AF776209

CATEGORY THEORY 1991

Titles in This Series

13 **R. A. G. Seely, Editor,** Category theory 1991

12 **A. Nicas and W. F. Shadwick, Editors,** Differential geometry, global analysis, and topology

11 **H. Tachikawa and V. Dlab, Editors,** Representations of finite dimensional algebras

10 **A. Bialynicki-Birula, J. Carrell, P. Russell, and D. Snow, Editors,** Group actions and invariant theory

9 **Joel S. Feldman and Lon M. Rosen, Editors,** Mathematical quantum field theory and related topics

8 **F. V. Atkinson, W. F. Langford, and A. B. Mingarelli, Editors,** Oscillation, bifurcation and chaos

7 **H. Kisilevsky and J. Labute, Editors,** Number theory

6 **J. Carrell, A. V. Geramita, and P. Russell, Editors,** Proceedings of the 1984 Vancouver Conference in algebraic geometry

5 **D. J. Britten, F. W. Lemire, and R. V. Moody, Editors,** Lie algebras and related topics

4 **I. Hambleton and C. R. Riehm, Editors,** Quadratic and hermitian forms

3 **Zeev Ditzian, Amram Meir, Sherman D. Riemenschneider, and Ambikeshwar Sharma, Editors,** Second Edmonton conference on approximation theory

2 **Richard M. Kane, Stanley O. Kochman, Paul S. Selick, and Victor P. Snaith, Editors,** Current trends in algebraic topology, (Parts 1 and 2)

1 **Carl Herz and R. Rigelhof, Editors,** 1980 Seminar on harmonic analysis

Canadian Mathematical Society
Société Mathématique du Canada

CONFERENCE PROCEEDINGS • Volume 13

CATEGORY THEORY 1991

Proceedings of an
International Summer Category Theory Meeting
held June 23–30, 1991

R. A. G. Seely, Editor

Published by the American Mathematical Society
for the Canadian Mathematical Society
PROVIDENCE, RHODE ISLAND

CMS CONFERENCE PROCEEDINGS
Series Editors
James B. Carrell and M. Ram Murty
Volume 13
1991 International Summer Category Theory Meeting
held in Montreal, Quebec, Canada,
June 23–30, 1991
Proceedings of a conference partially supported
by grants from Le Fonds FCAR, Quebec.
Accepted for publication by the Canadian Mathematical Society
Published by the American Mathematical Society
1991 *Mathematics Subject Classification*. Primary 18-06; Secondary 00B10, 18-02.

Library of Congress Cataloging-in-Publication Data

International Summer Category Theory Meeting (1991: Montréal, Québec)
Category theory 1991: proceedings of an International Summer Category Theory Meeting, held June 23–30, 1991/R.A.G. Seely, editor.
p. cm. – (Conference proceedings, ISSN 0731-1036; v. 13)
ISBN 0-8218-6018-6 (acid-free paper)
1. Categories (Mathematics)–Congresses. I. Seely, R. A. G. (Robert Andrew George), 1951– . II. Canadian Mathematical Society. III. Title. IV. Series: Conference proceedings (Canadian Mathematical Society); v. 13.
QA169.I58 1991 92-24186
512′.55–dc20 CIP

Printed in the United States of America.
The paper used in this book is acid-free and falls within the guidelines
established to ensure permanence and durability. ♾
Portions of this volume were printed directly from author-prepared copy.
Portions of this volume were typeset by the authors using AMS-TEX,
the American Mathematical Society's TEX macro system.
10 9 8 7 6 5 4 3 2 1 97 96 95 94 93 92

Contents

Preface

This volume is the published proceedings of the 1991 International Summer Category Theory Meeting (*Category Theory* 1991), which was held in Montréal, Canada during the week of June 23–30. As category theory approaches its first half-century it is clear that it continues to grow, finding new applications in domains that would have seemed inconceivable a generation ago, as well as in its traditional contexts. Some of this diversity is reflected in the papers in this volume.

The conference would not have been possible without the efforts of M. Barr, T. Fox, Raffaella Bruno, Monique Lemire, and the Scientific committee: Jim Lambek, M. Makkai, Andy Pitts, Gonzalo Reyes, Myles Tierney, and Richard Wood. In the production of this book I have been helped by the fine work (and especially the promptness) of the anonymous referees, of the authors themselves, and by Michael Barr, whose support was invaluable. The meeting was partially supported by grants from Le Fonds FCAR, Québec.

R. A. G. Seely

Canadian Mathematical Society
Conference Proceeding
Volume **13**, 1992

DeMorgan's Law and Related Identities in Classifying Topoi

ATISH BAGCHI

ABSTRACT. We relate model-theoretic properties of a geometric theory to the validity of certain equations in the Heyting algebra of truth-values of its classifying topos. The model-theoretic properties turn out to involve certain kinds of amalgamation.

Section I: Introduction

The research reported here is concerned with relating model-theoretic properties of a geometric theory (defined in Section IV) to algebraic properties (to be explained later) of the truth-value object (which, in general, is a Heyting algebra) of its classifying topos. The preceding considerations are closely related to the study of model-theoretic properties of a first-order theory T reflected in the lattice of equivalence classes (under T-provability) of existential positive formulas, a sublattice of the Lindenbaum algebra of T. This relationship is discussed in Section IV. The model-theoretic properties turn out to be the non-existence of pairwise contradictory finite homomorphic extensions of a sufficiently large part of the positive diagram of T, that are consistent with T. One can reformulate the preceding as an amalgamation property. This is discussed in Section V.

The algebraic structures that we need to consider are pseudocomplemented distributive lattices with 0 and 1. The equational subclasses of the class $\mathbf{B}_\omega$ of pseudocomplemented distributive lattices with 0 and 1 are exactly the members of the following ω-chain: $\mathbf{B}_{-1} \subset \mathbf{B}_0 \subset \mathbf{B}_1 \subset \ldots \subset \mathbf{B}_\omega$, where $\mathbf{B}_{-1}$ contains only the trivial algebra with $0 = 1, \mathbf{B}_0$ is the class of Boolean algebras and $\mathbf{B}_1$ is the class of Stone algebras, which are algebras satisfying De Morgan's law; by an algebraic property, we mean the defining equation I'_n for such a class $\mathbf{B}_n$. We note that every Heyting algebra is a particular kind of pseudocomplemented distributive lattice with 0 and 1. The specific examples of pseudocomplemented distributive lattices with 0 and 1 that we have considered are the following:

1991 Mathematics Subject Classification. Primary 03G30, 03C05.
This paper is in final form and no version of it will be submitted for publication elsewhere.
This is a part of the author's doctoral dissertation [B].

(1) $\mathrm{DE}_r(\mathrm{Mod(T)})$, which is the lattice of equivalence classes (with respect to T-provability) of arbitrary disjunctions of existential formulas for some theory T,

(2) $\mathrm{DE}_r^+(\mathrm{Mod(T)})$, which is the lattice of equivalence classes (with respect to T-provability) of arbitrary disjunctions of existential positive formulas for some theory T,

(3) Open(X), which is the Heyting algebra of open sets of a topological space X,

(4) $\Omega_{\mathcal{T}}$, which is the Heyting algebra of truth-values of the classifying topos $\mathcal{T}$ of a finitary geometric theory T, and

(5) $\Omega_{\mathcal{E}}$, which is the Heyting algebra of truth-values of an arbitrary topos $\mathcal{E}$.

The subscript r in $\mathrm{DE}_r(\mathrm{Mod(T)})$ and $\mathrm{DE}_r^+\mathrm{Mod(T)})$ indicates that the free variables of the formulas under consideration are among $x_1, \ldots, x_r$. We note that (4) is clearly a special case of (5). (3) may be subsumed under (5) by taking $\mathcal{E}$ to be the topos of sheaves on X. (2) and (4) are also closely related. Some of the theorems proved here extend earlier work of Blass and Scedrov, who considered Boolean classifying topoi [B&S]. It should be noted that the axioms characterizing $\mathbf{B}_n$, where $n \geq 2$, are weaker than De Morgan's law. Other non-equational conditions, weaker than De Morgan's law have been studied by Johnstone [DML]. This study also extends earlier work of Weispfenning, who considered the lattice of existential formulas, but not necessarily with a positive matrix [W]. The following chart summarizes what was already known.

	$\mathbf{B}_{-1}$	$\mathbf{B}_0$	$\mathbf{B}_1$	$\mathbf{B}_n$	
$\mathrm{DE}_r(\mathrm{Mod(T)})$	$\surd$	+	$\surd$	$\surd$	Weispfenning
$\mathrm{DE}_r^+(\mathrm{Mod(T)})$	$\surd$				
Open(X)	$\surd$	$\surd$	$\surd$		Gleason
$\Omega_{\mathcal{T}}$	$\surd$	$\surd$			Blass, Scedrov
$\Omega_{\mathcal{E}}$	$\surd$	$\surd$	$\surd$		Johnstone

A $\surd$ in the first row under $\mathbf{B}_n$ indicates that Welspfenning proved theorems that exactly describe conditions under which $\mathrm{DE}_r(\mathrm{Mod(T)}) \in \mathbf{B}_n$ for arbitrary $n \in \omega$. The other $\surd$'s are to be similarly Πinterpreted. The + indicates that what Welspfenning proved is a result about the lattice of equivalence classes of existential formulas and not equivalence classes of arbitrary disjunctions thereof. We have results that allow one to fill in all the remaining blank spots onΞΞ the chart. We shall concern ourselves here only with (1), (2) and (4). (3), (4), (5), various presheaf topoi and Open(X) for special topological spaces like the spectrum of a commutative ring will be the subject of a forthcoming paper.

We have already described the contents of Sections IV and V. We now briefly describe the contents of Sections II and III. In Section II we set up the model-theoretic framework necessary for the rest of the article. In Section III we prove results about distributive lattices that we use later. We list some open questions at the end of section V.

Throughout the paper we have omitted proofs that involve straightforward verification. This is particularly true of Section IV where the details of calculations are often rather tedious. We end the introduction by listing some notational conventions that we have adopted.

NOTATIONAL CONVENTIONS:

(1) Given integers m and $n, m..n$ denotes the set of integers i such that $m \leq i \leq n$. (2) Given a set $S, \mathrm{Fin}(S), \mathcal{P}(S)$ and $\#(S)$ denote respectively the set of finite subsets of S, the power-set of S and the cardinality of S.

(3) For all sets A, B such that B is included in A we denote by ι_A, the identity function on A, and by $\iota_{B \subset A}$, the inclusion mapping of B in A.

(4) Given a function $f : D \longrightarrow C$, a subset D' of D and a superset C' of C, we define: $\mathrm{Gr}(f) := \{(x, f(x)) | x \in D\}$;

$f|_{D'} := f \circ \iota_{D' \subset D}; f|^{C'} := \iota_{C \subset C'} \circ f$;

$f_> : \mathcal{P}(D) \longrightarrow \mathcal{P}(C) := S \mapsto \{f(s) \in C | s \in S\}$; and

$f^< : \mathcal{P}(C) \longrightarrow \mathcal{P}(D) := T \mapsto \{s \in D | f(s) \in T\}$.

(5) In any category $\mathcal{C}$, we have the maps Dom and Cod, that send a morphism to its domain and codomain respectively. Thus $\mathrm{Dom}^<$ and $\mathrm{Cod}^<$ are defined by (4).

(6) In any category $\mathcal{C}$, given a morphism f, we define $\mathrm{Lcomp}_f : \mathrm{Cod}^<(\mathrm{Dom}(f)) \longrightarrow \mathrm{Cod}^<(\mathrm{Cod}(f)) := g \mapsto f \circ g$. We shall have occasion to use $(\mathrm{Lcomp}_f)^< : \mathcal{P}(\mathrm{Cod}^<(\mathrm{Cod}(f))) \longrightarrow \mathcal{P}(\mathrm{Cod}^<(\mathrm{Dom}(f)))$. We note that given a family $F \in \mathcal{P}(\mathrm{Cod}^<(\mathrm{Cod}(f)))$ of morphisms with common codomain, the codomain of $f, (\mathrm{Lcomp}_f)^<(F) = \{g \in \mathrm{Cod}^<(\mathrm{Dom}(f)) | f \circ g \in F\}$.

(7) Boldface letters are used to denote finite sequences; the *ith* term of $\mathbf{a}$ is a_i. Given a map $h : A \longrightarrow B, h\mathbf{a} = (h(a_i) | i \in 1..n)$. We shall occasionally write $\mathbf{a}'$ to denote $(a'_i | i \in 1..n)$ and $(h(\mathbf{a}))'$ to denote $((h(a_i))' | i \in 1..n)$. We also define for every $n \in \omega$ and $\mathbf{a} \in A^n$, $\mathrm{Rng}(\mathbf{a}) := \{a_i | i \in 1..n\}$.

(8) $\wedge$ and $\bigwedge$ respectively denote binary and finitary of arity ≥ 3 or infinitary conjunctions. We adopt similar conventions for disjunctions.

(9) If $\mathbf{x}$ is $x_1, .., x_n$ and $\mathbf{y}$ is $y_1, .., y_n$ then we sometimes use the abbreviations $\exists\mathbf{x}$ for $\exists x_1 \ldots \exists x_n, \forall\mathbf{x}$ for $\forall x_1 \ldots \forall x_n$ and $\mathbf{x} = \mathbf{y}$ for the conjunction of the n formulas $x_i = y_i, i \in 1..n$.

(10) Given a category $\mathcal{C}, \mathrm{Op}(\mathcal{C})$ denotes the opposite category. $\mathrm{Obj}(\mathcal{C})$ and $\mathrm{Mor}(\mathcal{C})$ denote respectively the collection of objects of $\mathcal{C}$ and the collection of morphisms of $\mathcal{C}$.

Section II: Preliminaries

We discuss in this section the general model-theoretic framework within which we work. We assume given a countable relational language L. $\mathrm{Str}(L), \mathrm{Fml}(L)$ and $\mathrm{Sent}(L)$ denote respectively the collection of structures, the set of formulas and the set of sentences for the language L. Script letters, for instance $\mathcal{A}$, are used to denote a structure and the corresponding roman letter, A to denote the underlying set of the structure. Given $\phi \in \mathrm{Fml}(L), \mathrm{Fv}(\phi)$ denotes the set of free variables of ϕ. Given a theory T in $L, \mathrm{Mod}(\mathrm{T})$ denotes the collection of models of T.

DEFINITION II.1. Given $\mathcal{A} \in \mathrm{Str}(L), B \subset A, \underline{L(B)} := L \cup \{b' | b \in B\}$. We may define $(\mathcal{A}, B) \in \mathrm{Str}(L(B))$ by interpreting all symbols in L as they were interpreted in $\mathcal{A}$ and interpreting each b' in $(\mathcal{A}, B)$ as the corresponding b.

DEFINITION II.2. $\underline{\mathrm{Diag}^+(\mathcal{A})}$: = the $\underline{\text{positive diagram of } \mathcal{A}}$:= $\{\sigma \in \mathrm{Sent}(L(A)) | \sigma$ is atomic and $(\mathcal{A}, A) \models \sigma\}$.

DEFINITION II.3. For every $p \in \mathrm{Fin}(\mathrm{Diag}^+(\mathcal{A}))$, $\underline{\mathrm{Name}}\ (p) := \{a' \in L(A) | a'$ occurs in $p\}$.

DEFINITION II.4. Given the set of variables $\mathbf{x} := \{x_1, x_2, .., x_n\}, \mathrm{Fml}_{n;\mathbf{x}}(L) := \{\phi \in \mathrm{Fml}(L) | \mathrm{Fv}(\phi) \subset \mathbf{x}, \#(\mathbf{x}) = n\}$.

DEFINITION II.5 $\mathrm{E}^+_{n;\mathbf{x}} := \{\phi \in \mathrm{Fml}_{n,\mathbf{x}}(L) | \phi$ is existential positive, i.e., ϕ is an existential formula with a positive matrix$\}$.

DEFINITION II.6. $\mathrm{DE}^+_{n;\mathbf{x}} := \{\bigvee \Phi | \Phi \subset \mathrm{E}^+_{n;\mathbf{x}}\}$.

DEFINITION II.7. $\mathrm{Cond}^+_{n,\mathbf{x}}(T) := \{\eta \in \mathrm{E}^+_{n;x} | \eta$ is a finite conjunction of formulas consistent with T$\}$.

REMARK II.8. If there is no danger of confusion, the subscript $\mathbf{x}$ will be omitted from the three preceding definitions.

DEFINITION II.9. $\underline{\mathrm{Cond}^+}(\mathcal{A}) := \mathrm{Fin}(\mathrm{Diag}^+(\mathcal{A}))$.

REMARK II.10 Let $\mathcal{A} \in \mathrm{Str}(L)$, $\mathbf{a} \in A^n$ and $p \in \mathrm{Cond}^+(\mathcal{A})$ be given such that $\mathrm{Rng}(\mathbf{a}') \subset \mathrm{Name}(p)$. Then we may associate $\eta_{\mathbf{a},p} \in E^+_{n;x}$ with $(\mathbf{a}, p)$ as follows:

(1) Replace every name b' other than $a_i'(i \in 1..n)$ that occur in $\wedge p$ by a new distinct variable $y_{b'}$, and quantify existentially over this variable.
(2) For every $i \in 1..n$ replace a'_i in $\bigwedge p$ by the free variable x_i. For example, if $p = \{R_1(c'), R_2(b', a_1', a_2'), d' = a_2'\}$, then $\eta_{\mathbf{a},p} = \exists y_{c'}, \exists y_{b'}, \exists y_{d'}, (R_1(y_{c'}) \wedge R_2(y_{b'}, x_1, x_2) \wedge y_{d'} = x_2)$.

PROPOSITION II.11. $\mathcal{A} \in \mathrm{Mod(T)} \Rightarrow \eta_{\mathbf{a},p} \in \mathrm{Cond}^+_{n,x}(T)$.

PROOF. This is immediate as $\mathcal{A} \models \eta_{\mathbf{a},p}(\mathbf{a})$. ■

PROPOSITION II.12. *Let* $\psi(\mathbf{x}) \in \mathrm{DE}^+_{n;\mathbf{x}}$ *be given such that* $\psi(\mathbf{x})$ *is consistent with* T. *Then given* $\mathcal{A} \in \mathrm{Mod(T)}$ *such that* $\mathcal{A} \models \psi(\mathbf{a})$, *we may chooses* $p \in \mathrm{Cond}^+(\mathcal{A})$ *such that*

(1) $\mathrm{Rng}(\mathbf{a}') \subset \mathrm{Name}(p)$, *and*

(2) $\vdash \forall \mathbf{x}(\eta_{\mathbf{a},p}(\mathbf{x}) \to \psi(\mathbf{x}))$.

PROOF. We may, (up to logical equivalence), take $\psi(x)$ to be of the following form.

$$\psi(x) := \bigvee_{i \in I'} \exists \mathbf{y}_i (\bigvee_{j \in n_i'} (\bigwedge_{k \in m_{ij}} \psi_{ijk}(\mathbf{y}_i, \mathbf{x})).$$

Using distributivity, this may be rewritten in the form

$$\psi(\mathbf{x}) := \bigvee_{i \in I} \exists \mathbf{y}_i (\bigwedge_{j \in n_i} \psi_{ij}(\mathbf{y}_i, \mathbf{x})).$$

As $\mathcal{A} \models \psi(\mathbf{a})$, we may choose $i \in I$ such that $\mathcal{A} \models \exists \mathbf{y}_i (\bigwedge_{j \in n_i'} \psi_{ij}(\mathbf{y}_i, \mathbf{a})))$. Hence we may choose $\mathbf{b}_i \in A^{\mathrm{Length}(\mathbf{y}_i)}$ such that

$$\mathcal{A} \models \bigwedge_{j \in n_i} \psi_{ij}(\mathbf{b}_i, \mathbf{a}).$$

Define $p := \bigwedge_{j\in n_i} \psi_{ij}(\mathbf{b}_i, \mathbf{a})$. Then clearly $\mathrm{Rng}(\mathbf{a}') \subset \mathrm{Name}(p)$ and $\vdash \forall x(\eta_{\mathbf{a},p}(x) \to \psi(x))$ and $p \in \mathrm{Cond}^+(\mathcal{A})$.

DEFINITION II.13. For every $\mathcal{A}, \mathcal{B} \in \mathrm{Str}(L)$, such that $\mathcal{B}$ is a substructure of $\mathcal{A}$, we denote the inclusion homomorphism by $\iota_{\mathcal{B}\subset\mathcal{A}}$.

We shall need to consider finitely presented structures for L that admit a homomorphism into some $\mathcal{A} \in \mathrm{Mod(T)}$. Each finitely presented structure $\mathcal{G}$ is specified by a finite set of generators; $\underline{\mathrm{Gen}(\mathcal{G})}$:= the set of generators of $\mathcal{G}$ = the set of elements of G (as L is relational) and a finite set of atomic relations (in the vocabulary of L, augmented by names for the generators of $\mathcal{G}$) $\underline{\mathrm{Rel}(\mathcal{G})}$:= the set of atomic relations with appropriate lists of arguments as they appear in the presentation of $\mathcal{G}$. A typical element of $\mathrm{Gen}(\mathcal{G})$ is g_i, while a typical element of $\mathrm{Rel}(\mathcal{G})$ is $R_\alpha(\mathbf{g}_\alpha)$, where $\mathrm{Rng}(\mathbf{g}_\alpha) \subset \mathrm{Gen}(\mathcal{G})$ and Length $(g_\alpha) = \mathrm{Arity}(R_\alpha)$. We write $\mathcal{G} = \langle \mathrm{Gen}(\mathcal{G}) | \mathrm{Rel}(\mathcal{G}) \rangle$. We need the following definitions.

DEFINITION II.14. $\underline{\mathrm{FPS}(L)} := \{\mathcal{G} \in \mathrm{Str}(L) | \mathcal{G}$ is finitely presented$\}$.

DEFINITION II.15. $\underline{\mathrm{FP}_T(L)} := \{\mathcal{G} \in \mathrm{FPS}(L) | \exists \mathcal{A} \in \mathrm{Mod(T)}$ and a homomorphism $h : \mathcal{G} \longrightarrow \mathcal{A}\}$.

FPS is an abbreviation for finitely presented structures, while the subscript T is supposed to indicate that such a structure admits a homomorphism into some model of T.

REMARK II.16. Let $\mathcal{G} \in \mathrm{FPS}(L)$ and $h : \mathcal{G} \to \mathcal{A}$ be given. ($\mathcal{A}$ may or may not be a member of Mod(T).) We wish to associate, with $h, \mathcal{G}$ and $\mathcal{A}$ a set $p_{\mathcal{G}h\mathcal{A}} \in \mathrm{Cond}^+(\mathcal{A})$. We describe an example first.

EXAMPLE II.17. Suppose $\mathcal{G} := \langle a, b, c | R(a, a), S(a, b, c) \rangle$. Then $\mathrm{Gen}(\mathcal{G}) = \{a, b, c\}$ and $\mathrm{Rel}(\mathcal{G}) = \{R(a, a), S(a, b, c)\}$. Let $h : \mathcal{G} \longrightarrow \mathcal{A}$ be given. Suppose $h(a) = k \in A$ and $h(b) = h(c) = m \in A$. Then we define

$$p_{\mathcal{G}h\mathcal{A}} := \{R((h(a))', (h(a))'), S((h(a))', (h(b))', (h(b))')\},$$

and

$$\langle p_{\mathcal{G}h\mathcal{A}} \rangle := \langle (h(a))', (h(b))' | p_{\mathcal{G}h\mathcal{A}} \rangle.$$

We note that $< p_{\mathcal{G}h\mathcal{A}} >$ is essentially a quotient of $\mathcal{G}$ obtained by identifying some generators of $\mathcal{G}$, and that the embedding $\iota_{\langle p_{\mathcal{G}h\mathcal{A}} \rangle \hookrightarrow \mathcal{A}}$ is a homomorphism. The general description is as follows.

DEFINTION II.18. For every $\alpha \in 1..\#(\mathrm{Rel}(\mathcal{G})), \mathrm{F}_{R_\alpha} := \{\mathbf{F} \in \mathrm{Gen}(\mathcal{G}))^{\mathrm{Arity}(R_\alpha)} | R_\alpha(\mathbf{F})$ appears in $\mathrm{Rel}(\mathcal{G})\}$.

DEFINITION II.19. Given $\mathbf{F} \in \mathrm{F}_{R_\alpha}, \mathbf{F}' := ((h(F(i)))' | i \in 1..(\mathrm{Arity}(\mathbf{R}_\alpha))$.

For instance, in Example II.22, $\mathbf{F} = (a, b, c) \in \mathrm{F_S}$ and $\mathbf{F}' = ((h(a))', (h(b))', (h(c))')$.

DEFINITION II.20. We define

$p_{\mathcal{G}h\mathcal{A}} := \cup_{\alpha \in 1..\#\mathrm{Rel}(\mathcal{G})} \{\mathrm{R}_\alpha(\mathbf{F}') | \mathbf{F} \in \mathrm{R}_\alpha\}$.

DEFINITION II.21. $\langle p_{\mathcal{G}h\mathcal{A}} \rangle :=< \mathrm{Rng}((h\mathbf{g})') | p_{\mathcal{G}h\mathcal{A}} >$, where all the elements of $\mathcal{G}$ are in $\mathbf{g}$.

DEFINITION II.22. Let $\mathcal{G}_0, \mathcal{G}_i, \mathcal{G}_j \in \mathrm{Str}(L)$, the homomorphisms $h_{0i} : \mathcal{G}_0 \longrightarrow \mathcal{G}_i, h_{0j} : \mathcal{G}_0 \longrightarrow \mathcal{G}_j$ and the list $\mathbf{g} \in \mathrm{Gen}(\mathcal{G}_0))^n$ be given. We say that $\langle \mathcal{G}_0, \mathbf{g}, \mathcal{G}_i, \mathcal{G}_j, h_{0i}, h_{0j} \rangle$ may be amalgamated in Mod(T) :$\Leftrightarrow$ there exist $\mathcal{A} \in \mathrm{Mod(T)}$ and homomorphisms $h_{i\mathcal{A}}, h_{j\mathcal{A}}$ as shown such that the diagram commutes over $\mathbf{g}$ i.e. $h_{i\mathcal{A}} h_{0i}(\mathbf{g}) = h_{j\mathcal{A}} h_{0j}(\mathbf{g})$.
The diagram illustrates the situation.

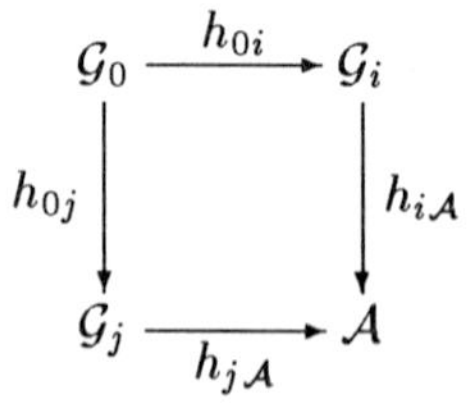

REMARK II.23. Definition II.22 refers to all structures, not just finitely presented ones.

REMARK II.24. We define

$$\eta_i(\mathbf{x}) := \eta_{(h(\mathbf{g}))', p_{\mathcal{G}_i} h_{i\mathcal{A}} \mathcal{A}} \quad \text{and} \quad \eta_j(\mathbf{x}) := \eta_{(h(\mathbf{g}))', p_{\mathcal{G}_j} h_{j\mathcal{A}} \mathcal{A}},$$

If $\mathcal{G}_0$, $\mathcal{G}_i$ and $\mathcal{G}_j$ are finitely presented, then, $< \mathcal{G}_0, g, \mathcal{G}_i, \mathcal{G}_j, h_{0i}, h_{0j} >$ cannot be amalgamated in Mod(T) $\Leftrightarrow \exists \mathbf{x}(\eta_i(\mathbf{x}) \wedge \eta_j(\mathbf{x}))$ is inconsistent in T. In this situation we shall also occasionally write that $p_{\mathcal{G}_i h_{i\mathcal{A}} \mathcal{A}}$ is inconsistent with T.

We describe in the following how certain quotients of E_n^+ and DE_n^+ may be endowed with the structure of distributive lattices. These will be our principal objects of study in later sections.

DEFINITION II.25. $\phi^{\mathrm{Mod(T)}} := \{\psi \in \mathrm{DE}_n^+ | \mathrm{Mod(T)} \models \phi \leftrightarrow \psi\}$. $\mathrm{Mod(T)} \models \theta$ means that θ holds in every model of T.

DEFINITION II.26. For every $\Phi \subset \mathrm{DE}_n^+(\mathrm{Mod(T)})$,

$$\Phi^{\mathrm{Mod(T)}} := \{\phi^{\mathrm{Mod(T)}} | \phi \in \Phi\}.$$

We also define $\mathrm{E}_n^+(\mathrm{Mod(T)}) := (\mathrm{E}_n^+)^{\mathrm{Mod(T)}}$, and $\mathrm{DE}_n^+(\mathrm{Mod(T)}) := (\mathrm{DE}_n^+)^{\mathrm{Mod(T)}}$.

DEFINITION II.27. The operations $\wedge$, $\vee$, $\rightarrow$, $\leftrightarrow$ without subscripts refer to the logical operations of L or the corresponding operations in the Lindenbaum algebra of T, where we allow infinitary conjunctions and disjunctions. We assume in the following that distributivity has been used to transform $\phi \wedge \psi$ and $\phi \vee \psi$ into logically equivalent forms that are explicitly members of $\mathrm{DE}_n^+(\mathrm{Mod(T)})$, if ϕ, $\psi \in \mathrm{DE}_n^+(\mathrm{Mod(T)})$. The subscript D refers to the lattice $\mathrm{DE}_n^+(\mathrm{Mod(T)})$.

Let $\phi, \psi \in \mathrm{DE}_n^+(\mathrm{Mod(T)})$ be given. We define $0_D := \bot^{\mathrm{Mod(T)}}$; $1_D := \top^{\mathrm{Mod(T)}}$; $\phi^{\mathrm{Mod(T)}} \wedge_D \psi^{\mathrm{Mod(T)}} := (\phi \wedge \psi)^{\mathrm{Mod(T)}}$; $\phi^{\mathrm{Mod(T)}} \vee_D \psi^{\mathrm{Mod(T)}} := (\phi \vee \psi)^{\mathrm{Mod(T)}}$; $\phi^{\mathrm{Mod(T)}} \leq_D \phi^{\mathrm{Mod(T)}} := \mathrm{Mod(T)} \models \phi \rightarrow \psi$.

REMARK II.28. With the operations and the ordering defined in the preceding, $\mathrm{DE}_n^+(\mathrm{Mod(T)})$ and $\mathrm{E}_n^+(\mathrm{Mod(T)})$ are both distributive lattices with 0 and 1. Moreover for every $n \in \omega$, $\mathrm{E}_n^+(\mathrm{Mod(T)})$ is a sublattice of $\mathrm{DE}_n^+(\mathrm{Mod(T)})$.

DEFINITION II.29. For every $\Phi \subset \mathrm{E}_n^+(\mathrm{Mod(T)})$,

Isp $(\Phi) :=$ ideal spanned by Φ in $\mathrm{E}_n^+(\mathrm{Mod(T)})$.

Idl (Mod(T))): $= \{\Phi^{\mathrm{Mod(T)}} \subset \mathrm{E}_n^+(\mathrm{Mod(T)}) \mid \Phi^{\mathrm{Mod(T)}}$ is an ideal$\}$.

PROPOSITION II.30. Given $\Phi^{\mathrm{Mod(T)}}, \Psi^{\mathrm{Mod(T)}} \subset \mathrm{E}_n^+(\mathrm{Mod(T)})$,

$\mathrm{Isp}(\Phi^{\mathrm{Mod(T)}}) \subset \mathrm{Isp}(\Psi^{\mathrm{Mod(T)}}) \Leftrightarrow (\bigvee \Phi)^{\mathrm{Mod(T)}} \leq_D (\bigvee \Psi)^{\mathrm{Mod(T)}}$.

PROOF $\mathrm{Isp}(\Phi^{\mathrm{Mod(T)}}) \subset \mathrm{Isp}(\Psi^{\mathrm{Mod(T)}})$

$\Leftrightarrow \forall \phi^{\mathrm{Mod(T)}} \in \Phi^{\mathrm{Mod(T)}}$ we may choose $n \in \omega$ and $\psi_1^{\mathrm{Mod(T)}}, .., \psi_n^{\mathrm{Mod(T)}} \in \Psi^{\mathrm{Mod(T)}}$ such that $\phi^{\mathrm{Mod(T)}} \leq \bigvee_{i \in 1..n} \psi_i^{\mathrm{Mod(T)}}$.

$\Leftrightarrow \quad \forall \phi^{\mathrm{Mod(T)}} \in \Phi^{\mathrm{Mod(T)}}, \exists n \in \omega$ and $\psi_1^{\mathrm{Mod(T)}}, .., \psi_n^{\mathrm{Mod(T)}} \in \Psi^{\mathrm{Mod(T)}}$ such that $\mathrm{Mod(T)} \models \phi \to \bigvee_{i \in 1..n} \psi_i$

$\Leftrightarrow \forall \phi^{\mathrm{Mod(T)}} \in \Phi^{\mathrm{Mod(T)}}, \mathrm{Mod(T)} \models \phi \to \bigvee \Psi$ (by compactness)

$\Leftrightarrow \mathrm{Mod(T)} \models \bigvee \Phi \to \bigvee \Psi$

$\Leftrightarrow (\bigvee \Phi)^{\mathrm{Mod(T)}} \leq_D (\bigvee \Psi)^{\mathrm{Mod(T)}}$.

PROPOSITION II.31. *In view of Proposition II.30, we may define the map*

$$h : \mathrm{DE}_n^+(\mathrm{Mod(T)}) \to \mathrm{Idl}(\mathrm{E}_n^+(\mathrm{Mod(T)}) := (\bigvee \Phi)^{\mathrm{Mod(T)}} \mapsto \mathrm{Isp}(\Psi^{\mathrm{Mod(T)}}).$$

h is a lattice isomorphism with inverse

$$h^{\leftarrow} : \mathrm{Idl}(\mathrm{E}_n^+(\mathrm{Mod(T)})) \to \mathrm{DE}_n^+(\mathrm{Mod(T)}) := I \mapsto \bigvee I.$$

PROOF. h is clearly injective and surjective, and by Proposition II.30, both h and $h^{\leftarrow}$ preserve order. Moreover $h^{\leftarrow}$ is clearly the inverse of h. ∎

Section III: Distributive Lattices

In this section we gather for later use all the results on distributive lattices that we shall need.

DEFINITION III.1. D_{01} := the class of distributive lattices with 0 and 1 in the language with symbol-set $S := \{0, 1, \wedge, \vee, \leq\}$. $a \to b$:= the relative pseudocomplement of a with respect to b, which, if it exists, is the largest element $c \in L$ such that $a \wedge c \leq b$. a^*:= the pseudocomplement of a, which, if it exists, is $a \to 0$.

DEFINITION III.2. Given $L \in D_{01}$, if for all $a, b \in L$, $a \to b$ exists, then L is called a Heyting algebra. For the purposes of this article, we shall consider Heyting algebras to be particular kinds of pseudocomplemented distributive lattices.

DEFINITION III.4. Pseudocomplemented lattices, regarded as structures for the language L^* with symbol-set $S^* := \{0, 1, \wedge, \vee, \leq, ^*\}$ are called distributive p-algebras.

DEFINITION III.5 $\mathbf{B}_\omega :=$ the class of distributive p-algebras.

REMARK III.6. $\mathbf{B}_\omega$ is the equational class axiomatized by the axioms for D_{01} together with the following axioms [W].

$$(1)\quad 0^* = 1. \quad (2)\quad 1^* = 0. \quad (3)\quad \forall x, y(x \wedge (x \wedge y)^* = x \wedge y^*).$$

REMARK III.7. The equational subclasses of $\mathbf{B}_\omega$ are exactly $\mathbf{B}_{-1} \subset \mathbf{B}_0 \subset \mathbf{B}_1 \subset \ldots \subset \mathbf{B}_\omega$, where $\mathbf{B}_{-1}$ contains only the trivial p-algebra, $\mathbf{B}_0$ is the class of Boolean algebras and B_1 is the class of Stone algebras. $\mathbf{B}_0$ is axiomatized by the axioms for distributive p-algebras and $\forall x(x \vee x^* = 1)$. $\mathbf{B}_1$ is axiomatized by the axioms for distributive p-algebras and $\forall x(x^* \vee x^{**} = 1)$. In general for all $r \geq 1$, $\mathbf{B}_r$ is axiomatized by the axiom I_r, defined as follows, in addition to the axioms for distributive p-algebras [L].

$$I_r := \bigwedge_{i,j \in 0..r, i<j} (x_j \wedge x_j = 0) \rightarrow (\bigvee_{i \in 0..r} x_i^* = 1).$$

I_r may be rewritten as an equation I_r', namely

$$I_r' := \forall x_1, .., x_r((\bigwedge_{i \in 1..r} x_i)^* \vee (\bigvee_{i \in 1..r} (\bigwedge_{j \in 1..r \setminus \{i\}} x_j \wedge x_i^*)^*)) = 1.$$

If the Heyting algebra of open sets of a topological space satisfies $I_r (r \geq 2)$, then we call it an r-Lee space.

We note that in pseudocomplemented distributive lattices, and hence in Heyting algebras, the identity $\forall x(x^* \vee x^{**} = 1)$ is equivalent to the identity $\forall a, b((a \wedge b)^* = a^* \vee b^*)$ [B& D p 164]. In fact $\forall a, b((a^* \vee b^*) \leq (a \wedge b)^*)$ holds in all pseudocomplemented distributive lattices, so that it is the opposite inequality that may not be valid in all pseudocomplemented distributive lattices. On the other hand, the identity $\forall a, b((a \vee b)^* = a^* \wedge b^*)$ holds in all pseudocomplemented distributive lattices, and hence in all Heyting algebras [B& D p153]. Thus, when we talk about De Morgan's law in a pseudocomplemented distributive lattice, we mean the identity $\forall a, b((a \wedge b)^* = a^* \vee b^*)$ or, equivalently, the identity $\forall x(x^* \vee x^{**} = 1)$.

REMARK III.8. A complete Heyting algebra (CHA) is a complete lattice satisfying the join infinite distributive identity JID: $\forall x \forall (y_i | i \in I) x \wedge (\bigvee_{i \in I} y_i) = \bigvee_{i \in I} (x \wedge y_i)$. For the definition of a complete lattice, we refer the reader to [B & D p45].

REMARK III.9. Any CHA is a Heyting algebra with $x \rightarrow y := \bigvee\{x | x \wedge z \leq y\}$ and $x^* := \bigvee\{x | x \wedge z = 0\}$. In particular, it is a distributive p-algebra.

DEFINITION III.10. An element a in a lattice L is said to be compact $:\Leftrightarrow$ for every $A \subset L$, $((a \leq \bigvee A) \Rightarrow$ (there exists $A' \in \mathrm{Fin}(A)$ such that $a \leq \bigvee A'$)).

DEFINITION III.11. Comp(L) $:= \{a \in L | a$ is compact$\}$.

REMARK III.12. $0 \in$ Comp(L) and Comp(L) is closed under $\vee$.

DEFINITION III.13. $L \in D_{01}$ is called a distributive arithmetical lattice: $\Leftrightarrow$

(1) Comp(L) is a sublattice of L containing 1, and

(2) for every $a \in L$, $a = \bigvee\{b \in \text{Comp(L)} | b \leq a\}$.

PROPOSITION III.14. *$DE_n^+(\text{Mod(T)})$ is a distributive arithmetical lattice with $E_n^+(\text{Mod(T)})$ as the sublattice of compact elements. Suprema in $DE_n^+(\text{Mod(T)})$ are given by*

$$\bigvee_{i \in I} D\phi_i^{\text{Mod(T)}} = (\bigvee_{i \in I} \phi_i)^{\text{Mod(T)}}.$$

In particular, as any distributive arithmetical lattice is a CHA, so is $DE_n^+(\text{Mod(T)})$.

PROOF. Essentially by definition and Proposition II.31. ■

REMARK III.15. The lattice operations $\wedge$ and $\vee$ in $DE_n^+(\text{Mod(T)})$ correspond to the propositional connectives $\wedge$ and $\vee$. But as $DE_n^+(\text{Mod(T)})$ is not closed under negation, implication or arbitrary infinite conjunctions, $*_D, \bigwedge_D, \rightarrow_D$ are different from $\neg$, $\bigwedge$ and $\rightarrow$.

DEFINITION III.16. For every $a \in L \in D_{01}$, we define $L_{\leq a} := \{y \in L | y \leq a\}$.

LEMMA III.17. *Let L be a CHA. Then for every $a \in L$, $L_{\leq a}$ is a CHA with the symbols $0', 1', \wedge', \vee', \leq', *'$ defined as follows:*

$0' := 0; 1' := a;$

$\forall y, z \in L_{\leq a}, y \wedge' z := y \wedge z$ (as $(y \wedge z) \leq (a \wedge a) = a$);

$\forall y, z \in L_{\leq a}, y \vee' z := y \vee z$ (as $(y \vee z) \leq (a \vee a) = a$);

$\forall (y_i \in L_{\leq a} | i \in I), \quad \bigvee'_{i \in I} y_i := (\bigvee_{i \in I} y_i) \wedge a.$

$\forall y, z \in L_{\leq a}, \quad y \longrightarrow' z := \bigvee\{w \in L_{\leq a} | w \wedge y \leq z\} := (y \longrightarrow z) \wedge a.$

In particular, $y^{*'} := y \rightarrow' 0 = (y \rightarrow 0) \wedge a = y^* \wedge a$. ■

PROPOSITION III.18. *Let L be a distributive arithmetical lattice. Then, for every $a \in Comp(L)$, $L_{\leq a}$ satisfies $I_r \Leftrightarrow L$ satisfies I_r.*

PROOF. $\Rightarrow$: This is immediate as $1 \in \text{Comp(L)}$ and $L = L_{\leq 1}$.
$\Leftarrow$: Let $a \in \text{Comp(L)}$ and $(y_i \in L_{\leq a} | i \in 0..r)$ be given such that for all $i, j \in 0..r, i \neq j \Rightarrow y_i \wedge' y_j = 0$. Hence for all $i, j \in 0..r, i \neq j \Rightarrow y_i \wedge y_j = y_i \wedge' y_j = 0$. Hence by assumption $\bigvee_{i \in 0..r}(y_i^*) = 1$. Hence $\bigvee_{i \in 1..r} y_i^{*'} = \bigvee_{i \in 0..r}(y_i^* \wedge a) = (\bigvee_{i \in 0..r} y_i^*) \wedge a = 1 \wedge a = a = 1'$. As $a \in \text{Comp(L)}$ was arbitrary, this establishes the result. ■

COROLLARY III.19 *For every $n \in \omega$, ($DE_n^+(\text{Mod(T)})$ satisfies I'_r) $\Leftrightarrow$ (for every $\phi^{\text{Mod(T)}} \in E_n^+(\text{Mod(T)}), DE_n^+(\text{Mod(T)})_{\leq \phi^{\text{Mod(T)}}}$ satisfies I'_r)).*

PROOF. $DE_n^+(\text{Mod(T)})$ is a distributive arithmetical lattice with

$$E_n^+(\text{Mod(T)}) = \text{Comp}(DE_n^+(\text{Mod(T)})).$$

■

LEMMA III.20. *For every set I and every I-indexed family $\Phi \in (DE_n^+(\text{Mod(T)}))^I$,*

$$(\bigvee \Phi = 1_{DE_n^+(\text{Mod(T)})}) \Leftrightarrow \text{ (for every } \mathcal{A} \in \text{Mod(T) and every } \mathbf{a} \in A^n,$$

$$\mathcal{A} \models (\bigvee \Phi)(\mathbf{a})).$$

PROOF. As $\mathrm{DE}_n^+(\mathrm{Mod(T)})$ is a distributive arithmetical lattice, $\bigvee \Phi = \bigvee_{i\in I} \phi_i$

$= \bigvee_{i\in I} \bigvee_{j\in J_i} \phi_{ij}$ (where $\forall i \in I, \forall j \in J_i, \phi_{ij} \in \mathrm{E}_n^+(\mathrm{Mod(T)})$)

$= \bigvee_{i\in I} \bigvee_{j\in J_i} \phi_{ij}$ (where $\forall i \in I, \forall j \in J_i, \phi_{ij} \in \mathrm{E}_n^+(\mathrm{Mod(T)})$)

$= \bigvee_{k\in K} \phi_k$ (where $K := I \times J_i, k := \langle i, j\rangle$, and $\forall k \in K, \phi_k := \phi_{\langle i,j\rangle} := \phi_i$)
Hence by assumption,

$$\bigvee_{k\in K} \phi_k = 1_{\mathrm{DE}_n^+(\mathrm{Mod(T)})}. \tag{1}$$

We recall that we have the isomorphisms $h : \mathrm{DE}_n^+\mathrm{Mod(T)}) \longrightarrow \mathrm{Idl}(\mathrm{E}_n^+(\mathrm{Mod(T)}))$ and $h^{\leftarrow} : \mathrm{Idl}(\mathrm{E}_n^+(\mathrm{Mod(T)})) \to \mathrm{DE}_n^+(\mathrm{Mod(T)}))$. Using h, we may conclude from (1) that $\mathrm{Isp}(\{\phi_k | k \in K\}) = \mathrm{Isp}(\{1_{\mathrm{E}_n^+}(\mathrm{Mod(T)})\})$. Hence we may choose $K' \in \mathrm{Fin}(K)$ such that $\bigvee_{k\in K'} \phi_k = 1_{\mathrm{E}_n^+(\mathrm{Mod(T)})}$. Hence $\mathrm{Isp}(\{\phi_k | k \in K'\}) = \mathrm{Isp}(\{1_{\mathrm{E}_n^+(\mathrm{Mod(T)})}\})$.

By using $h^{\leftarrow}$ we have

$$\bigvee_{k\in K'} \phi_k = 1_{\mathrm{DE}_n^+(\mathrm{Mod(T)})}. \tag{2}$$

Hence $\mathrm{T} \vdash \forall \mathbf{x}(\bigvee_{k\in K'} \phi_k \leftrightarrow \top)$

Hence $\bigvee \Phi = 1_{\mathrm{DE}_n^+(\mathrm{Mod(T)}))}$

$\Leftrightarrow \exists K' \in \mathrm{Fin}(K) \bigvee_{k\in K'} \phi_k = 1_{\mathrm{DE}_n^+(\mathrm{Mod(T)})}$

$\Leftrightarrow \exists K' \in \mathrm{Fin}(K)\ \mathrm{T} \vdash \forall \mathbf{x}(\bigvee_{k\in K'} \phi_k \leftrightarrow \top)$

$\Leftrightarrow \exists K' \in \mathrm{Fin}(K)\ \forall \mathcal{A} \in \mathrm{Mod(T)} \forall \mathbf{a} \in A^n \mathcal{A} \models (\bigvee_{k\in K'} \phi_k)(\mathbf{a})$. ■

LEMMA III.21. *Let H be a Heyting algebra. If we adjoin a new top element 1, then, with the definitions given as follows, the disjoint union, $H \amalg 1$ is a Heyting algebra. The operations on $H \amalg 1$ are the same as in H except for the instances noted below. $\forall x \in H$ the following hold*:

$1 \geq x; 1 \vee x = 1; 1 \wedge x = x; x \longrightarrow x = 1; x \longrightarrow 1 = 1; 1 \longrightarrow x = x; 1 \longrightarrow 1 = 1.$

Also $\forall x, y \in H(x \leq y \text{ in } H) \Rightarrow (x \longrightarrow y) = 1.$ ■

THEOREM III.22. [K.B. Lee] *For every $n \in \omega$, let $\underline{B}_n$ be the p-algebra obtained from the 2^n-element Boolean algebra B_n by adjoining a new unit 1. Then $\forall n \in \omega, \underline{B}_n \in \mathbf{B}_{n+1} \backslash \mathbf{B}_n[L]$.* ■

REMARK III.23. As every B_n is a Heyting algebra, by Lemma III.21 so is every $\underline{B}_n.\underline{B}_n$, considered as a Heyting algebra, is denoted by H_n. Thus $\{H_n | n \in \omega\}$ is a set of finite Heyting algebras such that for every $n \in \omega, H_n \in \mathbf{B}_{n+1} \backslash \mathbf{B}_n$.

Section IV: Geometric Theories

The primary purpose of this section is to reduce the problem of determining the conditions under which the truth-value object of a clasifying topos of a geometric theory T satisfies the rth Lee identity I_r to the problem of determining the conditions under which, for every $n \in \omega$, $\mathrm{DE}_n^+(\mathrm{Mod}(\mathrm{T}))$ satisfies I_r. We first present the relevant definitions and then proceed to carry out the reduction. The preliminary description of syntactic sites closely follows [B &S].

We assume initially in this section that we are given a first-order language L with equality. We shall later impose the restrictions that L is countable and relational. The geometric formulas of L are those obtainable from atomic formulas by finite conjunction (including the empty conjunction, $\top$), finite disjunction (including the empty disjunction, $\bot$), and existential quantification. Geometric formulas can be characterized model-theoretically as the formulas whose satisfaction is preserved by arbitrary homomorphisms of L-structures; see ([C & K] pp 126-128). Topos-theoretically, their key property is that, if M is an L-structure in a topos $\mathcal{E}$, if $f : \mathcal{F} \longrightarrow \mathcal{E}$ is a geometric morphism, and if ϕ is a geometric formula, then the truth-values of ϕ in M and f^*M satisfy $||\phi||_{f^*M} = f^*||\phi||_M$, i.e. f^* preserves the truth-value of ϕ.

A geometric sequent is a sentence of the form $\forall\mathbf{x}(\phi \longrightarrow \psi)$, where ϕ and ψ are geometric formulas and every variable free in ϕ or ψ occurs in the set $\mathbf{x}$. A geometric theory is a theory aximoatized by geometric sequents. We mention in passing that the geometric sequents classically provable in a geometric theory T hold in all models of T in arbitrary (not necessarily Boolean) topoi ([TT] p249).

If T is a geometric theory, then there exists a topos $\mathcal{E}$, called the classifying topos of T ([TT] pp 243-246), and there exists a model M of T in $\mathcal{E}$, called the universal model of T, such that, for any topos $\mathcal{F}$, the category of geometric morphisms $f : \mathcal{F} \longrightarrow \mathcal{E}$ and natural transformations $\nu : f^* \longrightarrow g^*$ is equivalent to the category of models of T in $\mathcal{F}$ and homomorphisms, the identity morphism corresponding to the model M. Thus, M has the universal property that every model of T in any topos $\mathcal{F}$ can be obtained (up to isomorphism) as f^*M for a unique (up to isomorphism) geometric morphism $f : \mathcal{F} \longrightarrow \mathcal{E}$. We shall need the following construction, due to Joyal and Reyes, of the classifying topos ([TT] pp. 244-246).

Given a geometric theory T, we shall define its syntactic site $(\mathcal{C}, J)$; the classifying topos will be the topos $\mathrm{Shv}(\mathcal{C}, J)$ of sheaves on this site. An object of $\mathcal{C}$ is a formal class term $\{\mathbf{x}|\phi(\mathbf{x})\}$ where $\phi(\mathbf{x})$ is a geometric formula whose free variables are among $\mathbf{x}$. We adopt the convention that a formula or a class term is not changed if its bound variables (including the $\mathbf{x}$ in $\{\mathbf{x}|\phi(\mathbf{x})\}$) are renamed subject to the usual precautions for avoiding clashes. To define morphisms from $\{\mathbf{x}|\phi(\{\mathbf{x}\}$ to $\{\mathbf{y}|\psi(\mathbf{y})\}$, we may and indeed do assume that the sets $\mathbf{x}$ and $\mathbf{y}$ are disjoint; a morphism is then equivalence class, with respect to T provable equivalence, of geometric formulas $\theta(\mathbf{x}, \mathbf{y})$ such that the following geometric sequents are provable in T.

(1) $\forall\mathbf{x}\forall\mathbf{y}(\theta(\mathbf{x}, \mathbf{y}) \longrightarrow \phi(\mathbf{x}) \wedge \psi(\mathbf{y}))$.

(2) $\forall\mathbf{x}(\phi(\mathbf{x}) \longrightarrow \exists\mathbf{y}\theta(\mathbf{x}, \mathbf{y}))$.

(3) $\forall\mathbf{x}\forall\mathbf{y}\forall\mathbf{z}(\theta(\mathbf{x},\mathbf{y}) \wedge \theta(\mathbf{x},\mathbf{z}) \longrightarrow \mathbf{y} = \mathbf{z})$.

The morphism defined by $\theta(\mathbf{x},\mathbf{y})$ will be called $[\mathbf{x} \mapsto \mathbf{y}|\theta(\mathbf{x},\mathbf{y})]$. The composite of $[\mathbf{x} \mapsto \mathbf{y}|\theta(\mathbf{x},\mathbf{y})]$ and $[\mathbf{y} \mapsto \mathbf{z}\ \eta(\mathbf{x},\mathbf{y})]$ is defined to be $[\mathbf{x} \mapsto \mathbf{z}|\exists\mathbf{y}(\theta(\mathbf{x},\mathbf{y}) \wedge \eta(\mathbf{y},\mathbf{z}))]$. It may be checked that this indeed defines a category $\mathcal{C}$. It is endowed with the structure of a site by defining a sieve on $\{\mathbf{y}|\psi(\mathbf{y})\}$ to be J-covering if and only if it includes a finite family of morphisms

$([\mathbf{x}_i \mapsto \mathbf{y}|\theta_i(\mathbf{x}_i,\mathbf{y})] : \{\mathbf{x}_i|\phi_i(x_i)\} \longrightarrow \{\mathbf{y}|\psi(\mathbf{y})\}|i \in 1..n)$ for some $n \in \omega$, for which the geometric sequent $\forall\mathbf{y}(\psi(\mathbf{y}) \longrightarrow \bigvee_{i\in 1..n} \exists\mathbf{x}_i\theta_i(\mathbf{x}_i,\mathbf{y}))$ is provable in T; it may be verified that J is a Grothendieck topology. Finally, the topos of sheaves $\mathcal{E}(T)$ on the site $(\mathcal{C}, J)$ is a classifying topos for the theory T ([TT] p. 245).

We note that $\{\mathbf{y}|\psi(\mathbf{y})\}$ is covered by the empty sieve if and only if $\psi(\mathbf{y})$ is inconsistent with T, i.e, T proves the geometric sequent $\forall\mathbf{y}(\psi(\mathbf{y}) \longrightarrow \perp)$. Such objects can be deleted from the site without changing the topos of sheaves ([SGA] III.4.1). More precisely, let $\mathcal{C}'$ be the full subcategory of $\mathcal{C}$ whose obejcts are $\{\mathbf{x}|\phi(\mathbf{x})\}$ for geometric $\phi(\mathbf{x})$ consistent with T, and let J' be the topology on $\mathcal{C}'$ induced by J. Then $\mathcal{E}(T)$ is also topos of sheaves on $(\mathcal{C}', J')$. We also note that, if $[\mathbf{x} \mapsto \mathbf{y}|\theta(\mathbf{x},\mathbf{y})] : \{x|\phi(\mathbf{x})\} \longrightarrow \{\mathbf{y}|\psi(\mathbf{y})\}$ is any morphism of $\mathcal{C}'$, then $\theta(\mathbf{x},\mathbf{y})$ is consistent with T. This follows immediately from the facts that $\phi(x)$ is consistent with T and that T proves $\forall\mathbf{x}(\phi(\mathbf{x}) \longrightarrow \exists\mathbf{y}\theta(\mathbf{x},\mathbf{y}))$.

So far in this section we have not made any assumptions on the language L. We prove that for our purposes it may be assumed without any loss that L is a purely relational language. To this end, we assume given a language L. We wish to construct a relational language L^r, into which the formulas of L may be translated. For every n-place function-symbol $f_{i,n} \in$ Symbol(L) we introduce an $(n+1)$-place relation symbol $F_{i,n+1}$ in L^r. This includes constant symbols as they are 0-place function symbols The n-place predicate symbols and the variables in L and L^r are the same. The translation $\tau : \mathrm{Fml}(L) \longrightarrow \mathrm{Fml}(L^r)$ is defined in two stages as follows.

DEFINTION IV.1. We define first a translation for formulas of the form $t = v$ (v a variable) by induction.
$\tau_0(v_i = v_j) := v_i = v_j$.

$\tau_0(f_{i,n}t_i..t_n = v_j) := \exists x_1, ..\exists x_n(\tau_0(t_1 = x_1) \wedge .. \wedge \tau_0(t_n = x_n) \wedge F_{i,n+1}(t_1..t_nv_j))$,

where $x_1, .., x_n$ are variables not occurring in the formula being translated.

We define translations next for all other formulas.

$$\tau(t_1 = t_2) := \exists x(\tau_0(t_1 = x) \wedge \tau_0(t_2 = x)); \tau(P_{i,n}t_1, ..t_n) :=$$

$$\exists x_1..\exists x_n(\tau_0(t_1 = x_1) \wedge .. \wedge \tau_0(t_2 = x_n) \wedge P_{i,n}x_1..x_n).$$

$$\tau(\neg\phi) := \tau(\neg\phi); \tau(\phi_1 \wedge \phi_2) :=$$

$$\tau(\phi_1) \wedge \tau(\phi_1); \tau(\phi_1 \vee \phi_2) := \tau(\phi_1) \vee \tau(\phi_1);$$

$$\tau(\phi_1 \longrightarrow \phi_2) := \tau(\phi_1) \longrightarrow \tau(\phi_1);$$

$$\tau(\exists v\phi(v)) := \exists v\tau(\phi(v)); \tau(\forall v\phi(v)) := \forall v\tau(\phi(v)).$$

EXAMPLE IV.2. $\tau(f_{1,2}c_{1,0}f_{1,1}v = c_{1,0}) = \exists x_1 \exists x_2 \exists x_3 (C_{1,1}(x_2) \wedge F_{1,2}(vx_3) \wedge F_{1,3}(x_2x_3x_1) \wedge C_{1,1}(x_1))$.

REMARK IV.3. It is clear from the precceding that the translation sends geometric formulas to geometric formulas. We also note that, given a geometric sequent $\forall \mathbf{x}(\phi(\mathbf{x}) \longrightarrow \psi(\mathbf{x}))$,

$$\tau(\forall \mathbf{x}(\phi(\mathbf{x}) \longrightarrow \psi(\mathbf{x}))) = \forall \mathbf{x}(\tau(\phi(\mathbf{x})) \longrightarrow \tau(\psi(\mathbf{x}))).$$

Hence a geometric sequent in L translates to a geometric sequent in L^r.

REMARK IV.4. Thus, given a geometric theory T in language L, we may construct a geometric theory T^r in L^r, of which the axioms include, in addition to the L^r-translations of the axioms of T, the following axioms that are clearly geometric.

(1) $\forall \mathbf{x}(\top \longrightarrow \exists y F_{i,n+1}(\mathbf{x}, y))$.

(2) $\forall \mathbf{x} \forall y \forall z (F_{i,n+1}(\mathbf{x}, y) \wedge F_{i,n+1}(\mathbf{x}, z) \longrightarrow y = z)$.

REMARK IV.5. It should be noted that there is a tanslation σ from L^r to L which is inverse to τ in the sense that $\forall \phi \in \mathrm{Fml}(L), \phi$ is logically equivalent to $\sigma \circ \tau(\phi)$.

Given any L -structure $\mathcal{A}$, we may define an L^r-structure $\mathcal{A}^r$ as follows. $A^r := A$. For every predicate symbol P of $L, P^{\mathcal{A}^r} := P^{\mathcal{A}}$. For every function symbol f of $L, F^{\mathcal{A}^r} := \mathrm{Gr}(f^{\mathcal{A}})$. With these definitions, we have the following.

(1) For every $n \in \omega$ and every $\phi \in \mathrm{Fml}_n(L), \phi^r \in \mathrm{Fml}(L^r)$.

(2) For every $\mathcal{A} \in \mathrm{Str}(L)$ and every $\mathbf{a} \in A^n, \mathcal{A} \models \phi(\mathbf{a}) \iff \mathcal{A}^r \models \phi^r(\mathbf{a})$. Furthermore, every model of T^r is $\mathcal{A}^r$ for a unique model $\mathcal{A}$ of T.

REMARK IV.6. In view of the preceding, in what follows, we shall assume that all geometric theories are specified in a relational language L.

We develop next several results about syntactic categories which will be needed to carry out the reduction referred to at the beginning of this section. In most cases we omit the proofs as they involve straightforward computations that are rather tedious.

REMARK IV.7. It can be easily verified that diagram (1) commutes for all functions $\sigma : S \longrightarrow X$ in the category $\mathcal{S}$ of sets and functions. π_1, ϕ_2 and π_2' the usual projections. This motivates the following proposition, which is essentially a translation of this observation in the context of syntactic categories.

PROPOSITION IV.7. *All the triangles in diagram (2), with* $f, g, p_\sigma, p'_\sigma, m_\sigma, \imath_\sigma, S, X, \mathrm{Gr}(m_\sigma)$ *and* $\mathrm{Im}(m_\sigma)$ *defined as follows, commute.*

(1) 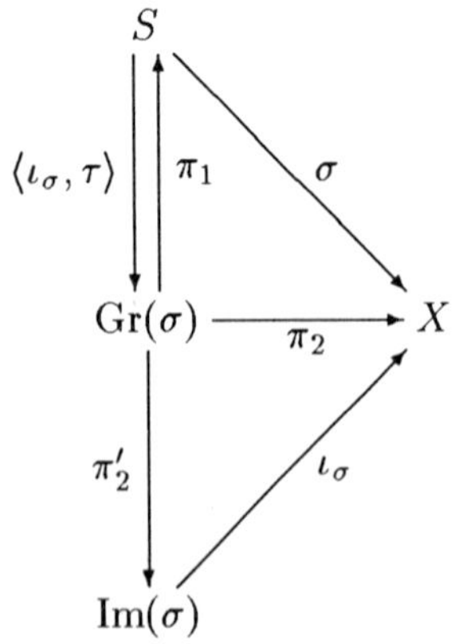

(2) 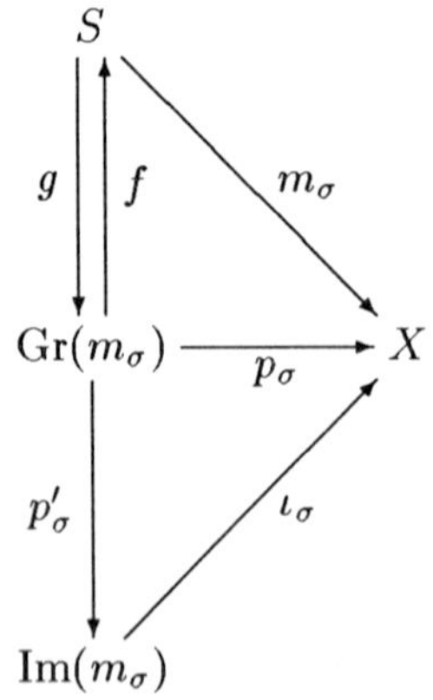

$S := \{\mathbf{s}|S(\mathbf{s})\}.X := \{\mathbf{x}|X(\mathbf{x})\}.m_\sigma := [\mathbf{s} \mapsto \mathbf{x}(\mathbf{s},\mathbf{x})] : S \longrightarrow X.$

$\mathrm{Gr}(m_\sigma) := \{(\mathbf{s}',\mathbf{x}')|\sigma(\mathbf{s}',\mathbf{x}')\} = \{(\mathbf{s}',\mathbf{x}')|\sigma(\mathbf{s}',\mathbf{x})' \wedge S(\mathbf{s}') \wedge X(\mathbf{x}')\}.$
$\mathrm{Im}(m_\sigma) := \{\mathbf{x}''|\exists \mathbf{s}\sigma(\mathbf{s},\mathbf{x}'')\}.$
$g := [(s',\mathbf{x}') \mapsto \mathbf{s} : \sigma(\mathbf{s}',\mathbf{x}') \wedge \mathbf{s} = \mathbf{s}'] : \mathrm{Gr}(m_\sigma) \longrightarrow S.$
$f := [\mathbf{s} \mapsto (\mathbf{s}',\mathbf{x}') : \sigma(\mathbf{s}',\mathbf{x}') \wedge s' = s] : S \longrightarrow \mathrm{Gr}(m_\sigma).$
$\imath_\sigma := [\mathbf{x}'' \mapsto \mathbf{x} : \exists \mathbf{s}\sigma(\mathbf{s},\mathbf{x}'') \wedge \mathbf{x} = \mathbf{x}''] : \mathrm{Im}(m_\sigma) \longrightarrow X.$
$p_\sigma := [(\mathbf{s}',\mathbf{x}') \mapsto \mathbf{x} : \sigma(\mathbf{s}',\mathbf{x}') \wedge \mathbf{x} = \mathbf{x}'] : \mathrm{Gr}(m_\sigma) \longrightarrow X.$
$p'_\sigma := [(\mathbf{s}',\mathbf{x}') \mapsto \mathbf{x}'' : \sigma(\mathbf{s}',\mathbf{x}') \wedge \mathbf{x}'' = \mathbf{x}'] : \mathrm{Gr}(m_\sigma) \longrightarrow \mathrm{Im}(m_\sigma).$ ∎

REMARK IV.8. Proposition IV.7 shows that every morphism m_σ and the corresponding projection p_σ factor through each other. Thus, given a sieve Σ on the object $\{\mathbf{x}|X(\mathbf{x})\}$, we may take the generators of Σ to be projections. We shall, in what follows, use $m_\sigma, p_\sigma, \imath_\sigma$ and p'_σ as defined in Proposition IV.7.

NOTATIONAL CONVENTIONS IV.9. We establish, for later use in this section, the following notational conventions. Let the sieve Σ on the object $\{\mathbf{x}|X(\mathbf{x})\}$ be given.

$\Omega(\{\mathbf{x}|X(\mathbf{x})\})$:= the collection of sieves on the object $\{\mathbf{x}|X(\mathbf{x})\}$.

$\Omega_J(\{\mathbf{x}|X(\mathbf{x})\})$:= the collection of J-closed sieves on the object $\{\mathbf{x}|X(\mathbf{x})\}$.

$\mathrm{Cl}_J(\Sigma)$:= the J-closure of Σ.

Given an object C of category $\mathcal{C}$ and a family of morphisms $(f_i \in \mathrm{Cod}^<(C)|i \in I)$ with the same codomain C, we define $\underline{\mathrm{Ssp}((f_i \in \mathrm{Cod}^<(C)|i \in I))}$:= the sieve spanned by $(f_i \in \mathrm{Cod}^<(C)|i \in I)$.

PROPOSITION IV.10. *Every J-closed sieve $\Sigma \in \Omega_J(\{\mathbf{x}|X\})$ is generated by the inclusions $\{\imath_\sigma|m_\sigma \in \Sigma\}$. (Recall IV.8 and* IV.9.)

PROOF. Let the closed sieve $\Sigma \in \Omega_J(\{\mathbf{x}|X\})$ be given.

It follows from Remark IV.8 that $\{p_\sigma|m_\sigma \in \Sigma\}$ generates Σ. We have from Proposition IV.7 that for every $m_\sigma \in \Sigma, p_\sigma$ factors through $\imath_\sigma$. Hence $\Sigma = \mathrm{Ssp}(\{p_\sigma|m_\sigma \in \Sigma\}) \subset \mathrm{Ssp}(\{\imath_\sigma|m_\sigma \in \Sigma\})$. To verify the reverse inclusion, it suffices to show that for every $m_\sigma \in \Sigma, \imath_\sigma \in \Sigma$. Hence as Σ is closed, it suffices to show that for every $m_\sigma \in \sigma, (\mathrm{Lcomp}_{\imath_\sigma})^{<}(\Sigma) \in J(\mathrm{Im}(\imath_\sigma))$, where $\mathrm{Im}(\imath_\sigma) = \{\mathbf{x}''|\exists \mathrm{s}\sigma(\mathrm{s}, \mathbf{x}'')\}$. The following diagram describes the situation.

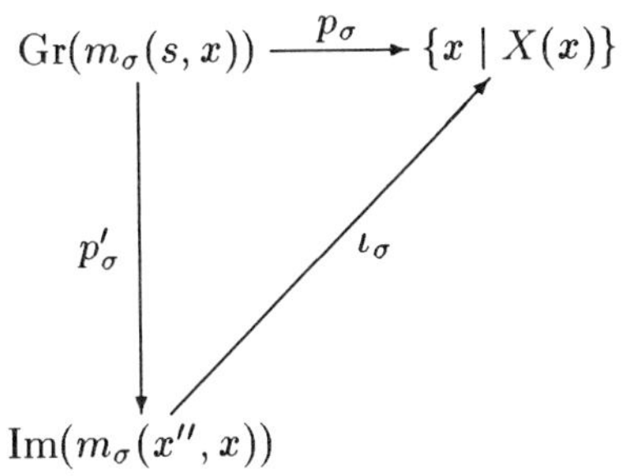

CLAIM. p'_σ covers $\mathrm{Im}(\imath_\sigma)$.

PROOF. We note that $\exists(\mathrm{s}', \mathbf{x}')p'_\sigma \equiv \exists(\mathrm{s}', \mathbf{x}')(\sigma(\mathrm{s}', \mathbf{x}') \wedge \mathbf{x}'' = \mathbf{x}') \equiv \exists \mathrm{s}(\sigma(\mathrm{s}, \mathbf{x}''))$. Hence $\mathrm{T} \vdash \forall \mathbf{x}''(\exists \mathrm{s}\sigma(\mathrm{s}, \mathbf{x}'') \longrightarrow \exists(\mathrm{s}', \mathbf{x}')p'_\sigma)$. Hence the claim. ∎

As $p_\sigma = \imath_\sigma \circ p'_\sigma$, it follows that $p'_\sigma \in (\mathrm{Lcomp}_{\imath_\sigma})^{<}(\Sigma)$. It follows from the above claim that $(\mathrm{Lcomp}_{\imath_\sigma})^{<}(\Sigma) \in J(\mathrm{Im}(\imath_\sigma))$. Hence, as $m_\sigma \in \Sigma$ was arbitrary, $\Sigma = \mathrm{Ssp}(\{\imath_\sigma | m_\sigma \in \Sigma\})$. ∎

LEMMA IV.11. *Let the closed sieve $\Sigma \in \Omega_J(\{\mathbf{x}|(\mathbf{x})\})$ be given. (Recall Notational Conventions IV.9.) Define $\Phi := \{(\exists \mathrm{s}\sigma(\mathrm{s}, \mathbf{x}))^{\mathrm{Mod(T)}} | m_\sigma \in \Sigma\}$. Then Φ is an ideal in $\mathrm{E}_n^+(\mathrm{Mod(T)})_{\leq (X(\mathbf{x}))^{\mathrm{Mod(T)}}}$.* ∎

LEMMA IV.12. *Let the ideal $\Phi \subset \mathrm{E}_n^+(\mathrm{Mod(T)})_{\leq (X(\mathbf{x}))^{\mathrm{Mod(T)}}}$ be given. We note that for every $(\sigma(\mathbf{x}))^{\mathrm{Mod(T)}} \in \Phi, \mathrm{T} \vdash \forall \mathbf{x}(\sigma(\mathbf{x}) \longrightarrow X(\mathbf{x}))$, and that we have (as in Lemma IV.11) an inclusion*

$$\imath_\sigma := [\mathbf{x}' \mapsto \mathbf{x} : \sigma(\mathbf{x}') \wedge \mathbf{x} = \mathbf{x}'] : \{\mathbf{x}'|\sigma(\mathbf{x}')\} \longrightarrow \{\mathbf{x}|X(\mathbf{x}\}.$$

Define

$$\Sigma := \mathrm{Ssp}(\{\imath_\sigma | (\sigma(\mathbf{x}))^{\mathrm{Mod(T)}} \in \Phi\}),$$

$$\Sigma' := C1_J(\Sigma) \in \Omega_J(\{\mathbf{x}|X(\mathbf{x})\}) \quad \textit{and}$$

$$\Phi' := \{(\exists \mathrm{s}\alpha(\mathrm{s}, \mathbf{x}))^{\mathrm{Mod(T)}} | m_\alpha \in \Sigma'\}.$$

$$\textit{Then} \quad \Phi' = \Phi.$$

∎

COROLLARY IV.13. *Σ' has no inclusions other than those already in Σ. As Σ' is closed, it is generated as a closed sieve, by all its inclusions.* ∎

PROPOSITION IV.14. *Let the closed sieve $\Sigma \in \Omega_J(\{\mathbf{x}|X(\mathbf{x})\})$ be given. Define $\Phi := \{(\exists \mathrm{s}\sigma(\mathrm{s}, \mathbf{x}))^{\mathrm{Mod(T)}} | m_\sigma \in \Sigma\}$. By Lemma IV.11, Φ is an ideal in $\mathrm{E}_n^+(\mathrm{Mod(T)})_{\leq (X(\mathbf{x}))^{\mathrm{Mod(T)}}}$. Define for every $(\phi(\mathbf{x}))^{\mathrm{Mod(T)}} \in \Phi$,*

$$\imath_\phi := [\mathbf{x}' \mapsto \mathbf{x} : \phi(\mathbf{x}') \wedge \mathbf{x} = \mathbf{x}'] : \{\mathbf{x}'|\phi(\mathbf{x}')\} \mapsto \{\mathbf{x}|X(\mathbf{x})\},$$

$$\Gamma := \mathrm{Ssp}(\{\imath_\phi | (\phi(\mathbf{x}))^{\mathrm{Mod(T)}} \in \Phi\}), \quad \textit{and}$$

$$\Gamma' := C1_J(\Gamma). \quad \textit{Then} \quad \Gamma' = \Sigma.$$

■

DEFINITION IV.15. In view of the preceding lemmas, we may define the maps

$$f : \Omega_J(\{\mathbf{x}|X(\mathbf{x})\}) \longrightarrow \mathrm{Idl}(\mathrm{E}_n^+(\mathrm{Mod(T)})_{\leq (X(\mathbf{x}))^{\mathrm{Mod(T)}}})$$

$$:= \Sigma \mapsto (\exists \mathbf{s}\sigma(\mathbf{s},\mathbf{x}))^{\mathrm{Mod(T)}}|m_\sigma \in \Sigma\} \quad \text{and}$$

$$f^{\leftarrow} : \mathrm{Idl}(\mathrm{E}_n^+(\mathrm{Mod(T)})_{\leq (X(\mathbf{x})^{\mathrm{Mod(T)}})}) \longrightarrow \Omega_J(\{\mathbf{x}|X(\mathbf{x})\})$$

$$:= \Phi \mapsto C1_J(\mathrm{Ssp}(\{\imath_\phi|(\phi(\mathbf{x})^{\mathrm{Mod(T)}} \in \Phi\})$$

PROPOSITION IV.16. *f and $f^{\leftarrow}$ are order-preserving.* ■

COROLLARY IV.19. $\mathrm{Idl}(\mathrm{E}_n^+(\mathrm{Mod(T)})_{\leq (X(\mathbf{x}))^{\mathrm{Mod(T)}}})^{\neg}$ *and* $\Omega_J(\{\mathbf{x}|X(\mathbf{x})\})$ *are isomorphic as lattices.*

PROOF. f and $f^{\leftarrow}$ are order-preserving bijections. ■

COROLLARY IV.20. $\mathrm{Shv}(\mathcal{C}, J) \models I_r \Leftrightarrow$ *for every* $n \in \omega, \mathrm{DE}_n^+(\mathrm{Mod(T)}) \models I_r$.

PROOF $\mathrm{Shv}(\mathcal{C}, J) \models I_r$

$\Leftrightarrow \mathrm{Shv}(\mathcal{C}, J) \models \forall x_0, .., \forall x_r (\bigwedge_{i,j \in 0..r, i<j} (x_i \wedge x_j = 0) \longrightarrow (\bigvee_{i \in 0..r} x_i^* = 1))$.

$\Leftrightarrow \forall X \in Obj(\mathcal{C}), \forall x_0, .., \forall x_r \in \Omega,$

$\forall x_0, .., \forall x_r (\bigwedge_{i,j \in 0..r, i<j} (x_i \wedge x_j = 0) \longrightarrow (\bigvee_{i \in 0..r} x_i^* = 1))$.

$\Leftrightarrow \forall X \in Obj(\mathcal{C}), \forall x_0, .., \forall x_r \in \Omega_J(X),$

$X \Vdash \forall x_0, .., \forall x_r (\bigwedge_{i,j \in 0..r, i<j} (x_i \wedge x_j = 0) \longrightarrow (\bigvee_{i \in 0..r} x_i^* = 1))$.

$\Leftrightarrow \forall n \in \omega, \forall X \in \mathrm{E}_n^+, \mathrm{Idl}(\mathrm{E}_n^+(\mathrm{Mod(T)})_{\leq (X(x_0,..,x_r))^{\mathrm{Mod(T)}}} \models I_r$.

(by Corollary III.19)

$\Leftrightarrow \forall n \in \omega, \forall X \in \mathrm{E}_n^+, \mathrm{DE}_n^+(\mathrm{Mod(T)})_{\leq (X(x_0,..,x_r))^{\mathrm{Mod(T)}}} \models I_r$.

(by Proposition II.36)

$\Leftrightarrow \forall n \in \omega, \mathrm{DE}_n^+(\mathrm{Mod(T)}) \models I_r$. (by Corollary III.19) ■

REMARK IV.21. This completes the reduction referred to at the begining of the section.

Section V: Model-Theoretic Considerations

We assume given a geometric theory T in a countable relational language L. In what follows, we derive results that describe precise model-theoretic conditions under which, given Ξsome $k \in \omega$, for every $n \in \omega$, $\mathrm{DE}_n^+(\mathrm{Mod(T)})$, i.e., the lattice of equivalence classes (up to T-provability) of arbitrary disjunctions of existential positive formulas with free variables among $x_1, .., x_n$, satisfies the kth Lee identity I_k. These results will, in particular, answer the question posed in Section IV, namely, given a geometric theory T, under what conditions does its classifying topos satisfy the Lee identity I_k. This extends the results of Blass and Scedrov, who studied the case $k = 0$, i.e., conditions under which the Boolean axiom is satisfied by the classifying topos [B&S]. They proved the following theorem.

THEOREM V.1. *The classifying topos $\mathcal{E}(T)$ of a geometric theory T is Boolean if and only if both the following conditions are satisfied:*
(1) Every formula in the language of T is T-provably equivalent to a geometric formula.
(2) *For any finite list* $\mathbf{x}$ *of variables, there are, up to T-provable equivalence, only finitely many formulas with free variables among* $\mathbf{x}$. Π We note that the case $k = 1$ is of particular interest. It characterizes those geometric theories whose classifying topoi satisfy De Morgan's law.

On the model-theoretic side these results generalize the results of Weispfenning [W], who considered the lattices of equivalence classes of existential formulas rather than existential positive ones. In fact, there are many purely model-theoretic questions that can be raised in this context, which we shall not have the occasion to consider, as our interest is to investigate topoi.

We shall, in what follows, need the notion of a c^+-resultant, which generalizes WeIspfenning's notion of a resultant, which, in turn, was a generalization of Robinson's notion of a resultant [R]. We shall use the c^+-resultant as a technical device for relating lattice-operations in $\mathrm{DE}_n^+(\mathrm{Mod(T)})$ with the usual lattice-operations in the Lindenbaum algebra of T. The other notions needed are those of a base and a base of conditions. Both of these are generalizations of similar notions developed by Weispfenning.

We begin by defining the c^+-resultant.

Let $\phi(\mathbf{x}) \in \mathrm{Fml}_{n,\mathbf{x}}(L)$ and $\rho(\mathbf{x}) \in \mathrm{DE}_n^+(\mathrm{Mod(T)})$ be given.

DEFINITION V.2. For every $\mathcal{A} \in \mathrm{Mod(T)}$ and every $\mathbf{a} \in A^n$, $\underline{\mathrm{FPS}_{\mathcal{A},\mathbf{a}}(L)} := \{\mathcal{G} \in \mathrm{FPS}_{\mathcal{A}}(L) |$ there exists a homomorphism $h : \mathcal{G} \longrightarrow \mathcal{A}$ and $\mathbf{g} \in (\mathrm{Gen}(\mathcal{G}))^n$ such that $h(\mathbf{g}) = \mathbf{a}\}$.

DEFINITION V.3. For every $\mathcal{C} \in \mathrm{Str}(L)$, $\underline{\mathrm{HomExtn}_T(\mathcal{C})} := \{\mathcal{A} \in \mathrm{Mod(T)}$ there exists a homomorphism $h : \mathcal{C} \longrightarrow \mathcal{A}\}$. The abbreivation $\mathrm{HomExtn}_T$ is supposed to indicate the collection of homomorphic extensions that are models of T.

DEFINTION V.4. $\rho(\mathbf{x})$ is called a $\overline{c^+\text{-resultant for } \phi(\mathbf{x}) \text{ in } \mathrm{Mod(T)}}$:$\Leftrightarrow$ for every $\mathcal{A} \in \mathrm{Mod(T)}$ and every $\mathbf{a} \in A^n$, $(\mathcal{A} \models \rho(\mathbf{a}) \Leftrightarrow$ there exists $\mathcal{G} \in \mathrm{FPS}_{\mathcal{A},\mathbf{a}}(L)$

and $\mathbf{g} \in (\mathrm{Gen}(\mathcal{G}))^n$ such that for every $\mathcal{B} \in \mathrm{HomExtn}_T(\mathcal{G})$, and every $h_{\mathcal{G}\mathcal{B}} : \mathcal{G} \longrightarrow \mathcal{B}, \mathcal{B} \models \phi(h_{\mathcal{G}\mathcal{B}}\mathbf{g})$).

REMARK V.5. The defining clause in the preceding, when expanded, means that there exist $\mathcal{G} \in \mathrm{FPS}_T(L), \mathbf{g} \in (\mathrm{Gen}(\mathcal{G}))^n$, and $h : \mathcal{G} \longrightarrow \mathcal{A}$ such that $\mathbf{a} = h\mathbf{g}$, and such that for every $\mathcal{B} \in \mathrm{HomExtn}_T(\mathcal{G})$ and every $h_{\mathcal{G}\mathcal{B}} : \mathcal{G} \longrightarrow \mathcal{B}, \mathcal{B} \models \phi(h_{\mathcal{G}\mathcal{B}}\mathbf{g})$. The following diagram describes the situation.

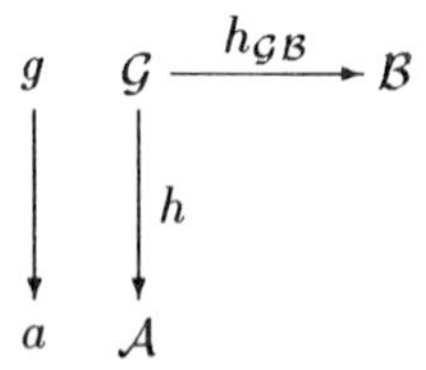

We first establish some properties of c^+-resultants.

THEOREM V.6. *Let $\phi, \psi, \theta \in \mathrm{DE}_n^+(\mathrm{Mod(T)})$ be given such that* $\theta^{\mathrm{Mod(T)}} = \phi^{\mathrm{Mod(T)}} \underset{D}{\longrightarrow} \psi^{\mathrm{Mod(T)}}$. *Then θ is a c^+-resultant for $\phi \longrightarrow \psi$ in* Mod(T).

PROOF. Let $\mathcal{A} \in \mathrm{Mod(T)}$ and $\mathbf{a} \in A^n$ be given.
($\Rightarrow$) Assume $\mathcal{A} \models \theta(\mathbf{a})$. By Proposition II.12, we may choose $p \in \mathrm{Cond}^+(\mathcal{A})$ such that

(1) $\mathrm{Rng}(\mathbf{a}') \subset$ Name (p), and

(2) $\vdash \forall \mathbf{x}(\eta_{\mathbf{a},p}(\mathbf{x}) \longrightarrow \theta(\mathbf{x})$, where $\eta_{\mathbf{a},p}(\mathbf{x}) \in \mathrm{Cond}^+_{n,\mathbf{x}}(T)$.

Let $\mathcal{B} \in \mathrm{HomExtn}_T(\langle p \rangle)$ and $h_{\langle p \rangle \mathcal{B}} : \langle p \rangle \longrightarrow \mathcal{B}$ be given. As $\eta_{\mathbf{a},p}(\mathbf{x}) \in \mathrm{E}_n^+(\mathrm{Mod(T)})$, we have $\mathcal{B} \models \eta_{\mathbf{a},p}(h_{\langle p \rangle \mathcal{B}}\mathbf{a})$. Hence by (2), $\mathcal{B} \models \theta(h_{\langle p \rangle \mathcal{B}}\mathbf{a})$. As $\mathrm{DE}_n^+(\mathrm{Mod(T)})$ is a CHA, it follows that $\theta^{\mathrm{Mod(T)}}$
$= \phi^{\mathrm{Mod(T)}} \longrightarrow_D \psi^{\mathrm{Mod(T)}}$
$= \bigvee\{\alpha^{\mathrm{Mod(T)}} | \phi^{\mathrm{Mod(T)}} \wedge \alpha^{\mathrm{Mod(T)}} \leq_D \psi^{\mathrm{Mod(T)}}\}$
$= \bigvee\{\alpha^{\mathrm{Mod(T)}} | (\phi \wedge \alpha)^{\mathrm{Mod(T)}} \leq_D \psi^{\mathrm{Mod(T)}}\}$
$= \bigvee\{\alpha^{\mathrm{Mod(T)}} | \mathrm{Mod(T)} \models \phi \wedge \alpha \longrightarrow \psi\}$
$= \bigvee\{\alpha^{\mathrm{Mod(T)}} | \mathrm{Mod(T)} \models \alpha \longrightarrow (\phi \longrightarrow \psi)\}$
$= (\bigvee\{\alpha | \mathrm{Mod(T)} \models \alpha \longrightarrow (\phi \longrightarrow \psi)\})^{\mathrm{Mod(T)}}$ (b)
Hence $\mathcal{B} \models (\bigvee\{\alpha | \mathrm{Mod(T)} \models \alpha \longrightarrow (\phi \longrightarrow \psi)\})(h_{\langle p \rangle \mathcal{B}}\mathbf{a})$. Hence, we may choose some disjunct α' such that $\mathcal{B} \models \alpha'(h_{\langle p \rangle \mathcal{B}}\mathbf{a})$. But, as $\mathrm{Mod(T)} \models \alpha' \longrightarrow (\phi \longrightarrow \psi)$, it follows that $\mathcal{B} \models (\phi \longrightarrow \psi)(h_{\langle p \rangle \mathcal{B}}\mathbf{a})$. We set $\mathcal{G} := \langle p \rangle, \mathbf{g} := \mathbf{a}$ and $h := \imath_{\langle p \rangle \subset \mathcal{A}}$. Then, clearly $\mathbf{a} = (\imath_{\langle p \rangle \subset \mathcal{A}})\mathbf{g} = h(\mathbf{g})$. As $\mathcal{B}, h_{\langle p \rangle \mathcal{B}}$ were arbitrary, we have established the $\Rightarrow$ part in Definition V.4.
($\Leftarrow$) Assume the right hand side in Definition V.4. Hence, we may choose $\mathcal{G} \in \mathrm{FPS}_T(L), \mathbf{g} \in (\mathrm{Gen}(\mathcal{G}))^n$) and $h : \mathcal{G} \longrightarrow \mathcal{A}$ such that

(1) $\mathbf{a} = h(\mathbf{g})$, and
(2) for every $\mathcal{B} \in \mathrm{HomExtn}_T(\mathcal{G})$ and every $h_{\mathcal{G}\mathcal{B}} : \mathcal{G} \longrightarrow \mathcal{B}, \mathcal{B} \models (\phi \longrightarrow \psi)(h_{\mathcal{G}\mathcal{B}}\mathbf{g})$.
Define $p := p_{\mathcal{G}h\mathcal{A}}$, and construct $\eta_{\mathbf{a},p}(\mathbf{x}) \in \mathrm{Cond}^+_{n,\mathbf{x}}(T)$ as in Remark II.10.

CLAIM. $\mathrm{Mod(T)} \models \forall \mathbf{x}(\eta_{\mathbf{a},p}(\mathbf{x}) \longrightarrow (\phi \longrightarrow \psi)(\mathbf{x}))$ (a)

PROOF. Suppose (a) is false.

Then, we may choose $\mathcal{C} \in \text{Mod(T)}$ and $\mathbf{c} \in C^n$ such that $\mathcal{C} \not\models \eta_{\mathbf{a},p}(\mathbf{c})$, and $\mathcal{C} \models (\phi \longrightarrow \psi)(\mathbf{c})$. As $\mathcal{C} \models \eta_{\mathbf{a},p}(\mathbf{c})$, it follows that $\mathcal{C} \in \text{HomExtn}_T(\mathcal{C})$, with a homomorphism sending $\mathbf{g}$ to $\mathbf{c}$. Hence, if $\mathcal{C} \not\models (\phi \longrightarrow \psi)(\mathbf{c})$, then this contradicts (2). Hence the claim. ∎

We note that $\mathcal{A} \in \text{HomExtn}_T(\mathcal{G})$. Hence we have $\mathcal{A} \models \eta_{\mathbf{a},p}(\mathbf{a})$. Hence by (a), $\eta_{\mathbf{a},p}(\mathbf{x})$ is one of the disjuncts of (b). Hence, as $\mathcal{A} \models \eta_{\mathbf{a},p}(\mathbf{a})$, it follows that $\mathcal{A} \models \theta(\mathbf{a})$. Hence the result. ∎

COROLLARY V.7. *Let $\phi, \theta \in \text{DE}_n^+(\text{Mod(T)})$ be given, and assume that $\theta^{\text{Mod(T)}} = (\phi^{\text{Mod(T)}})^{*_D}$. Then θ is a c^+-resultant for $\neg\phi$ in* Mod(T).

PROOF. $$(\phi^{\text{Mod(T)}})^{*_D} = \phi^{\text{Mod(T)}} \underset{D}{\longrightarrow} \bot^{\text{Mod(T)}}.$$
By Theorem V.6, θ is a c^+-resultant for $(\phi \longrightarrow 0) = \neg\phi$ in Mod(T). ∎

THEOREM V.8. *Let $\Phi \subset \text{DE}_n^+(\text{Mod(T)}), \psi \in \text{DE}_n^+(\text{Mod(T)})$ be given such that $(\psi)^{\text{Mod(T)}} = \bigwedge_D \Phi^{\text{Mod(T)}}$. Then ψ is c^+-resultant for $\bigwedge \Phi$.*

PROOF. We omit this proof as it is very similar to that of Theorem VI.6. ∎

We next define the notion of a base.

DEFINITION V.9. Let $\mathcal{G} \in \text{FPS}_T(L), \mathbf{g} \in (\text{Gen}(\mathcal{G}))^n$ and $\mathcal{B}_1, ..\mathcal{B}_r \in \text{Mod(T)}$ be given. $(\mathcal{B}_1, .., \mathcal{B}_r)$ is called a base (of length r) over $(\mathcal{G}, \mathbf{g})$ in Mod(T):⇔

(1) for every $i \in 1..r$, there exist homomorphisms $h_i : \mathcal{G} \longrightarrow \mathcal{B}_i$ such that

(2) for every pair of distinct elements $i, j \in 1..r, (\mathcal{G}, \mathbf{g}, \mathcal{B}_i, \mathcal{B}_j, h_i, h_j)$ cannot be amalgamated in Mod(T).

DEFINITION V.10. Let $\mathcal{G} \in \text{FPS}_T(L), \mathbf{g} \in (\text{Gen}(\mathcal{G}))^n$ and $\mathcal{G}_1, .., \mathcal{G}_r \in \text{FPS}_T(L)$ be given. $(\mathcal{G}_1, .., \mathcal{G}_r)$ is called a base of conditions (of length r) over $(\mathcal{G}, \mathbf{g})$ in Mod(T):⇔

(1) for every $i \in 1..r$, there exists a homomorphism $h_i : \mathcal{G} \longrightarrow \mathcal{G}_i$ such that

(2) for every pair of distinct elements $i, j \in 1..r, (\mathcal{G}, \mathbf{g}, \mathcal{G}_i, \mathcal{G}_j, h_i, h_j)$ cannot be amalgamated in Mod(T).

DEFINITION V.11. Let $\Phi \in \text{DE}_n^+(\text{Mod(T)})$ be given. A base $(\mathcal{B}_1, .., \mathcal{B}_r)$ over $(\mathcal{G}, \mathbf{g})$ in Mod(T) is called a test-base (of length r) for $\phi(g)$ over $(\mathcal{G}, \mathbf{g})$ in Mod(T) :⇔ for every $i \in 1..r$, every $\mathcal{B} \in \text{HomExtn}_T(\mathcal{B}_i)$, and every $h_{i\mathcal{B}} : \mathcal{B}_i \longrightarrow \mathcal{B}, \mathcal{B} \models \neg\phi(h_{i\mathcal{B}} h_i \mathbf{g})$.

DEFINITION V.12. Let $\phi \in \text{DE}_n^+(\text{Mod(T)})$ be given. A base of conditions $(\mathcal{G}_1, \ldots, \mathcal{G}_r)$ over $(\mathcal{G}, \mathbf{g})$ in Mod(T) is called a test-base of conditions (of length r) for $\phi(\mathbf{g})$ over $(\mathcal{G}, \mathbf{g})$ in Mod(T):⇔ for every $i \in 1..r$, ever $\mathcal{B} \in \text{HomExtn}_T(\mathcal{G}_i)$, and every $h_{i\mathcal{B}} : \mathcal{G}_i \longrightarrow \mathcal{B}, \mathcal{B} \models \neg\phi(h_{i\mathcal{B}} h_i \mathbf{g})$.

REMARK V.13. The following diagram summarizes the situation in the previous two definitions.

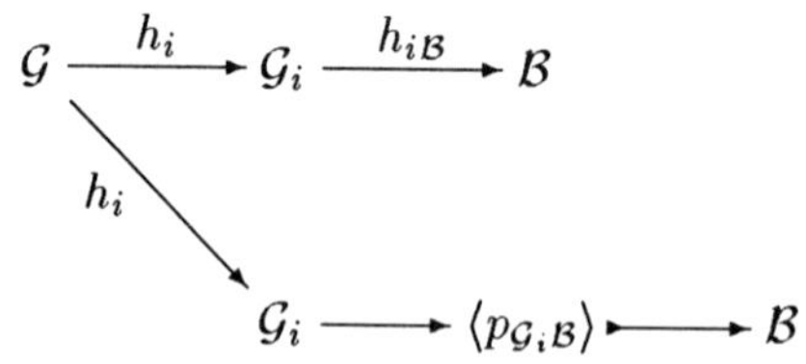

REMARK V.14. The following theorem elucidates the relationship between the notion of a test-base and the notion of a test-base conditions. In the following we assume that an integer r is given.

THEOREM V.15. *For every* $\mathcal{G}_0 \in \mathrm{FPS}_T(L)$ *and every* $\mathrm{g} \in (\mathrm{Gen}(\mathcal{G}_0))^n$, *and every* $\phi \in \mathrm{DE}_n^+$, *the following are equivalent:*

(1) There exists a test-base of length r *for* $\phi(\mathrm{g})$ *over* $(\mathcal{G}_0, \mathrm{g})$ in Mod(T).

(2) There exists a test-base of conditions of length r *for* $\phi(\mathrm{g})$ *over* $(\mathcal{G}_0, \mathrm{g})$ *in* Mod(T).

PROOF. Let $\mathcal{G}_0 \in \mathrm{FPS}_T(L), \mathrm{g} \in (\mathrm{Gen}(\mathcal{G}))^n$ and $\phi \in \mathrm{DE}_n^+$ be given.

(2)$\Rightarrow$(1) We choose a test-base of conditions $(\mathcal{G}_1, .., \mathcal{G}_r)$ for $\phi(\mathrm{g})$ over $(\mathcal{G}_0, \mathrm{g})$ in Mod(T). It follows from Definition V.12 that for every $i \in 1..r$, every $\mathcal{B} \in \mathrm{HomExt}n_T(\mathcal{G}_i)$, and every $h_{i\mathcal{B}} : \mathcal{G}_i \longrightarrow \mathcal{B}, \mathcal{B} \models \neg\phi(h_{i\mathcal{B}}h_{0i}\mathrm{g}).(a)$ Let $i \in 1..r$ be given. As $\mathcal{G}_i \in \mathrm{FPS}_T(L)$ we may choose $\mathcal{B}_i \in$ Mod(T) and a homomorphism $h_i : \mathcal{G}_i \longrightarrow \mathcal{B}_i$. Let $\mathcal{B} \in \mathrm{HomExt}n_T(\mathcal{B}_i)$ and $h_{\mathcal{B}_i\mathcal{B}} : \mathcal{B}_i \longrightarrow \mathcal{B}$ be given. As $\mathcal{B}_i \in \mathrm{HomExt}n_T(\mathcal{G}_i)$, it follows that $\mathcal{B} \in \mathrm{HomExt}n_T(\mathcal{G}_i)$. Hence it follows from (a) that $\mathcal{B} \models \neg\phi(h_{\mathcal{B}_i\mathcal{B}}h_ih_{0i}\mathrm{g})$. As $i, \mathcal{B}$ and $h_{\mathcal{B}_i\mathcal{B}}$ were arbitrary, we conclude that for every $i \in 1..r$, every $\mathcal{B} \in \mathrm{HomExt}n_T(\mathcal{B}_i)$ and every $h_{\mathcal{B}_i\mathcal{B}} : \mathcal{B}_i \longrightarrow \mathcal{B}, \mathcal{B} \models \neg\phi(h_{\mathcal{B}_i\mathcal{B}}h_ih_{0i}\mathrm{g})$. Thus the condition in Definition V.11 is satisfied.

CLAIM. $(\mathcal{B}_1, .., \mathcal{B}_r)$ is a base over $(\mathcal{G}, g)$ in Mod(T).

PROOF. For every $1 \in 1..r$ we have homomorphisms $h_ih_{0i} : \mathcal{G}_0 \longrightarrow \mathcal{B}_i$.

Assume that (2) in Definition V.10 is false. Then, we may choose distinct $i, j \in 1..r, \mathcal{B} \in$ Mod(T) and homomorphisms $h_{i\mathcal{B}}, h_{j\mathcal{B}}$ as shown in the following diagram such that $h_{i\mathcal{B}}(h_ih_{0i}(\mathrm{g})) = h_{j\mathcal{B}}(h_jh_{0j}(\mathrm{g}))$. Hence $h_{i\mathcal{B}}h_i(h_{0i}(\mathrm{g})) = h_{j\mathcal{B}}h_j(h_{0j}(\mathrm{g}))$. But then $(\mathcal{G}, \mathrm{g}, \mathcal{G}_i, \mathcal{G}_j, h_{0i}, h_{0j})$ can be amalgamated in Mod(T), which contradicts (2) in Definition V.8.

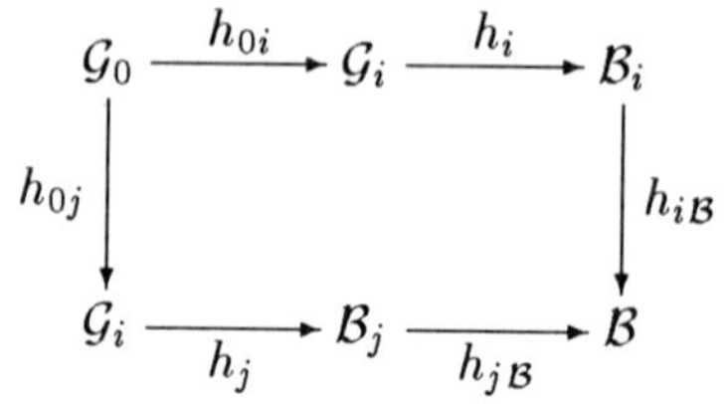

Hence (2) in Definition V.7 holds. Hence the claim. ■

Hence $(\mathcal{B}_1, .., \mathcal{B}_r)$ is a test-base for $\phi(\mathrm{g})$ over $(\mathcal{G}_0, \mathrm{g})$ in Mod(T). ■

(1)⇒(2) We choose a test-base $(\mathcal{B}_1, .., \mathcal{B}_r)$. Let $i \in 1..r$ be given. It follows from Definition V.11 that for every $\mathcal{B} \in \text{HomExtn}_T(\mathcal{B}_i)$ and every $h_{i\mathcal{B}} : \mathcal{B}_i \longrightarrow \mathcal{B}, \mathcal{B} \models \neg\phi(h_{i\mathcal{B}}h_{0i}\mathbf{g})$. Hence $\text{Diag}^+(\mathcal{B}_i) \cup \{\phi(h_{0i}\mathbf{g})\}$ is inconsistent with T. Hence, by compactness, we may choose $p'_i \in \text{Cond}^+(\mathcal{B}_i)$ such that $p'_i \cup \{\phi(h_{0i}\mathbf{g})\}$ is inconsistent with T. Hence $\exists\mathbf{x}(\eta_{h_{0i}(\mathbf{g}),p'_i}, \wedge\phi(\mathbf{x}))$ is inconsistent with T. (b)

We first choose names $(\xi_1, .., \xi_n)$ that do not occur in any $\mathcal{B}_k$, where $k \in 1..r$, and next define for every $k \in 1..r$,

$\text{Diag}^+_{\mathcal{G}_0,\mathbf{g}}(\mathcal{B}_k) :=$ (Set of all sentences of $\text{Diag}^+(\mathcal{B}_k), \forall s \in 1..n$, where every instance of $h_{0k}(g_s)$ has been replaced with $\xi_s) \cup \{\xi_m = \xi_{m'} | \forall m, m' \in 1..n$ such that $h_{0k}(g_m) = h_{0k}(g'_m)\}$. Let $j \in 1..r\backslash\{i\}$ be given. As $(\mathcal{B}_1, .., \mathcal{B}_r)$ is a base over $(\mathcal{G}_0, \mathbf{g})$ in Mod(T), it follows that, $(\mathcal{G}_0, \mathbf{g}, \mathcal{B}_i, \mathcal{B}_j, h_{0i}, h_{0j})$ cannot be amalgated in Mod(T). Hence $\text{Diag}^+_{\mathcal{G}_0,\mathbf{g}}(\mathcal{B}_i) \cup \text{Diag}^+_{\mathcal{G}_0\mathbf{g}}(\mathcal{B}_j)$ is inconsistent with T. Hence, we may choose $q_{ij} \in \text{Fin}(\text{Diag}^+_{\mathcal{G}_0,\mathbf{g}}(\mathcal{B}_i))$ and $q_{ji} \in \text{Fin}(\text{Diag}^+_{\mathcal{G}_0,\mathbf{g}}(\mathcal{B}_j))$ such that $q_{ij} \cup q_{ji}$ is inconsistent with T. (c)

Define $p_{ij} :=$ set of members of $\text{Cond}^+(\mathcal{B}_i)$ obtained by replacing for every $r \in 1..n$ every instance of $h_{0i}(g_r)$ in q_{ij} with ξ_r. Define $p_i := p'_i \cup \bigcup_{j=1..r\backslash\{i\}} p_{ij}$. Then $p_i \in \text{Cond}^+(\mathcal{B}_i)$.

Define $\mathcal{G}_i := \langle p_i \rangle$ and note that $h'_{0i} := \imath_{\langle p_i \rangle \hookrightarrow \mathcal{B}_i} \circ h_{0i} : \mathcal{G}_0 \longrightarrow \mathcal{G}_i$ is a homomorphism.

Let $\mathcal{B} \in \text{HomExtn}_T(\mathcal{G}_i)$ and $h_{i\mathcal{B}} : \mathcal{G}_i \longrightarrow \mathcal{B}$ be given.

CLAIM. $\mathcal{B} \models \neg\phi(h_{i\mathcal{B}}h'_{0i}(\mathbf{g}))$.

PROOF. We assume, towards a contradiction, that $\mathcal{B} \models \phi(h_{i\mathcal{B}}h'_{0i}(\mathbf{g}))$. Hence $\mathcal{B} \models \exists\mathbf{x}(\eta_{h'_{0i}(\mathbf{g}),p'_i}, \wedge\phi(\mathbf{x}))$. As $\mathcal{B} \in \text{Mod(T)}$, this contradicts (b). Hence $\mathcal{B} \not\models \phi(h_{i\mathcal{B}}h'_{0i}\mathbf{g})$, i.e. $\mathcal{B} \models \neg\phi(h_{i\mathcal{B}}h'_{0i}\mathbf{g})$. Hence the claim. ■

As $i \in 1..r, \mathcal{B}$ and $h_{i\mathcal{B}}$ were arbitrary, if we define, for every $i \in 1..r, \mathcal{G}_i$ as in the preceding, then $(\mathcal{G}_1, .., \mathcal{G}_r)$ satisfies the condition in Definition V.12.

It remains to show that $(\mathcal{G}_i, .., \mathcal{G}_r)$ is a base over $(\mathcal{G}_0, \mathbf{g})$.

To this end, let a pair of distinct elements $i, j \in 1..r$ be given, and assume that $(\mathcal{G}_0, \mathbf{g}, \mathcal{G}_i, \mathcal{G}_j, h'_{0i}, h'_{0j})$ can be amalgamated in Mod(T). Hence, we may choose $\mathcal{C} \in \text{Mod(T)}$ such that $\mathcal{C} \models \exists\mathbf{x}(\eta_{h'_{0i}(\mathbf{g}),p_i} \wedge \eta_{h'_{0i}(\mathbf{g}),p_j})$. Hence $\mathcal{C} \models \exists\mathbf{x}(\eta_{h'_{0i}(\mathbf{g}),p_{ij}} \wedge \eta_{h'_{0i}}(\mathbf{g}), p_{ji})$. This contradicts the fact (c) that $q_{ij} \cup q_{ji}$ is inconsistent with T. Hence the above assumption is untenable. Hence $(\mathcal{G}_1, .., \mathcal{G}_r)$ is a base over $(\mathcal{G}_0, \mathbf{g})$ and therefore $(\mathcal{G}_1, .., \mathcal{G}_r)$ is a test-base of conditions for $\phi(\mathbf{g})$ over $(\mathcal{G}_0, \mathbf{g})$ in Mod(T). ■

THEOREM V.16. *For every* $\phi_0 \in \text{DE}^+_n$, *every* $\mathcal{A} \in \text{Mod(T)}$, *and* $\mathbf{a} \in A^n$, *the following are equivalent:*

(α) *For every* $\mathcal{G} \in \text{FPS}_T(L)$ *and every* $\mathbf{g} \in (\text{Gen}(\mathcal{G}))^n$ *such that there exists a homomorphism* $h : \mathcal{G} \longrightarrow \mathcal{A}$ *satisfying* $\mathbf{a} = h(\mathbf{g})$, *there exists a test-base of conditions of length* r *for* $\phi_0(\mathbf{g})$ *over* $(\mathcal{G}, \mathbf{g})$ *in* Mod(T).

(β) *There exist* $\phi_1, .., \phi_r \in \text{DE}^+_n$ *such that* (i) $\phi_0^{\text{Mod(T)}}, \phi_1^{\text{Mod(T)}}, .., \phi_r^{\text{Mod(T)}}$ *are pairwise disjoint, and*

(ii) $\mathcal{A} \models \neg \bigvee_{i \in 1..r}(\phi_i{}^*(\mathbf{a}))$, *where* $\phi_i{}^*$ *is a representative of* $(\phi_i^{\text{Mod(T)}})^{*_D}$.

PROOF. Let $\phi_0 \in \mathrm{DE}_n^+(\mathrm{Mod(T)}), \mathcal{A} \in \mathrm{Mod(T)}$ and $\mathbf{a} \in A^n$ be given.

$(\beta) \Rightarrow (\alpha)$ As (β) holds, we may choose $\phi_1, .., \phi_r \in \mathrm{DE}_n^+(\mathrm{Mod(T)})$ such that (i) and (ii) in (β) are satisfied.

Let $\mathcal{G} \in \mathrm{FPS}_T(L), \mathbf{g} \in (\mathrm{Gen}(\mathcal{G}))^n$ and $h : \mathcal{G} \longrightarrow \mathcal{A}$ be given such that $\mathbf{a} = h(\mathbf{g})$. Let $i \in 1..r$ be given. It follows from (ii) that $\mathcal{A} \not\models \phi_i{}^*(\mathbf{a})$. By Corollary V.7, $\phi_i{}^*$ is a c^+-resultant for $\neg\phi_i$. Hence by Definition V.4, we may, for $\mathcal{G}, \mathbf{g}$ and h as given in the preceding, choose $\mathcal{B}_i \in \mathrm{HomExtn}_T(\mathcal{G})$ and $h_{\mathcal{G}i} : \mathcal{G} \longrightarrow \mathcal{B}_i$ such that

$$\mathcal{B}_i \not\models \neg\phi_i(h_{\mathcal{G}_i}\mathbf{g}), \quad \text{i.e.,} \quad \mathcal{B}_i \models \phi_i(h_{\mathcal{G}_i}\mathbf{g}).$$

By Proposition II.12, we may choose $p_i \in \mathrm{Cond}^+(\mathcal{B}_i)$ such that

(1) $(h_{\mathcal{G}_i}(\mathbf{g}))' \subset \mathrm{Name}(p_i)$, and

(2) $\vdash \forall\mathbf{x}(\eta_{h_{\mathcal{G}_i}(\mathbf{g}),p_i}(\mathbf{x}) \longrightarrow \phi_i(\mathbf{x}))$.

We may define, for every $i \in 1..r, \mathcal{G}_i := \langle p_i \cup p_{\mathcal{G}h_{\mathcal{G}_i}\mathcal{B}_i} \rangle$, where p_i has been chosen in the preceding manner. Note that as $i \in 1..r$ was arbitrary , we indeed have homomorphisms $h_{\mathcal{G}_i} : \mathcal{G} \longrightarrow \mathcal{G}_i$.

CLAIM. $(\mathcal{G}_1, .., \mathcal{G}_r)$ is a test-base of conditions for $\phi_0(\mathbf{g})$ over $(\mathcal{G}, \mathbf{g})$ in Mod(T).

PROOF. First we verify that (2) in Definition V.10 holds and so $(\mathcal{G}_1, .., \mathcal{G}_r)$ is a test-base of conditions.
We assume towards a contradiction that (2) in Definition V.10 does not hold. Hence we may choose distinct $i, j \in 1..r, \mathcal{C} \in \mathrm{Mod(T)}$ and homomorphisms as exhibited such that

$$h_{j\mathcal{C}}h_{\mathcal{G}j}(\mathbf{g}) = h_{i\mathcal{C}}h_{\mathcal{G}i}(\mathbf{g}). \qquad (a)$$

Hence $\mathcal{C} \models (\phi_i \wedge \phi_j)h_{i\mathcal{C}}h_{\mathcal{G}i}(\mathbf{g})$. This is a contradication, since by the choice of ϕ_i, we have

$$\phi_i^{\mathrm{Mod(T)}} \wedge \phi_j^{\mathrm{Mod(T)}} = (\phi_i \wedge \phi_j)^{\mathrm{Mod(T)}} = 0.$$

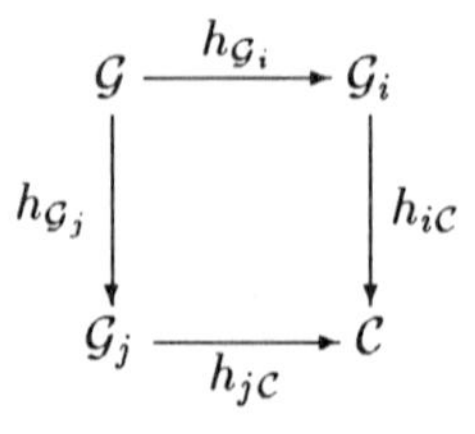

Hence (a) is untenable. Hence for every pair of distinct elements $i, j \in 1..r, (\mathcal{G}, \mathbf{g}, \mathcal{G}_i, \mathcal{G}_j, h_{\mathcal{G}i}, h_{\mathcal{G}j})$ cannot be amalgamated in Mod(T). Hence (2) in Definition V.10 holds. It remains to verify that the condition in Definition V.12 holds. Let $i \in 1..r$ and $\mathcal{B} \in \mathrm{HomExtn}_T(\mathcal{G}_i)$ and $h_{i\mathcal{B}} : \mathcal{G}_i \longrightarrow \mathcal{B}$ be given. Hence we have that $\mathcal{B} \models \eta_{h_{\mathcal{G}_i}(\mathbf{g}),p_i}(h_{i\mathcal{B}}h_{\mathcal{G}_i}\mathbf{g})$. Hence by choice of $p_i, \mathcal{B} \models \phi_i(h_{i\neg\mathcal{B}}h_{\mathcal{G}i}(\mathbf{g}))$.

Hence, as $\phi_0^{\mathrm{Mod(T)}}, \phi_i^{\mathrm{Mod(T)}}$ are disjoint, it follows that $\mathcal{B} \models \phi_0(h_{i\mathcal{B}}h_{\mathcal{G}i}\mathbf{g})$. Hence $\mathcal{B} \not\models \neg\phi_0(h_{i\mathcal{B}}h_{\mathcal{G}i}\mathbf{g})$.
As $i \in 1..r, \mathcal{B}$ and $h_{i\mathcal{B}}$ were arbitrary, the condition in Definition V.12 holds. Hence $(\mathcal{G}_1, .., \mathcal{G}_r)$ is a test-base of conditions for $\phi_0(\mathbf{g})$ over $(\mathcal{G}, \mathbf{g})$ in Mod(T). Hence the claim. ∎

This establishes (α). ■

$(\alpha) \Rightarrow (\beta)$ Under the assumption that (α) holds, we have the following.

CLAIM 1. For every $\mathcal{F} \in \mathrm{FPS}_T(L)$ and every $\mathbf{f} \in (\mathrm{Gen}(\mathcal{F}))^n$ such that there exists a homomorphism $h : \mathcal{F} \longrightarrow \mathcal{A}$ satisfying $\mathbf{a} = h(\mathbf{f})$, we may choose for every $i \in 1..r, \psi_{i\mathcal{F}} \in \mathrm{E}_n^+$ such that

(1) for every $i \in 1..r$, there exist $\mathcal{A}_i \in \mathrm{HomExt}n_T(\mathcal{F})$ and $h_{\mathcal{F}i} : \mathcal{F} \longrightarrow \mathcal{A}_i$ such that $\psi_{i\mathcal{F}}$ is satisfiable in $\mathcal{A}_i$ by $h_{\mathcal{F}i}\mathbf{f}$.

(2) $\phi_0^{\mathrm{Mod(T)}}, \psi_{1\mathcal{F}}^{\mathrm{Mod(T)}}, .., \psi_{r\mathcal{F}}^{\mathrm{Mod(T)}}$ are pairwise disjoint.

PROOF. Let $\mathcal{F} \in \mathrm{FPS_T}(L), \mathbf{f} \in (\mathrm{Gen}(\mathcal{F}))^n$ and a hmomorphism $h : \mathcal{F} \longrightarrow \mathcal{A}$ be given such that $\mathbf{a} = h(\mathbf{f}$. As (α) holds by assumption, we may choose a test-base of conditions $(\mathcal{F}_1, \ldots, \mathcal{F}_r)$ for $\phi_0(\mathbf{f})$ over $(\mathcal{F}, \mathbf{f})$ in Mod(T).

Hence the following hold by definition. For every $i \in i..r$ there exist homomorphisms $h_i : \mathcal{F} \longrightarrow \mathcal{F}_i$ such that

(1) for every pair of distinct elements $i, j \in 1..r, (\mathcal{F}, \mathbf{f}, \mathcal{F}_i, \mathcal{F}_j, h_i, h_j)$ cannot be amalgamated in Mod(T), and

(2) for every $i \in 1..r$, every $\mathcal{B} \in \mathrm{HomExtn_T}(\mathcal{F}_i)$ and every $h_{i\mathcal{B}} : \mathcal{F}_i \longrightarrow \mathcal{B}, \mathcal{B} \models \neg\phi_0(h_{i\mathcal{B}}h_i\mathbf{f})$.

The following diagram illustrates the situation.

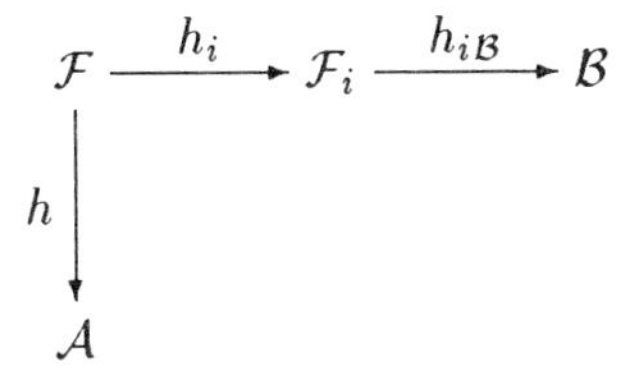

It follows from (2) that $\exists\mathbf{x}(\eta_i(\mathbf{x}) \wedge \phi_0(\mathbf{x})$ is inconsistent with T, where for every $i \in 1..r, \eta_i(\mathbf{x}) := \eta_{(h_i(\mathbf{x}))', p_{\mathcal{F}h_i}\mathcal{F}_i}$ as in Remark II.24. It follows from (1) that for every pair of distinct elements $i, j \in 1..r, \exists\mathbf{x}(\eta_i(\mathbf{x}) \wedge \eta_j(\mathbf{x}))$ is inconsistent with T.

We define for every $i \in 1..r, \psi_{i\mathcal{F}}(\mathbf{x}) := \eta_i(\mathbf{x})$. It follows from the preceding that for every $i \in 1..r, \exists\mathbf{x}(\psi_{i\mathcal{F}}(\mathbf{x}) \wedge \phi_0(\mathbf{x}))$ is inconsistent with T, and for every pair of distinct elements $i, j \in 1..r, \exists\mathbf{x}(\psi_{i\mathcal{F}}(\mathbf{x}) \wedge \psi_{j\mathcal{F}}(\mathbf{x}))$ is inconsistent with T.

Hence for every $i \in 1..r, \phi_0^{\mathrm{Mod(T)}}, \psi_{i\mathcal{F}}^{\mathrm{Mod(T)}}, .., \psi_{r\mathcal{F}}^{\mathrm{Mod(T)}}$ are pairwise disjoint.

Moreover, since for every $i \in 1..r, \mathcal{F}_i \in \mathrm{FPS}_T(L)$, we may choose $\mathcal{A}_i \in$ Mod(T) and a homomorphism $h_{ii} : \mathcal{F}_i \longrightarrow \mathcal{A}_i$ such that $\mathcal{A}_i \models \psi_{i\mathcal{F}}(h_{ii}h_i\mathbf{f})$. As $h_{ii}h_i$ is a homomorphism $\mathcal{F} \longrightarrow \mathcal{A}_i$, it follows that for every $i \in 1..r, \mathcal{A}_i \in \mathrm{HomExt}n_T(\mathcal{F})$.

We set for every $i \in 1..r, h_{\mathcal{F}i} = h_{ii}h_i$. Hence the claim. ■

We assume next, towards a contradiction that (α) holds and that (β) is false. Under the assumption that (β) is false, we have the following.

CLAIM 2. For all $\phi_1, .., \phi_r \in \mathrm{DE}_n^+$ such that $\phi_0^{\mathrm{Mod(T)}}, \phi_1^{\mathrm{Mod(T)}}, .., \phi_r^{\mathrm{Mod(T)}}$ are pairwise disjoint, there exist $\mathcal{G} \in \mathrm{FPS}_T(L)\mathbf{g} \in (\mathrm{Gen}(\mathcal{G}))^n$ and a homomorphism $h : \mathcal{G} \longrightarrow \mathcal{A}$ such that the following hold.

(1) $\mathbf{a} = h(\mathbf{g})$.

(2) There eixsts $i' \in 1..r$ such that for every $\mathcal{B} \in \text{HomExt}n_T(\mathcal{G})$, and every $h_{\mathcal{G}\mathcal{B}} : \mathcal{G} \longrightarrow \mathcal{B}, \mathcal{B} \models \neg\phi_i'(h_{\mathcal{G}\mathcal{B}}(\mathbf{g}))$.

(3) $p_{\mathcal{G}h\mathcal{A}} \cup \{\phi_i(\mathbf{a})\}$ is inconsistent with T.

PROOF. Let $\phi_1, \ldots, \phi_r \in DE_n^+(\text{Mod(T)})$ be given such that $\phi_0^{\text{Mod(T)}}, \phi_1^{\text{Mod(T)}}, \ldots, \phi_r^{\text{Mod(T)}}$ are pairwise disjoint. Since (β) is false by assumption, we have

$$\mathcal{A} \not\models \neg \bigvee_{i=1..r} (\phi_i)^*(\mathbf{a}), \text{ i.e., } \mathcal{A} \models \bigvee_{i\in 1..r} \phi_i^*(\mathbf{a}).$$

Hence we may choose $i' \in 1..r$ such that $\mathcal{A} \models \phi_{i'}^*(\mathbf{a})$. But $(\phi_{i'}^*)^{\text{Mod(T)}} = (\phi_{i'}^{\text{Mod(T)}})^{*D}$. Hence by Corollary V.7, $\phi_{i'}^*$ is a c^+-resultant for $\neg\phi_{i'}^*$.

Hence by Definition V.4, we may choose $\mathcal{G} \in \text{FPS}_T(L), \mathbf{g} \in (\text{Gen}(\mathcal{G}))^n$, and a homomorphism $h : \mathcal{G} \longrightarrow \mathcal{A}$ such that

(b1) $\mathbf{a} = h(\mathbf{g})$, and

(b2) for every $\mathcal{B} \in \text{HomExt}n_T(\mathcal{G})$ and every $h_{\mathcal{G}\mathcal{B}} : \mathcal{G} \longrightarrow \mathcal{B}, \mathcal{B} \models \neg\phi_{i'}(h_{\mathcal{G}\mathcal{B}}\mathbf{g})$. It follows from (b2) that $\exists\mathbf{x}(\eta_{(h_{\mathcal{G}\mathcal{B}}(\mathbf{g}))', p_{\mathcal{G}h_{\mathcal{G}\mathcal{B}}\mathcal{B}}(\mathbf{x}) \wedge \phi_{i'}(\mathbf{x})})$ is inconsistent with T. Hence as $\mathcal{A} \in \text{HomExtn}_\text{T}\mathcal{G}), \exists\mathbf{x}(\eta_{h(\mathbf{g}))', p_{\mathcal{G}h\mathcal{A}}(\mathbf{x}) \wedge \phi_{i'}(\mathbf{x})})$ is inconsistent with T. Hence, by (b1), $p_{\mathcal{G}h\mathcal{U}} \cup \{\phi_{i'}(\mathbf{a})\}$ is inconsistent with T. Hence the claim. ■

CLAIM 3. Let $\mathcal{F} \in \text{FPS}_T(L)$ and $\mathbf{f} \in (\text{Gen}(\mathcal{F}))^n$ be given such that there exists $h : \mathcal{F} \longrightarrow \mathcal{A}$ satisfying $\mathbf{a} = h(\mathbf{f})$. Then there exist $\phi_{1\mathcal{F}}, \ldots, \phi_{r\mathcal{F}} \in \text{E}_n^+, \mathcal{F}' \in \text{FPS}_T(L)$, and there exist homomorphisms $h_{\mathcal{F}\mathcal{F}'} : \mathcal{F} \longrightarrow \mathcal{F}'$ and $h' : \mathcal{F}' \longrightarrow \mathcal{A}$ such that the following hold.

(c0) $\mathbf{a} = h'h_{\mathcal{F}\mathcal{F}'}(\mathbf{f})$.

(c1) $\phi_0^{\text{Mod(T)}}, \phi_{1\mathcal{F}}^{\text{Mod(T)}}, \ldots, \phi_{r\mathcal{F}}^{\text{Mod(T)}}$ are pairwise disjoint.

(c2) For every $i \in 1..r, p_{\mathcal{F}h\mathcal{A}} \cup \{\phi_{i\mathcal{F}}(\mathbf{a})\}$ is consistent with T.

(c3) For every $1 \in 1..r, p_{\mathcal{F}'h'\mathcal{A}} \subset \{\phi_{i\mathcal{F}}(\mathbf{a})\}$ is inconsistent with T. The following diagram illustrates the situation.

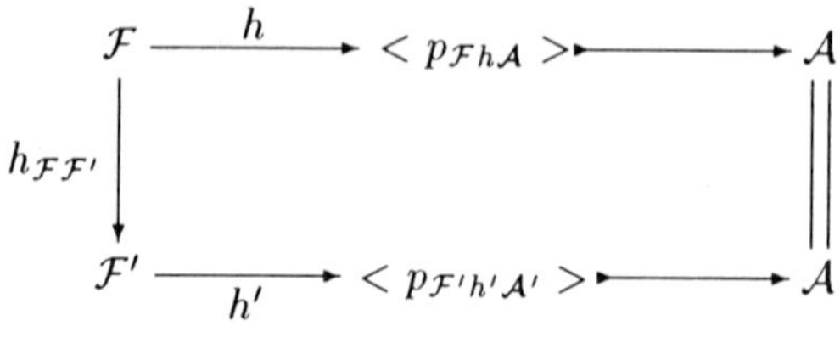

PROOF. Let $\mathcal{F} \in \text{FPS}_T(L)$ and $\mathbf{f} \in (\text{Gen}(\mathcal{F}))^n$ and $h : \mathcal{F} \longrightarrow \mathcal{A}$ be given such that $\mathbf{a} = h(\mathbf{f})$.

We construct, by induction on $s \in 0..r$, formulas $\phi_{1\mathcal{F}}^s, \ldots, \phi_{s\mathcal{F}}^s \in \text{E}_n^+$, structures $\mathcal{F}_s' \in \text{FPS}_T(L)$ and homomorphisms $h_s' : \mathcal{F}_s' \longrightarrow \mathcal{A}$ and $h_{\mathcal{F}\mathcal{F}_s'} : \mathcal{F} \longrightarrow \mathcal{F}_s'$ such that the analogues of (c0), (c1), (c2) and (c3) hold with $\mathcal{F}', h', h_{\mathcal{F}\mathcal{F}'}$ and r replaced with $\mathcal{F}'_s, h'_s, h_{\mathcal{F}\mathcal{F}'_s}$ and s respectively. As (c2) and (c3) are universally quantified over i, the induction may be taken to begin vacuously at $s = 0$ with $\mathcal{F}'_0 = \mathcal{F}, h'_0 = h$ and $h_{\mathcal{F}\mathcal{F}'_0} = \imath_{\mathcal{F}}$.

Induction step: We assume that $\phi^{s-1}_{1\mathcal{F}}, \ldots, \phi^{s-1}_{s-1\mathcal{F}}, \mathcal{F}'_{s-1}, h'_{s-1}$ and $h_{\mathcal{F}\mathcal{F}'_{s-1}}$ have been constructed.

We note that $\mathcal{F}'_{s-1}, h'_{s-1}$ and $(h_{\mathcal{F}\mathcal{F}'_{s-1}}\mathbf{f}) \in \mathrm{Gen}(\mathcal{F}'_{s-1}))^n$ satisfy the hypothesis in Claim 1.

As assumed that (α) holds, by Claim 1, we may choose $\psi_{1\mathcal{F}'_{s-1}}, \ldots, \psi_{r\mathcal{F}'_{s-1}} \in \mathrm{E}^+_n$ such that the following hold.

(α1) For every $i \in 1..r$, there exists $\mathcal{A}^{s-1}_i \in \mathrm{HomExt}n_T(\mathcal{F}'_{s-1})$ such that $\psi_{i\mathcal{F}'_{s-1}}$ is satisfiable in $\mathcal{A}^{s-1}_i$.

(α2) $\phi_0^{\mathrm{Mod(T)}}, \psi_{1\mathcal{F}'_{s-1}}^{\mathrm{Mod(T)}}, \ldots, \psi_{r\mathcal{F}'_{s-1}}^{\mathrm{Mod(T)}}$ are pairwise disjoint.

We note that $\phi_0^{\mathrm{Mod(T)}}, \psi_{1\mathcal{F}'_{s-1}}^{\mathrm{Mod(T)}}, \ldots, \psi_{r\mathcal{F}'_{s-1}}^{\mathrm{Mod(T)}}$ satisfy the hypothesis in Claim 2.

As (β) is false by assumption, by Claim 2, we may choose $\mathcal{G} \in \mathrm{FPS}_T(L), \mathbf{g} \in (\mathrm{Gen}(\mathcal{G}))^n$ and a homomorphism $h'' : \mathcal{G} \longrightarrow \mathcal{A}$ such that the following hold.

(β1) $\mathbf{a} = h''(\mathbf{g})$.

(β2) There exists $i' \in 1..r$ such that for every $\mathcal{B} \in \mathrm{HomExt}n_T(\mathcal{G})$ and $h_{\mathcal{G}\mathcal{B}} : \mathcal{G} \longrightarrow \mathcal{B}, \mathcal{B} \models \neg\psi_{i'\mathcal{F}'_{s-1}}(h_{\mathcal{G}\mathcal{B}}\mathbf{g})$.

(β3) $p_{\mathcal{G}h''\mathcal{A}} \cup \{\psi_{i'\mathcal{F}'_{s-1}}(\mathbf{a})\}$ is inconsistent with T.

We define

$$\mathcal{F}'_s := \langle p_{\mathcal{G}h''\mathcal{A}} \cup p_{\mathcal{F}'_{s-1}h'_{s-1}\mathcal{A}}\rangle,\ \phi^s_{s\mathcal{F}} := \psi_{i'\mathcal{F}'_{s-1}} \wedge \eta_{\mathbf{a},p_{\mathcal{F}'_{s-1}h'_{s-1}\mathcal{A}}} \in \mathrm{E}^+_n$$

and for every $i \in 1..(s-1), \phi^s_{i\mathcal{F}} := \phi^{s-1}_{i\mathcal{F}}$.

REMARK V.16.1. Thus $\phi^s_{i\mathcal{F}}$, once defined for some s does not change at later stages. (So the superscript can be omitted). We define $h'_s := \imath_{\langle p_{\mathcal{G}h''\mathcal{A}} \cup p_{\mathcal{F}'_{s-1}h'_{s-1}\mathcal{A}}\rangle} \hookrightarrow \mathcal{A}$. We note that $h'_{s-1} : \mathcal{F}'_{s-1} \longrightarrow \langle p_{\mathcal{F}'_{s-1}h'_{s-1}\mathcal{A}}\rangle$ remains a homomorphism if we extend the codomain to $\langle p_{\mathcal{G}h''\mathcal{A}} \cup p_{\mathcal{F}'_{s-1}h'_{s-1}\mathcal{A}}\rangle$ as $p_{\mathcal{G}h''\mathcal{A}} \in \mathrm{Cond}^+(\mathcal{A})$ can impose relations on the generators of $\langle p_{\mathcal{F}'_{s-1}h'_{s-1}\mathcal{A}}\rangle$, but cannot have sentences contradicting the sentences in $p_{\mathcal{F}'_{s-1}h'_{s-1}\mathcal{A}}$. We may thus define

$$h_{\mathcal{F}\mathcal{F}'_s} := (h'_{s-1}\,|^{\langle p_{\mathcal{G}h''\mathcal{A}} \cup p_{\mathcal{F}'_{s-1}h'_{s-1}\mathcal{A}}\rangle}) \circ (h_{\mathcal{F}\mathcal{F}'_{s-1}}).$$

Hence $h'_s h_{\mathcal{F}\mathcal{F}'_s}(\mathbf{f})$

$$\begin{aligned}
&= (\imath_{\langle p_{\mathcal{G}h''\mathcal{A}} \cup p_{\mathcal{F}'_{s-1}h'_{s-1}\mathcal{A}}\rangle \hookrightarrow \mathcal{A}}) \circ (h'_{s-1}\,|^{\langle p_{\mathcal{G}h''\mathcal{A}} \cup p_{\mathcal{F}'_{s-1}h'_{s-1}\mathcal{A}}\rangle}) \circ (h_{\mathcal{F}\mathcal{F}'_{s-1}})(\mathbf{f}) \\
&= (h'_{s-1}h_{\mathcal{F}\mathcal{F}'_{s-1}})(\mathbf{f}) \\
&= \mathbf{a}, \quad \text{by the induction hyothesis.}
\end{aligned}$$

Hence $(c0)$ holds.

As $\phi^s_{i\mathcal{F}}$, once defined at some stage remains unchanged at all later stages, to show that (c1) holds, it suffices to show that

(a) $\phi_0^{\mathrm{Mod(T)}}$ and $\phi^s_{s\mathcal{F}}$ are disjoint, and

(b) for every $i \in 1..(s-1)$, $\phi^s_{i\mathcal{F}}$ and $\phi^s_{s\mathcal{F}}$ are disjoint.

(a) follows from property ($\alpha 2$), as $\psi_{i'\mathcal{F}'_{s-1}}$ is a conjunct of $\psi^s_{s\mathcal{F}}$.
To show that (b) holds, let $i \in 1..(s-1)$ be given and assume that (b) is false. Hence we may choose $\mathcal{C} \in \mathrm{Mod(T)}$ and $\mathbf{c} = C^n$ such that $\mathcal{C} \models \phi^s_{i\mathcal{F}}(\mathbf{c}) \wedge \phi^s_{s\mathcal{F}}(\mathbf{c})$. Hence $\mathcal{C} \models \phi^s_{i\mathcal{F}}(\mathbf{c}) \wedge \psi_{i'\mathcal{F}'_{s-1}}(\mathbf{c}) \wedge \eta_{\mathbf{a},p_{\mathcal{F}'_{s-1}h'_{s-1}\mathcal{C}}}(\mathbf{c})$ (recall the definition of $\phi^s_{s\mathcal{F}}$).
Hence $p_{\mathcal{F}'_{s-1}h_{s-1}\mathcal{U}} \cup \{\phi^s_{i\mathcal{F}}\}$ is consistent with T. (c)

As $i \in 1..(s-1), \phi^s_{i\mathcal{F}}$ did not change at the sth stage, and therefore (c) contradicts (c3) in the induction hypothesis. As $i \in 1..(s-1)$ was arbitrary (b) holds and hence (c1) holds.

As for every $i \in 1..(s-1), \phi^s_{i\mathcal{F}}$ did not change at the sth stage, to show that (c2) holds, it suffices to verify that $p_{\mathcal{F}h\mathcal{A}} \cup \{\phi^s_{s\mathcal{F}}(\mathbf{a})\}$ consistent with T, i.e., $p_{\mathcal{F}h\mathcal{A}} \cup \{\psi_{i'\mathcal{F}'_{s-1}}(\mathbf{a}) \wedge \eta_{\mathbf{a},p'_{\mathcal{F}_{s-1}}h_{s-1}\mathcal{A}}(\mathbf{a})\}$ is consistent with T. This is true because of the following reasons.

(i) There is a homomorphism $h'_{s-1}h_{\mathcal{F}\mathcal{F}'_{s-1}} : \mathcal{F} \longrightarrow \langle p_{\mathcal{F}'_{s-1}h_{s-1}\mathcal{A}}\rangle$, and $\mathbf{a} = h(\mathbf{f}) = h'_{s-1}h_{\mathcal{F}\mathcal{F}'_{s-1}}(\mathbf{f})$.

(ii) By ($\alpha 2$), we may choose $\mathcal{A}_i^{s-1} \in \mathrm{HomExt}n_T(\mathcal{F}'_{s-1})$ such that $\psi_{i\mathcal{F}'_{s-1}}$ is satifiable in $\mathcal{A}_i^{s-1}$.

We next need to show that (c3) holds.
It follows from ($\beta 3$) that, $p_{\mathcal{G}h''\mathcal{A}} \cup \{\psi_{i'\mathcal{F}'_{s-1}}(\mathbf{a})\}$ is inconsistent with T.
As $\mathcal{F}'_s = \langle p_{\mathcal{G}h''\mathcal{A}} \cup p_{\mathcal{F}'_{s-1}h'_{s-1}\mathcal{A}}\rangle$ and $h'_s := \imath_{\langle p_{\mathcal{G}h''\mathcal{A}} \cup p_{\mathcal{F}'_{s-1}h'_{s-1}\mathcal{A}}\rangle \hookrightarrow \mathcal{A}}$, we have $p_{\mathcal{F}'_s h'_s \mathcal{A}} \cup \{\psi_{i'\mathcal{F}'_s}(\mathbf{a})\}$ is inconsistent with T. As $\psi_{i'\mathcal{F}'_{s-1}}$ is one of the conjuncts of $\phi^s_{i\mathcal{F}}$, it follows that $p_{\mathcal{F}'_s h'_s \mathcal{A}} \cup \{\phi^s_{i\mathcal{F}}(\mathbf{a})\}$ is inconsistent with T. By the induction hypothesis, for every $i \in 1..(s-1), p_{\mathcal{F}'_{s-1}h'_{s-1}\mathcal{A}} \cup \{\phi^{s-1}_{i\mathcal{F}}(\mathbf{a})\}$ is inconsistent with T. As $\mathcal{F}'_s = \langle p_{\mathcal{G}h''\mathcal{A}} \cup p_{\mathcal{F}'_{s-1}h'_{s-1}\mathcal{A}}\rangle$ and as for every $i \in 1..(s-1), \phi^s_{i\mathcal{F}}(\mathbf{a}) = \phi^{s-1}_{i\mathcal{F}}(\mathbf{a})$, it follows that for every $i \in 1..(s-1), p_{\mathcal{F}'_s h'_s \mathcal{A}} \cup \{\phi^s_{i\mathcal{F}}(\mathbf{a})\}$ is inconsistent with T. Hence (c3) holds.

This completes the proof of Claim 3. ∎

We may, in view of the Lowenheim-Skolem theorem, assume that $\mathcal{A}$ is countable.

Remark V.16.2. Since L and $\mathcal{A}$ are countable, there exists a sequence $(p_m \in \mathrm{Cond}^+(\mathcal{A}) | m \in \omega)$ such that the following hold.

(1) For every $p \in \mathrm{Cond}^+(\mathcal{A})$ such that $\mathrm{Rng}(\mathbf{a}) \subset \mathrm{Name}(p)$, there exists $m \in \omega$ such that $p \subset p_m$.

(2) For all $m, m' \in \omega, m < m' \Rightarrow p_m \subset p_{m'}$.

In what follows, we assume that such a sequence $(p_m \in \mathrm{Cond}^+(\mathcal{A}) | m \in \omega)$ has been chosen.

We define for every $m \in \omega, \mu_m := \eta_{\mathbf{a},p_m}$.

CLAIM 4. For every $j \in \omega$, there exist $m_j \in \omega$ and $\phi_i^j, \ldots, \phi_r^j \in \mathrm{E}_n^+$ such that the following hold.

(d1) For every $i \in 1..r$, $\{\phi_i^j(\mathbf{a})\}$ is inconsistent with T.

(d2) For every $i \in 1..r$, $(\phi_i^j)^{\mathrm{Mod(T)}} \leq (\mu_{m_j})^{\mathrm{Mod(T)}}$.

(d3) For every $i \in 1..r$, $p_{m_{j+1}} \cup \{\phi_i^j(\mathbf{a})\}$ is inconsistent with T.

(d4) $\phi_0^{\mathrm{Mod(T)}}, (\phi_i^j)^{\mathrm{Mod(T)}}, \ldots, (\phi_r^j)^{\mathrm{Mod(T)}}$ are pairwise disjoint.

(d5) The sequence $(m_j \,|\, j \in \omega)$ is strictly increasing.

PROOF. We prove Claim 4 by induction on j.
$j = 0$: We note that $\langle p_0 \rangle, \mathbf{a} \in (\mathrm{Gen}(\langle p_0 \rangle))^n$ and $\imath_{\langle p_0 \rangle \subset \mathcal{A}}$ satisfy the hypothesis in Claim 3. Hence by Claim 3, we may choose $\phi_{1\langle p_0 \rangle}, \ldots, \phi_{r\langle p_0 \rangle} \in \mathrm{E}_n^+, \mathcal{F}' \in \mathrm{FPS}_T(L), h_{\langle p_0 \rangle \mathcal{F}'} : \langle p_0 \rangle \longrightarrow \mathcal{F}'$, and $h' : \mathcal{F} \longrightarrow \mathcal{A}$ such that the following hold.

(c0) $\mathbf{a} = h' h_{\langle p_0 \rangle \mathcal{F}'}(\mathbf{a})$.

(c1) $\phi_0^{\mathrm{Mod(T)}}, \phi_{1\langle p_0 \rangle}^{\mathrm{Mod(T)}}, \ldots, \phi_{r\langle p_0 \rangle}^{\mathrm{Mod(T)}}$ are pairwise disjoint.

(c2) For every $i \in 1..r$, $p_0 \cup \{\phi_{i\langle p_0 \rangle}(\mathbf{a})\}$ is consistent with T.

(c3) For every $i \in 1..r$, $p_{\mathcal{F}'h'\mathcal{A}} \cup \{\phi_{i\langle p_0 \rangle}(\mathbf{a})\}$ is inconsistent with T.

By Remark V.16.2, we may choose $m_1 \in \omega$ such that $m_1 > 0$ and $p_{\mathcal{F}'h'\mathcal{A}'} \subset p_{m_1}$. We set $m_0 := 0$ and for every $i \in 1..r$, $\phi_i^0 := \phi_{i\langle p_0 \rangle} \wedge \mu_{m_0}$. Then it follows from (c2) that for every $i \in 1..r$, $\{\phi_i^0(\mathbf{a})\}$ is consistent with T. Hence (d1) holds.It follows from the defintion that for every $i \in 1..r$, $(\phi_i^0)^{\mathrm{Mod(T)}} \leq (\mu_{m_0})^{\mathrm{Mod(T)}}$. Hence (d2) holds.

As $p_{\mathcal{F}'h'\mathcal{A}} \subset p_{m_1}$, it follows from (c3) that for every $i \in 1..r$, $p_{m_1} \cup \{\phi_i^0(\mathbf{a})\}$ is inconsistent with T. Hence (d3) holds.

It follows from (c1) that $\phi_0^{\mathrm{Mod(T)}}, (\phi_1^0)^{\mathrm{Mod(T)}}, \ldots, (\phi_r^0)^{\mathrm{Mod(T)}}$ are pairwise disjoint. Hence (d4) holds.

$j > 0$:
We assume that m_{j-1}, m_j and ϕ_i^{j-1} for $i \in 1..r$ have been defined such that (d1)-(d4) hold.

We note that $\langle p_{m_j} \rangle, \mathbf{a} \in (\mathrm{Gen}(\langle p_{m_j} \rangle))^n$ and $\imath_{\langle p_{m_j} \rangle \subset \mathcal{A}}$ satisfy the hypothesis in Claim 3. Hence by Claim 3, we may choose $\phi_{1\langle p_{m_j} \rangle}, \ldots, \phi_{r\langle p_{m_j} \rangle} \in E_n^+, \mathcal{F}' \in \mathrm{FPS}_T(L), h_{\langle p_{m_j} \rangle \mathcal{F}'} : \langle p_{m_j} \rangle \longrightarrow \mathcal{F}'$ and $h' : \mathcal{F}' \longrightarrow \mathcal{A}$ such that the following hold.

(c0) $\mathbf{a} = h' h_{\langle p_{m_j} \rangle \mathcal{F}'}(\mathbf{a})$.

(c1) $\phi_0^{\mathrm{Mod(T)}}, \phi_{1\langle p_{m_j} \rangle}^{\mathrm{Mod(T)}}, \ldots, \phi_{r\langle p_{m_j} \rangle}^{\mathrm{Mod(T)}}$ are pairwise disjoint.

(c2) For every $i \in 1..r$, $p_{m_j} \cup \{\phi_{i\langle p_{m_j} \rangle}(\mathbf{a})\}$ is consistent with T.

(c3) For every $i \in 1..r$, $p_{\mathcal{F}'h'\mathcal{A}} \cup \{\phi_{i\langle p_{m_j} \rangle}(\mathbf{a})\}$ is inconsistent with T.

By Remark V.16.2. we may choose $m_{j+1} \in \omega$ such that $m_{j+1} > m_j$ and $p_{\mathcal{F}'h'\mathcal{A}} \subset p_{m_{j+1}}$.

We set for every $i \in 1..r, \phi_i^j := \phi_{i(p_{m_j})} \wedge \mu_{m_j}$, where $\mu_{m_j} = \eta_{\mathbf{a},p_{m_j}}$. Then it follows from (c2) that for every $i \in 1..r, \{\phi_i^j(\mathbf{a})\}$ is consistent with T. Hence (d1) holds.

It follows from the defintion that for every $i \in 1..r, (\phi_i^j)^{\mathrm{Mod(T)}} \leq \mu_{m_j}^{\mathrm{Mod(T)}}$. Hence (d2) holds.

As $p_{\mathcal{F}'h'\mathcal{A}} \subset p_{m_{j+1}}$, it follows from (c3) that for every $i \in 1..r, p_{m_{j+1}} \cup \{\phi_i^j(\mathbf{a})\}$ is inconsistent with T. Hence (d3) holds.

It follows from (c1) that $(\phi_0)^{\mathrm{Mod(T)}}, (\phi_1^j)^{\mathrm{Mod(T)}}, \ldots, (\phi_r^j)^{\mathrm{Mod(T)}}$ are pairwise disjoint. Hence (d4) holds.

The choice of m_j guarantees that (d5) holds. Hence we have Claim 4. ■

Finally, we define for every $i \in 1..r, \phi_i := \bigvee_{j\in\omega} \phi_i^j$.

REMARK V.16.3. As we assumed that (β) is false, we shall obtain a contradiction by showing that $\phi_0^{\mathrm{Mod(T)}}, \phi_1^{\mathrm{Mod(T)}}, \ldots, \phi_r^{\mathrm{Mod(T)}}$ satisfy clauses (i) and (ii) in (β).

We recall that $\mathrm{DE}_n^+(\mathrm{Mod(T)})$ is a complete Heyting algebra and hence satisfies the join infinite distributive identity.

(i) For every $i \in 1..r$, we have $(\phi_0)^{\mathrm{Mod(T)}} \wedge (\phi_i^j)^{\mathrm{Mod(T)}} = 0$. Hence for every $i \in 1..r, (\phi_0)^{\mathrm{Mod(T)}} \wedge_D \phi_i^{\mathrm{Mod(T)}} = \phi_0^{\mathrm{Mod(T)}} \wedge_D (\bigvee_{j\in\omega} \phi_i^{j\,\mathrm{Mod(T)}}) = \bigvee_{j\in\omega}(\phi_0^{\mathrm{Mod(T)}} \wedge (\phi_i^j)^{\mathrm{Mod(T)}}) = 0$. Next let the pair of distinct elements $i, i' \in 1..r$ be given. To show that $\phi_i^{\mathrm{Mod(T)}} \wedge_D \phi_{i'}^{\mathrm{Mod(T)}} = 0$, it suffices to show that for every $j, j' \in \omega, (\phi_i^j)^{\mathrm{Mod(T)}} \wedge (\phi_{i'}^{j'})^{\mathrm{Mod(T)}} = 0$. Without loss, we may assume that $j \leq j'$. If $j = j'$, then by (d4) $(\phi_i^j)^{\mathrm{Mod(T)}} \wedge (\phi_{i'}^j)^{\mathrm{Mod(T)}} = 0$. If $j < j'$, then by (d2) and (d3)

$$(\phi_i^j)^{\mathrm{Mod(T)}} \wedge_D (\phi_{i'}^{j'})^{\mathrm{Mod(T)}} \leq (\phi_i^j)^{\mathrm{Mod(T)}} \wedge_D (\mu_{m_{j'}})^{\mathrm{Mod(T)}}$$

$$\leq (\phi_i^j)^{\mathrm{Mod(T)}} \wedge_D (\mu_{m_{j+1}})^{\mathrm{Mod(T)}} = 0.$$

(ii) Let $\mathcal{G} \in \mathrm{FPS}_T(L), \mathbf{g} \in (\mathrm{Gen}(\mathcal{G}))^n$ and $h : \mathcal{G} \longrightarrow \mathcal{A}$ be given such that $\mathbf{a} = h(\mathbf{g})$.

By Remark V.16.2 and (d5), we may choose $j \in \omega$ such that $p_{\mathcal{G}h\mathcal{A}} \subset p_{m_j}$. Let $i \in 1..r$ be given.

It follows from (d1) that for every $i \in 1..r, \{\phi_i^j(\mathbf{a})\}$ is consistent with T, and from (d2) that for every $i \in 1..r, (\phi_i^j)^{\mathrm{Mod(T)}} \leq (\mu_{m_j})^{\mathrm{Mod(T)}}$. Hence $p_{m_j} \cup \{\phi_i^j(\mathbf{a}\}$ and hence $p_{\mathcal{G}h\mathcal{A}} \cup \{\phi_i^j(\mathbf{a})\}$ are consistent with T. Hence we may choose $\mathcal{B} \in \mathrm{HomExt}n_T(\mathcal{G})$ and $h_{\mathcal{G}\mathcal{B}} : \mathcal{G} \longrightarrow \mathcal{B}$ such that $\mathcal{B} \models \phi_i^j(h_{\mathcal{G}\mathcal{B}}\mathbf{g})$; hence $\mathcal{B} \models \phi_i(h_{\mathcal{G}\mathcal{A}}\mathbf{g})$, i.e., $\mathcal{B} \not\models \neg\phi_i(h_{\mathcal{G}\mathcal{A}}\mathbf{g})$. Hence, as ϕ_i^* is a c^+-resultant for $\neg\phi_i$, by Definition V.4, we have $\mathcal{A} \not\models \phi_i^*(\mathbf{a})$, i.e., $\mathcal{A} \models \neg\phi_i^*(\mathbf{a})$.

As $i \in 1..r$ was arbitrary, $\mathcal{A} \models \neg\bigvee_{i=1..r} \phi_i^*(\mathbf{a})$. Hence $\phi_0^{\mathrm{Mod(T)}}, \phi_1^{\mathrm{Mod(T)}}, \ldots, \phi_r^{\mathrm{Mod(T)}}$ satisfy clauses (i) and (ii) in (β).

This completes the proof of Theorem V.16. ■

COROLLARY V.17. *For every* $\mathcal{A} \in \mathrm{Mod(T)}$ *and* $\mathbf{a} \in A^n$, *the following are equivalent*:

(α) *(For every $\mathcal{G} \in \mathrm{FPS}_T(L)$ and every $\mathbf{g} \in (\mathrm{Gen}(\mathcal{G}))^n$ such that there exists a homomorphism $h : \mathcal{G} \longrightarrow \mathcal{A}$ satisfying $\mathbf{a} = h(\mathbf{g})$), there exists a base of conditions of length r over $(\mathcal{G}, \mathbf{g})$ in* Mod(T).

(β) *There exist $\phi_1, \ldots, \phi_r \in \mathrm{DE}_n^+$ such that*
(i) $\phi_1^{\mathrm{Mod(T)}}, \ldots, \phi_r^{\mathrm{Mod(T)}}$ *are pairwise disjoint, and*

(ii) $\mathcal{A} \models \neg \bigvee_{i \in 1..r} \phi_i^*(\mathbf{a})$.

PROOF. We note that every base of conditions over $(\mathcal{G}, \mathbf{g})$ in Mod(T) is a test-base of conditions for $\phi := \perp$, as the condition in Definition V.10 is automatically satisfied.

Also ϕ, is defined in the preceding, is disjoint from every $\psi^{\mathrm{Mod(T)}}$. Theorem V.16 now yields Corollary V.17 if we take $\phi_0 = \perp$. ∎

We proceed to define the relevant amalgamation properties. Let $\mathcal{A} \in$ Mod(T), $\mathbf{a} \in A^n$ be given.

DEFINITION V.18. $(\mathcal{A}, \mathbf{a})$ is said to allow r^+-amalgamation over conditions in Mod(T) :$\Leftrightarrow$ there exist $\mathcal{G} \in \mathrm{FPS}_T(L), \mathbf{g} \in (\mathrm{Gen}(\mathcal{G}))^n$ and there exists $h : \mathcal{G} \longrightarrow \mathcal{A}$ such that

(1) $\mathbf{a} = h(\mathbf{g})$, and

(2) there is no base of conditions of length $r + 1$ over $(\mathcal{G}, \mathbf{g})$ in Mod(T).

DEFINITION V.19. $\mathcal{A} \in$ Mod(T) is said to allow r^+-amalgamation over conditions in Mod(T) :$\Leftrightarrow$ for every $n \in \omega$ and $\mathbf{a} \in A^n, (\mathcal{A}, \mathbf{a})$ allows r^+-amalgamation over conditions in Mod(T).

THEOREM V.20. *For every $n \in \omega$, $\mathrm{DE}_n^+(\mathrm{Mod(T)})$ satisfies $I_r \Leftrightarrow$ every $\mathcal{A} \in$* Mod(T) *allows r^+-amalgamation over conditions in* Mod(T).

PROOF. The right hand side holds

$\Leftrightarrow$ for every $\mathcal{A} \in$ Mod(T) and $\mathbf{a} \in A^n((\alpha)$ with $r + 1$ in place of r in Corollary V.17 fails)

$\Leftrightarrow$ for every $\mathcal{A} \in$ Mod(T) an $\mathbf{a} \in A^n((\beta)$ with $r + 1$ in place of r in Corollary V.17 fails)

$\Leftrightarrow$ for every $\phi_1, \ldots, \phi_{r+1} \in \mathrm{DE}_n^+(\mathrm{Mod(T)})$, if the $\phi_i^{\mathrm{Mod(T)}}$ are disjoint, then every $(\mathcal{A}, \mathbf{a})$ satisfies some ϕ_i^*

$\Leftrightarrow$ for every $\phi_1, \ldots, \phi_{r+1} \in \mathrm{DE}_n^+(\mathrm{Mod(T)})$, if the $\phi_i^{\mathrm{Mod(T)}}$ are disjoint, then $\bigvee_i \phi_i^* = 1$

$\Leftrightarrow I_r$ holds. ∎

COROLLARY V.21. *The classifying topos of a geometric theory T satisfies $I_r \Leftrightarrow$ every $\mathcal{A} \in$* Mod(T) *allows r^+-amalgamation over conditions in* Mod(T).

PROOF. This is immediate from Theorem V.20 and Corollary IV.20. ∎

COROLLARY V.22. *In particular, the classifying topos of a geometric theory T satisfies De Morgan's law*

$\Leftrightarrow$ every $\mathcal{A} \in \mathrm{Mod(T)}$ allows 1^+-amalgamation over conditions in Mod(T)

$\Leftrightarrow$ for every $n \in \omega, \mathcal{A} \in \mathrm{Mod(T)}$ and $\mathbf{a} \in A^n$, there exist $\mathcal{G} \in \mathrm{FPS}_T(L), \mathbf{g} \in (\mathrm{Gen}(\mathcal{G}))^n$ and a homomorphism $h : \mathcal{G} \longrightarrow \mathcal{A}$ such that

(1) $\mathbf{a} = h(\mathbf{g})$, and

(2)there is no base of length 2 over $(\mathcal{G}, \mathbf{g})$ in Mod(T). ■

EXAMPLE V.23. We construct, for every $r \in \omega$, examples of finitary geoemtric theories T_r, such that the corresponding classifying topos $\mathcal{T}_r$ satisfies I_r but not I_s for $s \in 1..(r-1)$.

T_r has exactly r propositional variables $p_1, \ldots, p_r$. The axioms for T_r are $p_i \wedge p_j \longrightarrow \perp$, for every pair of distinct elements $i, j \in 1..r$.

T_r has exactly $r+1$ models, all of which are finite. These are as follows: $\mathcal{F}_0$, with every p_j interpreted as $\perp$, and, for every $i \in 1..r, \mathcal{F}_i$, with p_i interpreted as $\top$, and every other variable p_j as $\perp$.

Thus $\mathcal{F}_0$ is the only model that admits any homomorphic extension other than itself, and each such non-trivial extension is $\mathcal{F}_i$, for some $i \in 1..r$. Thus any set of $r+1$ distinct homomorphic extensions of $\mathcal{F}_0$ must include $\mathcal{F}_0$, and hence fails to be a base. There being no base of length $r+1$, it follows from Corollary V.21 that I_r is satisfied, and hence, so is I_n, for every $n > 4$.

Moreover, no two distinct $\mathcal{F}_i$'s can be amalgamated, and thus there are bases of all lengths less than or equal to r. It therefore follows from Corollary V.21 that none of the identities I_n, with $n < r$, is satisfied.

Open Questions

Johnstone has studied the validity of the identity $\forall x, y((x \longrightarrow y) \vee (y \longrightarrow x) = 1)$ called strong De Morgan's law in arbitrary topoi [DML]. One may use the results of this article to deduce conditions under which the classifying topos of a geometric theory T satisfies strong De Morgan's law. However the conditions are somewhat unsatisfactory as they involve extensions of T rather than T itself. It remains to find a "nicer" description.

Also, the study of the validity of equational axioms stronger than strong De Morgan's law ([B& D] p. 176) in $\mathrm{DE}_n\mathrm{Mod(T)}), \mathrm{DE}_n^+(\mathrm{Mod(T)})$ or topoi ([B& D] p. 176) has not yet been undertaken.

REFERENCES

[B] A. Bagchi, Topoi satisfying De Morgan's law and related identities, Ph.D. dissertation, (University of Michigan (August 1991)).

[B&D] R. Balbes and P. Dwinger, *Distributive Lattices*, University of Missouri Press, Columbia, 1974.

[B&S] A. Blass and A. Scedrov, Boolean classifying topoi, J. Pure and Appl. Alg. **28** (1983), 15-30.

[CK] C.C. Chang and H.J. Keisler, *Model Theory*, North Holland, Amsterdam, 1973.

[DML] P.T. Johnstone, Conditions related to De Morgan's Law, Lecture Notes in Mathematics, **753** (1979), 479-491.

[MDML] P.T. Johnstone, Another Condition Equivalent to De Morgan's Law, Commum. Alg. **7** (1979), 1309-1312.

[G] A.M. Gleason, Projective Topological Spaces, Ill. J. Math. **2** (1958) 482-489.

[GC1] P.T. Johnstone, The Gleason Cover of a Topos I, J. of Pure and Appl. Alg. **19** (1980), 171-192.

[GC2] P.T. Johnstone, The Gleason of Cover of a Topos II, J. of Pure and Appl. Alg. **22** (1981), 229-247.

[G&J] L. Gillman and M. Jerison, *Rings of Continuous Functions*, Van Nostrand, Princeton, 1960.

[L] K.B. Lee, Equational Classes of Distributive Pseudocomplemented Lattices, Can. J. Math., Vol. XXII, **4** (1970), 881-891.

[NR1] S.B. Niefield and K.I. Rosenthal, Strong De Morgan's Law and the Spectrum of a Commutative Ring, J. of Alg. **93** (1985), 169-181.

[NR2] S.B. Niefield and K.I. Rosenthal, Note on the Algebraic De Morgan's Law, Cahiers de topologie et geometrie differentielle categorique, **XXVI-2** (1985), 1-6.

[R] A. Robinson, A Decision Method for Elementary Algebra and Geometry Revisited, Proceedings of Symposia in Pure Mathematics, 25 (1974), 139-152.

[S] A. Scedrov, Forcing and Classifying Topoi, *Memoirs of the AMS* 295, Vol. 48, (1984).

[SGA] A. Grothendieck and J.L. Verdier, *Séminaire de géométrie algébrique IV, Tome 1, Lecture Notes In Mathematics*, Springer, Berlin, New York, 1972.

[SS] P.T. Johnstone, *Stone Spaces*, Cambridge University Press, Cambridge, 1982.

[TT] P.T. Johnstone, *Topos Theory*, Academic Press, New York, 1977.

[W] V. Weispfenning, Model Theory and Lattices of Formulas, *Patras Logic Symposium*, North Holland Publishing Company, 1982.

Department of Mathematics, Case Western Reserve University, Cleveland, Ohio 44106

Current address: Department of Mathematics, Case Western Reserve University, Cleveland, Ohio 44106

E-mail address: axb32@po.cwru.edu

Canadian Mathematical Society
Conference Proceeding
Volume **13**, 1992

A Stone Duality for Metric Spaces

B. BANASCHEWSKI AND A. PULTR

ABSTRACT. A dual equivalence is established between Boolean metric frames and complete metric spaces, involving frame homomorphisms on one side and weakly open maps on the other, each subject to analogous metric requirements, providing uniformly continuous, Lipschitz, and contraction maps on the space side. Prior to this, metric frames are considered in general, dealing with the dual adjointness to metric spaces, completion, and Booleanization.

This note presents a duality between the category of complete metric spaces and certain maps and the category of metric Boolean frames and analogous homomorphisms. The relevant maps $f:X \longrightarrow Y$ between the metric spaces can be uniform, Lipschitz, or contractive but are subject to the topological restriction that, for any non–void open $U \subseteq X$, int $f[U] \neq \emptyset$. On the other hand, a metric Boolean frame is a Boolean frame (= complete Boolean algebra) equipped with a metric diameter (Definition 1), and the homomorphisms involved are frame homomorphisms (= complete Boolean homomorphisms) subject to metric restrictions which correspond to uniform, Lipschitz, and contraction maps of spaces (Definition 2). We obtain our duality

---

1991 Mathematics Subject Classification: 006D99, 18B30, 54E35, 54H99
Supported by an operating grant from the Natural Sciences and Engineering Research Council of Canada.
This paper is in final form and no version of it will be submitted for publication elsewhere.

(III) by putting together two antecedent results, the dual equivalence between complete metric frames and complete metric spaces (I) and the equivalence between Boolean metric frames and complete metric frames, with suitable maps (II).

1. Pointfree Metric Topology

Recall that pointfree topology deals with certain lattices, called frames, and their homomorphisms, where a frame is a complete lattice L in which

$$x \wedge \bigvee S = \bigvee\{x \wedge t \mid t \in S\} \quad (x \in L, S \subseteq L)$$

and a frame homomorphism is a map $h:L \longrightarrow M$ between frames preserving all finitary meets, including the unit e, and arbitrary joins, including the zero 0. A principal example of these notions is the frame $\mathfrak{O}X$ of open sets of a topological space X, and the frame homomorphism $\mathfrak{O}f: \mathfrak{O}Y \longrightarrow \mathfrak{O}X$, determined by any continuous map $f:X \longrightarrow Y$ between topological spaces, taking $U \in \mathfrak{O}Y$ to $f^{-1}[U] \in \mathfrak{O}X$. Indeed, the resulting correspondence constitutes a contravariant functor $\mathfrak{O}$:**Top** $\longrightarrow$ **Frm** from the category **Top** of topological spaces and continuous maps to the category **Frm** of frames and their homomorphisms. On the other hand, there is a contravariant functor Σ:**Frm** $\longrightarrow$ **Top** such that, for any frame L, ΣL is the space of all frame homomorphisms $\xi:L \longrightarrow 2$ (2 the two-element frame), with open sets $\Sigma_a = \{\xi \in \Sigma L \mid \xi(a) = 1\}$ for $a \in L$, and for any frame homomorphism $h:L \longrightarrow M$ $\Sigma h:\Sigma M \longrightarrow \Sigma L$ acts by composition with h. Moreover, $\mathfrak{O}$ and Σ are adjoint on the right, with adjunction maps

$$\eta_L:L \longrightarrow \mathfrak{O}\Sigma L, \quad a \rightsquigarrow \Sigma_a$$

and

$$\varepsilon_X:X \longrightarrow \Sigma\mathfrak{O}X, \quad x \rightsquigarrow \hat{x}, \ \hat{x}(U) = \mathrm{card}(U \cap \{x\}).$$

For any L, the space ΣL is called the spectrum of L, and L is called spatial whenever η_L is an isomorphism. A general reference for frames is Johnstone [7].

Now, pointfree metric topology is concerned with frames equipped with a suitable kind of metric structure (Pultr [10]) as described by the following

DEFINITION 1. A metric diameter on a frame L is a map $d:L \longrightarrow \overline{R_+}$ such that

(M1) $d(0) = 0$.
(M2) If $a \leq b$ then $d(a) \leq d(b)$.
(M3) If $a \wedge b \neq 0$ then $d(a \vee b) \leq d(a) + d(b)$.
(M4) For any $\alpha < d(a)$ and $\epsilon > o$, there exist $b,c \leq a$ such that $d(b),d(c) < \epsilon$ and
$\alpha < d(b \vee c)$.
(M5) Each $D_\epsilon = \{x \in L \mid d(x) < \epsilon\}$ is a cover, that is, $\bigvee D_\epsilon = e$.
(M6) Each $a \in L$ is the join of all $x \in L$ such that there exist $\epsilon > 0$ for which $D_\epsilon x = \bigvee\{s \in D_\epsilon \mid s \wedge x \neq 0\} \leq a$.

A metric frame will then be a frame equipped with a specified metric diameter. We write L,M,... for metric frames, $d_L, d_M, \ldots$ for their metric diameters, and $D^L_\epsilon, D^M_\epsilon, \ldots$ for the covers consisting of the elements of diameter less than ϵ.

The basic fact about metric diameters is (Pultr [10]) that in any (metrizable) space X, each metric ρ of X determines a metric diameter on the frame $\mathfrak{O}X$ in the obvious way, while each metric diameter d on $\mathfrak{O}X$ gives rise to a metric ρ of X such that

$$\rho(x,y) = \inf\{d(U) \mid x,y \in U \in \mathfrak{O}X\}.$$

Moreover, these correspondences are inverse to each other.

In particular, this means that, with each metric space X, one has the associated metric frame $\mathfrak{O}X$, the frame of open subsets of X equipped with the metric diameter resulting from the metric of X. On the other hand, any metric frame L determines its metric spectrum ΣL whose points and open sets are as for mere frames and whose metric is given by

$$\rho(\xi,\zeta) = \inf\{d(a) \mid \xi(a) = 1 = \zeta(a)\}.$$

Just as for metric spaces, one considers certain types of metrically restricted maps in the frame context, as follows:

DEFINITION 2. A frame homomorphism $h:L \longrightarrow M$ between metric frames is called

(U) uniform if, for each $\epsilon > 0$, there exist $\delta > 0$ such that D^M_δ refines $h[D^L_\epsilon]$, that is, for any $x \in D^M_\delta$, there exist $z \in D^L_\epsilon$ with $x \leq h(z)$,

(L) Lipschitz if there exist k such that, for all $\epsilon > 0$, $D^M_{\epsilon/k}$ refines $h[D^L_\epsilon]$, and

(C) contractive if (L) holds with $k = 1$.

The crucial result then is (Pultr [11]) that a continuous map $f:X \longrightarrow Y$ between metric spaces is uniform, Lipschitz, or a contraction in the usual sense iff the associated frame homomorphism $\mathfrak{O}f:\mathfrak{O}Y \longrightarrow \mathfrak{O}X$ satisfies the corresponding condition (U), (L) or (C). Moreover, for any homomorphism of one of these types the induced map between the corresponding metric spectra will be of analogous type. Thus, if **MFrm** and **MSp** are the categories of metric frames and metric spaces, respectively, each with one of these three corresponding types of maps, then $\mathfrak{O}$ and Σ determine contravariant functors between these categories. Moreover, these functors are again adjoint on the right with adjunctions as in the case of mere frames, while $\mathfrak{O}$ actually provides a full dual embedding of **MSp** into **MFrm**.

It should be noted that Isbell [6] uses the term "metric" for uniform frames with countably generated uniformity. That is, of course, not quite the same thing, but close: In any metric frame L, the covers D^L_ϵ, $\epsilon > 0$, define a uniformity, obviously countably generated (take $\epsilon = \frac{1}{n}$), while any uniform frame of the stated kind has a metric diameter d such that its given uniformity is the same as the uniformity determined by d (Pultr [10]).

2. Completion

The fact that a metric space X is complete iff any dense embedding $X \longrightarrow Y$ is an isomorphism points to an obvious notion of completeness for metric frames, as follows:

DEFINITION 3. (i) A homomorphism $h:L \longrightarrow M$ between metric frames is called a dense surjection if $h(x) = 0$ implies $x = 0$, $d_M(h(a)) = d_L(a)$, and h is onto.

(ii) A metric frame M is called complete if any dense surjection $L \longrightarrow M$ is an isomorphism.

(iii) For any metric frame L, a complete metric frame M together with a dense surjection $M \longrightarrow L$ is called a completion of L.

Now we have:

Any metric frame L has an essentially unique completion $\gamma_L:CL \longrightarrow L$, and completion is functorial in the category **MFrm**.

A convenient approach to proving this is to make use of the <u>uniform</u> completion $h:M \longrightarrow L$ (Isbell [6], Kříž [9], Banaschewski–Pultr [1]) of L relative to its metric uniformity by showing that $\delta(x) = d_L(h(x))$, $x \in M$, defines a metric diameter on M.

(M1) and (M2) hold trivially for δ, as does (M5) since $\delta(r(x)) = d_L(x)$ for the right adjoint $r:L \longrightarrow M$ of h, and $r[D^L_\epsilon]$ is a cover of M by basic facts concerning uniform completions (Banaschewski–Pultr [1]). Regarding (M4), if $\alpha < \delta(a)$ for some $a \in M$ and $\epsilon > 0$, then also $\alpha < d_L(h(a))$, hence there exist $b,c \leq h(a)$ in D^L_ϵ for which $\alpha < d(b \vee c)$ by (M4) for d_L, and one readily checks that $s = a \wedge r(b)$ and $t = a \wedge r(c)$ are the desired elements to prove (M4) for δ. The crucial point concerning (M6) is that, by the fact that M is the stated uniform completion, each $a \in M$ is the join of the $x \in M$ such that $r[D^L_\epsilon]x \leq a$ for some $\epsilon > 0$. Since $D^M_\epsilon = h^{-1}[D^L_\epsilon]$ is a refinement of $r[D^L_\epsilon]$ because $s \leq rh(s)$ for all $s \in M$, this immediately implies the condition (M6) for δ.

That the resulting metric frame provides a metric completion is obvious, and the uniqueness is an immediate consequence of the familiar uniqueness of uniform completions. Regarding functoriality, this is clear in the case of uniform homomorphisms, again by the known properties of completions of uniform frames. Concerning the more subtle metric conditions, one proves first, for any homomorphism $h:L \longrightarrow M$ between metric frames, that D^{CM}_δ refines $Ch[D^{CL}_\epsilon]$ whenever D^M_δ refines $h[D^L_\epsilon]$, as follows: For any $x \in D^{CM}_\delta$, $\gamma_M(x) \in D^M_\delta$ and therefore $\gamma_M(x) \leq h(y)$ for some $y \in D^L_\epsilon$ by hypothesis. Next, by basic properties of metric frames, $d_L(z) \leq d_L(y) + 2\alpha$ for $z = D^L_\alpha y$ and thus also $d_L(z) < \epsilon$ for sufficiently small α. Now, by the properties of uniform completions, $r_L(y) \prec r_L(z)$ for the right adjoint r_L of γ_L, where $s \prec t$ means $s \wedge u = 0$, $t \vee u = e$ for some u. Consequently, $(Ch)r_L(y) \prec (Ch)r_L(z)$, and this implies for the right adjoint r_M of γ_M that

$$r_M\gamma_M(Ch)r_L(y) \leq (Ch)r_L(z)$$

by the density of γ_M. Finally, since

$$\gamma_M(Ch)r_L(y) = h\gamma_L r_L(y) = h(y),$$

we have

$$x \leq r_M\gamma_M(x) \leq r_M h(y) \leq (Ch)r_L(z)$$

which completes the proof because $r_L(z) \in D_\epsilon^{DM}$.

Evidently, it now follows that Ch is Lipschitz whenever h is, with the same Lipschitz constant k, and that proves the result.

REMARK. While there are other ways of obtaining this result which avoid the use of the completion of uniform frames, the advantage of the present approach is that it is choice free and that it provides a method applicable in a constructive treatment of this subject.

A further basic result is, assuming the Axiom of Choice, that any complete metric frame is spatial since any complete uniform frame with countably generated uniformity is spatial (Isbell [6], also Banaschewski–Pultr [1]). As an immediate consequence one then has that a metric space X is complete iff its metric frame $\mathfrak{O}X$ is complete.

Using C to indicate restriction to the complete case, one therefore has, assuming the Axiom of Choice:

I. Σ and $\mathfrak{O}$ induce a dual equivalence bewteen **CMFrm** and **CMSp**.

3. Booleanization

In any frame L, let $a^* = \bigvee\{x \in L \mid x \wedge a = 0\}$ be the pseudocomplement of $a \in L$,
$\mathfrak{B}L = \{a \in L \mid a = a^{**}\}$, and $\beta_L : L \longrightarrow \mathfrak{B}L$ the map $a \leadsto a^{**}$. Then, it is well–known that $\mathfrak{B}L$ is a complete Boolean algebra, and β_L is dense and factors through any dense onto homomorphism $L \longrightarrow M$.

Clearly, $\mathfrak{B}L$ is, in some sense, "L made Boolean" but it is a familiar fact that this Booleanization fails to be functorial. This suggests considering those frame homomorphimsms for which it is functorial:

DEFINITION 4. A frame homomorphism is called $\mathfrak{B}$–compatible if there exist $\overline{h}:\mathfrak{B}L \longrightarrow \mathfrak{B}M$ such that $\beta_M h = \overline{h}\beta_L$.

Of course, this $\overline{h}$ is necessarily unique, and it is then trivial that the correspondence $L \rightsquigarrow \mathfrak{B}L$ is functorial on the category of all frames and $\mathfrak{B}$–compatible homomorphisms. A useful characterization of $\mathfrak{B}$–compatible frame homomorphisms is that these are exactly the $h:L \longrightarrow M$ such that $h(a)$ is dense whenever a is dense, dense meaning that the pseudocomplement is zero. An alternative condition is that $h(a^{**}) \leq h(a)^{**}$ for all $a \in L$ (Banaschewski–Pultr [3]).

Turning now from frames to metric frames, we have the following important fact:

In any metric frame L, $d_L(a) = d_L(a^{**})$ for all $a \in L$.

We have to exclude the possibility that $d_L(a) < d_L(a^{**})$. Suppose this were the case for some $a \in L$. Then, take $\epsilon > 0$ such that $d_L(a) + 2\epsilon < d_L(a^{**})$ and $x,y \leq a^{**}$ such that $d_L(x),d_L(y) < \epsilon$ while $d_L(a) + 2\epsilon < d_L(x \vee y)$. Then both, x and y are non–zero, hence also $a \wedge x \neq 0 \neq a \wedge y$ since $x,y \leq a^{**}$, and using (M3) we obtain

$$d_L(a \vee y \vee x) \leq d_L(a)+d_L(x)+d_L(y) < d(a) + 2\epsilon,$$

yielding the contradiction $d_L(x \vee y) < d_L(x \vee y)$.

As a consequence, we now have:

In any metric frame L, the restriction of d_L to $\mathfrak{B}L$ is a metric diameter.

Here, (M1) and (M2) are trivial, and for the remaining conditons one variously uses the above result, the fact that binary join in $\mathfrak{B}L$ is $(x \vee y)^{**}$ and binary meet as in L, and $x \leq x^{**}$.

It follows that we may consider $\mathfrak{B}L$ as a metric frame for any metric frame L. Further, we let $\mathbf{MFrm}_{\mathfrak{B}}$ be the subcategory of **MFrm** given by the $\mathfrak{B}$–compatible maps in **MFrm**. Then we obviously have:

The correspondence $L \rightsquigarrow \mathfrak{B}L$ is functorial on $\mathbf{MFrm}_{\mathfrak{B}}$.

Note that $\beta_L:L \longrightarrow \mathfrak{B}L$ is a dense surjection. Hence, for complete L, it is the completion of $\mathfrak{B}L$. On the other hand, if a metric frame M is Boolean then we have a commuting triangle

$$\begin{array}{ccc} CM & \xrightarrow{\gamma_M} & M \\ & {\scriptstyle \beta_{CM}} \searrow \quad \swarrow {\scriptstyle h} & \\ & \mathfrak{B}CM & \end{array}$$

where h results from the fact that γ_M is dense. Now, h is onto and dense, which makes it a frame isomorphism since M is Boolean, and then it is obviuosly a metric frame isomorphism.

Considering the full subcategory $\mathbf{M\mathfrak{B}Frm}$ of **MFrm** given by the metric frames with underlying Boolean frame, it is obvious from these observations that $\mathfrak{B}$ and C induce functors from $\mathbf{CMFrm}_{\mathfrak{B}}$ to $\mathbf{M\mathfrak{B}Frm}$ and back, and that on objects these functors are inverse to each other. Regarding maps, $\mathfrak{B}$ is full because, for any $h:\mathfrak{B}L \longrightarrow \mathfrak{B}M$, $\mathfrak{B}f = h$ for

$$f:L \xrightarrow{\sim} C\mathfrak{B}L \xrightarrow{Ch} C\mathfrak{B}M \xrightarrow{\sim} M$$

On the other hand, $\mathfrak{B}$ is faithful since $\mathfrak{B}h = \mathfrak{B}k$ for $h,k:L \longrightarrow M$ implies $\beta_M h = \beta_M k$ and therefore $h = k$ because β_M is dense and the frames involved are regular.

As a result, we now have:

II. $\mathfrak{B}$ induces an equivalence between $\mathbf{CMFrm}_{\mathfrak{B}}$ and $\mathbf{M\mathfrak{B}Frm}$, with inverse given by C.

4. Duality

In order to put I and II together properly, a topological characterization of $\mathfrak{B}$–compatibility is needed. This is given by the following general result (Banaschewski–Pultr [3]):

For any continuous map $f:X \longrightarrow Y$ between topological spaces, $\mathfrak{O}f:\mathfrak{O}Y \longrightarrow \mathfrak{O}X$ is $\mathfrak{B}$–compatible iff, for any non–void $U \subseteq X$, $\text{int}\, f[U] \neq \emptyset$.

We call these maps weakly open since open continuous maps obviously are of this kind. Elsewhere, they have also been called demi–open (Herrlich–Strecker [5]) or skeletal (Johnstone [8]).

Indicating the restriction to weakly open maps by subscript, we now obtain from I and II:

III. ΣC and $\mathfrak{B}\mathfrak{O}$ induce a dual equivalence between $\mathbf{M\mathfrak{B}Frm}$ and $\mathbf{CMSp}_{wo}$.

In conclusion, we add that the functor ΣC has a suggestive equivalent which can be described without any reference to completions. It follows from the corresponding result of Banaschewski–Pultr [1] that $\Sigma C \cong \Psi$ where ΨL, the Cauchy spectrum of L, is the metric space of minimal Cauchy filters $P \subseteq L$, "Cauchy" meaning that $P \cap D^L_\epsilon \neq \emptyset$ for each $\epsilon > 0$, with metric given by $\rho(P,Q) = \inf\{d(a) \mid a \in P \cap Q\}$, and the topology generated by the basic open sets $\Psi_a = \{P \in \Psi L \mid a \in P\}$, $a \in L$. An alternative description of the Cauchy spectrum of a metric frame is given in Banaschewski–Pultr [2], and a treatment of the duality III in terms of that occurs in Banaschewski–Pultr [4].

As a final remark, we point out that there are results analogous to those presented here for uniform frames – but with some significant differences resulting from the fact that the uniform frame associated with a complete uniform space need not be complete (Banaschewski–Pultr [3]).

REFERENCES

1. B. Banaschewski and A. Pultr, Samuel compactification and completion of uniform frames. Math. Proc. Cambridge Phil. Soc. **108** (1990), 63–78.

2. ————, Cauchy points of metric locales. Can. J. Math. **41** (1989), 830–854.

3. ————, Booleanization as reflection. Unpublished manuscript.

4. ————, Universal categories of metric and uniform frames. Unpublished manuscript.

5. H. Herrlich and G. Strecker, H–closed spaces and reflective subcategories. Math. Ann. **177** (1968), 302–309.

6. J.R. Isbell, Atomless parts of spaces Math. Scand. **31** (1972), 5–32.

7. P.T. Johnstone, *Stone spaces.* Cambridge University Press, Cambridge 1982.

8. ————, *Factorization theorems for geometric morphisms II.* Springer Lecture Notes in Mathematics 915, 216–233. Springer–Verlag, Berlin Heidelberg, New York 1982.

9. I. Kříž, A direct description of uniform completion in locales and a characterization of LT groups. Cahiers Top. et Géom. Diff. Cat. **27** (1986), 19–34.

10. A. Pultr, Pointless uniformities II. (Dia)metrization. Comm. Math. Univ. Carolinae **25** (1984), 91–104.

11. ———, Categories of diametric frames. Math. Proc. Cambridge Phil. Soc. **105** (1989), 285–297.

B. BANASCHEWSKI
MATHEMATICS & STATISTICS, MCMASTER UNIVERSITY, HAMILTON, ONTARIO CANADA L8S 4K1
E–mail address: Math@McMaster

A. PULTR
APPLIED MATHEMATICS, CHARLES UNIVERSITY, MALOSTRANSKE NAM. 25, 11800 PRAHA 1, CZECHOSLOVAKIA
E–mail address: PULTR@CSPGUK 11

Canadian Mathematical Society
Conference Proceeding
Volume **13**, 1992

Sheaves in Cocomplete Categories

RENATO BETTI

ABSTRACT. The paper deals with sheaves in categories relative to a base topos. The main result consists of a characterization of Grothendieck topoi in terms of cocomplete categories.

1. Introduction

This work deals with sheaves for topologies in a topos $\mathcal{E}$, taking values in categories defined, and locally internal, over the topos. The motivation arises mainly from the study of locally internal categories, when one tries to understand these categories in terms of special structures. Moreover, many sheaf theoretic arguments become more transparent and many constructions are unified.

The main definition is due to Paré [6], who developed the notion for indexed categories (not necessarily locally internal). Here we adopt a different description: categories over $\mathcal{E}$ are regarded as $(\mathrm{Span}\,\mathcal{E})$-enriched categories, as in Betti and Walters [3], and the module calculus which has been developed in that setting is constantly applied. Moreover, a slight extension of the definition is given, in order to encompass the case of sheaves indexed in internal categories.

In the paper, we consider cocomplete categories with an object of generators and show that also in such a general case they are categories of sheaves in $\mathcal{E}^{\mathbf{C}}$ for a suitable internal category $\mathbf{C}$.

1991 Mathematics Subject Classification. Primary 18D20, 18B25.

This paper is in final form and no version of it will be submitted for publication elsewhere.

2. Sheaves

Following Betti and Walters [3], recall that a locally internal category $\mathcal{X}$ over the elementary topos $\mathcal{E}$ is a category enriched in the bicategory Span $\mathcal{E}$, satisfying the *substitution property*

$$\mathcal{X}(x,y)\cdot f \simeq \mathcal{X}(f^{*}x,y)$$

or equivalently

$$f^{o}\cdot \mathcal{X}(x,y) \simeq \mathcal{X}(x,f^{*}y)$$

for every pair of objects x, y and every map f in $\mathcal{E}$ from the underlying object of x to that of y (recall that a *map* is an arrow of $\mathcal{E}$ regarded in Span $\mathcal{E}$, and that f^{o} denotes its right adjoint in Span $\mathcal{E}$).

This means in particular that $\mathcal{X}$ has objects $x, y, \ldots$ over objects of $\mathcal{E}$ (say over $I, J, \ldots$). So an object should be regarded as an indexed family $x = (x_i)_{i\in I}$, and the hom $\mathcal{X}(x,y)$ as the $(I\times J)$-matrix $\mathcal{X}(x,y)_{ij}$ of arrows from x_i to y_j, i.e. a span $I \not\longrightarrow J$.

Let $\mathcal{T}$ be a pullback stable family of monos in $\mathcal{E}$ (called *dense monos*) and let $\mathbf{C}$ be an internal category (i.e. a one object category enriched in Span $\mathcal{E}$).

DEFINITION. *A functor $F:\mathbf{C}\to\mathcal{X}$ is suid to be a sheaf in $\mathcal{X}$ over $\mathbf{C}$ if for any dense mono $x : X' \rightarrowtail X$, for any object B of $\mathcal{X}$ over X and any module $X \overset{\alpha}{\not\longrightarrow} \mathbf{C}$ as in the following diagram*

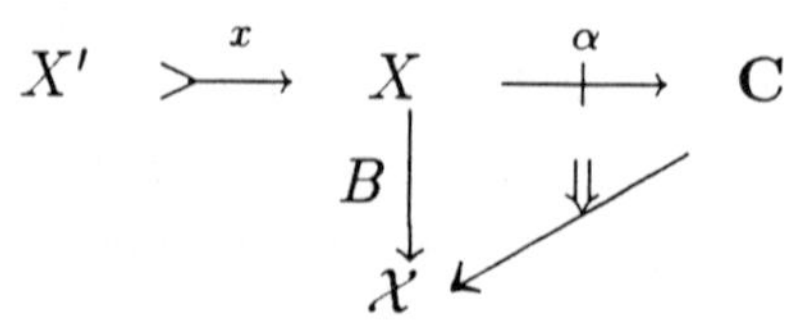

composition with x provides a bijection of 2-cells

$$\frac{\alpha \Rightarrow \mathcal{X}(B,F)}{\alpha\cdot x \Rightarrow \mathcal{X}(B,F)\cdot x}$$

Remarks

2.1 The definition of Paré [6] regards the case when $\mathbf{C} = I$ is a discrete (internal) category, i.e. an object of $\mathcal{E}$, and a map is used instead of the module α. Stability of $\mathcal{T}$ under pullbacks and the substitution property allow one to show that this second generalization is apparent: if the sheaf condition of the previous definition is satisfied with respect to maps, then it is satisfied also for arbitrary spans.

2.2 The sense of the definition is that "homs of $\mathcal{X}$ extend along dense monos". To see this, the formal rôle of substitution property should be considered: taking $\mathbf{C} = X$ and $\alpha = 1_X$ in the definition, one obtains the bijection

$$\frac{1_{X'} \Rightarrow \mathcal{X}(B, F)}{1_X \Rightarrow \mathcal{X}(x^*B, x^*F)}$$

2.3 Let $Sh_{\mathcal{T}}\mathcal{X}$ denote the full subcategory of $\mathcal{X}$ whose objects are sheaves over objects of $\mathcal{E}$. This is a locally internal subcategory. In other words, $Sh_{\mathcal{T}}\mathcal{X}$ admits substitutions: if A is a sheaf of $\mathcal{X}$ over I and $f : J \to I$ is any map of $\mathcal{E}$, then f^*A is again a sheaf: this depends on the fact that f^*A is a particular indexed limit, and the sheaf condition is stable for indexed limits (see later).

2.4 The sheaf condition for the object A of $\mathcal{X}$can be expressed in terms of right Kan extensions in the bicategory Mod $\mathcal{E}$ of internal categories and modules between them, in the following way:

$$\mathcal{X}(B, A) \simeq [x, \mathcal{X}(B, A) \cdot x]$$

for any dense mono x and any B in $\mathcal{X}$. In fact, for any module α, we have the bijections:

$$\frac{\dfrac{\alpha \Rightarrow [x, \mathcal{X}(B, A) \cdot x]}{\alpha \cdot x \Rightarrow \mathcal{X}(B, A) \cdot x}}{\alpha \Rightarrow \mathcal{X}(B, A).}$$

3. Complete and cocomplete categories

In this section we first consider the case when $\mathcal{X}$ is a *cocomplete category*. This means that for any functor $F : \mathbf{C} \to \mathcal{X}$ and any module $\alpha : \mathbf{C} \not\to I$ (where $\mathbf{C}$ is an internal category), the right Kan extension $hom^{\mathbf{C}}(\alpha, \mathcal{X}(1, F))$ of the module $\mathcal{X}(1, F) : \mathbf{C} \not\to \mathcal{X}$ through α is represented by an object $\alpha * F$ in $\mathcal{X}$:

$$hom^{\mathbf{C}}(\alpha, \mathcal{X}(1, F)) \simeq \mathcal{X}(\alpha * F, 1) : I \not\longrightarrow \mathcal{X}.$$

The universal property of $\alpha * F$ can be expressed by the bijection

$$\frac{\gamma \Rightarrow \mathcal{X}(\alpha * F, B)}{\alpha \cdot \gamma \Rightarrow \mathcal{X}(F, B)}$$

for any module $\gamma : I \nrightarrow J$ and any object B over J. As usual, we have the isomorphisms:

$$\mathcal{X}(\alpha * F, B) \simeq \mathcal{E}^{\mathbf{C}^{op}}(\alpha, \mathcal{X}(F, B))$$

where $\mathcal{E}^{\mathbf{C}^{op}}$ has, as objects over I, the modules $\mathbf{C} \nrightarrow I$ and homs are given by right Kan extensions:

$$\mathcal{E}^{\mathbf{C}^{op}}(\alpha, \beta) = hom^{\mathbf{C}}(\alpha, \beta).$$

In particular, $\mathcal{E}^{\mathbf{C}^{op}}$ is cocomplete, and it holds true the following bijection:

$$\frac{\delta \Rightarrow \mathcal{E}^{\mathbf{C}^{op}}(\alpha, \beta)}{\delta \cdot \alpha \Rightarrow \beta}$$

for any module δ.

Dually, $\mathcal{X}$ is a *complete category* if for any $F : \mathbf{C} \to \mathcal{X}$ and any indexing module $\beta : I \nrightarrow \mathbf{C}$, the right lifting $hom_{\mathbf{C}}(\beta : \mathcal{X}(1, F))$ of the module $\mathcal{X}(1, F) : \mathcal{X} \nrightarrow \mathbf{C}$ through β is represented by an object $\{\beta, F\}$ in $\mathcal{X}$:

$$hom_{\mathbf{C}}(\beta, \mathcal{X}(1, F)) \simeq \mathcal{X}(1, \{\beta, F\}).$$

Again, the universal property of $\{\beta, F\}$ can be expressed by the bijection

$$\frac{\gamma \Rightarrow \mathcal{X}(B, \{\beta, F\})}{\beta \cdot \gamma \Rightarrow \mathcal{X}(B, F)}$$

but, in terms of functor categories, this property involves the symmetric structure of Span $\mathcal{E}$. More precisely, Span $\mathcal{E}$ is a *symmetric bicategory* in the sense that to each arrow α is associated an *opposite arrow* α^o (interchanging the components of the span) with the properties:

$$1_u^o \simeq 1_u$$
$$(\alpha \cdot \beta)^o \simeq \beta^o \cdot \alpha^o$$
$$(\alpha^o)^o \simeq \alpha.$$

Moreover, any 2-cell $\alpha \Rightarrow \beta$ corresponds exactly to one 2-cell $\alpha^o \Rightarrow \beta^o$.

The symmetry of Span $\mathcal{E}$ allows to define *opposite categories* by

$$\mathcal{X}^{op}(B, C) = \mathcal{X}(C, B)^o$$

and to consider the functor category $\mathcal{E}^{\mathbf{C}}$. Objects of $\mathcal{E}^{\mathbf{C}}$ over I are modules $I \nrightarrow \mathbf{C}$ and homs are given by opposite arrows of right liftings:

$$\mathcal{E}^{\mathbf{C}}(\beta, \gamma) = hom_{\mathbf{C}}(\beta, \gamma)^o.$$

In particular, the universal property of indexed limits $\{\beta, F\}$ in $\mathcal{X}$ can be expressed by the bijection

$$\frac{\delta \Rightarrow \mathcal{E}^{\mathbf{C}}(\beta, \gamma)}{\beta \cdot \delta^o \Rightarrow \gamma}$$

and for a complete category $\mathcal{X}$ it holds:

$$\mathcal{E}^{\mathbf{C}}(\beta, \mathcal{X}(F, B)) \simeq \mathcal{X}(B, \{\beta, F\})^o.$$

Remark

When the internal category $\mathbf{C}$ is discrete, the notation $\sum_\alpha F$ is commonly used for $\alpha * F$.

When $\mathcal{X}$ is cocomplete, with the notation of the previous definition, we have the bijection

$$\frac{1_{\mathbf{C}} \Rightarrow \mathcal{X}(\Sigma_\alpha B, F)}{\alpha \Rightarrow \mathcal{X}(B, F)}$$

hence the sheaf condition for F can be expressed by:

$$\mathcal{X}(\Sigma_\alpha B, F) \simeq \mathcal{X}(\Sigma_{\alpha \cdot x} x^* B, F)$$

for any dense mono $x : X' \rightarrowtail X$ and any object B of $\mathcal{X}$over X.

Examples

3.1 When $\mathcal{X}$ is the base topos $\mathcal{E}$ regarded as locally internal over itself, consider $\mathbf{C} = 1$ and let $t : X \to 1$ be the terminal map in $\mathcal{E}$. Then, an object B of $\mathcal{E}$ over X is a family $B_0 \to X$. We have the situation

$$\begin{array}{ccccc} B'_0 & \overset{y}{\rightarrowtail} & B_0 & \overset{f}{\longrightarrow} & F_0 \\ {\scriptstyle x^*B}\downarrow & & {\scriptstyle B}\downarrow & & {\scriptstyle F}\downarrow \\ X' & \overset{x}{\rightarrowtail} & X & \overset{t}{\longrightarrow} & 1 \end{array}$$

and the sheaf condition now becomes

$$\mathcal{E}/1(t \cdot B, F) \simeq \mathcal{E}/1(t \cdot x \cdot x^*B, F)$$

where x^*B is the pullback of B along x: in other words, any map like $f : B_0 \to F_0$ in the above diagram can be uniquely extended along the dense mono y. But all dense monos are like y, hence:

F is a sheaf in $\mathcal{E}$ over 1 if and only if it admits the unique extension property with respect to all dense monos.

In particular, when $\mathcal{T}$ is a topology in $\mathcal{E}$, sheaves in $\mathcal{E}$ coincide with the usual ones.

3.2 Again consider $\mathcal{X}$=$\mathcal{E}$, but take for $\mathbf{C}$ any discrete category I. The sheaf condition for F becomes

$$\mathcal{E}/I(\alpha \cdot B, F) \simeq \mathcal{E}/I(\alpha \cdot y \cdot y^*B, F)$$

where $y = x \times I : X' \times I \rightarrowtail X \times I$ is a mono in $\mathcal{E}/I$. Regard now F as an object over 1 in $\mathcal{E}/I$ and take for α the projection $X \times I \to I$. It is not difficult to show that F can be extended along y. The converse is also true and we have:

F is a sheaf in $\mathcal{E}$ over I if and only if, regarded as an object of $\mathcal{E}/I$, it satisfies the unique extension property with respect to all monos (in $\mathcal{E}/I$) which are pullbacks of monos of the type I^*x, with x dense in $\mathcal{E}$.

Again, if $\mathcal{T}$ is a topology in $\mathcal{E}$, pullbacks of monos of the type I^*x generatete a topology $I^*\mathcal{T}$, and a sheaf in $\mathcal{E}$ over I is exactly a sheaf in $\mathcal{E}/I$ for this topology.

As a consequence, by the Diaconescu theorem, we have that the inverse image of $\mathcal{E}/I$ (regarded as a locally internal cocomplete category over $\mathcal{E}$) along a sheaf subcategory of $\mathcal{E}$ consists precisely of sheaves in $\mathcal{E}/I$.

3.3 Analogous results hold in the general case:

F is a sheaf in $\mathcal{E}$ over the internal category $\mathbf{C}$ *if and only if, regarded as an object in* $\mathcal{E}^{\mathbf{C}}$, *satisfies the unique extension property with respect to all monos which are pullbacks of monos of the type* $\mathbf{C}^*x$, *with x dense.*

Here, $\mathbf{C}^*$ denotes the inverse image of the canonical geometric morphism $\mathcal{E}^{\mathbf{C}} \to \mathcal{E}$ and again, when $\mathcal{T}$ is a topology in $\mathcal{E}$, this means that sheaves in $\mathcal{E}^{\mathbf{C}}$ (regarded as a locally internal cocomplete category over $\mathcal{E}$) coincide with sheaves (in the usual sense) for the topology $\mathbf{C}^*\mathcal{T}$. Moreover, the inverse image of $\mathcal{E}^{\mathbf{C}}$ along a sheaf subcategory of $\mathcal{E}$ consists precisely of sheaves in the topos $\mathcal{E}^{\mathbf{C}}$ with respect to the pullback topology.

Here is the general case: suppose $\mathcal{F} \xrightarrow{p} \mathcal{E}$ is a geometric morphism, and consider $\mathcal{F}$ as a locally internal cocomplete category $\sum_p \mathcal{F}$ over $\mathcal{E}$ (the *direct image* of $\mathcal{F}$, see Betti [1]).

This means that there is a bijection of functors

$$(+) \qquad \frac{\mathbf{C} \xrightarrow{F} \Sigma_p \mathcal{F}}{p^*\mathbf{C} \xrightarrow{G} \mathcal{F}.}$$

The main property we use is that $\sum_p \mathcal{F}$ is cocomplete, and colimits in $\sum_p \mathcal{F}$ can be computed by means of colimits in $\mathcal{F}$ (regarded as locally internal over itself). Namely, the colimit $\alpha * F$ in $\sum_p \mathcal{F}$:

$$\begin{array}{ccc} \mathbf{D} & \xrightarrow{F} & \Sigma_p \mathcal{F} \\ {\scriptstyle\alpha}\downarrow & \nearrow_{\alpha * F} & \\ \mathbf{C} & & \end{array}$$

under the above bijection corresponds to the colimit $p^*\alpha * G$ in $\mathcal{E}$.

So, one has that the sheaf condition for a functor F in $\sum_p \mathcal{F}$

$$(\Sigma_p\mathcal{F})(\Sigma_\alpha B, F) \simeq (\Sigma_p\mathcal{F})(\Sigma_{\alpha x} x^* B, F)$$

is equivalent with

$$\mathcal{F}(\Sigma_{p^*\alpha} B', G) \simeq \mathcal{F}(\Sigma_{p^*\alpha\cdot y} y^* B', G)$$

where B' correspond to B in a bijection like (+) above.

The last equivalence expresses the sheaf condition for G in $\mathcal{F}$, and so it is proved the following:

THEOREM. *F is a sheaf in $\mathcal{F}$, regarded as the locally internal category $\sum_p \mathcal{F}$ over $\mathcal{E}$, if and only if it satisfies the unique extension property with respect to monos which are pullbacks of monos of the type*

$$p^*X' \overset{p^*x}{\rightarrowtail} p^*X$$

where x is a dense mono.

Again, when $\mathcal{T}$ is a topology in $\mathcal{E}$, the above monos generate the pullback topology $p^*\mathcal{T}$ in $\mathcal{F}$, and it holds:

COROLLARY. *Let $p : \mathcal{F} \longrightarrow \mathcal{E}$ be a geometric morphism. The inverse image of $\mathcal{F}$, regarded as a locally internal category over $\mathcal{E}$, along a sheaf subcategory of $\mathcal{E}$ consists precisely of sheaves in $\mathcal{F}$ with respect to the pullback topology.*

Remark

The previous result appears already in Paré [6], with a different proof.

THEOREM. *If $\mathcal{X}$ is a complete category, also $Sh_\mathcal{T}\mathcal{X}$ is complete.*

PROOF. Suppose $F : \mathbf{C} \to Sh_\mathcal{T}\mathcal{X}$ is a functor and $\beta : I \,+\!\!\!\longrightarrow \mathbf{C}$ is a module. The fact that the indexed limit $\{\beta, F\}$ is a sheaf in $\mathcal{X}$ (over I) is obtained by the following bijections:

$$\frac{\dfrac{\dfrac{\alpha \cdot x \Rightarrow \mathcal{X}(B, \{\beta, F\}) \cdot x \simeq \mathcal{X}(x^* B, \{\beta, F\})}{\beta \cdot \alpha \cdot x \Rightarrow \mathcal{X}(x^* B, F) \simeq \mathcal{X}(B, F) \cdot x}}{\beta \cdot \alpha \Rightarrow \mathcal{X}(B, F)}}{\alpha \Rightarrow \mathcal{X}(B, \{\beta, F\})}$$

THEOREM. *Let $\mathcal{X}$ be a complete category and let $\mathbf{C}$ be the internal full subcategory generated by the object A. Then A is a sheaf in $\mathcal{X}$ over $\mathbf{C}$ if and only if the representable functor $\mathcal{X}(-,A)$: $\mathcal{X}^{op} \to \mathcal{E}^{\mathbf{C}}$ factors through* $\mathrm{Sh}_{\mathcal{T}}(\mathcal{E}^{\mathbf{C}})$:

$$\begin{array}{ccc} \mathcal{X}^{op} & \xrightarrow{\mathcal{X}(-,A)} & \mathcal{E}^{\mathbf{C}} \\ \downarrow & \nearrow & \\ \mathrm{Sh}_{\mathcal{T}}(\mathcal{E}^{\mathbf{C}}) & & \end{array}$$

PROOF. For any object B in $\mathcal{X}$, we have that $\mathcal{X}(B,A)$ is a sheaf, as results by the following bijections:

$$\begin{array}{c} f\cdot x \Rightarrow \mathcal{E}^{\mathbf{C}}(\beta,\mathcal{X}(B,A))\cdot x \simeq \mathcal{E}^{\mathbf{C}}(\beta\cdot x,\mathcal{X}(B,A)) \\ \hline \beta\cdot x\cdot x^o\cdot f^o \Rightarrow \mathcal{X}(B,A) \\ \hline \beta\cdot x \Rightarrow \mathcal{X}(f^*B,A)\cdot x \\ \hline \beta \Rightarrow \mathcal{X}(f^*B,A) \simeq \mathcal{X}(B,A)\cdot f \\ \hline f \Rightarrow \mathcal{E}^{\mathbf{C}}(\beta,\mathcal{X}(B,A)) \end{array}$$

for any dense mono $x : X' \rightarrowtail X$, any module $\beta : X \to \mathbf{C}$ (i.e. an object of $\mathcal{E}^{\mathbf{C}}$ over X) and any map f.

Conversely, suppose that $\mathcal{X}(-,A)$ factors through $\mathrm{Sh}_{\mathcal{T}}(\mathcal{E}^{\mathbf{C}})$. Then A is a sheaf, as it results by the following bijections:

$$\begin{array}{c} \alpha\cdot x \Rightarrow \mathcal{X}(B,A)\cdot x \\ \hline x^o \Rightarrow \mathcal{E}^{\mathbf{C}}(\alpha,\mathcal{X}(B,A)\cdot x) \simeq x^o\cdot \mathcal{E}^{\mathbf{C}}(\alpha,\mathcal{X}(B,A)) \\ \hline 1_{X'} \Rightarrow x^o\cdot \mathcal{E}^{\mathbf{C}}(\alpha,\mathcal{X}(B,A))\cdot x \\ \hline x \Rightarrow \mathcal{E}^{\mathbf{C}}(\alpha,\mathcal{X}(B,A))\cdot x \\ \hline 1_X \Rightarrow \mathcal{E}^{\mathbf{C}}(\alpha,\mathcal{X}(B,A)) \\ \hline \alpha \Rightarrow \mathcal{X}(B,A) \end{array}$$

□

4. Cocomplete categories with generators

Let $F : \mathbf{C} \to \mathcal{X}$ be a functor. Then, consider the largest family $\mathcal{T}_F$ of monos such that F satisfies the sheaf condition in $\mathcal{X}$. This is the *family induced by* F: a mono $x : X' \rightarrowtail X$ is in $\mathcal{T}_F$ provided for any $f : Y \to X$, any object B in $\mathcal{X}$ over X and any module $\alpha : Y \mathrel{+\!\!\!\longrightarrow} \mathbf{C}$, there is a bijection

$$\frac{\alpha \cdot y \Rightarrow \mathcal{X}(B,F)\cdot y}{\alpha \Rightarrow \mathcal{X}(B,F)}$$

where

$$\begin{array}{ccc} Y' & \stackrel{y}{\rightarrowtail} & Y \\ \downarrow & & \downarrow{\scriptstyle f} \\ X' & \stackrel{x}{\rightarrowtail} & X \end{array}$$

is a pullback.

Clearly, the family $\mathcal{T}_F$ is stable under pullbacks.

THEOREM (Paré [6]). *When $\mathcal{X}$ is a complete category, $\mathcal{T}_F$ is a topology in $\mathcal{E}$ and it is the largest topology such that F is a sheaf.* □

Suppose now that the object A *strongly generates* the cocomplete category $\mathcal{X}^{op}$. This means (see Betti [2]) that the functor

$$\mathcal{X}(-,A) : \mathcal{X}^{op} \to \mathcal{E}^{\mathbf{C}}$$

is fully faithful, where $\mathbf{C}$ denotes the internal full subcategory generated in $\mathcal{X}$ by A.

THEOREM. *Let $\mathcal{X}^{op}$ be a cocomplete category with a strong generator A. Let $\mathbf{C}$ denote the internal full subcategory generated by A and $\mathcal{T}_A$ the topology generated by A. Then $\mathcal{X}^{op}$ is a reflective subcategory of the category $Sh_{\mathcal{T}_A}(\mathcal{E}^{\mathbf{C}})$ of sheaves in $\mathcal{E}^{\mathbf{C}}$ for the topology $\mathcal{T}_A$.*

PROOF. By the previous theorem, the functor

$$\mathcal{X}(-,A) : \mathcal{X}^{op} \to \mathcal{E}^{\mathbf{C}}$$

factors through sheaves. Moreover, it admits a left adjoint. Namely the functor "taking the limit of A in $\mathcal{X}$":

$$\{-,A\} : \mathcal{E}^{\mathbf{C}} \to \mathcal{X}^{op}.$$

The family $\mathcal{T}_A$ is a topology hence the embedding

$$Sh_{\mathcal{T}_A}(\mathcal{E}^{\mathbf{C}}) \hookrightarrow \mathcal{E}^{\mathbf{C}}$$

admits as a left adjoint the "associated sheaf functor". By standard arguments on adjoint functors, the embedding $\mathcal{X}^{op} \hookrightarrow Sh_{\mathcal{T}_A}(\mathcal{E}^{\mathbf{C}})$ admits as a left adjoint the composition

$$Sh_{\mathcal{T}_A}(\mathcal{E}^{\mathbf{C}}) \hookrightarrow \mathcal{E}^{\mathbf{C}} \xrightarrow{\{-,A\}} \mathcal{X}^{op}.$$

□

COROLLARY. *A cocomplete category $\mathcal{X}^{op}$ is equivalent with a category of sheaves in $\mathcal{E}^{\mathbf{C}}$ for a suitable (internal) category $\mathbf{C}$ if and only if it admits a strong generator A such that $\mathcal{X}(\text{-},A)$ preserves limits of $A : \mathbf{C} \to \mathcal{X}$ indexed by any sheaf for the topology $\mathcal{T}_A$.*

PROOF. When the strong generator A is such that $\mathcal{X}(\text{-},A)$ preserves the limits $\{\beta, A\}$ (with $\beta : I \longrightarrow\!\!\!\!\!\!+\;\; \mathbf{C}$ a sheaf in $\mathcal{E}^{\mathbf{C}}$ over the object I), we have that the unity in the adjunction

$$\mathcal{X}^{op} \rightleftarrows \mathrm{Sh}_{\mathcal{T}_A}(\mathcal{E}^{\mathbf{C}})$$

of the previous theorem is invertible:

$$\mathcal{X}(\{\beta, A\}, A) \simeq \{\beta, \mathcal{X}(A, A)\} \simeq \beta$$

where the first isomorphism depends on the fact that $\mathcal{X}(\text{-},A)$ preserves the limit $\{\beta, A\}$ and the second one because the Yoneda embedding $\mathcal{X}(A, A)$ is the identity under module composition.

Conversely, if $\mathcal{X}^{op} \simeq \mathrm{Sh}_{\mathcal{T}}(\mathcal{E}^{\mathbf{C}})$, then $\mathcal{X}^{op}$ is a reflective subcategory of $\mathcal{E}^{\mathbf{C}}$ and we obtain a strong generator $A : \mathbf{C}^{op} \to \mathcal{X}^{op}$.

We have that, for a β in $\mathrm{Sh}_{\mathcal{T}}(\mathcal{E}^{\mathbf{C}})$, the limit $\{\beta, A\}$ is the object of $\mathcal{X}^{op}$ which corresponds to β in the equivalence, hence $\{\beta, A\} \simeq \beta$.

Now, $\mathcal{T}_A$ is the largest topology for which A is a sheaf, and A is a sheaf also for the topology $\mathcal{T}$ (because $\mathcal{X}(-, A)$ factors through sheaves). So there is a reflection

$$\mathrm{Sh}_{\mathcal{T}_A}(\mathcal{E}^{\mathbf{C}}) \rightleftarrows \mathrm{Sh}_{\mathcal{T}}(\mathcal{E}^{\mathbf{C}})$$

and $\{\beta, A\} \simeq \beta$ also for β in $\mathrm{Sh}_{\mathcal{T}_A}(\mathcal{E}^{\mathbf{C}})$.

□

Remark

In the particular case when $\mathcal{T}$ consists only of isomorphisms, any β in $\mathcal{E}^{\mathbf{C}}$ is a sheaf and it is not difficult to see that $\mathcal{X}(-,A)$ preserves the limits $\{\beta, A\}$ if and only if it preserves any indexed limit, i.e. A is *internal projective.*

In this case we recover a known characterization of presheaves categories:

$\mathcal{X}^{op}$ is equivalent with a presheaves category if and only if it admits a strong generator which is internal projective.

The above characterization is due to Bunge [4] and [5], who developed it both for internal categories and categories enriched in a monoidal one. It appears unified in Betti [2], in terms of categories enriched in a bicategory.

References

[1] R. Betti, Change of base for locally internal categories, J.Pure App. Alg. 61 (1989), 233-242.

[2] R. Betti, An elementary characterization of presheaves categories, Rend. Circolo Mat. Palermo 24 (1990), 257-266.

[3] R. Betti and R.F.C. Walters, On completeness of locally internal categories, J. Pure App. Alg. 47 (1987), 105-117.

[4] M. Bunge, Relative functor categories and categories of algebras, J. of Alg. 11 (1969), 64-101.

[5] M. Bunge, Internal presheaves toposes, Cahiers top. géom. diff. 18 (1977), 291-329.

[6] R. Paré, Indexed categories and generated topologies, J. Pure App. Alg. 19 (1980), 385-400.

Renato Betti
Dipartimento di Matematica
Politecnico di Milano
via Bonardi, 9
20133 Milano
Italy

Canadian Mathematical Society
Conference Proceeding
Volume **13**, 1992

Low dimensional geometry of the notion of choice

Dominique BOURN

ABSTRACT : In the same way as the category of groupoids is monadic above the category of split epimorphisms, i.e. epimorphisms equipped with the choice of a point in every inverse image, the category of 2-groupoids is shown to be monadic above the category of normalized groupoids, i.e. groupoids equipped with the choice of a point and certain morphisms in every connected components.

There is a way of realizing cohomology groups by means of classes of n-groupoids [3] which makes it very likely that the tower of n-groupoids has a universal property with respect to cohomology theories. Trying to understand it and coming back to the significant fact (cohomologically speaking) that the category of groupoids is monadic above the category of split epimorphisms [5], I was led to the following result : the category of 2-groupoids, via the inevitable notion of full or oriented [16] triangles, is again monadic but above the category of normalized groupoids : i.e. groupoids equipped with the choice of an object in each connected component and a morphism between this object and every other object in this component. This normalization of groupoid is strongly connected to the "equipement d'un groupoïde" [11] which a recent conference gave me the opportunity to learn and which is a determining tool for the general formulation of the Van Kampen theorem.

The groupoid monadicity clearly dealt with the notion of choice since the splitting of an epimorphism is nothing but the choice of a point in every inverse image. The new monadicity, through the concept of normalization of a groupoid, delineates what could be thought as the 1-geometry of the notion of choice and is the first step of a higher order process.

1991 Mathematics Subject Classification. Primary 18D05, 18G30.
This paper is in final form and no version of it will be submitted for publication elsewhere.

There is a lot of interest about the combinatorial geometry of the pasting for n-categories and the comparison with simplicial sets (nerves) [14], [6], [16], [10], [4], [13], [1]. The monadicities presented here show that, in the case of (n-)groupoids, the situation is simpler and the combinatorial-geometric aspect can be entirely described in a categorical way. In particular the normalization provides a substitute for what is dealing with hollowness in [9] and [16].

This paper contains a brief recall of the monadicity of the groupoids, the introduction of the notion of normalized groupoid, the proof of the monadicity of the 2-groupoids, the link with the nerve of a 2-groupoid and a sketch of the problem for the higher dimensional orders.

0 - Groupoids and split epimorphisms.

Let $\mathbb{E}$ be a left exact category and Pt $\mathbb{E}$ denote the category whose objects are the epimorphism in $\mathbb{E}$ with a given splitting and whose morphisms are the commutative squares between such data. $\mathbb{E}$ being left exact, the left exact forgetful functor q_0 : Pt $\mathbb{E} \to \mathbb{E}$ associating to each split epimorphisms its codomain is a fibered reflection (i.e. a fibration with a right inverse right adjoint). The cartesian morphisms in Pt $\mathbb{E}$ are the commutative squares which are pullbacks.

Let us denote by Grd $\mathbb{E}$ the category of internal groupoids in $\mathbb{E}$. Given a groupoid X_1 :

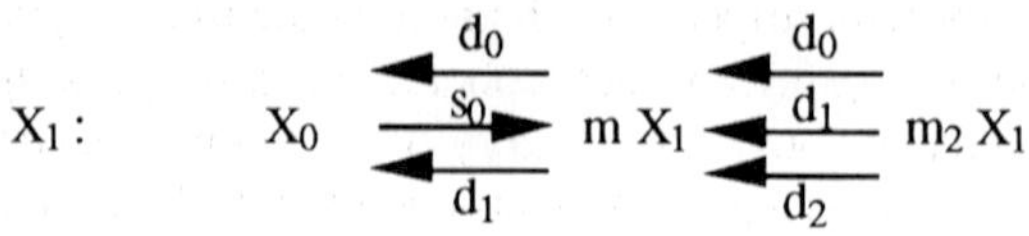

we shall denote by $(\ )_0$: Grd $\mathbb{E} \to \mathbb{E}$ the functor associating X_0 to X_1 and by U_0 : Grd $\mathbb{E} \to$ Pt $\mathbb{E}$ the functor associating the pair (d_0, s_0) to X_1 :

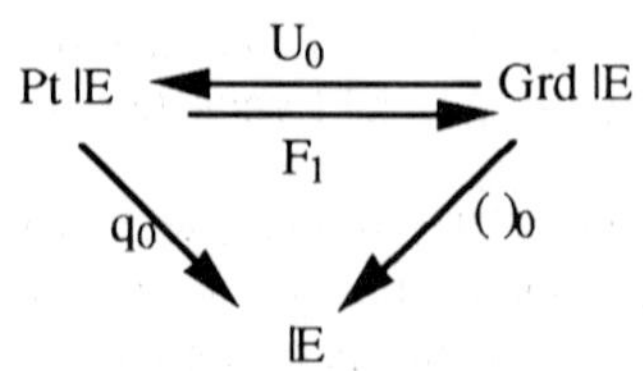

The functor $(\ )_0$ is again a fibered reflection (we shall denote by G_1 its right inverse right adjoint). The functor U_0 does not preserve the cartesian maps.
The functor U_0 has a left adjoint F_1 associating to (p_0, s_0) its kernel groupoid :

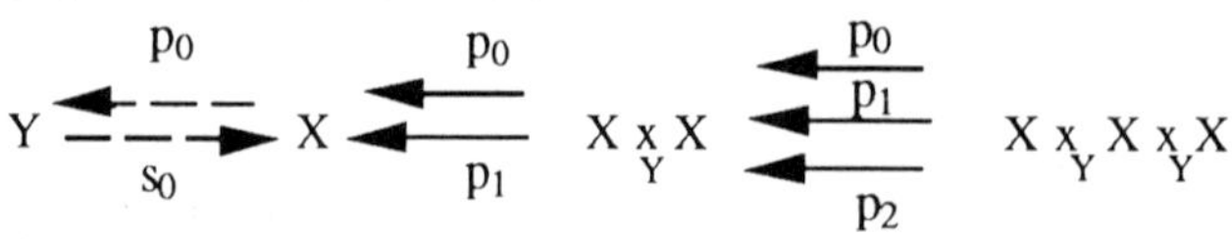

Let us briefly recall how U_0 is monadic.

The monad (Γ_0, γ, ψ) generated by (U_0, F_1) on Pt |E is given by the following diagram :

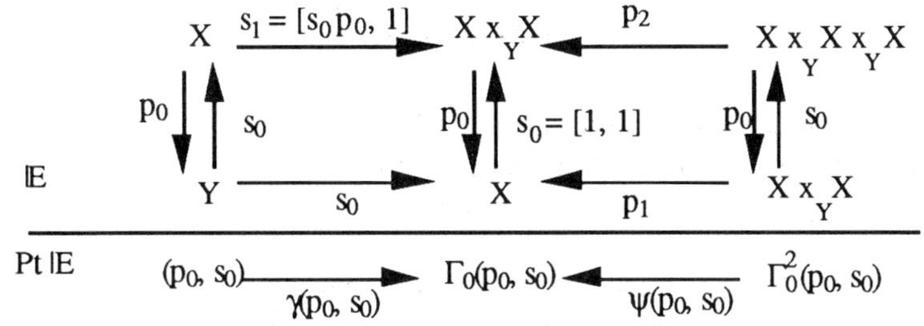

Clearly the functor Γ_0 preserves the cartesian morphisms of Pt |E and for every object (p_0, s_0) in Pt |E, both $\gamma(p_0, s_0)$ and $\psi(p_0, s_0)$ are cartesian.

<u>Definition 1</u>. In such a situation, in the case of any fibration, we shall say that the monad is *transversely cartesian* with respect to the fibration.

The comonad generated on Grd |E is denoted by (Dec, ε, υ) in reference to the simplicial notation. The following diagram :

$$X_1 \xleftarrow{\varepsilon X_1} \text{Dec } X_1 \underset{\text{Dec } \varepsilon X_1}{\overset{\varepsilon \text{ Dec } X_1}{\leftleftarrows}} \text{Dec}^2 X_1$$

being a componentwise split fork in |E, the functor εX_1 is a coequilizer in Grd |E. Consequently the comparison functor K_1 : Grd |E $\rightarrow$ Alg Γ_0 is fully faithful since :

<u>Proposition 1</u>. *Let* $(F, G, \eta, \varepsilon)$: /A $\rightarrow$ |B *be an adjunction,* $T = (G F, \eta, G \varepsilon F)$ *the monad it defines in* |B. *The comparaison functor* K : /A $\rightarrow$ Alg T *is fully faithful if and only if for every object* A *in* /A, *the map* ε A *is the cokernel of* ε F U A *and* F U ε A.

Proof : Straightforward.

Now we must show that K_1 is essentially surjective. Any algebra determines obviously a 3-truncated simplicial object. Is it underlying to an internal groupoid ? Thanks to the Yoneda embedding, it is sufficient to prove it in the category $ of sets.

1. Monadic adjustment of the 0-geometry.

Let (d_0, s_0) be a split epimorphism : X $\rightarrow$ Y in $. We shall suppose Y a subset of X and shall write d_0 instead of $d_0 \,.\, s_0$.

An algebra of Γ_0 is given by a pair d_2, d_1 of maps :

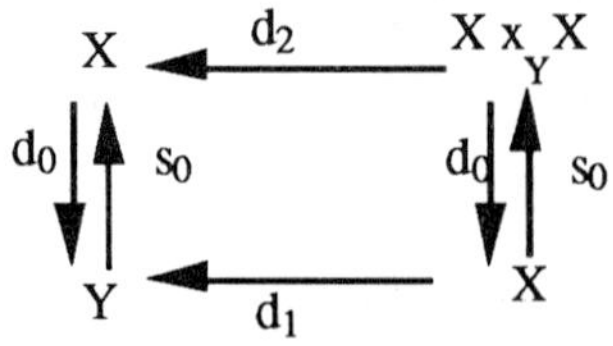

Let us rather consider d_2 as an operation $* : X \times_Y X \to X$.
The pair $(*, d_1)$ must be a morphism in Pt \$, then :

0] $x * x = d_1 x$, $d_0(x * y) = d_1 x$

satisfying

1] $d_1 x = x$ for any x in Y , $d_0 x * x = x$

2] $d_1(x * y) = d_1 y$, $(x * y) * (x * z) = y * z$.

Therefore d_0 and d_1 are two projections on Y, and the first part of the axioms can be pictured by the following representation where d_0 is the vertical projection from X to Y and d_1 the oblique one :

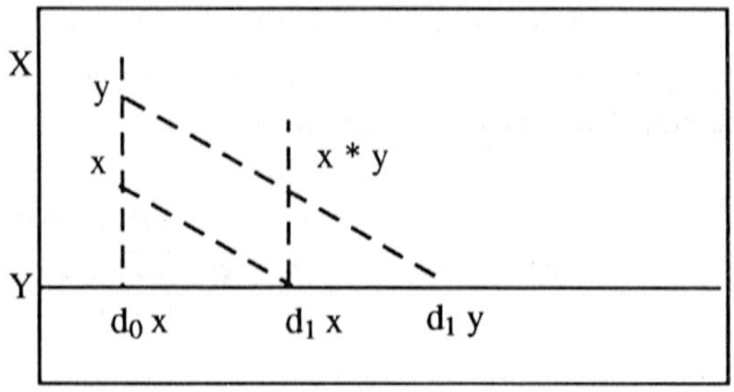

while the second part is pictured by the following one :

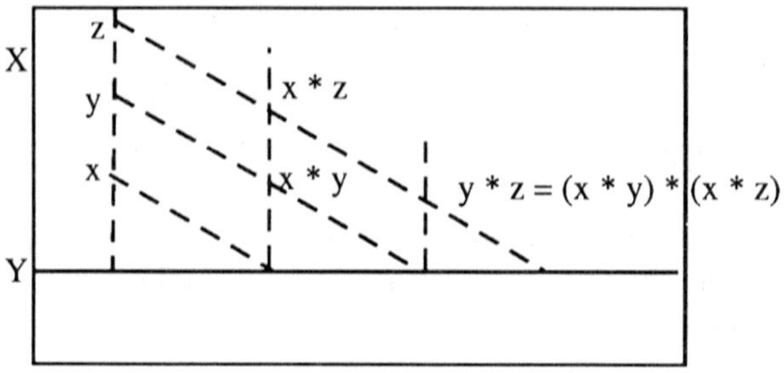

Now, let us set $x^{-1} = x * d_0 x$. If $d_0 y = d_1 x$, we can check that $d_0 x^{-1} = d_0(x * d_0 x) = d_1 x = d_0 y$ and then set $y . x = x^{-1} * y$.

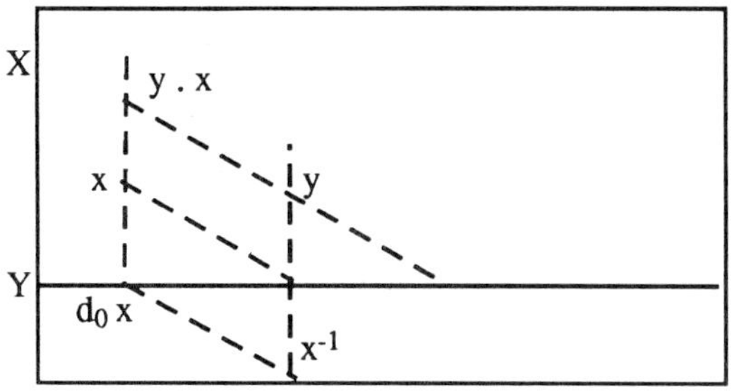

Now t = y . x if and only if x * t = y. Then it is not difficult to check that $(d_0, d_1, .)$ determines a groupoid whose associated algebra is $(*, d_1)$ [5].
As a corollary, we get that any fork in Grd IE whose image by U_0 is split, determines a cokernel in Grd IE.

2. Normalized groupoids.

Definition 2 : A *normalized groupoid* is a groupoid X_1 endowed with the choice of an object x_0 in each connected component and of a morphism between this object and every other object in this component, with the only coherence condition that the morphism associated to x_0 is 1_{x_0}. This particular distribution of choices for a groupoid, justified by the main result of this paper, could be thought as the 1-geometry of choice.

Internally speaking, it is then a groupoid X_1 endowed with an extra map $s_1 : X_0 \to m\, X_1$, satisfying :

1] $d_1\, s_1 = 1_{X_0}$

2] $d_0\, s_1\, d_0 = d_0\, s_1\, d_1$

3] $s_1\, d_0\, s_1 = s_0\, d_0\, s_1$

Let us denote by N-Grd IE the category whose objects are the normalized groupoids and morphisms the functors preserving the s_1.

Examples

[0] The commutative triangles of a groupoid X_1

Given any groupoid X_1, the groupoid Dec X_1 is normalized since there is an initial object in each connected component.

[1] The non commutative triangles of a groupoid X_1

Given any groupoid X_1, let us denote by $\Gamma_1\, X_1$ the groupoid whose objects are the maps of X_1 and morphisms between two objects f and g are the non commutative triangles of the following form :

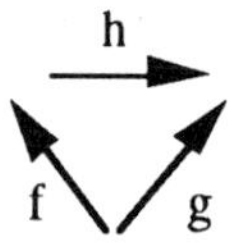

the composition being given by the composition of the upper maps.
This groupoid is canonically normalized since in every connected component there is a specific object of the form 1_X, and for every object f a specific map :

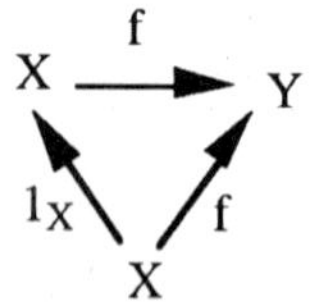

[2] The oriented triangles of a 2-groupoid X_2

An internal 2-groupoid in IE is an internal groupoid in the fibers of $(\ )_0 : \mathrm{Grd}\ \mathbb{E} \to \mathbb{E}$, that is a groupoid X_2 in Grd IE

$$X_2 : \quad X_1 \underset{d_1}{\overset{d_0}{\leftleftarrows}} m\,X_2 \;\Lleftarrow\; m_2\,X_2$$

such that the image by $(\ )_0$ of any of its structural maps is invertible. Let us denote by $(\ )_1 : 2\text{-}\mathrm{Grd}\ \mathbb{E} \to \mathrm{Grd}\ \mathbb{E}$ the functor associating X_1 to X_2. It is a fibered reflection and we denote by G_2 its right inverse right adjoint, where $G_2\, X_1$ is the kernel groupoid associated to the terminal map in the fiber of X_1 :

$$G_2\,X_1 : \quad G_1\,X_0 \dashleftarrow X_1 \underset{p_1}{\overset{p_0}{\leftleftarrows}} X_1 \times_0 X_1 \;\underset{p_2}{\overset{p_0,\ p_1}{\Lleftarrow}}\; X_1 \times_0 X_1 \times_0 X_1$$

We shall denote by $U_1(X_2)$ the following normalized groupoid : it has the 1-morphisms of X_2 as objects and the oriented triangles as maps between f and g :

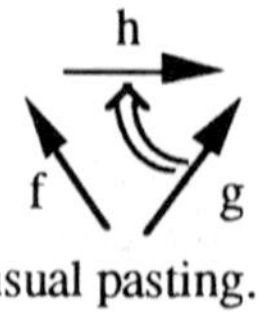

the composition being given by the usual pasting.

The normalization is as in the example 1. In fact the example 1 is nothing but $U_1\, G_2\, X_1$. Internally speaking, we obtain $U_1\, X_2$ by the following pullback :

$$\begin{array}{ccc} U_1\,X_2 & \longrightarrow & m\,X_2 \\ \delta \downarrow \uparrow \sigma & (*) & d_0 \downarrow \uparrow s_0 \\ \mathrm{Dec}\,X_1 & \underset{\varepsilon X_1}{\longrightarrow} & X_1 \end{array}$$

whence a functor $U_1 : 2\text{-}\mathrm{Grd}\ \mathbb{E} \to \mathrm{N}\text{-}\mathrm{Grd}\ \mathbb{E}$. The principal aim of this paper is to show that U_1 is monadic.

3. N-Grd IE is fibered above Pt IE.

We are now going to show that N-Grd IE is fibered above Pt IE in a way that makes it an extension of Pt IE.

Given (X_1, s_1) a normalized groupoid, then the kernel $s : X_{-1} \to X_0$ of s_0 and s_1 (where X_{-1} is the object of the distinguished objects in the components) appears to be also the cokernel of d_0 and d_1 by a morphism $d : X_0 \to X_{-1}$ in a way that makes (d, s) the splitting of the idempotent $d_0 . s_0$. We shall denote by $q_1(X_1, s_1)$ the split epimorphism (d, s) and by q_1 : N-Grd IE $\to$ Pt IE the associated left exact functor.

It is clear that $q_1 . U_1 = U_0 . (\,)_1$:

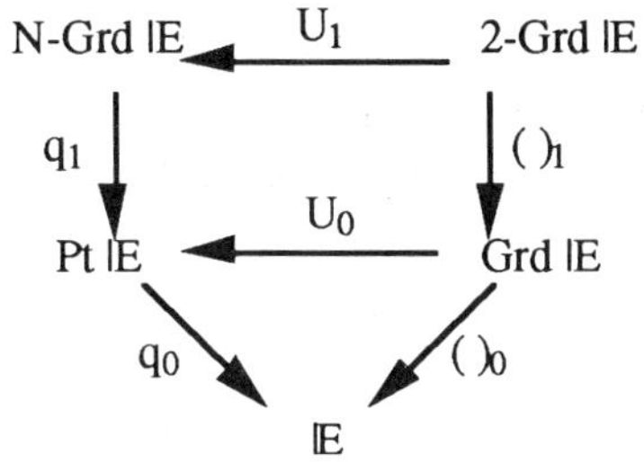

This functor q_1 has a right adjoint right inverse r_1 which makes it a fibered reflexion . The q_1-cartesian maps are the functors which are "fully faithful" but only with respect to the connected components.

The groupoid $r_1(p_0, s_0)$ is just the kernel groupoid of (p_0, s_0), the normalization being generated by s_0. This functor r_1 precises the way how the 1-geometry is an extension of the 0-geometry.

4. The terminal map in the fibers of q_1 always splits.

Let (X_1, s_1) be a normalized groupoid and $\tau_1 : (X_1, d_0) \to r_1\, q_1(X_1, s_1)$ be the terminal map in the fiber of q_1. The functor τ_1 is determined by the canonical factorization :

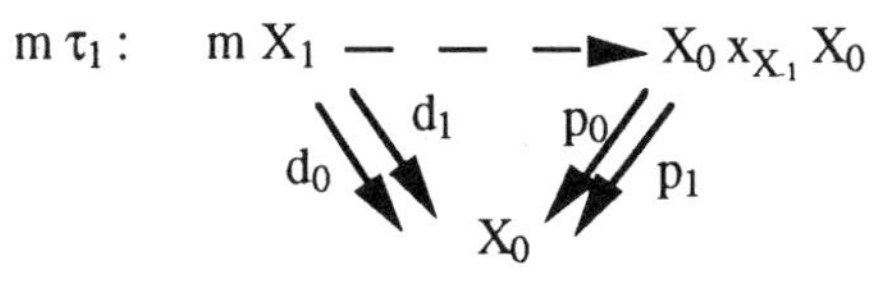

Proposition 2 : *The functor τ_1 is canonically split.*

Proof : Thanks to the Yoneda embedding, it is sufficient to show it in \$. Now, given two points X and X' in the same connected component, the normalization produces a map between X and X' which determines the splitting functor σ_1 :

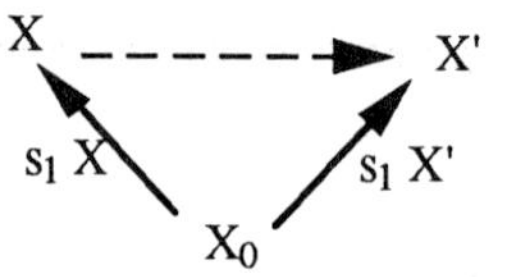

$\sigma_1(X, X') = s_1 X' \,.\, (s_1 X)^{-1}$

Conversely a normalized groupoid can be seen as a pair of a groupoid X_1 and of an objet (d, s) in Pt $\mathbb{E}$ together with a $(\,)_0$-invertible split epimorphism

$$X_1 \underset{\sigma_1}{\overset{\tau_1}{\rightleftarrows}} F_1(d, s).$$

Remark : Thus the whole structure of $U_1 X_2$ as a normalized groupoid (see 2.) necessarily involves the data of the splitting $U_1 X_2 \underset{\sigma}{\overset{\delta}{\rightleftarrows}} \mathrm{Dec}\, X_1$.

Now, if we denote by $Pt_{(\,)_0}(Grd\,\mathbb{E})$ the full subcategory of Pt(Grd $\mathbb{E}$) whose objects are the $(\,)_0$-invertible split epimorphisms (that is the split epimorphisms in the fibers of the fibration $(\,)_0$), the category N-Grd $\mathbb{E}$ is up to equivalence, the category defined by the following pullback :

$$\begin{array}{ccc} S F_0 & \xrightarrow{\tilde{F}_1} & Pt_{(\,)_0}(Grd\,\mathbb{E}) \\ {\scriptstyle k_0}\big\downarrow & & \big\downarrow{\scriptstyle q_0} \\ Pt\,\mathbb{E} & \xrightarrow[F_1]{} & Grd\,\mathbb{E} \end{array}$$

and the fibration q_1 is up to equivalence the fibration k_0.

5. The functor U_1 has a left adjoint F_2

Given (X_1, s_1) a normalized groupoid, $F_2(X_1, s_1)$ is just the kernel groupoid associated to τ_1. Indeed, the functor τ_1 being $(\,)_0$-invertible, the structural maps of the kernel groupoid are $(\,)_0$-invertible and consequently produce a 2-groupoid.

When $\mathbb{E}$ = \$, the 2-groupoid $F_2(X_1, s_1)$ has the same objects and 1-morphisms as X_1 and a unique 2-cell between two distinct 1-morphisms. It is clear then that any functor $f_1 : X_1 \to U_1 Y_2$ can be uniquely extended into a 2-functor $f_2 : F_2(X_1, s_1) \to Y_2$. Indeed any morphism $g : X \to X'$ determines an oriented triangle

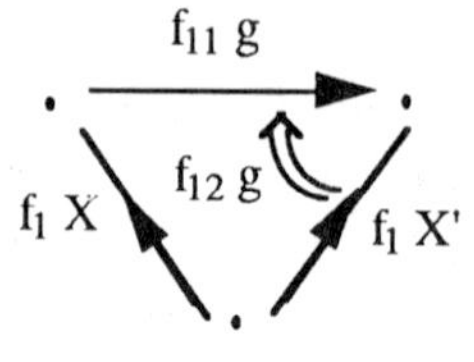

Let us set $f_2\, x = d_1\, f_1\, X$, $f_2\, g = f_{11}\, g$. Now given two parallel maps g and g', the two oriented triangles produce a 2-cell between $f_{11}\, g$ and $f_{11}\, g'$:

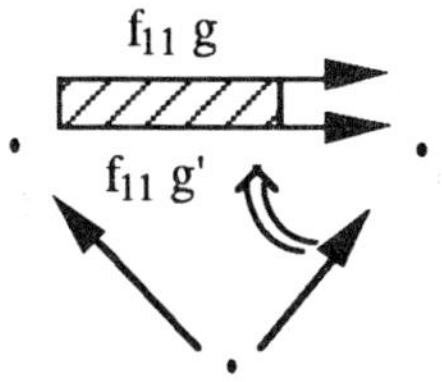

$$f_2(g, g') = (f_{12}\, g' \,.\, f_{12}^{-1}\, g) \,.\, f_1^{-1}\, X \,.$$

For a strict internal proof of this adjunction, we are going to decompose the functor U_1 and, for that, to wind about a little through a general construction which will be useful later on.

6. Split free-extensions

Let $\mathbb{C}$ and $\mathbb{D}$ be two categories and $U : \mathbb{D} \to \mathbb{C}$ a functor having a left adjoint F. Let us then consider the following pullback :

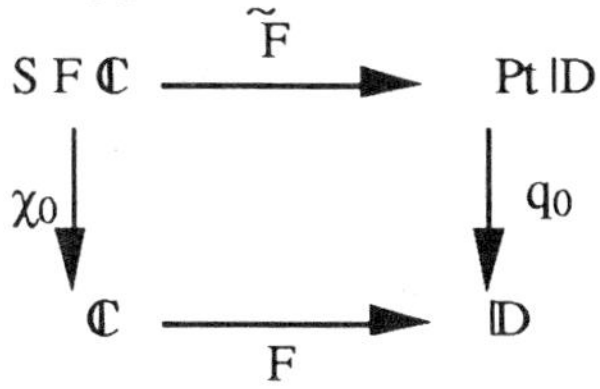

The objects of $SF\mathbb{C}$ (split free-extensions) are the pairs (Y, (p, s)) where Y is an object of $\mathbb{C}$ and (p, s) a split epimorphism in $\mathbb{D}$, such that $q_0(p, s) = FY$. When $\mathbb{D}$ is left exact, q_0 is a fibered reflection and then χ_0 is a fibered reflection. A morphism in $SF\mathbb{C}$ is χ_0-cartesian if and only if its image by $\tilde{F}$ is q_0-cartesian and thus $\tilde{F}$ preserves and reflects the cartesian morphisms. On the other hand, $\tilde{F}$ obviously preserves the χ_0-invertible morphisms.

<u>Example</u> : The functor $\tilde{F}_1 : S\, F_0 \to Pt_{(\ )_0}(Grd\, \mathbb{E})$, defined in 4, is a full restriction of the functor $\tilde{F}_1 : S\, F\, (Pt\, \mathbb{E}) \to Pt\, (Grd\, \mathbb{E})$.

<u>Proposition 3</u> : *The functor* $\tilde{F}$ *has a right adjoint* $\tilde{U}$.

<u>Proof</u>. Let (p, s) be an object of Pt $\mathbb{D}$ and let us consider the following pullback :

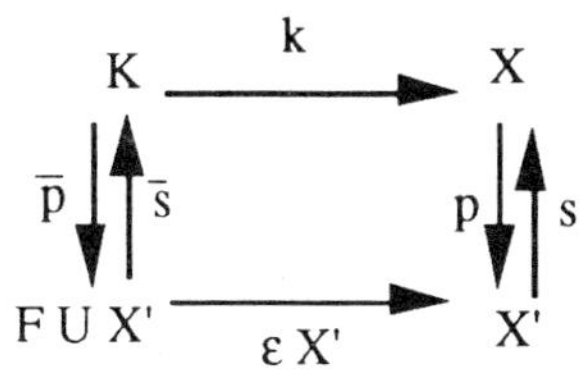

Then let us set $\tilde{U}(p, s) = (U X', (\bar{p}, \bar{s}))$. Now when $(Y, (p', s'))$ is an object in $SF\mathbb{C}$ and (h, l) a morphism : $\tilde{F}(Y, (p', s')) \to (p, s)$:

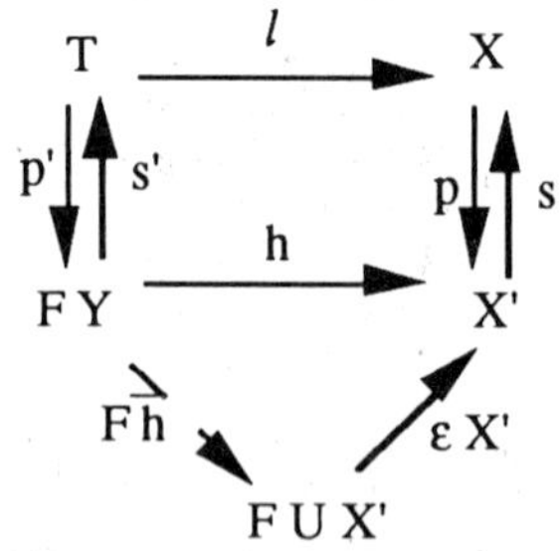

then there is a unique $\bar{h} : Y \to U X'$ such that $\varepsilon X' . F\bar{h} = h$, whence a unique $\bar{l} : T \to K$ such that $k . \bar{l} = l$. Now $\bar{s} . F\bar{l} = \bar{l} . s'$ if and only if $s . h = l . s'$.

Remark [1] Clearly $\tilde{U}$ preserves the cartesian maps (box lemma) and the q_0-inversible maps.

[2] Actually we are in the following situation :

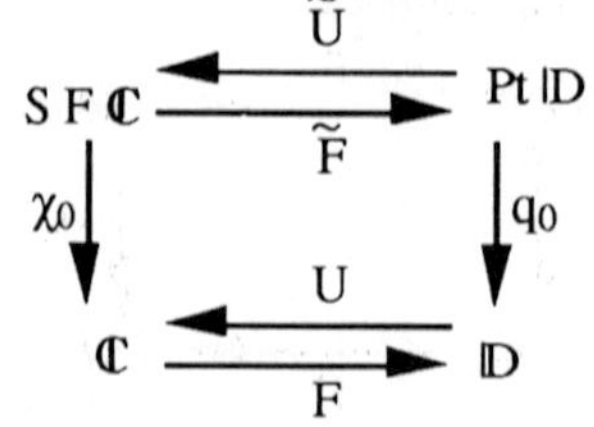

where $\tilde{U}$ and $\tilde{F}$ are cartesian liftings of U and F. Furthermore the unit and co-unit of this adjunction are cartesian, according to the following pullbacks :

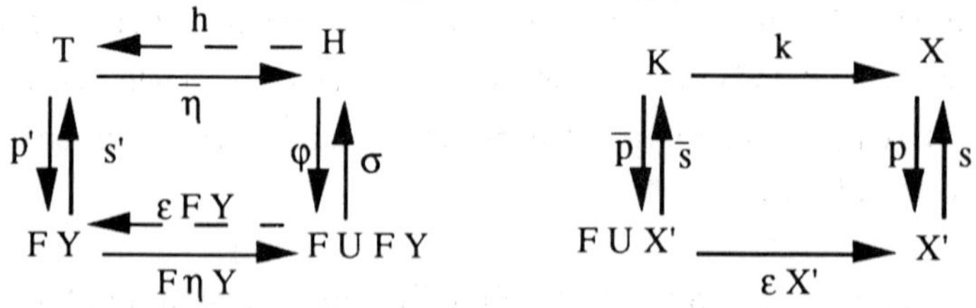

So the adjunction $(\tilde{U}, \tilde{F})$ is a transversely cartesian lifting of the adjunction (U, F). As a consequence the associated monad $(\tilde{U} . \tilde{F}, \tilde{\eta}, \tilde{\mu})$ is transversely cartesian.

Now, the functor U_1 is, up to equivalence, the composite of the two following functors :

$$2\text{-Grd } \mathbb{E} = \mathrm{Grd}_{(\ )_0}(\mathrm{Grd}\ \mathbb{E}) \xrightarrow{U_0} \mathrm{Pt}_{(\ )_0}(\mathrm{Grd}\ \mathbb{E}) \xrightarrow{\tilde{U}_1} S F_0$$

where $\tilde{U}_1$ is the right adjoint of $\tilde{F}_1$ (cf 4) and U_0 the restriction of the general functor $U_0 : \mathrm{Grd}(\mathbb{C}) \to \mathrm{Pt}(\mathbb{C})$ (cf 0) to the full subcategory 2-Grd $\mathbb{E}$ of Grd(Grd $\mathbb{E}$). This restriction has for left adjoint the restriction of the functor F_1 to the full subcategory

$Pt_{(\ )_0}$ (Grd IE) of Pt(Grd IE). Furthermore this restriction U_0 is clearly a monadic restriction of the general U_0. It is not difficult to check that the left adjoint $F_1 . \tilde{F}_1$ of $U_1 \cong \tilde{U}_1 . U_0$ is up to equivalence the functor F_2 described in 5. Now when IE = \$, we have $U_1 . F_2 = \Gamma_1$ (see 2 example 1).

More generally, we shall denote by (Γ_1, γ, ψ) the monad generated on N-Grd IE (or equivalently S F_0) by the adjunction (U_1, F_2)

$$\text{N-Grd IE} \underset{F_2}{\overset{U_1}{\rightleftarrows}} \text{2-Grd IE}$$

7. Some results about monadicity

<u>Proposition 4</u>. *Let* $\mathbb{H} \xrightarrow{\tilde{U}} \mathbb{G} \xrightarrow{U'} \mathbb{F}$ *be two functors having* $\tilde{F}$ *and* F' *as left adjoints. If* U' . $\tilde{U}$ *is monadic, then the comparison functor* K : $\mathbb{H} \rightarrow$ Alg $\tilde{T}$ *is fully faithful and has a left adjoint* M.

<u>Proof</u>. The functor U' . $\tilde{U}$ being conservative, such is $\tilde{U}$. Now, if (f, g) is a parallel pair in $\mathbb{H}$ such that its image by $\tilde{U}$ has a split coequalizer, it is a fortiori the case for its image by U' . $\tilde{U}$. Now U' . $\tilde{U}$ being monadic, the pair (f, g) has a coequalizer in $\mathbb{H}$ (which is preserved by U' . $\tilde{U}$). Consequently (Beck's theorem) the comparison functor K has a left adjoint M.

Now starting from an object H in $\mathbb{H}$, the object M . K(H) is the following coequalizer in $\mathbb{H}$:

$$(\tilde{F}\tilde{U})^2 H \underset{\tilde{\varepsilon}\tilde{F}\tilde{U} H}{\overset{\tilde{F}\tilde{U}\tilde{\varepsilon} H}{\rightrightarrows}} \tilde{F}\tilde{U} H \xrightarrow{m} MKH \xrightarrow{\gamma} H, \qquad \gamma \cdot m = \tilde{\varepsilon} H$$

The morphism $\tilde{\varepsilon}$ H coequalizes the pair, whence a factorization γ. Now the coequalizer m is preserved by U' . $\tilde{U}$. On the other hand U' . $\tilde{U}(\tilde{\varepsilon}$ H) is a split coequalizer since it is the case for $\tilde{U}(\tilde{\varepsilon}$ H). So U' . $\tilde{U}$(m) and U' . $\tilde{U}(\tilde{\varepsilon}$ H) are coequalizers of the same pair and U' . $\tilde{U}(\gamma)$ is an isormorphism. The functor U' . $\tilde{U}$ being conservative, γ it is itself invertible.

<u>Proposition 5</u>. *If furthermore* U' *is conservative then* K *is an equivalence.*

<u>Proof</u>. Given an algebra (X, δ) of $\tilde{T}$ in G, the object M(X, δ) in $\mathbb{H}$ is given by the following cokernel :

$$\tilde{F}\tilde{U}\tilde{F}X \underset{\tilde{\varepsilon}\tilde{F}X}{\overset{\tilde{F}\delta}{\rightrightarrows}} \tilde{F}X \xrightarrow{z} Z$$

The unit $(X, \delta) \to K\,M(X, \delta)$ is determined by the following factorization ξ in G :

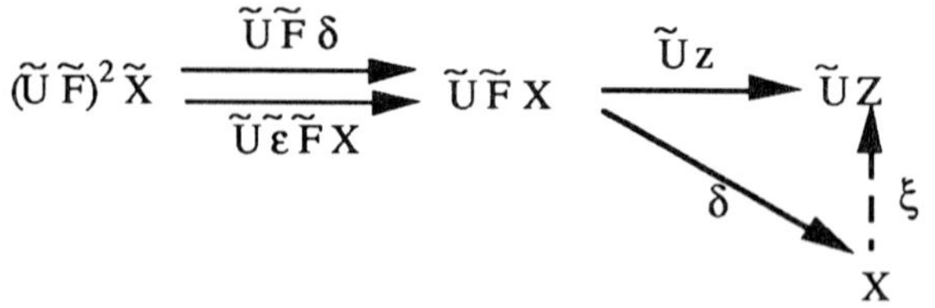

We know that z, as a cokernel above a split pair, is preserved by the monadic functor $U' \, . \, \tilde{U}$. On the other hand, δ is preserved by U' as a coequalizer, then $U'(\xi)$ is necessarily an isomorphism. If U' is conservative, ξ itself is an iso.

8. The comparison functor K_1 : 2-Grd $\mathbb{E} \to$ Alg Γ_2 is fully faithful

We shall get now our result in two steps.
The first one comes from an observation about the split free extensions :

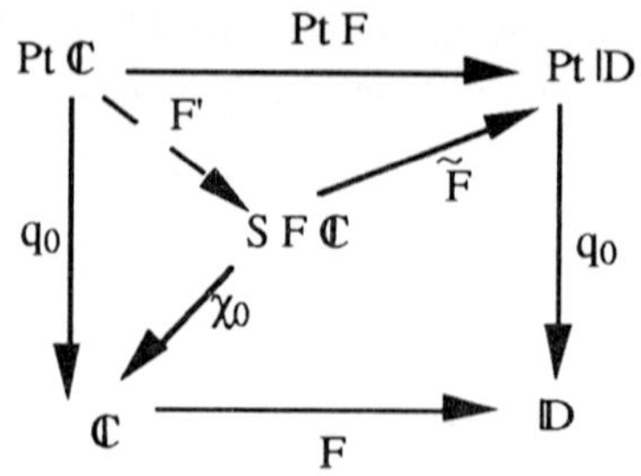

The definition of S F $\mathbb{C}$ determines a factorization F' : Pt $\mathbb{C} \to$ S F $\mathbb{C}$ defined by $F'((p, s)) = (q_0(p, s), (Fp, Fs))$. It has a right adjoint U', where, for every object (Y, (p, s)) in S F $\mathbb{C}$, its image in $\mathbb{C}$ is given by the following pullback :

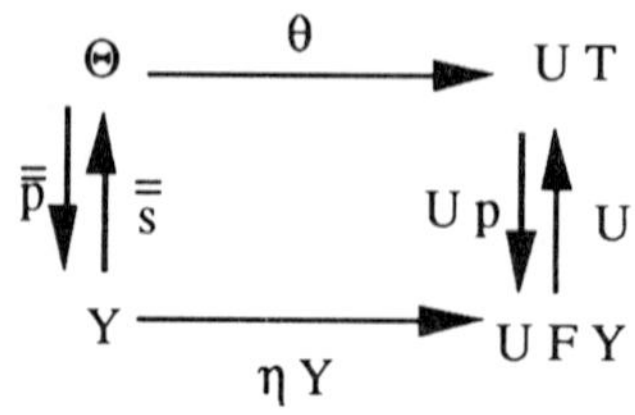

Further more $U' \, . \, \tilde{U}$ = Pt(U) the extension of the functor U to the categories of split epimorphisms.

The second steps consists in the following remark. When U is left exact, it can be not only extended into a functor Pt U : Pt $\mathbb{D} \to$ Pt $\mathbb{C}$ but also into a functor Grd U : Grd $\mathbb{D} \to$ Grd $\mathbb{C}$, with the commutative diagram :

$$\begin{array}{ccc} \text{Grd } \mathbb{C} & \xleftarrow{\text{Grd U}} & \text{Grd } \mathbb{D} \\ {\scriptstyle U_0}\downarrow & & \downarrow{\scriptstyle U_0} \\ \text{Pt } \mathbb{C} & \xleftarrow[\text{Pt U}]{} & \text{Pt } \mathbb{D} \end{array}$$

Proposition 6 : *When* U *is monadic, then* Pt U . $U_0 = U_0$. Grd U *is monadic.*

Proof. Let us denote by (T = U F, η, U ε F) the monad generated on $\mathbb{C}$ by the adjunction (U, F). Then Pt T is the monad generated on Pt $\mathbb{C}$ by the adjunction (Pt U, Pt F).

The functor U being left exact, the functor Pt U always commutes with the monads Γ_0 : Pt U . $\Gamma_0(\mathbb{C}) = \Gamma_0(\mathbb{D})$. Pt U. Following [2], this commutation produces a distributive law between Pt T and $\Gamma_0(\mathbb{C})$.

If furthermore U is monadic, Pt U is clearly monadic. The category Pt $\mathbb{D}$ is up to equivalence the category Alg Pt T, and $\Gamma_0(\mathbb{D})$ the lifting of $\Gamma_0(\mathbb{C})$ to this category of algebras. Again following [2], Pt U. U_0 is monadic.

Corollary. If U is monadic and Grd U admits a left adjoint, then Grd U is monadic.

Proof. Following the proposition 6, the functor U_0 . Grd U is monadic.
If Grd U admits a left adjoint we are under the hypothesis of the proposition 4. Now U_0 is monadic and thus conservative, so following the proposition 5, Grd U is monadic.

We are now in position to assert that

Proposition 7. *The comparison functor* K_2 : 2-Grd $\mathbb{E} \to$ Alg Γ_1, *is fully faithfull and has a left adjoint.*

Proof. Let us consider the following diagram :

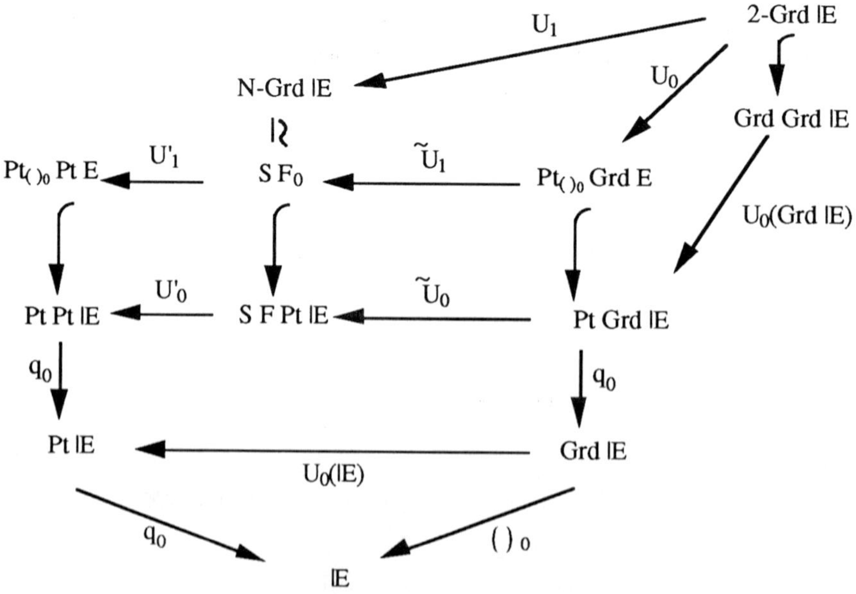

We saw that $U_0(\mathbb{E})$ is monadic. Then by proposition 6

$$Pt(U_0) \,.\, U_0(Grd\ \mathbb{E}) = U'_0 \,.\, \tilde{U}_0 \,.\, U_0\ Grd(\mathbb{E})$$

is again monadic. It is also the case for its full restriction $U'_1 \,.\, \tilde{U}_1 \,.\, U_0 \cong U'_1 \,.\, U_1$. Then following proposition 4, the comparison functor K_2 : 2-Grd $\mathbb{E} \to$ Alg Γ_1 is fully faithful and has a left adjoint M_2.

9. The comparison functor K_2 is an equivalence of categories.

Proposition 9. *The functor* U_1 : 2-Grd $\mathbb{E} \to$ N-Grd $\mathbb{E}$ *is monadic.*

Proof. Following prop. 5, it is sufficient to prove that the functor U'_1 is conservative. Given (X_1, s_1) a normalized groupoid, $U'_1(X_1, s_1)$ is given by the following diagram :

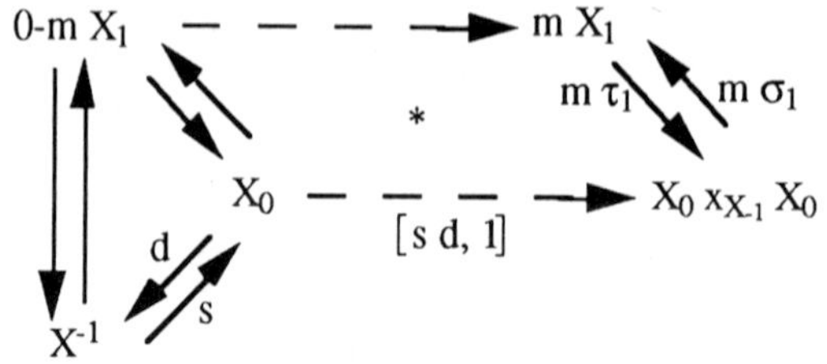

where the square (*) is a pullback. Then 0-m X_1 represents the set of morphisms of X_1 whose domain is a distinguished object in X_1.

Now thanks to the Yoneda embedding it is sufficient to prove that U'_1 is conservative when $\mathbb{E} = \$$. Let $F_1 : X_1 \to Y_1$ be a functor between two normalized groupoids such that $U'(F_1)$ is a isomorphism. In particular F_0 is an isomorphism and F_1 is bijective on the objects. We must show that is it fully faithful. Given two maps $f, g : X \to X'$ such that $F_1(f) = F_1(g)$. Then $s_1 X = s_1 X'$ and :

$$F_1(f) . s_1 F X = F_1 g . s_1 FX' .$$

That means that $F_1(f . s_1 X) = F_1(g . s_1 X)$. Whence $f . s_1 X = g . s_1 X$ and $f = g$. Conversely given a map $h : FX \to FX'$, then $h. s_1 FX$ comes from a unique map $\tau : s X \to X'$. That is $F \tau = h . s_1 FX$. Then $F(\tau . s_1^{-1} X) = F \tau . F(s_1^{-1} X) = h$.

10. Description of the inverse equivalence M_2. Monadic adjustment of the 1-geometry.

Let X_1 be a normalized groupoid and $\delta_1 : \Gamma_1 X_1 \to X_1$ an algebra. The objects of the 2-groupoid $Z_2 = M_2(X_1, \delta_1)$ are the distinguished elements of X_1, in other words $Z_0 = X_1$. The 1-morphisms of Z_2 are the objects of X_1 (whence $m Z_1 = X_0$). The 2-morphisms of Z_2 are the elements of 0-m X_1 (whence $m(m Z_2) =$ 0-m X_1).

Now let X be an object of X_1, as a 1-morphism in Z_2 its 0-domain is s X and its 0-codomain is $\delta_1(1_X)$. If $f : s X \to X$ is a map in X_1, as a 2-cell in Z_2, its 1-domain is X and its 1-codomain is $\delta_1(f)$. The quickest way to describe the composition in Z_1 is to describe its structure of algebra on Γ_0. Given two 1-maps X and Y having the same 0-domain (that means s X = s Y and thus that X and Y are in the same connected component), then $X * Y = \delta_1(\sigma_1(X, Y))$ where $\sigma_1(X, Y)$ is seen as an object of $\Gamma_1 X_1$:

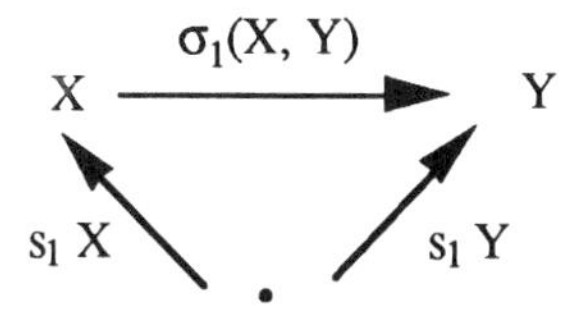

In the same way, given two 2-cells f and g with the same 1-domain, then f * g is the image by δ_1 of the following morphism of $\Gamma_1 X_1$:

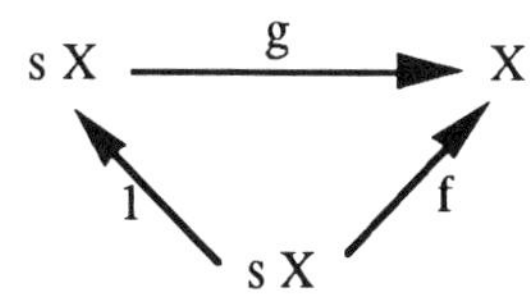

Thus any map $f : X \to Y$ in X_1 determines an oriented triangle in Z_2 :

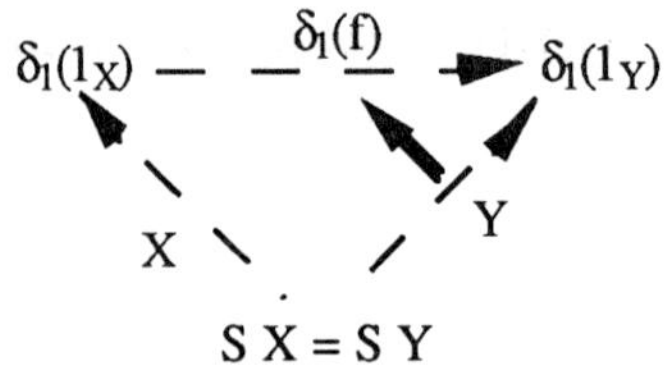

the filler being given by $f . s_1 X : sX \longrightarrow X \xrightarrow{f} Y$
Conversely an oriented triangle in Z_2 :

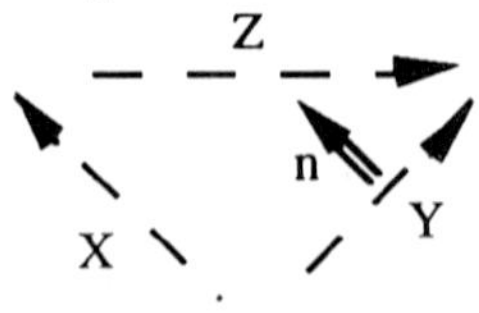

produces a map $n . s_1^{-1} X : X \longrightarrow sX \xrightarrow{n} Y$ in X_1 .

11. The nerve of a 2-groupoid.

A normalized groupoid produces a 1-truncated split augmented simplicial object:

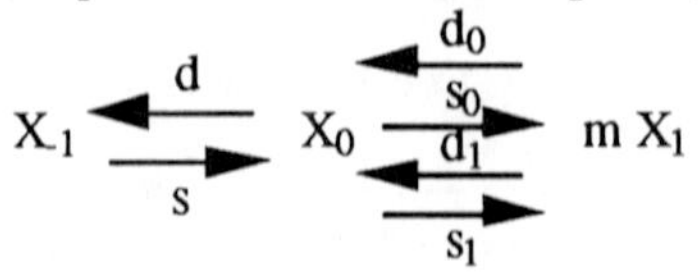

It can be completed into a 2-truncated split augmented simplicial object by its simplical kernel K_2, which is nothing but the object of the non commutative triangles of X_1, i.e. $m\, \Gamma_1\, X_1$. Via the iterated simplicial kernels, we define a functor :

$$n : \text{N-Grd}\, \mathbb{E} \to \text{Spl Aug Simpl}\, \mathbb{E}$$

Let us denote by δec the endofunctor on Spl Aug Simpl $\mathbb{E}$ obtained by cancelling X_{-1} and the upper indexed face and degeneracy maps. It is underlying to a monad whose category of algebras [9] is the category Simpl $\mathbb{E}$ of simplicial objects in $\mathbb{E}$.

Now, following our previous remark on K_2, the object δec . $n(X_1, s_1)$ is nothing but $n . \Gamma_1(X_1, s_1)$, and thus n defines a strict morphism of monads : $(\Gamma_1, \gamma, \psi) \to (\delta\text{ec}, \lambda, \mu)$. Consequently it produces a functor at the level of the category of algebras.

This functor N : 2-Grd $\mathbb{E} \to$ Simpl $\mathbb{E}$ is the "nerve" functor for the 2-groupoids [6], [16] [10] [4]. When $\mathbb{E}$ = \$, $N_0(X_2)$ is the set of objects (or 0-cells) $N_1(X_2)$ the set of 1-morphisms (or 1-cells), $N_2(X_2)$ the set of oriented triangles, $N_3(X_2)$ the set of "commutative" tetrahedrons, the upper levels being only simplicial kernels.

12. The higher dimensional groupoids.

Let us denote by $(T_2, \varepsilon_2, \nu_2)$ the comonad generated on 2-Grd $\mathbb{E}$ by the adjunction (U_1, F_2).

Now let us define a normalized 2-groupoid as a 2-groupoid X_2 together with a given normalization s_1 of its underlying groupoid X_1 and a given splitting σ_2 of its terminal map in the fiber $\tau_2 : X_2 \to G_2 X_1$. A normalized 2-functor is a 2-functor which preserves the normalization data. Whence a category denoted by N-2 Grd $\mathbb{E}$.

Examples :

[1] Given X_2 a 2-groupoid, then $T_2 X_2 = F_2 U_1 X_2 = G_2 U_1 X_2$. Now $U_1 X_2$ is a normalized groupoid, and then $T_2 X_2$ is a normalized 2-groupoid.

[2] The normalized 2-groupoid associated to a 3-groupoid.

Given a 3-groupoid X_3 in $\mathbb{E}$, the following pullback in 2-Grd $\mathbb{E}$ will define a normalized 2-groupoid :

$$\begin{array}{ccc} U_2 X_3 & \dashrightarrow & m X_2 \\ {\scriptstyle \delta_2}\downarrow \uparrow{\scriptstyle \sigma_2} & & {\scriptstyle d_0}\downarrow \uparrow{\scriptstyle s_0} \\ T_2 X_2 & \overset{\varepsilon_2 X_2}{\dashrightarrow} & X_2 \end{array}$$

since $T_2 X_2$ is the terminal object in the fiber of X_3.

Whence a functor U_2 : 3-Grd $\mathbb{E} \to$ N-2-Grd $\mathbb{E}$.

Actually the category of normalized 2-groupoids is equivalent to the category obtained by the following pullback :

$$\begin{array}{ccc} S F_1 & \xrightarrow{\tilde{F}_2} & Pt_{(\ )_1}(2\text{-Grd } \mathbb{E}) \\ {\scriptstyle k_1}\downarrow & & \downarrow{\scriptstyle q_0} \\ \text{N-Grd } \mathbb{E} & \xrightarrow{F_2} & 2\text{-Grd } \mathbb{E} \end{array}$$

The general construction of 6 shows us that U_2 has a left adjoint F_3.

More generally, let us suppose we have defined the notion of normalized (n - 1)-groupoid, then a normalized n-groupoid is a n-groupoid X_n together with a given normalization of its underlying (n - 1)-groupoid X_{n-1} and a given splitting σ_n of the terminal map $\tau_n : X_n \to G_n X_{n-1}$. We get then the category N-n-Grd $\mathbb{E}$ of normalized n-groupoids.

Furthermore let us suppose we have an adjunction

$$\text{N-(n - 1)-Grd }\mathbb{E} \underset{F_n}{\overset{U_{n-1}}{\rightleftarrows}} \text{n-Grd }\mathbb{E}$$

and we have denoted by $(\Gamma_{n-1}, \gamma, \psi)$ the monad generated on N-(n - 1)-Grd $\mathbb{E}$ and by $(T_n, \varepsilon_n, \nu_n)$ the comonad generated on n-Grd $\mathbb{E}$. We are now in position to define the functor U_n : (n + 1)-Grd $\mathbb{E} \rightarrow$ N-n-Grd $\mathbb{E}$ by the following pullback :

$$\begin{array}{ccc} U_n(X_{n+1}) & \dashrightarrow & m\,X_{n+1} \\ \delta_n \downarrow \uparrow \sigma_n & & d_0 \downarrow \uparrow s_0 \\ T_n\,X_n & \overset{\varepsilon_n}{\dashrightarrow} & X_n \end{array}$$

Again, by 6, it has a left adjoint F_{n+1}.

That U_n is monadic is not easy to show. The proof of it will be developped elsewhere. This monadicity will allow a description of the nerve functor for the n-groupoids in the same way as the one given here for the 2-groupoids.

REFERENCE

1. F.A. Agl and R. Steiner, Nerves of multiple categories, Preprint, Univ. Glasgow (1991).

2. J. Beck, Distributive laws., Springer, Lecture Notes in Math. 80 (1969), 119-140.

3. D. Bourn, The tower of n-groupoids and the long cohomology sequence., J. Pure Appl. Algebra 62 (1989) 137-183.

4. D. Bourn, Another denormalization theorem for the abelian chain complexes., J. Pure Appl. Algebra 66 (1990) 229-249.

5. D. Bourn, The shift functor and the comprehensive factorization for internal groupoids, Cahiers Topol. Géom. Diff. 28(3) (1987) 197-226.

6. R. Brown and P.J. Higgins, The equivalence of ∞-groupoids and crossed complexes, Cahiers Topol. Géom. Diff. 22(4) (1981) 371-386.

7. A. Burroni, Higher dimensional word problem, Springer L.N. in Computer Sciences 530 (1991) 94-105.

8. M.K. Dakin, Kan complexes and multiple groupoid structures, Ph D. Thesis, Univ. of Wales (1974).

9. J.W. Duskin, Simplicial methods and the interpretation of triple cohomology, Mem. Amer. Math. Soc. vol. 3, Issue 2, n° 163 (1975).

10. J.W. Duskin, Private communications.

11. A. Gramain, Le théorème de Van Kampen, Lecture at the Conference "Point de vue des catégories", Tours 2,3,4 octobre 1991.

12. M. Johnson, The combinatorics of n-categorical pasting, J. Pure Appl. Algebra 62 (1989) 211-225.

13. M.M. Kapranov and V.A. Voevodsky, Combinatorial geometric aspects of polycategory theory : pasting schemes and higher Bruhat orders, Cahiers Topol. Géom. Diff. 32(1) (1991) 11-27.

14. J.E. Roberts, Mathematical aspects of local cohomology. Proc. Colloquium on Operator Algebras and their appl. to Math. Physics, Marseille (1977).

15. J. Stasheff, Homotopy associativity of H-spaces, Trans. Amer. Math. Soc. 108 (1963) 275-312.

16. R. Street, The algebra of oriented simplexes, J. Pure Appl. Algebra 49 (1987) 283-335.

17. R. Street, Parity complexes. Macquarie Math. Report (1988).

FACULTE DE MATHEMATIQUES ET D'INFORMATIQUE - UNIVERSITE DE PICARDIE - 33,RUE SAINT LEU - 80039 AMIENS CEDEX 1 - FRANCE.

Canadian Mathematical Society
Conference Proceeding
Volume **13**, 1992

CLASSIFYING TOPOSES AND FUNDAMENTAL LOCALIC GROUPOIDS

MARTA BUNGE

ABSTRACT. The fundamental group of a pointed Grothendieck topos was defined by Grothendieck, Artin and Mazur as a progroup. The localic approach to the subject of the fundamental group of a topos was initiated by Moerdijk in the case of a pointed connected locally connected topos $\mathcal{E}$ bounded over a base topos $\mathcal{S}$, using the theory of classifying toposes of localic groupoids originating in the work of Joyal and Tierney. In this paper we introduce the notion of a totally disconnected topos and use it to define the fundamental localic groupoid of a topos $\mathcal{E}$ bounded over a base topos $\mathcal{S}$, for a topos $\mathcal{E}$ with a locale of path components in $\mathcal{S}$ (among these are the locally connected toposes), but not assuming that $\mathcal{E}$ has points. The fundamental localic groupoid of $\mathcal{E}$ is a totally disconnected open etale complete localic groupoid in $\mathcal{S}$, representing first degree cohomology of $\mathcal{E}$ with coefficients in discrete groups. It is also locally, i.e., on an open cover of the base topos $\mathcal{S}$, Morita equivalent to a prodiscrete localic groupoid.

0. Introduction

For a pointed Grothendieck topos, the fundamental group was defined by Grothendieck (cf. [Gro], also [A/M]) as a progroup, i.e., as an inversely filtered system of groups and and homomorphisms. Associated with a progroup $\mathcal{G}$ is a Grothendieck topos $B\mathcal{G}$, said to be the classifying topos of $\mathcal{G}$. The objects of $B\mathcal{G}$ are called $\mathcal{G}$-sets by analogy with the case of G a group, but the notion of a $\mathcal{G}$-action is not a straightforward generalization of the notion of a G-action (cf. [Ken_3]).

The work of Joyal and Tierney [J/T] on toposes and localic groupoids provided the natural framework for a drastic simplification of the subject, initiated by Moerdijk [M_4] for the case of a pointed connected locally connected topos bounded over a base topos $\mathcal{S}$. The starting point was the remark_ also made by Tierney_ that a strict progroup $\mathcal{G}$ (i.e., one in which all transition maps are surjective homomorphisms) has a localic group G associated with it, namely the limit of the inversely filtered category of discrete groups in the category of localic groups and that the complicated notion of a $\mathcal{G}$-action is equivalently described by the notion of a G-action, the latter being just the localic version of the usual notion. Specifically, to any given localic group(oid) G is associated a topos BG of discrete G-locales or G_0-sheaves (in the terminology of [J/T]). For a localic groupoid G

1991 Mathematics Subject Classification. Primary, 18B25; Secondary, 20L15. Supported by the National Science & Engineering Council of Canada. This paper is in final form and no version of it will be submitted for publication elsewhere.

with locale of objects G_0 and locale of morphisms G_1, a G_0-sheaf is a locale X equipped with a local homeomorphism $X \to G_0$ and with a G-action in the sense of a continuous map $X \times_{G_0} G_1 \to X$ which satisfies the usual conditions. A morphism of G_0-sheaves is a continuous map which preserves the action. There results a category BG which is a topos, said to be the classifying topos of G (cf. [M$_2$], [M$_3$]). It is shown in [Bu$_1$] (cf. also [M$_5$]) that BG classifies principal localic $\hat{G}$-bundles, for $\hat{G}$ the etale completion of G. In the case of a strict progroup $\mathcal{G}$, the associated localic groupoid G is such that the toposes $B\mathcal{G}$ and BG are equivalent.

The purpose of this paper is to extend the localic approach to the question of the fundamental "group" of a topos from the pointed to the case of an arbitrary topos which (cf.[Ba]) need not have any points. Although one cannot expect the assignment of the fundamental "group" of an unpointed topos to be functorial (cf. [Gra]), it seems of interest to investigate to what extent such a notion can be defined so as to retain some of the usual properties of the fundamental "group" of a pointed topos.

We approach the construction of π_1 of a topos $\mathcal{E}$ guided by a Čech-style definition of topos cohomology (cf.[Ja]). In §1, the notion of a totally disconnected topos is introduced and investigated for its own sake as well as for applications in §3. The notion of a locale of path components of a topos, defined in §1, is naturally linked with the result of Moerdijk and Wraith in [M/W] whereby any connected locally connected locale(or topos) is path connected. The class of toposes with a locale of path components properly includes that of all locally connected toposes. In §2, using the theory of classifying toposes for localic groupoids from [J/T], we associate with any cover U in a topos $\mathcal{E}$ with a locale of path components, a totally disconnected localic groupoid Π^U which classifies U-split torsors in $\mathcal{E}$ for discrete groups. The local data is then patched up in §3 by taking the limit in $\mathrm{Top}_{\mathcal{S}}$ of an inversely filtered system arising from the toposes $B\Pi^U$ in order to exhibit a totally disconnected localic groupoid Π in $\mathcal{S}$ which represents first degree cohomology of $\mathcal{E}$ with coefficients in discrete groups. Such a localic groupoid Π (the fundamental localic groupoid of $\mathcal{E}$) is, locally (i.e, on an open cover of the base topos $\mathcal{S}$) Morita equivalent to a prodiscrete localic groupoid.

I am grateful to J.Kennison for valuable discussions on the subject of the fundamental group during the academic year 1988-89. In the meantime, Kennison has also given (cf.[Ken$_4$]) a definition of the fundamental localic groupoid of an unpointed Grothendieck topos by different methods and in the spirit of his own previous work on the subject (cf.[Ken$_1$],[Ken$_2$] and [Ken$_3$]). I am also grateful to P.T.Johnstone and I.Moerdijk for useful suggestions on how to improve a preliminary version [Bu$_2$] of this paper. Moreover, Moerdijk provided me with the correct version of total disconnectedness for geometric morphisms and outlined some of its basic properties.

1. Totally Disconnected Toposes and Locales of Path Components.

Let $\mathcal{S}$ be an elementary topos with natural numbers object. Denote by $\mathrm{Top}_{\mathcal{S}}$ the 2-category of bounded $\mathcal{S}$-toposes, geometric morphisms and **iso** 2-cells. All constructions performed are to be understood in a 2-categorical sense and relative to $\mathcal{S}$. Moreover, we will often work as if $\mathcal{S}$ were the category of sets as well as employ terminology accordingly. This is justified (cf.[Bo/J]) so long as the arguments be explicit and constructive.

The following notion of total disconnectedness for toposes and geometric morphisms is the analogue in $\mathrm{Top}_{\mathcal{S}}$ of the localic version for locales and continuous maps from [J/M] and has been suggested by Moerdijk, who also outlined Propositions 1.2, 1.3 ,1.4 and Corollary 1.5, below.

1.1. Definition Let I be the unit interval constructed as a locale in $\mathcal{S}$ (cf. [H]). Let $\mathcal{S}[I]$ be the topos of sheaves on I . The latter is exponentiable (cf. [J/J]) and there is, for each $\mathcal{E}$ in $\mathrm{Top}_{\mathcal{S}}$, a geometric morphism

$$c_{\mathcal{E}} : \mathcal{E} \to \mathcal{E}^{\mathcal{S}[I]}$$

which is the transpose of the projection. Call $\mathcal{E}$ a *totally disconnected topos* if $c_{\mathcal{E}}$ is an equivalence.

A geometric morphism $f : \mathcal{F} \to \mathcal{E}$ is said to be a *totally disconnected geometric morphism* if $\mathcal{F}$ is totally disconnected as a topos over $\mathcal{E}$. Explicitly, consider the diagram

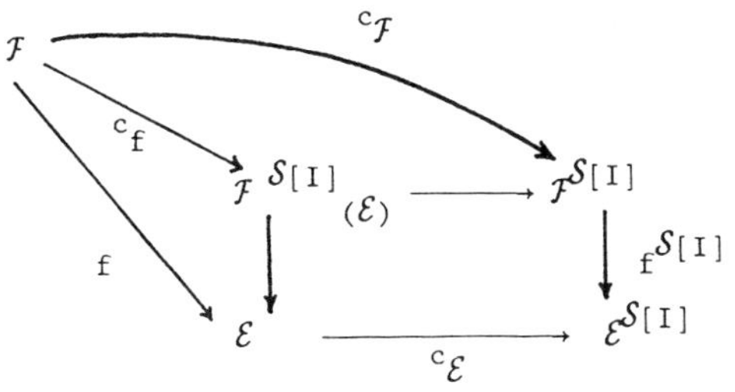

where the inner square is a pullback in $\mathrm{Top}_{\mathcal{S}}$ and c_f is the induced geometric morphism. Then, f is totally disconnected if and only if c_f is an equivalence.

1.2 Proposition. *(i) Let $\mathcal{E} \in \mathrm{Top}_{\mathcal{S}}$ with structural map $\gamma : \mathcal{E} \to \mathcal{S}$. Then $\mathcal{E}$ is a totally disconnected topos if and only if γ is a totally disconnected geometric morphism.*

(ii) Let X be a locale in $\mathcal{S}$ (respectively, $f : X \to Y$ be a continuous map of locales in $\mathcal{S}$). Then X is totally disconnected (respectively f is totally disconnected, both in the sense of [J/M]) if and only if the topos $\mathcal{S}[X]$ (respectively the geometric morphism $f : \mathcal{S}[X] \to \mathcal{S}[Y]$) is totally disconnected. □

1.3. Proposition.(Characterizations of total disconnectedness). *Let $f : \mathcal{F} \to \mathcal{E} \in \mathrm{Top}_{\mathcal{S}}$. Then: (i) f is totally disconnected if and only if the commutative diagram*

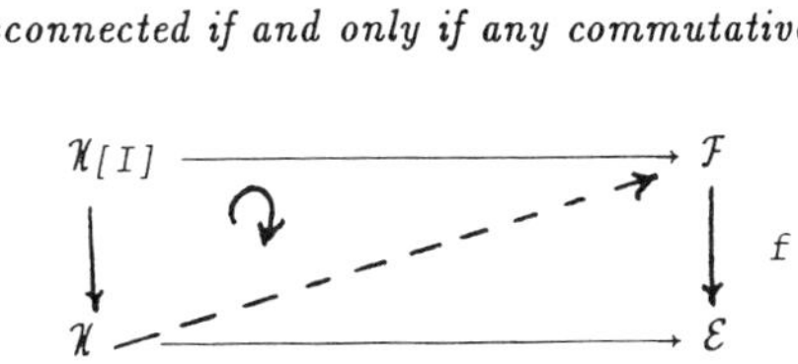

is a pullback;

(ii) f is totally disconnected if and only if any commutative diagram of solid arrows

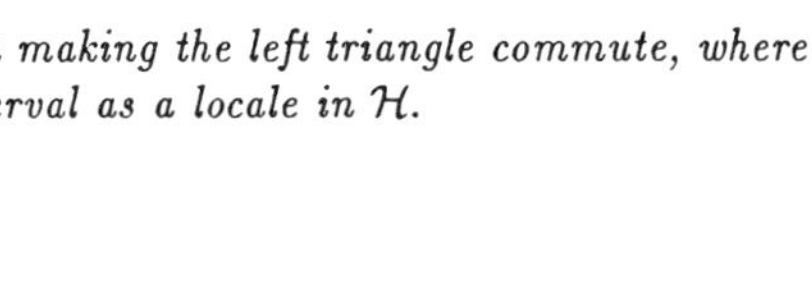

admits a diagonal fill-in making the left triangle commute, where $\mathcal{H}[I]$ denotes the topos of sheaves on the unit interval as a locale in $\mathcal{H}$.

Proof. (i) This is immediate from Def. 1.1. (ii) Since $\mathcal{H}[I] \to \mathcal{H}$ is connected (cf.[H]), if a diagonal fill-in exists making the left triangle commute then it is unique and also makes the right triangle commute. But the existence of a diagonal making the entire diagram commute is equivalent to the characterization of total disconnectedness given in (i), after passing to transposes.□

1.4. Proposition. (Stability properties of totally disconnected geometric morphisms). *(i) Totally disconnected geometric morphisms are stable under composition. (ii) If, in a commutative diagram*

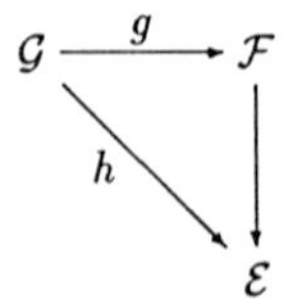

in $Top_{\mathcal{S}}$, h is totally disconnected then so is g. (iii) Totally disconnected geometric morphisms in $Top_{\mathcal{S}}$ are pullback stable. (iv) Any equivalence of toposes is totally disconnected.

Proof. (i) This is immediate from the first characterization given in Prop.1.3.
(ii) Assume that h is totally disconnected and let us consider a diagram of solid arrows

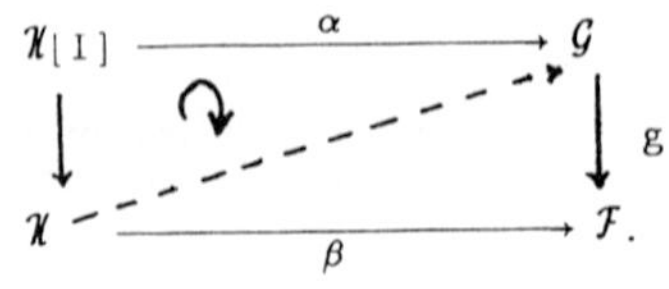

Since h is totally disconnected, a diagonal fill-in making the left triangle commute exists for the diagram

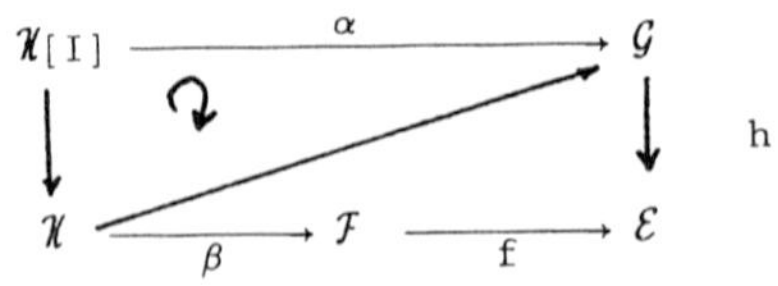

and the same diagonal arrow may be used to show that g is totally disconnected, by the second characterization given in Prop. 1.3.
(iii) If, in a diagram

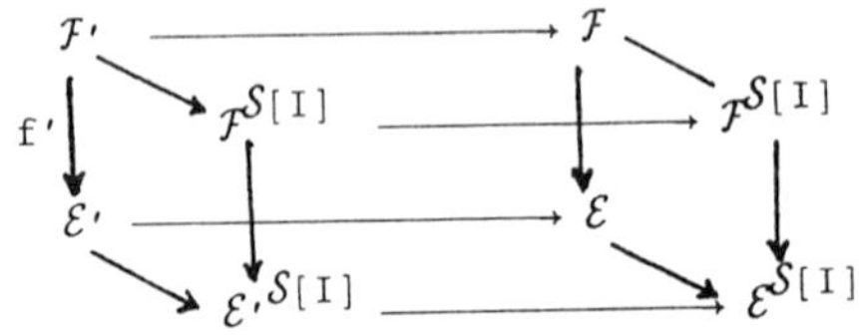

f is totally disconnected and the back square is a pullback, then, using that exponentiating to $\mathcal{S}[I]$ preserves pullbacks, it follows by a standard categorical argument (cf. [CWM]) that the left vertical square is a pullback. Thus, by Prop. 1.3(i), f' is totally disconnected. (iv) If $f : \mathcal{F} \to \mathcal{E}$ is an equivalence, so is $f^{\mathcal{S}[I]}$. It follows that the commutative square of Prop. 1.3(i) is a pullback. □

1.5. Corollary. *(i) Any point $p : \mathcal{S} \to \mathcal{E}$ of a topos $\mathcal{E} \in Top_{\mathcal{S}}$ is totally disconnected. (ii) Any localic point $p : \mathcal{S}[X] \to \mathcal{E}$ in $Top_{\mathcal{S}}$ with X a totally disconnected locale is a totally disconnected geometric morphism. (iii) Any local homeomorphism is totally disconnected. (iv) Totally disconnected toposes are closed under limits of inversely filtered systems in $Top_{\mathcal{S}}$.*

Proof. (i) & (ii). These are immediate consequences of Prop. 1.4 (ii) & (iv). (ii) Discrete locales are totally disconnected. (iii) The argument is similar to that of Prop. 1.4 (iii) using that exponentiating to $\mathcal{S}[I]$ preserves those inverse limits which exist in $\mathrm{Top}_{\mathcal{S}}$, e.g., those of inversely filtered systems (cf. [M_1]).□

In the localic case (cf. [J/M]), the class of path connected continuous maps is orthogonal to the class of totally disconnected continuous maps. We recall explicitly the relevant definitions and results.

1.6. Definition. (cf [M/W]). A locale X in $\mathcal{S}$ is said to be *path connected* if both the unique map $X \to 1$ and evaluation at the endpoints $\epsilon : X^I \to X \times X$ are open surjections. Similarly, a topos $\mathcal{E}$ is said to be *path connected* if both its structural map $\gamma : \mathcal{E} \to \mathcal{S}$ and $\epsilon : \mathcal{E}^{\mathcal{S}[I]} \to \mathcal{E} \times \mathcal{E}$ are open surjections in $\mathrm{Top}_{\mathcal{S}}$. The corresponding notions for continuous maps and geometric morphisms are given by relativization.

By a theorem of Moerdijk and Wraith[M/W], any connected locally connected locale (or topos) is path connected. Their notion of path connectedness does not seem to be strong enough to imply orthogonality to the class of arbitrary totally disconnected geometric morphisms, but the stronger notion of connected locally connected is, as shown in Theorem 1.9. We begin by recalling the localic version of orthogonality.

1.7. Theorem [J/M]. *In the category $Loc(\mathcal{S})$ of locales in $\mathcal{S}$ and continuous maps, any path connected morphism is orthogonal to any totally disconnected morphism.*

Proof. Since the notions involved are relative to the base, it is enough to prove that any continuous map $f : X \to Y$, with X path connected and Y totally disconnected locales, is constant. Since X is path connected, $X \to 1$ is an open surjection in $\mathrm{Loc}(\mathcal{S})$ hence (cf. [M_3]) a coequalizer of its kernel pair. Since path connectedness of X also says that $\epsilon : X^I \to X \times X$ is an open surjection (hence surjective in $\mathrm{Loc}(\mathcal{S})$), the diagram

$$X^I \underset{\epsilon_1}{\overset{\epsilon_0}{\rightrightarrows}} X \longrightarrow 1$$

is a coequalizer in $\mathrm{Loc}(\mathcal{S})$, where $\epsilon_i = \pi_i \epsilon$ for $i = 0, 1$ and where π_0, π_1 are the projections $X \times X \rightrightarrows X$. Since Y is totally disconnected and both ϵ_0 and ϵ_1 (defined for Y) are left inverses to the constant paths map c_Y, they are equal. From the commutativity of the diagrams

$$\begin{array}{ccc} X^I & \xrightarrow{\quad \epsilon_i \quad} & X \\ {\scriptstyle f^I}\big\downarrow & & \big\downarrow{\scriptstyle f} \\ Y^I & \xrightarrow[\quad \epsilon_i \quad]{} & Y \end{array}$$

follows that f coequalizes ϵ_0 and ϵ_1 (defined for X) and so f admits a factorization

$$\begin{array}{ccc} X & \longrightarrow & 1 \\ {\scriptstyle f}\big\downarrow & \swarrow & \\ Y & & \end{array}$$

as claimed.□

That connected locally connected geometric morphisms are regular epis in $\mathbf{Top}_{\mathcal{S}}$ follows from the following result which is basically due to Pitts (lecture, McGill University, 1984).

1.8. Theorem. *Let*

$$\begin{array}{ccc} \mathcal{A} & \xrightarrow{\quad f \quad} & \mathcal{C} \\ {\scriptstyle g}\big\downarrow & & \big\downarrow{\scriptstyle q} \\ \mathcal{B} & \xrightarrow[\quad p \quad]{} & \mathcal{D} \end{array}$$

be a pullback in $\mathbf{Top}_{\mathcal{S}}$, *with p and q both connected and locally connected. Then, f and g are both connected and locally connected and the diagram is a pushout.*

Proof. That connected locally connected geometric morphisms are pullback stable and the BCC ("Beck-Chevalley condition") holds was shown independently by Pitts[Pi] and Paré[Pa]. Since the diagram of inverse image parts is now one of inclusions, to show that it is a pullback (so that the given diagram be a pushout in $\mathbf{Top}_{\mathcal{S}}$) it is enough to verify that it is an intersection. This is done as follows.

Let $C \in \mathcal{C}$ and $B \in \mathcal{B}$ be given with $\alpha : g^*B \to f^*C$ an iso in $\mathcal{A}$. We claim that defining $D = q_!C$, where $q_!$ is the left adjoint of q^*, has the property that $p^*D \cong B$ and $q^*D \cong C$. Since g is connected, the counit of adjunction $\epsilon : g_!g^* \to \mathrm{id}$ is an iso natural transformation. Also, by the BCC, the canonical natural transformation $\omega : p^*q_! \to g_!f^*$ is iso. First, we have the composite isomorphism

$$p^*D = p^*q_!C \xrightarrow{\omega_C} g_!f^*C \xrightarrow{g_!\alpha} g_!q^*B \xrightarrow{\epsilon_B} B\ .$$

Next, there is also an iso

$$f^*q^*q_!C \xrightarrow{\theta_{q_!C}} g^*p_{q!}C \xrightarrow{g^*\omega_C} g^*g_!f^*C \xrightarrow{g^*g_!\alpha^{-1}} g^*g_!g^*B \xrightarrow{\epsilon_{g^*B}} g^*B \xrightarrow{\alpha} f^*C$$

so that by the fully faithfulness of f^* gives rise to an iso $q^*D \to C$.□

1.9. Theorem. *In $Top_{\mathcal{S}}$, any connected locally connected morphism is orthogonal to any totally disconnected morphism.*

Proof. As in the proof of Theorem 1.7, we reduce to the case of showing that any morphism $f : \mathcal{E} \to \mathcal{F}$, with $\mathcal{E}$ connected locally connected and $\mathcal{F}$ totally disconnected, is constant. By Theorem 1.8, the structural map $\gamma : \mathcal{E} \to \mathcal{S}$ is the coequalizer of its kernel pair. Since $\mathcal{F}$ is totally disconnected, $c_{\mathcal{F}}$ is an equivalence hence $\epsilon_0 = \epsilon_1$ (for $\mathcal{F}$). It follows from the commutativity in $\mathrm{Top}_{\mathcal{S}}$ of the diagrams

$$\begin{array}{ccc} \mathcal{E}^I & \xrightarrow{\epsilon_i} & \mathcal{E} \\ {\scriptstyle f^I}\downarrow & & \downarrow{\scriptstyle f} \\ \mathcal{F}^I & \xrightarrow[\epsilon_i]{} & \mathcal{F} \end{array}$$

that $f\epsilon_0 = f\epsilon_1$ and, since ϵ (for $\mathcal{E}$) is a surjection, that $f\pi_0 \cong f\pi_1$ hence a factorization

$$\begin{array}{ccc} \mathcal{E} & \xrightarrow{\gamma} & \mathcal{S} \\ {\scriptstyle f}\downarrow & \swarrow & \\ \mathcal{F} & & \end{array}$$

exists.□

1.10. Proposition. *(i) Factorizations in $Top_{\mathcal{S}}$ into connected locally connected followed by totally disconnected are unique if they exist. (ii) Locally connected geometric morphisms admit factorizations into connected locally connected followed by totally disconnected (in fact, discrete) morphisms. (iii) A geometric morphism which is both locally connected and totally disconnected is a local homeomorphism.*

Proof. (i) This is a consequence of orthogonality and the surjectivity in $\mathrm{Top}_{\mathcal{S}}$ of connected morphisms. (ii) It is well known (cf.[TT]) that any locally connected morphism $f : \mathcal{E} \to \mathcal{F}$ in $\mathrm{Top}_{\mathcal{S}}$ factors canonically as

$$\mathcal{E} \xrightarrow{q} \mathcal{F}/f_!1 \xrightarrow{t} \mathcal{F}$$

with q connected locally connected and t the canonical map φ_U induced by $U \to 1$, where $U = f_!1$. Thus, t is a local homeomorphism and since it is localic it is determined by a discrete locale in $\mathcal{F}$. (iii) Let $f : \mathcal{E} \to \mathcal{F}$ be locally connected, totally disconnected. In the factorization given by q and t as in (ii), q is then also totally disconnected (use Prop.1.4). As q is connected locally connected, it is an equivalence. Thus, f is naturally isomorphic to t, hence a local homeomorphism.□

1.11. Definition. A topos $\mathcal{E} \in \mathrm{Top}_{\mathcal{S}}$ is said to have *a locale of path components over $\mathcal{S}$* if its structural map factors into a connected locally connected followed by a totally disconnected localic morphism.

1.12. Proposition. *(i)Let $\mathcal{E} \in \mathrm{Top}_{\mathcal{S}}$ have a locale of path components over $\mathcal{S}$. Let U be any object of $\mathcal{E}$. Then the slice topos $\mathcal{E}/U$ also has a locale of path components over $\mathcal{S}$. (ii) Any locally connected topos $\mathcal{E}$ bounded over $\mathcal{S}$ has a locale of path components over $\mathcal{S}$.*

Proof. (i) Let $\gamma : \mathcal{E} \to \mathcal{S}$ factor as $\gamma = tq$ with q connected locally connected and t totally disconnected localic. Since the canonical map $\varphi_U : \mathcal{E}/U \to \mathcal{E}$ is locally connected, by Prop. 1.10 it factors as $t_U q_U$ with q_U connected locally connected and t_U a local homeomorphism. Since t_U is locally connected, the composite qt_U is locally connected hence factors as $t'q'$ with q' connected locally connected and t' a local homeomorphism, hence totally disconnected localic. Thus, in the factorization of $\gamma_U : \mathcal{E}/U \to \mathcal{E}$ as $\gamma_U = \gamma\varphi_U = tqt_Uq_U = tt'q'q$, we have that $q'q$ is connected locally connected and that tt' is totally disconnected (by Prop.1.4) and localic. (ii) If $\mathcal{E}$ is locally connected, then the assertion follows directly from Proposition 1.10(ii).□

2. Classification of Torsors Split by a Cover.

Our ultimate goal is to associate with an unpointed topos $\mathcal{E} \in \mathrm{Top}_\mathcal{S}$ with a locale of path components (in the sense of Definition 1.11) a totally disconnected localic groupoid $\Pi = \pi_1(\mathcal{E})$ which represents the cohomology functor $H^1(\mathcal{E}; -) :$ Groups($\mathcal{S}$) $\to \mathcal{S}$ of $\mathcal{E}$. The strategy for so doing is based on a Čech-style definition of cohomology for Grothendieck toposes (cf.[Ja]).

If $\mathcal{E}$ is a Grothendieck topos, then for any given generating filtered poset Cov(1) of covers U in $\mathcal{E}$ and a group K, $H^1(\mathcal{E}; K)$ is the filtered colimit of a family of subsets $H^1(\mathcal{E}; K)^U$ consisting of the isomorphism classes of U-split K-torsors in $\mathcal{E}$. The situation is no different over an arbitrary base topos $\mathcal{S}$, as we now observe.

Let $\mathcal{E} \in \mathrm{Top}_\mathcal{S}$ with structural map $\gamma : \mathcal{E} \to \mathcal{S}$ and let $(\mathbf{C}, J)$ be a site of definition (with a terminal object). Let Cov(1) be the "set" with $U \in$ Cov(1) if and only if $U = \Sigma\{1C_i | C_i \in R\}$ for some $R \in J(1)$ and $1 : \mathbf{C} \to \mathcal{E}$ the canonical continuous map (cf.[TT]).

Let Tors($\mathcal{E}; K$) denote the category of K-torsors in $\mathcal{E}$ (cf. [TT]). A K-torsor $T = \langle T, \zeta \rangle$ is an object T of $\mathcal{E}$ with global support (i.e., $T \to 1$ epi) and ζ a K-action on T such that $\alpha = \langle \zeta, \pi_1 \rangle : T \times \gamma^* K \to T \times T$ is iso. A morphism $T \to T'$ of K-torsors is a morphism $T \to T'$ which commutes with the K-actions.

Let U be a cover in $\mathcal{E}$, i.e., $U \to 1$ is epi. A K-torsor T is said to be U-split if there is an isomorphism $\alpha : U \times \gamma^* K \to U \times T$ in $\mathcal{E}/U$. Note that any K-torsor $T = \langle T, \zeta \rangle$ is T-split for the cover T. Denote by Tors $(\mathcal{E}; K)^U$ the full subcategory of Tors($\mathcal{E}; K$) whose objects are the U-split K-torsors. It will be convenient to work with an equivalent category, obtained by letting the objects be pairs $\langle T, \alpha \rangle$ where T is a U-split K-torsor and α is a splitting but where a morphism $\langle T, \alpha \rangle \to \langle T', \alpha' \rangle$ is just a morphism $T \to T'$ of K-torsors, not necessarily compatible with the splittings. We use the same notation for this equivalent category.

Since any K-torsor T is T-split, it is also split by any cover U for which there is a morphism $U \to T$. This implies that the canonical map

$$\varinjlim_{U \in \mathrm{Cov}(1)} H^1(\mathcal{E}; K)^U \longrightarrow H^1(\mathcal{E}; K)$$

is a isomorphism.

2.1. Theorem.(Diaconescu). *Let $\mathcal{E} \in \mathrm{Top}_\mathcal{S}$, K a group. Let $\mathcal{S}^K$ be the topos of K-sets. Then there exists an equivalence of categories*

$$\phi : \mathrm{Top}_\mathcal{S}(\mathcal{E}, \mathcal{S}^K) \longrightarrow \mathrm{Tors}(\mathcal{E}; K)$$

which assigns, to a geometric morphism $f : \mathcal{E} \to \mathcal{S}^K$ the K-torsor $f^*(K)$. *The value of ϕ on a 2-cell $\delta : f \to g$ is its K-component $\delta_K : f^* \to g^*$. (Cf. [TT] for a proof.)* □

2.2. Definition. For U a cover in $\mathcal{E}$, the topos $\mathcal{H}_U$ "of U-split objects of $\mathcal{E}$" is defined via the pushout in $\mathrm{Top}_{\mathcal{S}}$

$$\begin{array}{ccc} \mathcal{E}/U & \xrightarrow{\varphi_U} & \mathcal{E} \\ {\scriptstyle \gamma_U}\downarrow & & \downarrow{\scriptstyle h_U} \\ \mathcal{S} & \xrightarrow[k_U]{} & \mathcal{H}_U \end{array}$$

By the nature of colimits in $\mathrm{Top}_{\mathcal{S}}$ (cf. [M_1]), an object of $\mathcal{H}_U$ is a 3-tuple (S, A, χ) with S a set (i.e., an object of $\mathcal{S}$), A an object of $\mathcal{E}$ and $\chi : \gamma_U^* S \to \varphi_U^* A$ an iso in $\mathcal{E}/U$. A morphism in $\mathcal{H}_U$ from (S, A, χ) to (S', A', χ') is a pair (α, f) with $\alpha : S \to S'$, $f : A \to A'$ such that $\varphi_U^* f \cdot \chi = \chi' \cdot \gamma_U^* \alpha$. The geometric morphisms h_U and k_U are defined via their inverse images by the corresponding forgetful functors.

2.3. Lemma. *Let*

$$\begin{array}{ccc} \mathcal{C} & \xrightarrow{q} & \mathcal{B} \\ {\scriptstyle p}\downarrow & & \downarrow{\scriptstyle h} \\ \mathcal{A} & \xrightarrow[k]{} & \mathcal{D} \end{array}$$

be a pushout in $\mathrm{Top}_{\mathcal{S}}$. Then: (i) if p and q are both open (respectively, locally connected) then so are each of k and h. (ii) if p is connected then so is h (and similarly for q and k).

Proof. (i) An argument for the open case is given in [M_3]. The locally connected case can be handled similarly, reducing it first to the preservation of exponentiation. Given objects (A, B, χ) and (A', B', χ') in $\mathcal{D}$, define $\chi^{\chi'}$ to be the composite

$$p^*\left(A^{A'}\right) \to p^*A^{p^*A'} \xrightarrow{\chi^{p^*A}} q^*B^{p^*A'} \xrightarrow{q^*B^{\chi'}} q^*B^{q^*B'} \to q^*\left(B^{B'}\right) \quad ,$$

with canonical isomorphisms given from the preservation of exponentials by p^* and q^*. It can be checked that $\left(A^{A'}, B^{B'}, \chi^{\chi'}\right) \cong (A, B, \xi)^{(A', B', \chi')}$ in $\mathcal{D}$.

(ii) Assume that p is connected and let $g : B \to B'$ in $\mathcal{B}$, where (A, B, χ) and (A', B', χ') are objects of $\mathcal{D}$. Consider the composite

$$p^* \xrightarrow{\chi} q^*B \xrightarrow{q^*g} q^*B' \xrightarrow{\chi'^{-1}} p^*A'$$

and use the fully faithfulness of p^* to obtain $f : A \to A'$ such that p^*f equals the above composite. Clearly, $(f, g) : (A, B, \chi) \to (A', B', \chi')$ in $\mathcal{D}$.□

2.4. Definition. A cover U in $\mathcal{E}$ is said to be *connected* if the structure map $\gamma_U : \mathcal{E}/U \to \mathcal{S}$ is connected.

2.5. Propostion. *Let $\mathcal{E} \in \mathrm{Top}_{\mathcal{S}}$ be locally connected and let U be a connected cover in $\mathcal{E}$. Then, with notation as in Def.2.2, h_U is connected locally connected and k_U is a surjection and a local homeomorphism.*

Proof. Since φ_U and γ_U are both locally connected, it follows from Lemma 2.3 that h_U and k_U are both locally connected. Since γ_U is also connected, so is h_U by the same lemma. That k_U is localic and totally disconnected follows since it is a point of a topos (cf. Corollary 1.5 for the latter), hence defined by a discrete locale (by Corollary 1.10).□

We are now ready to apply methods of Joyal and Tierney[J/T] on classifying toposes for localic groupoids. We briefly recall some terminology. Further details are to be found in [J/T], [M₂],[M₃],[M₅] and [Bu₁]. A localic groupoid G is a groupoid in the category Loc($\mathcal{S}$), i.e., a category object

$$G_1 \times_{G_0} G_1 \;\begin{array}{c}\xrightarrow{\pi_0}\\ \xrightarrow{m}\\ \xrightarrow[\pi_1]{}\end{array}\; G_1 \;\begin{array}{c}\xrightarrow{d_0}\\ \xrightarrow[d_1]{}\end{array}\; G_0$$

with $(m, \pi_0) : G_1 \times G_0 G_1 \to G_1 \times G_1$ an isomorphism of locales. The localic groupoid G is said to be open (respectively, locally connected, totally disconnected, discrete) if d_0 and d_1 have these properties as continuous maps of locales. By the stability properties which these classes of maps have, it follows that the rest of the structure is given by maps of the same type.

The notation $p_G : \mathcal{S}[G_0] \to BG$ will be employed for the topos BG of G-locales (cf. Introduction) equipped with its canonical localic point p_G. A localic groupoid G is said to be etale complete if G is isomorphic to the localic groupoid of automorphisms of the localic point p_G. The representation theorem of Joyal and Tierney [J/T] states that any $\mathcal{S}$-bounded topos is equivalent to a topos of the form BG for an etale complete open localic groupoid G. For a pair (BG, p_G) and (BH, p_H), denote by $\mathrm{Top}_{\mathcal{S}}(BG, BH)_+$ the following category. An object is a 3-tuple (f, χ, α) with $f : BG \to BH$ in $\mathrm{Top}_{\mathcal{S}}$, $\chi : G_0 \to H_0$ a continuous map and α a 2-cell in $\mathrm{Top}_{\mathcal{S}}$ (thus, an iso 2-cell) as in the diagram

$$\begin{array}{ccc} \mathcal{S}[G_0] & \xrightarrow{\quad p_G \quad} & BG \\ {\scriptstyle\chi}\big\downarrow & & \big\downarrow{\scriptstyle f} \\ \mathcal{S}[H_0] & \xrightarrow[\quad p_H \quad]{} & BH \end{array}$$

A morphism $\delta : (f, \chi, \alpha) \to (f', \chi', \alpha')$ is any 2-cell $\delta : f \to f'$ in $\mathrm{Top}_{\mathcal{S}}$ **not** required to be compatible with the rest of the data).

2.6.Theorem. (Morita equivalence for localic groupoids). *Let G and H be open localic groupoids in $\mathcal{S}$ with H etale complete and let Hom(G, H) denote the category of continuous homomorphisms. Then there is an equivalence of categories induced by the functor B,*

$$Hom(G, H) \longrightarrow Top_{\mathcal{S}}(BG, BH)_+ \ .$$

Proof. This is Corollary 5.6 of [Bu₁]. The same result for localic groups appears in [M₂].□

2.7. Proposition. *Let $\mathcal{E} \in Tops$ be locally connected and let U be a connected cover in $\mathcal{E}$. Then there is an open etale complete discrete localic group Π^U in $\mathcal{S}$ such that $\mathcal{H}_U \cong B\Pi^U$ under an equivalence of categories which identifies k_U with the canonical (localic) point $p_U : \mathcal{S} \to B\Pi^U$.*

Proof. This follows readily from [J/T] using the fact (cf. Prop. 2.5) that k_U is a discrete localic open surjection.□

2.8. Theorem.(Classification of U-split torsors,I). *Let $\mathcal{E} \in Tops$ and let U be a connected cover. Then, with notation as in Prop.2.7, there is an equivalence of categories*

$$\Omega : Tops_{\mathcal{S}}\left(B\Pi^U, BK\right)_+ \longrightarrow Tors(\mathcal{E}; K)^U$$

natural in K.

Proof. We revert to the notation $\mathcal{H}_U$ for $B\Pi^U$ and $\mathcal{S}^K$ for BK, when convenient. Let $g : \mathcal{H}_U \to \mathcal{S}^K$ be a morphism in $\text{Top}_{\mathcal{S}}$ and let $\beta : p_k \to g k_U$ be a 2-cell in $\text{Top}_{\mathcal{S}}$. Define $\Omega(g, \beta) = (f, \alpha)$ with $f = g h_U$ and α the composite

$$p_K \gamma_U \xrightarrow{\beta\gamma_U} g k_U \gamma_U \xrightarrow{g\theta} g h_u \varphi_U = f\varphi.$$

We now check that this is well defined. First, α fits into a diagram

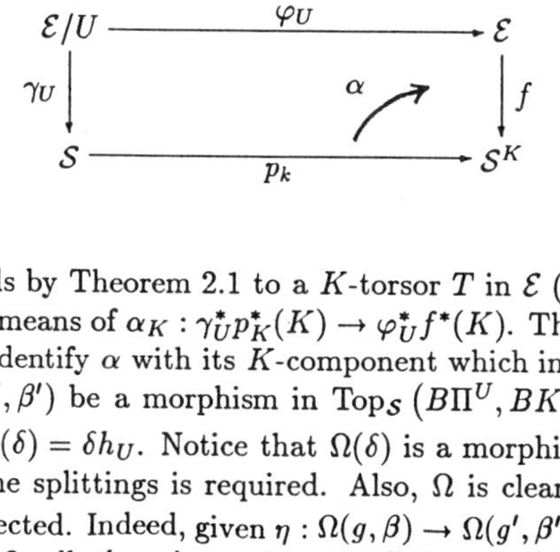

and, as such, corresponds by Theorem 2.1 to a K-torsor T in $\mathcal{E}$ (associated with $f : \mathcal{E} \to \mathcal{S}^K$) which is U-split by means of $\alpha_K : \gamma_U^* p_K^*(K) \to \varphi_U^* f^*(K)$. Thus, (f, α) is an object of $\text{Tors}(\mathcal{E}, K)^U$, where we identify α with its K-component which in any case determines it.

Let $\delta : (g, \beta) \to (g', \beta')$ be a morphism in $\text{Top}_{\mathcal{S}}\left(B\Pi^U, BK\right)_+$, i.e., $\delta : g \to g'$ is a 2-cell in $\text{Top}_{\mathcal{S}}$. Define $\Omega(\delta) = \delta h_U$. Notice that $\Omega(\delta)$ is a morphism in $\text{Tors}(\mathcal{E}; K)^U$ since no compatibility with the splittings is required. Also, Ω is clearly a functor. It is fully faithful since h_U is connected. Indeed, given $\eta : \Omega(g, \beta) \to \Omega(g', \beta')$ in in $\text{Tors}(\mathcal{E}; K)^U$, i.e., given $\eta : g h_U \to g' h_U$, a 2-cell, there is a unique 2-cell $\delta : g \to g'$ such that $\delta h_U = \eta$.

Finally, Ω is dense. Given $(f, \alpha) \in \text{Tors}(\mathcal{E}; K)^U$, by the pushout property defining $\mathcal{H}_U$, we get the existence of g, β, ν as in the diagram

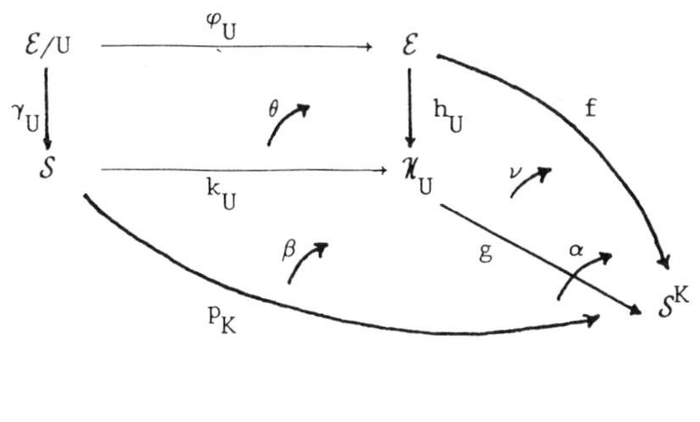

with $\alpha = \nu\varphi_U \cdot g\theta \cdot \beta\gamma_U$. We claim that $\Omega(g,\beta) \cong (f,\alpha)$ in the category $\mathrm{Tors}(\mathcal{E};K)^U$. Since $\Omega(g,\beta) = (gh_U, g\theta \cdot \beta\gamma_U)$, all that is needed is an (iso) 2-cell such as $\nu : gh_U \to f$ and this is given. □

2.9. Corollary. *Let $\mathcal{E} \in Top_{\mathcal{S}}$ be locally connected, U a connected cover and K a group. Then there is an isomorphism*

$$[\Pi^U, K] \to H^1(\mathcal{E};K)^U$$

natural in K, where square brackets indicate the set of connected components of the corresponding category of continuous homomorphisms and where Π^U is the discrete open localic group defined in Prop.2.7.

Proof. By Theorem 2.8 combined with Theorem 2.6 (where the version for localic groups given by [M$_2$] suffices here), we get an equivalence

$$\mathrm{Hom}\,(\Pi^U, K) \cong \mathrm{Tors}(\mathcal{E};K)^U.$$

The result now follows by taking connected components.□

The next goal is to provide a more general set up than that of $\mathcal{E}$ locally connected and U a connected cover under which a similar result (with Π^U a groupoid) can be shown.

2.10 Definition. Let $\mathcal{E} \in \mathrm{Top}_{\mathcal{S}}$ be a topos having a locale of path components over $\mathcal{S}$ in the sense of Def. 1.9. Define a *topos* $\mathcal{G}_U$ together with a connected locally connected geometric morphism $\sigma_U : \mathcal{E} \to \mathcal{G}_U$ and a totally disconnected localic point p_U as in the pushout diagram below, obtained by constructing the pushout of Def. 2.2 in two steps along the (given) factorization of γ_U into a connected locally connected geometric morphism ρ_U followed by a totally disconnected localic geometric morphism λ_U with locale Y^U of definition:

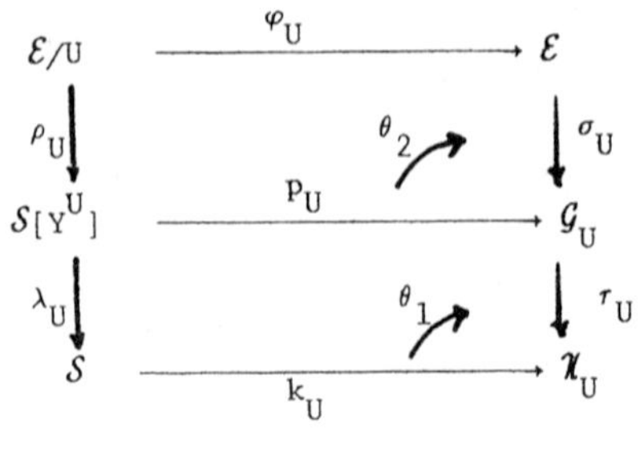

That σ_U is connected locally connected follows from Lemma 2.3 and that p_U is totally disconnected and localic follows from similar properties of the composite $k_U\lambda_U$ by stability properties (Proposition 1.4 (ii) & Corollary 1.5 for totally disconnected; [J/T] for localic).

2.11 Proposition. *Let $\mathcal{E} \in Top_{\mathcal{S}}$ be a topos having a locale of path components in the sense of Def. 1.11. Let U be any cover in $\mathcal{E}$ and let $\mathcal{G}_U, \sigma_U, p_U$ be as in Def. 2.10. Then $\mathcal{G}_U \cong B\Pi^U$ for Π^U a discrete open localic groupoid, by means of an equivalence which identifies p_U with the canonical localic point of $B\Pi^U$.*

Proof. We have already remarked (in Def. 2.10) that p_U is localic and totally disconnected. It is also a surjection since the composite $\sigma_U \rho_U$ is one. Furthermore, p_U is locally connected since in the top pushout both φ_U and ρ_U are locally connected (use Lemma 2.3). It follows from Proposition 1.10 (iii) that p_U is necessarily a surjective local homeomorphism. By the Joyal and Tierney theorem mentioned earlier, $\mathcal{G}_U$ is equivalent to the classifying topos $B\Pi^U$ of an etale complete discrete open localic groupoid Π^U, with locale of objects given by Y^U by an equivalence which identifies p_U with the canonical localic point of $B\Pi_U$.□

2.12. Theorem. (Classification of U-split torsors,II). *Let $\mathcal{E} \in Top_\mathcal{S}$ be a topos with a locale of path components over $\mathcal{S}$. Let U be any cover in $\mathcal{E}$ and K a group. Then there is an equivalence of categories*

$$\Omega : Top_\mathcal{S}\left(B\Pi^U, BK\right)_+ \longrightarrow Tors(\mathcal{E};K)^U$$

natural in K.

Proof. Recall that classifying toposes are equipped with their canonical localic points, in this case these are p_U and p_K. We also recall that an object of the left hand side category is a 3-tuple (g, χ_U, β) with $g : \mathcal{G}_U \to \mathcal{S}^U$ and $\beta : p_K \lambda_U \to g p_U$ a 2-cell. With any such, let Ω associate the U-split K-torsor (f, α) given by $f = g\sigma_U$ and $\alpha : p_K \gamma_U \to f \varphi_U$ the splitting given by the composite

$$p_K \gamma_U = p_K \lambda_U \rho_U \xrightarrow{\beta \rho_U} g p_u \rho_U \xrightarrow{g\theta_2} g \sigma_U \varphi_U = f \varphi_U.$$

To $\delta : (g, \lambda_U, \beta) \to (g', \lambda_U, \beta')$, Ω assigns $\delta \rho_U$. As in the proof of Theorem 2.8, it is shown that Ω is fully faithful (this time on account of the connectedness of σ_U) and that it is dense (this time using the pushout defining $\mathcal{G}_U$).□

2.13. Corollary. *Let $\mathcal{E} \in Top_\mathcal{S}$ be a topos with a locale of path components, U a cover in $\mathcal{E}$ and K a group. Then there is an isomorphism*

$$\left[\Pi^U, K\right] \to H^1(\mathcal{E};K)^U$$

natural in K, where Π^U is the discrete localic open groupoid defined in Prop.2.11. Moreover, if $\mathcal{E}$ is locally connected and U a connected cover, then Π^U is a localic group.

Proof. By Theorem 2.12 and Theorem 2.6, there is an equivalence

$$\mathrm{Hom}\left(\Pi^U, K\right) \to \mathrm{Tors}(\mathcal{E};K)^U.$$

Take connected components. The second assertion is clear : if γ_U is connected locally connected (already) then in the factorization into ρ_U followed by λ_U, the latter is an equivalence. In particular Y^U is here isomorphic to the terminal locale, i.e., $\mathcal{S}\left[Y^U\right] \cong \mathcal{S}$.□

3. The Fundamental Localic Groupoid of a Topos with Locale of Path Components.

We deal first with the case of a connected locally connected topos $\mathcal{E} \in \mathrm{Top}_\mathcal{S}$. In this case there is a site of definition with a terminal object and in which every covering crible is non-empty and connected (cf. [B/Pa] and [M₁]). Associated with such a site as in §2 is a generating inversely filtered poset Cov(1) of connected covers in $\mathcal{E}$ (in the sense of

Def.2.4). Our immediate concern is to render (pseudo-) functorial the assignment of a discrete localic group Π^U to each $U \in \mathrm{Cov}(1)$, in this case. For this purpose we first make Cov(1) into a category $\mathcal{C}$, the full subcategory of $\mathcal{E}$ whose objects are the elements U of Cov(1).

3.1. Proposition. *Let $\mathcal{E} \in \mathcal{T}op_{\mathcal{S}}$ be connected locally connected and let $\mathcal{C}$ be any category of connected covers as defined above. Then, the assignment $U \mapsto \Pi^U$ of Prop.2.7 is pseudo functorial on $\mathcal{C}$ and defines an inversely bifiltered biordered 2-category $\mathcal{I}$ (in the sense of [Ken$_4$], cf. also [K]) of discrete localic groups and open continuous homomorphisms.*

Proof. Let $\alpha : U \to V$ be a morphism in $\mathcal{C}$. From the pushout diagrams defining $\mathcal{H}_U$ and $\mathcal{H}_V$ as in Def.2.2, we derive the existence of a geometric morphism $f_\alpha : \mathcal{H}_U \to \mathcal{H}_V$ and of 2-cells μ_α and ν_α, as in the diagram below:

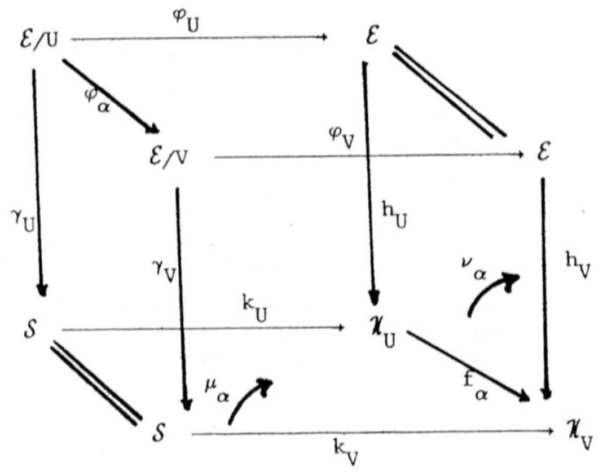

The geometric morphism f_α is an open surjection (in fact, a locally connected surjection) uniquely determined (up to iso 2-cell) by the existence of an iso 2-cell $\nu_\alpha : f_\alpha h_U \to h_V$, since h_U is connected. Moreover, the pair $(f_\alpha, \mu_\alpha) \in \mathrm{Top}_{\mathcal{S}}\left(B\Pi^U, B\Pi^V\right)_+$ (cf.§2), where we have identified $\mathcal{H}_U$ and $\mathcal{H}_V$ with the classifying toposes of the discrete localic groups Π^U and Π^V.

By Theorem 2.6 (where the version for localic groups given in [M$_2$] suffices here), (f_α, μ_α) determines an open continuous homomorphism χ_a with $B\chi_\alpha = f_\alpha$. Let $\mathcal{I}$ be the 2-category whose objects are the Π^U for $U \in \mathrm{Cov}(1)$, the 1-cells are the $\chi_\alpha \in \mathrm{Hom}\left(\Pi^U, \Pi^U\right)$ for $\alpha : U \to V$, and a 2-cell $\lambda : \chi_\alpha \to \chi_{\alpha'}$ is any one such induced by a 2-cell $\lambda : f_\alpha \to f_{\alpha'}$ in $\mathrm{Top}_{\mathcal{S}}$ such that $\lambda h_U = \nu_{\beta'}^{-1} \nu_\beta$. In particular, the assignment is pseudo functorial. We show that $\mathcal{I}$ is inversely bifiltered and biordered.

Given any two objects U, V in $\mathcal{C}$, there is $W \in \mathcal{C}$ and $\alpha : W \to U, \beta : W \to V$ since Cov(1) is inversely filtered as a poset. This gives that for any Π^U, Π^V in $\mathcal{I}$ there exists Π^W and 1-cells $\chi_\alpha : \Pi^W \to \Pi^U$ and $\chi_\beta : \Pi^W \to \Pi^V$ in $\mathcal{I}$. Next, given any two 1-cells $\chi_\beta, \chi_{\beta'} : \Pi^V \to \Pi^U$ in $\mathcal{I}$, there is a unique 2-cell $\lambda : \chi_\beta \to \chi_{\beta'}$ defined as above. This also says that $\mathcal{I}$ is biordered. □

There is a more general case where the work done goes through with virtually no modifications. This is the case of a topos $\mathcal{E} \in \mathrm{Top}_{\mathcal{S}}$ which is assumed to have the structure of a bounded locally connected topos over a totally disconnected localic topos over $\mathcal{S}$, rather than over $\mathcal{S}$ itself. In the terminology of Def.1.9, this is equivalent with requiring that $\mathcal{E}$ have a locale of path components over $\mathcal{S}$. We let Cov(1) be any generating inversely filtered poset of covers and $\mathcal{C}$ the corresponding full subcategory of $\mathcal{E}$ as before. For each $U \in$ Cov(1) we obtain, as in Def. 2.10 , a topos $\mathcal{G}_S \cong B\Pi^U$ with Π^U a discrete localic groupoid with locale of objects Y^U (cf.Prop.2.11) and a canonical localic point $p_U : \mathcal{S}[Y^U] \to \mathcal{G}_U$ as well as a connected locally connected geometric morphism $\sigma_U : \mathcal{E} \to \mathcal{G}_U$ in $\mathrm{Top}_\mathcal{S}$. We now approach the question of pseudofunctoriality in this context.

3.2. Proposition. *Let $\mathcal{E} \in Top_\mathcal{S}$ be a topos with a locale of path components over $\mathcal{S}$. Then the assignment $U \mapsto \Pi^U$ of Prop. 2.11 is pseudofunctorial on $\mathcal{C}$ and defines an inversely bifiltered biordered 2-category $\mathcal{I}$ of discrete open localic groupoids.*

Proof. The proof is analogous to that of Prop.3.1, except that the relevant diagrams, for any given $\alpha : U \to V$ in $\mathcal{C}$ now look as follows:

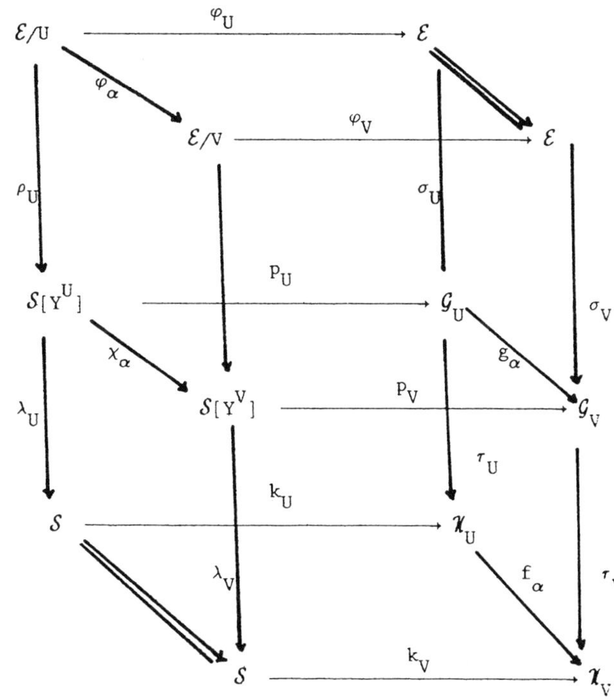

where φ_α is induced by α, χ_α arises from orthogonality (cf.Theorem 1.9) since ρ_U is connected locally connected and λ_V is totally disconnected localic. The morphism g_α arises from the pushout defining $\mathcal{G}_U$ and does not depend on α since σ_U is connected, just as in Prop. 3.1. It must now be argued that the same is true of the χ_α. This requires a different argument since the latter are only determined (up to unique 2-cell) by the existence of a 2-cell $\omega_\alpha : \chi_\alpha \rho_U \to \rho_V \varphi_\alpha$, hence, a priori, depend on α.

For this purpose, let $\alpha, \alpha' : U \to V$ induce $\varphi_{\alpha'}\varphi_{\alpha'}$ as well as $\chi_{a'}\chi_{\alpha'}$ as in the above diagram. By the theorem of Moerdijk and Wraith[M/W], ρ_V is path connected. Thus, evaluation at the endpoints

$$(\mathcal{E}/V)^{\mathcal{S}[I]} \xrightarrow{\epsilon} \mathcal{E}/V \times \mathcal{E}/V$$

is an open surjection in $\mathrm{Top}_{\mathcal{S}}$. Since $\mathcal{S}[Y^V]$ is localic and totally disconnected, it follows as in the proof of Theorem 1.9 that $\rho_V \pi_0 = \rho_V \pi_1$. Consider the diagram in $\mathrm{Top}_{\mathcal{S}}$

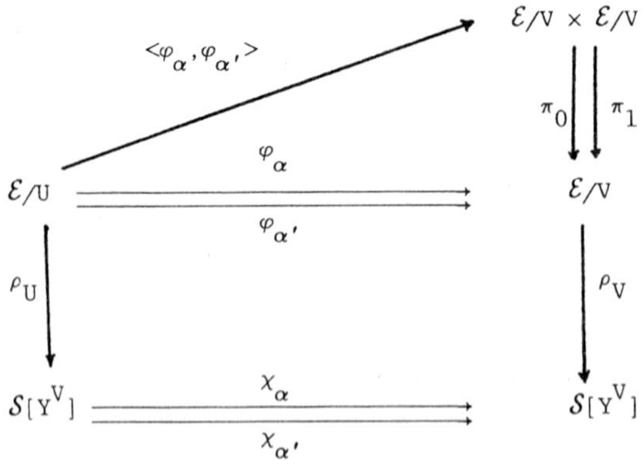

It follows from the above that $\rho_V \varphi_\alpha = \rho_V \varphi_{\alpha'}$, hence that there is an iso 2-cell

$$\chi_\alpha \rho_U \xrightarrow{\omega_\alpha} \rho_U \varphi_\alpha = \rho_V \varphi_{\alpha'} \xrightarrow{\omega_{\alpha'}^{-1}} \chi_\alpha \rho_U.$$

Since ρ_U is connected, there is an iso 2-cell $\lambda : \chi_\alpha \to \chi_{\alpha'}$ unique so that $\lambda \rho_U = \omega_{\alpha'}^{-1} \omega_\alpha$. The rest is exactly as in Prop. 3.1.□

The notion of internal or essential equivalence for category objects in a topos is studied in [Bu/P]. For localic groupoids, this notion specializes as follows (cf. [M_3]). Let $\varphi : G \to H$ be an open continuous homomorphism of open localic groupoids. Call φ an essential equivalence if it is fully faithful and essentially surjective, where φ is fully faithful if the diagram

$$\begin{array}{ccc} G_1 & \xrightarrow{\varphi_1} & H_1 \\ {\scriptstyle \langle d_0, d_1 \rangle} \downarrow & & \downarrow {\scriptstyle \langle d_0, d_1 \rangle} \\ G_0 \times G_0 & \xrightarrow[\varphi_0 \times \varphi_0]{} & H_0 \times H_0 \end{array}$$

is a pullback of locales and where φ is essentially surjective if the composite

$$G_0 \times_{H_0} H_1 \xrightarrow{\pi_1} H_1 \xrightarrow{d_1} H_0$$

is an open surjection.

3.3. Lemma. *Let $\varphi : G \to K$ be an open continuous homomorphism of open localic group(oids) with G and K etale complete. Then there exists an open localic groupoid H*

and a (not necessarily commutative) diagram

$$\begin{array}{ccc} H & \xrightarrow{\epsilon} & G \\ {\scriptstyle \xi}\downarrow & \swarrow{\scriptstyle \varphi} & \\ K & & \end{array}$$

with ϵ an essential equivalence, ξ an open surjective homomorphism and $B\varphi \cdot B\epsilon \cong B\xi$.

Proof. (cf. [M$_3$]). Explicitly, H is obtained from the pullbacks

$$\begin{array}{ccccc} \mathcal{S}[H_0] & & \xrightarrow{\quad \xi_0 \quad} & & \mathcal{S}[K_0] \\ {\scriptstyle \epsilon_0}\downarrow & & & & \downarrow{\scriptstyle p_K} \\ \mathcal{S}[G_0] & \xrightarrow[p_G]{} & BG & \xrightarrow[B\varphi]{} & BK \end{array} \qquad \begin{array}{ccccc} \mathcal{S}[H_1] & & \xrightarrow{\quad \xi_1 \quad} & & \mathcal{S}[K_1] \\ {\scriptstyle \epsilon_1}\downarrow & & & & \downarrow{\scriptstyle p_K d_1} \\ \mathcal{S}[G_1] & \xrightarrow[p_G d_1]{} & BG & \xrightarrow[B\varphi]{} & BK \end{array}$$

in $\mathrm{Top}_{\mathcal{S}}$.□

3.4. Definition. Let $\mathcal{E} \in \mathrm{Top}_{\mathcal{S}}$. An open localic groupoid Π in $\mathcal{S}$ is said to be *a fundamental localic groupoid of* $\mathcal{E}$ and denoted $\pi_1(\mathcal{E})$ if Π is totally disconnected and represents first degree cohomology of $\mathcal{E}$ with coefficients in discrete groups, in the sense that there exist isomorphisms

$$H^1(\mathcal{E}; K) \cong [\Pi, K]$$

natural in K for each discrete group K in $\mathcal{S}$.

3.5 Theorem. *Let $\mathcal{E} \in Top_{\mathcal{S}}$ with $\mathcal{E}$ a topos with a locale of path components. Let $\mathcal{G}$ be the topos $\varprojlim_{\mathcal{I}} B\Pi^U$, the limit in $Top_{\mathcal{S}}$ (cf. [M$_1$]) of the inversely filtered system of toposes $B\Pi^U$, geometric morphisms f_α and 2-cells ν_α, as in Prop.3.2, and let $p : \varprojlim_{\mathcal{I}} \mathcal{S}[Y^U] \to \varprojlim_{\mathcal{I}} B\Pi^U$ be the geometric morphism induced by the $p_U : \mathcal{S} \to B\Pi^U$. Then, if p is an open surjection, the topos $\mathcal{E}$ has a fundamental localic groupoid Π in the sense of Definition 3.4., with the topos $\mathcal{G} = \varprojlim_{\mathcal{I}} B\Pi^U$ equivalent to the classifying topos $B\Pi$ by an equivalence which identifies the geometric morphism p with the canonical localic point p_Π of $B\Pi$.*

Proof. Notice that p is localic since the localic reflection preserves filtered inverse limits in $\mathrm{Top}_{\mathcal{S}}$ (cf.[M$_1$]). Similarly, p is totally disconnected by Corollary 1.5. By assumption, p is also an open surjection. Hence a totally disconnected open etale complete localic groupoid Π exists by the representation theorem of [J/T] with $\mathcal{G} \cong B\Pi$ by an equivalence identifying p with p_Π. We need only verify that Π represents cohomology of $\mathcal{E}$. For this, we have the following string of equivalences:

$$\varinjlim_{\mathcal{I}} \mathrm{Tors}(\mathcal{E}; K)^U \cong \varinjlim_{\mathcal{I}} \mathrm{Top}_{\mathcal{S}} \left(B\Pi^U, BK\right)_+ \cong$$

$$\mathrm{Top}_{\mathcal{S}} \left(\varprojlim_{\mathcal{I}} B\Pi^U, BK\right)_+ \cong \mathrm{Top}_{\mathcal{S}}(B\Pi, BK)_+ \cong \mathrm{Hom}(\Pi, K),$$

the latter by Theorem 2.6., since Π is open and K is etale complete. Taking connected components is a functor $\prod_0$ which commutes with filtered colimits. Hence we have:

$$H^1(\mathcal{E};K) \cong \varinjlim_{\mathrm{Cov}(1)} H^1(\mathcal{E};K)^U \cong \varinjlim_{\mathrm{Cov}(1)} \prod_0 \mathrm{Tors}(\mathcal{E};K)^U \cong$$

$$\prod_0 \varinjlim_{\mathcal{I}} \mathrm{Tors}(\mathcal{E};K)^U \cong \prod_0 \mathrm{Hom}(\Pi,K) \cong [\Pi,K].\square$$

A localic groupoid G will be said to be *prodiscrete* if it is the inverse limit (in locales) of a filtered 1-system of discrete localic group(oid)s and surjective homomorphisms. By the results of §1, a prodiscrete localic groupoid is open and totally disconnected.

3.6. Theorem. *Let $\{\Pi^i\}_{i\in I}$ be an inversely filtered 2-system of open discrete localic group(oid)s in a topos $\mathcal{S}$. Let $\mathcal{G} = \varprojlim_{\mathcal{I}} B\Pi^i$ and let p be the geometric morphism $p : \varprojlim_{\mathcal{I}} \mathcal{S}[Y^i] \longrightarrow \mathcal{G}$ induced by the canonical localic points $p_i : \mathcal{S}[Y_i] \to B\Pi^i$, each of which is an open surjection. Then, $\mathcal{G}$ is atomic and p is a localic open surjection.*

Proof. As in [J/M] one can construct an open surjection $q : \mathcal{S}' \to \mathcal{S}$ as well as a sequence

$$\cdots q^{\#}\Pi^3 \xrightarrow{\chi_2} q^{\#}\Pi^2 \xrightarrow{\chi_1} q^{\#}\Pi^1$$

in $\mathcal{S}$, cofinal to the inversely filtered 2-system $\{q^{\#}\Pi^i\}$ of discrete localic group(oid)s and continuous homomorphisms. Using Lemma 3.3, we now replace inductively the above by an essentially equivalent sequence

$$\cdots \longrightarrow H^3 \xrightarrow{\zeta_2} H^2 \xrightarrow{\zeta_1} H^1$$

of discrete localic groupoids H^n in $\mathcal{S}'$ and open surjective homomorphisms ζ_n. Explicitly, having constructed H^n as well as

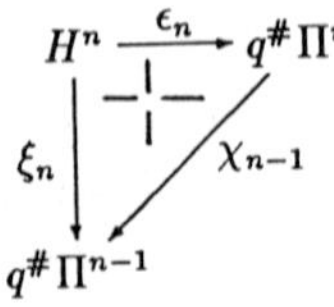

with ϵ_n an essential equivalence and ξ_n an open surjective homomorphism as in Lemma 3.3, proceed by applying the same construction to the top horizontal map in the pullback

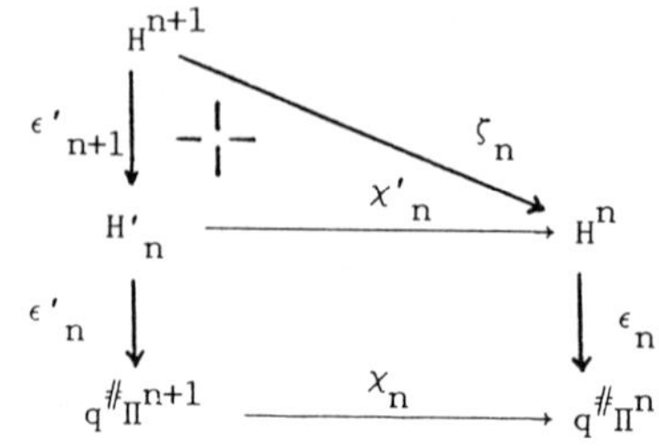

letting $\epsilon_{n+1} : H^{n+1} \to q^{\#}\Pi^{n+1}$ be the composite $\epsilon_{n+1} = \epsilon'_n \epsilon'_{n+1}$, an essential equivalence by the stability properties (under pullback and composition) of the latter.

For the sequence $\{H^n\}$ of open localic groupoids and open surjective homomorphisms, it can be shown as in [M3] that the canonical geometric morphism

$$\varprojlim BH^n \longrightarrow B\left(\varprojlim H^n\right)$$

is an equivalence, where $H = \varprojlim H^n$ is the prodiscrete localic groupoid obtained as the limit of the 1-system of the discrete groupoids H^n and the surjective homomorphisms χ_n. The atomic site S^∞ for the limit topos $\mathcal{G}'$ is obtained as an inverse limit of a sequence of sites S^n and adjoint pairs

$$\cdots \rightleftarrows S^3 \underset{T_{\xi_2}}{\overset{P_{\xi_2}}{\rightleftarrows}} S^2 \underset{T_{\xi_1}}{\overset{P_{\xi_1}}{\rightleftarrows}} S^1$$

with S^n the full subcategory of BH^n whose objects are the H^n-sheaves of the form $(H^n)_1 \cap d_1^{-1}(U)/N$, where $U \subset (H^n)_0$ is an open sublocale and $N \subset (H^n)_1$ is an open U-congruence. That the atomic site S^∞ is $H = \varprojlim H^n$ follows by considering the cofinal system consisting of those objects of the above form with $U = \xi_n^{-1}(V)$ for some open sublocale $V \subset (H^n)_0$ and $N = \xi_n^{-1}(M)$ for some open congruence $M \subset (H^n)_1$.

By Moerdijk's Stability Theorem (cf. [M3],[M5]), there are equivalences

$$B\left(q^{\#}\Pi^n\right) \cong B\Pi^n \times_{\mathcal{S}} \mathcal{S}'$$

since each Π^n is an open localic groupoid, these equivalences identifying the appropriate localic points. Further, since $\{H^n\}$ is a 1-system essentially equivalent to the 2-system $\{q^{\#}\Pi^n\}$, it follows that there is also an equivalence of toposes

$$\varprojlim BH^n \cong \varprojlim B\left(q^{\#}\Pi^n\right).$$

We then have the following string of equivalences

$$B\left(\varprojlim H^n\right) \cong \varprojlim (BH^n) \cong \varprojlim B\left(q^{\#}\Pi^n\right) \cong \varprojlim (B\Pi^n \times_{\mathcal{S}} \mathcal{S}') \cong$$
$$\left(\varprojlim B\Pi^n\right) \times_{\mathcal{S}} \mathcal{S}' \cong \left(\varprojlim_{\mathcal{I}} B\Pi^i\right) \times_{\mathcal{S}} \mathcal{S}' \cong \mathcal{G} \times_{\mathcal{S}} \mathcal{S}'.$$

by an equivalence which identifies p_H with $p \times_{\mathcal{S}} \mathcal{S}'$ in $\mathrm{Top}_{\mathcal{S}'}$.

As pointed out earlier, since the localic reflection preserves limits of filtered systems of toposes (cf. [M1] Corollary 5.4), the topos $\varprojlim_{\mathcal{I}} \mathcal{S}[Y^i]$ is localic over $\mathcal{S}$. Let Π_0 be the locale in $\mathcal{S}$ with $\mathcal{S}[\Pi_0]$ equivalent to the limit topos above.

Consider now the following pullback in $\mathrm{Top}_{\mathcal{S}}$

$$\begin{array}{ccc}
\mathcal{S}[\Pi_0] \times_{\mathcal{S}} \mathcal{S}' & \longrightarrow & \mathcal{S}[\Pi_0] \\
{\scriptstyle p \times_{\mathcal{S}} \mathcal{S}'} \downarrow & & \downarrow {\scriptstyle p} \\
\mathcal{G} \times_{\mathcal{S}} \mathcal{S}' & \xrightarrow{\quad r \quad} & \mathcal{G} \\
{\scriptstyle \gamma_{\mathcal{G}'}} \downarrow & & \downarrow {\scriptstyle \gamma_{\mathcal{G}}} \\
\mathcal{S}' & \xrightarrow{\quad q \quad} & \mathcal{S}
\end{array}$$

where q (and therefore r) is an open surjection. By a result of Moerdijk (cf.[M1] Theorem 5.2), pulling back along open surjections in $\mathrm{Top}_{\mathcal{S}}$ reflects several types of geometric morphisms, among which are the open surjections and the (connected) atomic geometric morphisms. Since $p \times_S S'$ is naturally equivalent with p_H, it is an open surjection – hence so is p. Similarly, since $\gamma_{\mathcal{G}} \times_S S' = \gamma_{\mathcal{G}'}$ is the structural map of the classifying topos BH of a prodiscrete localic groupoid, it is atomic – hence, so is $\gamma_{\mathcal{G}}$. This establishes the claim.□

3.7. Theorem. *Let $\mathcal{E} \in Top_S$ be a topos with a locale of path components in the sense of Definition 1.11. Then $\mathcal{E}$ has a fundamental localic groupoid in the sense of Def.3.4. Moreover, the latter may be taken to be locally (i.e., on an open cover of S) Morita equivalent to a prodiscrete localic groupoid.*

Proof. The result is a consequence of applying Theorem 3.6.to the system $\{\Pi^U\}$ of Theorem 3.5. This yields the existence of a totally disconnected open etale complete localic groupoid Π which – by "construction" is locally Morita equivalent to a prodiscrete localic groupoid. □

We end by making a few remarks about our construction of the fundamental localic groupoid as compared with some related results in the literature.

In [M/W], Moerdijk and Wraith propose a notion of $\pi_1(\mathcal{E})$ for an arbitrary topos $\mathcal{E} \in$ Top$_S$, which is defined as a groupoid topos. However, this is not pursued therein beyond the remark that if $\mathcal{E}$ is connected locally connected, then this groupoid topos is connected atomic. Notice that this conclusion also applies,in our case, to the topos $\mathcal{G}$ which is the classifying topos of the fundamental localic groupoid $\Pi = \pi_1(\mathcal{E})$, by Theorems 3.6 and 3.7.

In [M$_4$], Moerdijk defines the fundamental localic group of a pointed connected locally connected topos over an arbitrary base topos $\mathcal{S}$. His methods, which inspired ours, are primarily based in the work of Joyal and Tierney [J/T]. However, in order to prove the existence and prodiscrete nature of the the fundamental localic group, arguments which involve the point of the topos are employed in [M$_4$]. In the absence of a point, we were forced to resort to different arguments. Notice also that – for the same reason – in the unpointed case we settle for total disconnectedness rather than prodiscreteness, although the latter does hold in our case as well, if only just locally and up to Morita equivalence.

As mentioned in the Introduction, Kennison [Ken$_4$] has also associated, by different methods, a prodiscrete localic groupoid π to any Grothendieck topos, (i.e., bounded over Sets) and he has done so under no other particular assumptions except for connectedness. But it is only in the (connected) locally connected case that the prodiscrete localic groupoid π defined in [Ken$_4$] is shown to have nice properties – namely, the comparison functor $\kappa : \mathcal{E} \to B\pi$ is connected, the image of κ^* is the full subcategory of $\mathcal{E}$ generated by the split objects and π classifies torsors, hence represents cohomology.

REFERENCES

[A/M] M.Artin and B.Mazur. Etale Homotopy. Lecture Notes in Mathematics 100 (Springer-Verlag, Berlin 1972).

[B] M.Barr, Toposes without points. J.Pure and Applied Algebra 5 (1974)265-280.

[B/Pa] M.Barr and R.Paré, Molecular toposes. J.Pure and Applied Algebra 17 (1980) 127-152.

[Bo/J] A.Boileau and A.Joyal, La logique des topos. J. Symb.Logic 46 (1981) 6-16.

[Bu$_1$] M.Bunge, An application of descent to a classification theorem for toposes. Math. Proc. Cambridge Phil.Soc. 107 (1990) 59-79.

[Bu_2] M.Bunge, Open covers and the fundamental localic groupoid of a topos. Report of the Department of Mathematics #89-33, McGill University, 1989.

[Bu/Pa] M.Bunge and R.Paré, Stacks and equivalences of indexed categories. Cahiers Topologie et Géométrie Différentielle Catégoriques 20 (1979) 401-436.

[CWM] S.Mac Lane. Categories for the working mathematician. Graduate Texts in Mathematics 5. (Springer-Verlag, Berlin, 1971).

[Gra] B.Gray. Homotopy Theory (Academic Press, 1975).

[Gro] A.Grothendieck. Revêtements étales et groupe fondamental (SGA1). Lecture Notes in Mathematics 224 (Springer-Verlag, Berlin, 1981).

[H] M.Hyland, Function spaces in the category of locales, in : B.Banaschewski and R.-E.Hoffman (eds.), Continuous lattices. Lecture Notes in Mathematics 871 (Springer-Verlag, Berlin, 1981).

[Ja] J.P.Jardine, Universal Hasse-Witt classes. Preprint, University of Western Ontario, 1988.

[J/J] P.T.Johnstone and A.Joyal, Continuous categories and exponentiable toposes. J.Pure and Applied Algebra 25 (1982) 255-296.

[J/M] A.Joyal and I.Moerdijk, Toposes as homotopy groupoids. Advances in Mathematics 80 (1990) #1, 22-38.

[J/T] A.Joyal and M.Tierney. An extension of the Galois theory of Grothendieck. Memoirs of the American Mathematical Society 58 #30, 1984.

[K] G.M.Kelly, Elementary observations on 2-categorical limits. Bulletin of the Australian Mathematical Society 39 (1988), 301-317.

[Ken_1] J.Kennison, The fundamental group of a topos. J.Pure and Applied Algebra 30 (1983) 23-38.

[Ken_2] J.Kennison, The fundamental group of a molecular topos. J.Pure and Applied Algebra 46 (1987) 187-215.

[Ken_3] J.Kennison, Progroup actions and fundamental progroups. J.Pure and Applied Algebra 66 (1990) 185-218.

[Ken_4] J.Kennison, The fundamental localic groupoid of a topos. To appear: J. Pure and Applied Algebra.

[M_1] I.Moerdijk, Continuous fibrations and inverse limits of toposes. Comp. Math. 58 (1986) 45-72.

[M_2] I.Moerdijk, Morita equivalence for continuous groups. Math. Proc. Cambridge Philosophical Society 103 (1988) 97-115.

[M_3] I.Moerdijk, The classifying topos of a continuous groupoid I. Trans. Amer. Math. Soc. 310 #2 (1988) 629-668.

[M_4] I.Moerdijk, Prodiscrete groups and Galois toposes. Proc. Kon. Nederl. Akad. van Wetenschappen, series A, vol92 #2 (1989) 219-234.

[M_5] I.Moerdijk, Toposes and groupoids, in : F.Borceux (ed.), Catgorical algebra and its applications. Proceedings Louvain-la-Neuve 1987. Lecture Notes in Mathematics 1348 (Springer-Verlag, Berlin, 1988) ,280-298.

[M/W] I.Moerdijk and G.Wraith, Connected locally connected toposes are path connected. Trans. Amer. Math. Soc. 295 (1986) 849-859.

[Pa] R.Paré, Indexed categories and generated topologies. J.Pure and Applied Algebra 19 (1980) 305-400.

[Pi] A.Pitts, On product and change of base for toposes. Cahiers de Topologie et Geometrie Differentielle Categoriques 26-1 (1985) 43-61.

[TT] P.T. Johnstone. Topos theory.(Academic Press, 1977).

Department of Mathematics and Statistics, McGill University, Montreal, Quebec H3A 2K6, Canada.

E-mail address: inbe@musicb.mcgill.ca

Canadian Mathematical Society
Conference Proceeding
Volume **13**, 1992

Internal Graphs and Internal Groupoids in Mal'cev Categories

A. CARBONI M.C. PEDICCHIO N. PIROVANO

Abstract. We show that the Mal'cev semantical condition in a left exact category, not necessarily varietal, allows to prove the *generalized Eckmann-Hilton argument*: every (internal) reflexive–multiplicative graph is uniquely a groupoid. The internal reflexive graphs which are underlying graphs of a groupoid are also characterized.

Introduction

Mal'cev's result about varieties of algebras says that the semantical property that every reflexive relation in the category of algebras is already a congruence, is equivalent to the syntactical one of the definability of a ternary operation satisfying certain equations (for some historical details, see [CLP]). Then, a natural question arises to know how much remains when we drop the algebraicity condition and just ask the semantical property ("*Mal'cev categories*"). Clearly, there are examples of such categories; for instance a large class is given by the dual categories of topoi (and in fact much less than a topos is needed): one can easily show that in such categories the semantical property holds, and that they are never algebraic. Another example is provided by the dual category of the category of topological spaces.

An evidence of the need for an axiomatic study of the semantical axiom is mentioned for instance in the introduction to [B], where is stated that for an exact category to be Mal'cev is equivalent to every simplicial object being Kan; further, the classical result about the algebraic model of 2-types (see e.g. [Lo]) being the (homotopy) category of groupoids in the Mal'cev category of groups, also indicates the nature of the Mal'cev axiom.

In the present note we pursue the study of internal category theory in a Mal'cev category, where we require only finite limits, since both category theory and the Mal'cev axiom are of a lex nature. Such a study was initiated in

---

[0]1991 Mathematics Subject Classification: primary 20L17; secondary 08B05, 18G30.
This paper is in final form and no version of it will be submitted for publication elsewhere.
Typeset by LaTeX; thanks to Paul Taylor for his package *Commutative Diagrams in TeX*.

the paper [Ja] of G. Janelidze, where the generalized Eckmann–Hilton kind of argument is proved to be valid in the *varietal* Mal'cev categories: every reflexive-multiplicative internal graph is in fact a groupoid. The author also characterizes those internal graphs which admit a (unique) reflexive-multiplicative structure. Here whe show that both results still can be proved by dropping the algebraicity of the category. We also show that reflexive-multipicative internal graphs with just one object are *abelian* groups. As for the characterization of those graphs that admit a (unique) reflexive–multiplicative structure, and hence are the underlying graphs of a groupoid, we should mention that it already appears in [Jo2] in terms of the notion of *herdoid* in the context of just a lex category; what we have more in the context of a Mal'cev category is the uniqueness of the structure and the fact that associativity and inverse follow.

The proofs are very simple and transparent, provided we use a property (*difunctionality* of relations) that in a category with finite limits can be proved to be equivalent to the Mal'cev property. The simplicity of the proofs indicates that the axiomatic method not only has a wider range of models, but that it should also be used in the varietal case, and that the work already done in the subject of Mal'cev varieties (as for instance [S]) should be revisited in the same way.

While working in a category with finite limits, we freely use elements as far as Horn sentences built up from atomic ones and the equality are concerned; we use existential quantification only when we can show uniqueness. It is well known that in such a way we can always interpret our statements in a lex category in a valid way.

1 Remarks on the Mal'cev Axiom

Let $\mathcal{E}$ be a category with finite limits (*"left exact", "lex"*), and recall that a *relation* from an object X to an object Y of $\mathcal{E}$ is a subobject R of the product $X \times Y$. As we remarked in the Introduction, we will use a set theoretical notation; for a given relation R, we will write xRy instead than $(x, y) \in R$.
Observe that for a relation $R \hookrightarrow X \times X$ to be an equivalence relation, is just a lex statement.

DEFINITION **1.1** *A lex category $\mathcal{E}$ is a Mal'cev category if and only if it satisfies the following axiom:*

(M) *for any X in $\mathcal{E}$, any reflexive relation $R \hookrightarrow X \times X$ is an equivalence relation.*

Observe that the notion of Mal'cev category considered in [CLP] is stronger than Definition 1.1: in [CLP] $\mathcal{E}$ had also to be a *regular* category; under this assumption one can define an associative composition of relations and show that the Mal'cev axiom **(M)** is equivalent to equivalences commutativity.

PROPOSITION **1.2** *Let $\mathcal{E}$ be a lex category, then the following are equivalent:*

1) $\mathcal{E}$ satisfies the axiom **(M)**;

2) for any X in $\mathcal{E}$, any reflexive relation on X is symmetric;

3) for any X in $\mathcal{E}$, any reflexive relation on X is transitive;

4) for any X and for any Y in $\mathcal{E}$, any relation $R \hookrightarrow X \times Y$ is difunctional, *i.e.*

$$\text{(D)} \qquad \forall a \forall c \in X \text{ and } \forall b \forall d \in Y, \quad \left.\begin{matrix} aRb \\ cRb \\ cRd \end{matrix}\right\} \Rightarrow aRd\,;$$

5) for any X in $\mathcal{E}$, any relation on X is difunctional.

PROOF. For any relation $R \hookrightarrow X \times Y$, define [F] a new relation

$$S \hookrightarrow R \times R$$

as follows:

$$(a,b)S(c,d) :\Longleftrightarrow aRd$$

which is reflexive by definition:

1) $\Rightarrow$ 2) trivial;

2) $\Rightarrow$ 3) if aRb and bRc, then $(b,c)S(a,b)$, and by symmetry $(a,b)S(b,c)$, i.e. aRc;

3) $\Rightarrow$ 4) if aRb, cRb and cRd, then $(a,b)S(c,b)$ and $(c,b)S(c,d)$, so by transitivity $(a,b)S(c,d)$, i.e. aRd;

4) $\Rightarrow$ 5) trivial;

5) $\Rightarrow$ 1) symmetry:
if aRb, then bRb, aRb and aRa, hence bRa;

transitivity:
if aRb and bRc, then aRb, bRb and bRc, hence aRc. □

In [CLP] a relation $R \hookrightarrow X \times Y$ in a regular category is said to be *difunctional* if it satisfies the condition

$$\text{(D')} \qquad R \circ R^{\circ} \circ R \subseteq R\,,$$

where $\circ$ denotes the usual composition of relations and $(\ )^{\circ}$ the opposite relation; observe that, in any regular category, the inclusion $R \subseteq R \circ R^{\circ} \circ R$ is always satisfied, so that difunctionality means $R \circ R^{\circ} \circ R = R$.
Notice that **(D')** is also just a lex statement; it is easy to show that the property **(D)** is the interpretation, in the internal logic, of **(D')**.

2 Internal graphs and internal groupoids

Given a lex category $\mathcal{E}$, consider the category $\mathrm{rg}(\mathcal{E})$ of *internal reflexive graphs* in $\mathcal{E}$, i.e. diagrams

$$M \underset{c}{\overset{d}{\rightrightarrows}} \xleftarrow{\;i\;} O$$

satisfying

(R) $\qquad di = 1_O = ci\ .$

We recall that an *internal category* in $\mathcal{E}$ is a reflexive graph equipped with a *multiplication*

$$M * M \xrightarrow{\ m\ } M$$

where $M * M$ denotes the pull-back

$$\begin{array}{ccc} M * M & \longrightarrow & M \\ \big\downarrow & \lrcorner & \big\downarrow{\scriptstyle d} \\ M & \xrightarrow{\ c\ } & O \end{array}$$

satisfying the following axioms:

(C1) $\quad \forall \alpha \in M\ , \qquad \alpha 1_{d(\alpha)} = \alpha = 1_{c(\alpha)}\alpha$

(C2) $\quad \forall (\alpha, \beta) \in M * M\ , \qquad \begin{array}{l} d(\beta\alpha) = d(\alpha) \\ c(\beta\alpha) = c(\beta) \end{array}$

(C3) $\quad \forall (\alpha, \beta)\ \forall (\beta, \gamma) \in M * M\ , \qquad \gamma(\beta\alpha) = (\gamma\beta)\alpha$

where $\quad 1_x := i(x)$ and $\beta\alpha := m(\alpha, \beta).$

A reflexive graph with a multiplication m verifying **(C1)** will be called a *reflexive–multiplicative graph.*

We denote by $\mathrm{cat}(\mathcal{E})$ the category of internal categories in $\mathcal{E}$ and by $\mathrm{rmg}(\mathcal{E})$ that one of internal reflexive–multiplicative graphs, where the morphisms are the canonical ones, namely, any pair $(f_O\ ,\ f_M)$ of arrows of $\mathcal{E}$ as in the diagram

$$\begin{array}{ccc} M_1 * M_1 & & M_2 * M_2 \\ \big\downarrow{\scriptstyle m_1} & & \big\downarrow{\scriptstyle m_2} \\ M_1 & \xrightarrow{\ f_M\ } & M_2 \\ {\scriptstyle d_1}\big\downarrow\ \big\uparrow{\scriptstyle i_1}\ \big\downarrow{\scriptstyle c_1} & & {\scriptstyle d_2}\big\downarrow\ \big\uparrow{\scriptstyle i_2}\ \big\downarrow{\scriptstyle c_2} \\ O_1 & \xrightarrow{\ f_O\ } & O_2 \end{array}$$

such that

(M1) $\quad \forall x \in O_1\ , \qquad f_M(1_x) = 1_{f_O(x)}$

(M2) $\quad \forall \alpha \in M_1\ , \qquad \begin{array}{l} d_2 f_M(\alpha) = f_O d_1(\alpha) \\ c_2 f_M(\alpha) = f_O c_1(\alpha)\ ; \end{array}$

in the case of reflexive graphs; otherwise, we also require the *functoriality*, which means that

(M3) $\forall(\alpha,\beta) \in M_1 * M_1$, $\qquad f_M(\beta\alpha) = f_M(\beta)f_M(\alpha)$.

Furthermore, we write $\mathrm{grpd}(\mathcal{E})$ for the category of *internal groupoids* in $\mathcal{E}$, i.e. internal categories with an *inversion*

$$M \xrightarrow{\ s\ } M$$

such that

(G1) $\forall\alpha \in M$, $\qquad d(s(\alpha)) = c(\alpha)$, $\ c(s(\alpha)) = d(\alpha)$

(G2) $\forall\alpha \in M$, $\qquad s(\alpha)\alpha = 1_{d(\alpha)}$, $\ \alpha s(\alpha) = 1_{c(\alpha)}$.

There are forgetful functors

$$\mathrm{grpd}(\mathcal{E}) \overset{U}{\hookrightarrow} \mathrm{cat}(\mathcal{E}) \overset{V}{\hookrightarrow} \mathrm{rmg}(\mathcal{E}) \overset{W}{\hookrightarrow} \mathrm{rg}(\mathcal{E})$$

where only W is not in general full.

PROPOSITION **2.1** *If $\mathcal{E}$ is a Mal'cev category, then the functor W is full.*

PROOF. Let $(f_O\ ,\ f_M)$ be a morphism of reflexive graphs, and consider the relation $R \hookrightarrow M_1 \times M_1$ defined by

$$\beta R\alpha :\Longleftrightarrow \left\{ \begin{array}{rcl} d_1(\beta) & = & c_1(\alpha) \\ f_M(\beta\alpha) & = & f_M(\beta)f_M(\alpha) \end{array} \right. :$$

$\forall\,(\alpha,\beta) \in M_1 * M_1$, if $1 := 1_{d_1(\beta)} = 1_{c_1(\alpha)}$, then

$$\begin{array}{rcl} \beta R1 & : & f_M(\beta 1) = f_M(\beta) = f_M(\beta)f_M(1) \\ 1R1 & : & \text{trivial} \\ 1R\alpha & : & f_M(1\alpha) = f_M(\alpha) = f_M(1)f_M(\alpha) \\ \hline \beta R\alpha & , & \text{by axiom } \mathbf{(D)}, \end{array}$$

i.e. $f_M(\beta\alpha) = f_M(\beta)f_M(\alpha)$. □

In [Jol] Johnstone proves some results about the functors W , V and U; in fact, his proof shows that the functor W is an isomorphism iff $\mathcal{E}$ is a *naturally Mal'cev category*, i.e. a lex category equipped with a natural transformation μ from the functor $A \longrightarrow A^3$ to the identity functor on $\mathcal{E}$ such that for each object A and for each x , $y \in A$

(*) $\qquad \mu_A(x,x,y) = y = \mu_A(y,x,x)$;

under this assumption, he also proves that the functors V and U are isomorphisms.

On the other hand, in [Ja] Janelidze shows that V and U are isomorphisms also in the case in which $\mathcal{E}$ is a *Mal'cev variety*, i.e. a category of algebras whose signature contains a ternary operator μ with the identities **(*)**.
Note that both the naturally Mal'cev categories and the Mal'cev varieties are Mal'cev categories, by the following argument (which must be interpreted in the internal logic and in the underlying set structure, respectively): any (homomorphic) relation R in these categories is difunctional, since if aRb, cRb and cRd then $\mu(a,c,c)R\mu(b,b,d)$, so aRd, by **(*)**.

We can now state our main result:

THEOREM 2.2 *If $\mathcal{E}$ is a Mal'cev category, then*

1) any internal reflexive graph admits at most one structure of reflexive–multiplicative graph;

2) the forgetful functors $\mathrm{cat}(\mathcal{E}) \xhookrightarrow{V} \mathrm{rmg}(\mathcal{E})$ *and* $\mathrm{grpd}(\mathcal{E}) \xhookrightarrow{U} \mathrm{cat}(\mathcal{E})$ *are isomorphisms.*

PROOF. 1) Let

$$M * M \overset{m}{\underset{m'}{\rightrightarrows}} M \underset{c}{\overset{d}{\rightrightarrows}} O, \qquad i : O \to M$$

be a reflexive graph equipped with two structures of multiplicative–reflexive graph; we define a relation $R \hookrightarrow M \times M$ by

$$\beta R\alpha :\Longleftrightarrow \begin{cases} d(\beta) &= c(\alpha) \\ \beta\alpha &= \beta \circ \alpha \end{cases}$$

where $\beta \circ \alpha$ denotes the composite $m'(\alpha, \beta)$.
Then, $\forall (\alpha, \beta) \in M * M$, if we pose $1 := 1_{d(\beta)} = 1_{c(\alpha)}$, we get:

$$\begin{array}{rcl} \beta R1 & : & \beta 1 = \beta = \beta \circ 1 \\ 1R1 & : & \text{trivial} \\ 1R\alpha & : & 1\alpha = \alpha = 1 \circ \alpha \\ \hline \beta R\alpha & , & \text{by axiom } \mathbf{(D)}, \end{array}$$

i.e. in particular $\beta\alpha = \beta \circ \alpha$.

2) Let

$$M * M \xrightarrow{m} M \underset{c}{\overset{d}{\rightrightarrows}} O, \qquad i : O \to M$$

be a reflexive–multiplicative graph: we must show that it is an internal category which is in fact a groupoid.
To show **(C2)**, define a relation $R \hookrightarrow M \times M$ by

$$\beta R\alpha :\Longleftrightarrow \begin{cases} d(\beta) &= c(\alpha) \\ d(\beta\alpha) &= d(\alpha) \end{cases}$$

$\forall (\alpha,\beta) \in M * M$, as above we get:

$$\begin{array}{rcl} \beta R 1 & : & d(\beta 1) = d(\beta) \\ 1 R 1 & : & \text{trivial} \\ 1 R \alpha & : & d(1\alpha) = d(\alpha) \\ \hline \beta R \alpha & , & \text{by axiom } \mathbf{(D)}, \end{array}$$

hence $d(\beta\alpha) = d(\alpha)$.
Similarly, it holds $c(\beta\alpha) = c(\beta)$.
To show **(C3)**, consider the relation $R \hookrightarrow M \times (M * M)$ defined by

$$\gamma\, R\, (\beta,\alpha) :\Longleftrightarrow \left\{ \begin{array}{rcl} d(\gamma) & = & c(\beta) \\ \gamma(\beta\alpha) & = & (\gamma\beta)\alpha \end{array} \right. ;$$

if (α,β) and (β,γ) are in $M * M$ and 1 denotes $1_{d(\gamma)} = 1_{c(\beta)}$, then we get:

$$\begin{array}{rcl} \gamma R(1,1) & : & \gamma(11) = \gamma = (\gamma 1)1 \\ 1R(1,1) & : & \text{trivial} \\ 1R(\beta,\alpha) & : & 1(\beta\alpha) = \beta\alpha = (1\beta)\alpha \\ \hline \gamma R(\beta,\alpha) & , & \text{by axiom } \mathbf{(D)}, \end{array}$$

so $\gamma(\beta\alpha) = (\gamma\beta)\alpha$.

To define the inversion $M \stackrel{s}{\longrightarrow} M$ we first show that the multiplication m satisfies the *left-cancellation property*:

$$\beta\gamma = \beta\alpha \text{ implies } \gamma = \alpha.$$

Let $R \hookrightarrow (M \times M) \times M$ be the relation defined by

$$(\gamma,\alpha)\, R\, \beta :\Longleftrightarrow \left\{ \begin{array}{rcccl} d(\beta) & = & c(\gamma) & = & c(\alpha) \\ \beta\gamma & = & \beta\alpha & & \end{array} \right.$$

then, if $\beta\gamma = \beta\alpha$ and $1 := 1_{d(\beta)} = 1_{c(\gamma)} = 1_{c(\alpha)}$, we get:

$$\begin{array}{rcl} (\gamma,\alpha)R\beta & : & \beta\gamma = \beta\alpha \\ (1,1)R\beta & : & \beta 1 = \beta 1 \\ (1,1)R1 & : & 11 = 11 \\ \hline (\gamma,\alpha)R1 & , & \text{by axiom } \mathbf{(D)}, \end{array}$$

so $\gamma = 1\gamma = 1\alpha = \alpha$.
Similarly, the right-cancellation property for m is true.
To show **(G1)** and **(G2)** consider the relation $S \hookrightarrow M \times M$ defined by

$$\beta S \alpha :\Longleftrightarrow \text{ there exists } \gamma \in M \text{ such that } \left\{ \begin{array}{rcl} d(\gamma) & = & c(\beta) \\ \gamma\beta & = & \alpha \end{array} \right. ;$$

note that γ *is unique by the previous remark.*
Since $\forall\, \alpha \in M$, $\alpha S \alpha$ — put $\gamma := 1_{c(\alpha)}$ —, by **(M)** S is an equivalence relation, so from $1_{d(\alpha)} S \alpha$ — put $\gamma := \alpha$ — it follows $\alpha S 1_{d(\alpha)}$, i.e.

$$\forall\, \alpha \in M \ , \ \exists !\ {}^{\bullet}\alpha \ \text{ such that } \left\{ \begin{array}{lcl} d({}^{\bullet}\alpha) & = & c(\alpha) \\ {}^{\bullet}\alpha\alpha & = & 1_{d(\alpha)} \ . \end{array} \right.$$

Similarly, one can show that

$$\forall\, \alpha \in M \ , \ \exists !\ \alpha^{\bullet} \ \text{ such that } \left\{ \begin{array}{lcl} c(\alpha^{\bullet}) & = & d(\alpha) \\ \alpha\alpha^{\bullet} & = & 1_{c(\alpha)} \ . \end{array} \right.$$

Since the composition is associative it follows $\alpha^{\bullet} = {}^{\bullet}\alpha$ for any $\alpha \in M$; then we can define $s(\alpha) = \alpha^{\bullet}$. □

COROLLARY **2.3** *For any Mal'cev category* $\mathcal{E}$*:*

1) any internal semimonoid (i.e. an object with multiplication m and unit 1) *is an abelian group in a unique way;*

2) any arrow of $\mathcal{E}$ *which preserves* 1 *also preserves* m.

PROOF. Since an internal semimonoid is a graph of the form

$$M * M = M \times M \xrightarrow{\ m\ } M \underset{t_M}{\overset{t_M}{\rightrightarrows}} 1 \qquad (1 : 1 \to M)$$

we can apply the previous results to show 1) and 2), but for the commutativity of m : let $R \hookrightarrow M \times M$ be a relation defined by

$$\beta R \alpha :\Longleftrightarrow \beta\alpha = \alpha\beta \ .$$

Since $\forall \beta \forall \alpha \in M$ $\beta R 1$, $1 R 1$ and $1 R \alpha$, we get $\beta R \alpha$, by axiom **(D)**. □

3 Centralizers for equivalence relations and partial Mal'cev operations

Given two equivalence relations $R \hookrightarrow M \times M \hookleftarrow S$ in a lex category $\mathcal{E}$, we may state [S] the following

DEFINITION **3.1** *A* double relation $C = C(R, S)$ *is a subobject* $C \hookrightarrow M^4$ *such that*

1) $\forall (x, y, z, u) \in M^4$, $(x, y, z, u) \in C$ *implies* xRy , uRz , xSu *and* ySz ; we will denote these conditions pictorially, by $\begin{array}{ccc} x & R & y \\ S & & S \\ u & R & z \end{array}$;

2) the relations $H \hookrightarrow R \times R$ *and* $V \hookrightarrow S \times S$ *defined by*

$$(x,y)H(u,z) \text{ iff } (x,y,z,u) \in C \text{ iff } (x,u)V(y,z)$$

are equivalence relations.

Note that any double relation $C = C(R,S)$ is a subobject of the object $R\Box S$ defined by

$$(x,y,z,u) \in R\Box S \text{ iff } \begin{array}{ccc} x & R & y \\ S & & S \\ u & R & z \end{array}$$

which also is a double relation. For example, given an arrow $X \xrightarrow{f} Y$ in $\mathcal{E}$, let $Ker\,(f) \hookrightarrow X \times X$ be the *kernel pair of* f, which is always an equivalence relation: then, if we consider any graph $M \overset{d}{\underset{c}{\rightrightarrows}} O$ on $\mathcal{E}$, the double relation $Ker\,(d)\Box Ker\,(c)$ consists of all "diagrams" of the form

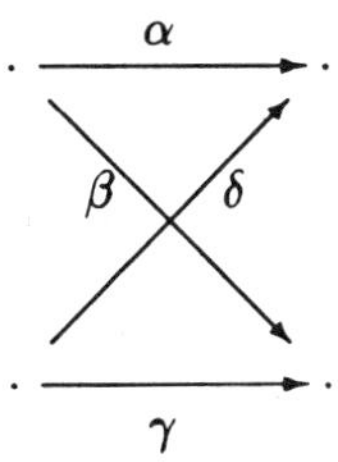

DEFINITION **3.2** *A double relation* $C = C(R,S)$ *is a* centralizer *for* R *and* S *iff it satisfies the following condition*

(C) $\forall (x,y,z) \in M^3$ *such that* $xRySz$*, there exists a unique* $u \in M$ *such that* $(x,y,z,u) \in C$.

If we denote the double relation C by

$$\begin{array}{ccc} C & \overset{h_0}{\underset{h_1}{\rightrightarrows}} & R \\ {\scriptstyle v_0}\downarrow\downarrow{\scriptstyle v_1} & & {\scriptstyle r_0}\downarrow\downarrow{\scriptstyle r_1} \\ S & \overset{s_0}{\underset{s_1}{\rightrightarrows}} & M \end{array}$$

then the above condition **(C)** can be described in terms of pull-backs by saying that the following square

$$\begin{array}{ccc} C & \xrightarrow{h_0} & R \\ {\scriptstyle v_1}\downarrow & & \downarrow{\scriptstyle r_1} \\ S & \xrightarrow{s_0} & M \end{array}$$

is a pull–back.

In particular a relation $R \overset{<d,c>}{\hookrightarrow} X \times Y$ is difunctional iff $Ker\,(d)\square Ker\,(c)$ is a centralizer and $Ker\,(d)\square Ker\,(c) = S$ [PV], where S is the relation defined in 1.2.

PROPOSITION 3.3 *Given two equivalence relations R and S on M in a Mal'-cev category $\mathcal{E}$, consider a subobject $C \hookrightarrow M^4$: then*

1) C is a double relation iff:

i) $(x,y,z,u) \in C$ implies $\begin{matrix} x & R & y \\ S & & S \\ u & R & z \end{matrix}$,

ii) xRy implies $(x,y,y,x) \in C$,

iii) xSu implies $(x,x,u,u) \in C$;

2) if C is a double relation, then the following are equivalent:

i) C is a centralizer;

ii) $(x,y,y,z) \in C$ implies $x = z$;

iii) $(x,y,z,u) \in C$ and $(x,y,z,v) \in C$ implies $u = v$.

PROOF. 1) *ii*) and *iii*) imply that the relations H and V defined in 3.1 are equivalence relations, since they are reflexive;

2) *i*) $\Rightarrow$ *ii*) since $xRySy$ and the relation H is reflexive, $(x,y,y,x) \in C$;
ii) $\Rightarrow$ *iii*) observe that $(x,y)H(u,z)$ and $(x,y)H(v,z)$ imply $(u,z)H(v,z)$ since H is an equivalence, so $(u,z,z,v) \in C$, hence $u = v$;
iii) $\Rightarrow$ *i*) consider the relation $T \hookrightarrow M^2 \times M$ defined by

$(x,y)\,T\,z :\Longleftrightarrow$ there exists u (necessarily unique) with $(x,y,z,u) \in C$:

for any (x,y,z) such that $xRySz$, we get

$$\begin{array}{rl} (x,y)Ty & : \; (x,y,y,x) \in C \\ (y,y)Ty & : \; (y,y,y,y) \in C \\ (y,y)Tz & : \; (y,y,z,z) \in C \\ \hline (x,y)Tz & , \; \text{by axiom } \mathbf{(D)}. \end{array}$$

□

COROLLARY 3.4 *In a Mal'cev category $\mathcal{E}$, the centralizer of two equivalence relations $R \hookrightarrow M \times M \hookleftarrow S$, if it exists, is unique (say, $R|S$).*

PROOF. Suppose C and C' are two centralizers for R and S: then, as in 3.3, we can define a relation $W \hookrightarrow M^2 \times M$ by

$(x,y)\,W\,z :\Longleftrightarrow$ there exists u (necessarily unique) such that $(x,y,z,u) \in C$ and $(x,y,z,u) \in C'$;

difunctionality of W implies $C = C'$. □

We recall [Jo2] that, given a span $X \xleftarrow{f} A \xrightarrow{g} Y$ in a lex category, a *partial Mal'cev operation* on A is an arrow $A \times_X A \times_Y A \xrightarrow{p} A$ which satisfies the identities (*), and that p is said to be *associative* iff

$$p(x, y, p(z, u, v)) = p(p(x, y, z), u, v)$$

if either side is defined.

PROPOSITION 3.5 *Let* $A \times_X A \times_Y A \xrightarrow{p} A$ *be a (partial) Mal'cev operation in a Mal'cev category, then:*

1) *if* p' *is another (partial) Mal'cev operation on* A, *then* $p = p'$;

2) p *is associative.*

PROOF. 1) follows from difunctionality of the relation $R \hookrightarrow A^2 \times A$ defined by

$$(x, y)Rz :\Longleftrightarrow \begin{cases} x, y, z \text{ are compatible} \\ p(x, y, z) = p'(x, y, z) \end{cases},$$

via $(x, y)Ry$, $(y, y)Ry$ and $(y, y)Rz$.
Similarly, 2) follows from difunctionality of the relation $T \hookrightarrow A^3 \times A^2$ defined by

$$(x, y, z)T(u, v) :\Longleftrightarrow \begin{cases} x, y, z, u, v \text{ are compatible} \\ p(x, y, p(z, u, v)) = p(p(x, y, z), u, v) \end{cases},$$

via $(x, y, z)T(u, u)$, $(x, x, z)T(u, u)$ and $(x, x, z)T(u, v)$. □

We can now state the following

THEOREM 3.6 *Let* $M \underset{c}{\overset{d}{\rightrightarrows}} O$ *(with* $i: O \to M$*) be a reflexive graph in a Mal'cev category* $\mathcal{E}$, *then the following are equivalent:*

1) *the reflexive graph is a reflexive-multiplicative graph;*

2) *there exists the centralizer* $Ker\,(d)|Ker\,(c)$;

3) *there exists the partial Mal'cev operation on* M, *defined for any* α, β, γ *such that* $d(\alpha) = d(\beta)$ *and* $c(\beta) = c(\gamma)$.

PROOF. 1) ⇒ 2) By Theorem 2.2, M is a groupoid, so we can define an object C on $\mathcal{E}$ by

$$(\alpha, \beta, \gamma, \delta) \in C :\Longleftrightarrow \delta = \alpha\beta^{-1}\gamma\,,$$

for any $(\alpha, \beta, \gamma, \delta) \in M^4$ such that $d(\alpha) = d(\beta)$, $d(\gamma) = d(\delta)$, $c(\alpha) = c(\delta)$ and $c(\beta) = c(\gamma)$; by Proposition 3.3, C is the centralizer $Ker\,(d)|Ker\,(c)$, since $\alpha = \alpha\beta^{-1}\beta$, $\delta = \alpha\alpha^{-1}\delta$, and $\delta = \alpha\beta^{-1}\beta$ implies $\delta = \alpha$.

2) ⇒ 3) We know that for any $(\alpha, \beta, \gamma) \in M^3$ such that $d(\alpha) = d(\beta)$ and $c(\beta) =$

$c(\gamma)$ there exists a unique δ with $(\alpha, \beta, \gamma, \delta) \in Ker(d)|Ker(c)$ [so $d(\delta) = d(\gamma)$ and $c(\delta) = c(\alpha)$]; thus we can define a partial operation p on M by $p(\alpha, \beta, \gamma) = \delta$: the identities $p(\alpha, \alpha, \beta) = \beta$ and $p(\alpha, \beta, \beta) = \alpha$ correspond to the conditions $(\alpha, \alpha, \beta, \beta) \in Ker(d)|Ker(c)$ and $(\alpha, \beta, \beta, \alpha) \in Ker(d)|Ker(c)$.

3) $\Rightarrow$ 1) Given a partial Mal'cev operation p, we can define a multiplication on M by

$$\beta\alpha := p(\beta, 1_{d(\beta)}, \alpha)$$

for any $(\alpha, \beta) \in M * M$: the axiom **(C1)** follows by the Mal'cev identities. □

Observe that if $\mathcal{E}$ is only a lex category, internal groupoids still admit a description in terms of centralizers or of a suitable partial Mal'cev operation [Jo2], in the sense that the following are equivalent:

1) a reflexive graph $M \underset{c}{\overset{d}{\rightrightarrows}} O$ (with $i: O \to M$) has a groupoid structure;

2) there exists a centralizer for $Ker(d)$ and $Ker(c)$;

3) there exists an associative partial Mal'cev operation on M, etc.

In this case one has to define the groupoid inversion by

$$\alpha^{-1} := p(1_{d(\alpha)}, \alpha, 1_{c(\alpha)}).$$

References

[B] M. Barr, P. A. Grillet, D. H. van Osdol *Exact Categories*, Springer Lecture Notes in Mathematics 236 (1971).

[CLP] A. Carboni, J. Lambek, M. C. Pedicchio *Diagram chasing in Mal'cev Categories*, JPAA 69(1990) 271–284.

[F] E. Faro *On a conjecture of Lawvere*, Preprint, SUNY Buffalo, 1989.

[Ja] G. Janelidze *Internal Categories in Mal'cev varieties*, Preprint, Toronto, 1990.

[Jo1] P. T. Johnstone *Affine Categories and Naturally Mal'cev Categories*, JPAA 61(1989) 251–256.

[Jo2] P. T. Johnstone *The "closed subgroup Theorem" for localic herds and pregroupoids*, JPAA 70(1991) 97–106.

[K] A. Kock *Generalised fibre bundles*, LNM 1348 Proceed. LLN 1987.

[Lo] J. L. Loday *Spaces with finitely many non-trivial homotopy groups*, JPAA 24(1982) 179–202.

[PV] M. C. Pedicchio, J. Vermeulen *Remarks on centralizer relations*, in preparation.

[S] J. D. H. Smith *Mal'cev Varieties*, Springer Lecture Notes in Mathematics 554 (1976).

A. Carboni
Dipartimento di Matematica, Universitá degli Studi di Milano, v. C. Saldini 50 - 20133 Milano (Italy)
e-mail: carboni @imiucca.csi.unimi.it

M. C. Pedicchio
Dipartimento di Matematica, Universitá degli Studi di Trieste, v. Europa 1 – 34100 Trieste (Italy)
e-mail: pedicchi@univ.trieste.it

N. Pirovano
Dipartimento di Matematica, Universitá degli Studi di Milano, v. C. Saldini 50 - 20133 Milano (Italy)

Canadian Mathematical Society
Conference Proceeding
Volume **13**, 1992

HEREDITARY AND MODAL CLOSURE OPERATORS

G. CASTELLINI, J. KOSLOWSKI, G. E. STRECKER

ABSTRACT: We first investigate a symmetrized generalization of the notion of categorical closure operator on an arbitrary class $\mathcal{M}$ of morphisms. For this we introduce the concept of $\mathcal{Z}$-modality which presents a common generalization of hereditariness and modality. Applications are indicated concerning the Cassidy-Hébert-Kelly Galois connection linking certain pre-factorizations with reflective subcategories of a finitely complete category that has arbitrary intersections of strong monos. In the case that $\mathcal{M}$ is part of a factorization system $\langle E, \mathcal{M}\rangle$ for sinks and that E is sufficiently well-behaved, we present constructions for $\mathcal{Z}$-modal cores and hulls, and investigate how these constructions interact with closure operators that are weakly hereditary and with those that are idempotent. If E is not well-behaved, but the category is $\mathcal{M}$-well-powered, it is shown that iterative methods can be used to construct various hulls.

1991 Mathematics Subject Classification. Primary 18A32. Secondary 06A15, 54B30.

The first author was supported in part by the Research Office of the Faculty of Arts and Sciences – University of Puerto Rico – Mayagüez Campus.

The third author was supported in part by the U.S. Office of Naval Research, grant N00014-888-K-0455.

This paper is in final form and no version of it will be submitted for publication elsewhere.

0. Introduction

Closure operators on a category $\mathcal{X}$ were originally introduced relative to classes $\mathcal{M}$ of monos in the setting of categorical topology. This notion, however, makes sense for other classes $\mathcal{M}$ of morphisms as well. A particularly useful symmetric version of this concept that we call *cd-operators* is introduced in Section 1. Many properties of closure operators in the topological framework are present in general cd-operators, and the point of view of [4] that closure operators are essentially fixed points of a Galois connection induced by a simple relation still applies in the new context. Factorization systems for morphisms turn out to be special cases of cd-operators.

Two properties of closure operators on a class of monos that are closely related are the notions of hereditariness and of modality. Both can be viewed as special cases of $\mathcal{Z}$-modality. We introduce this concept for cd-operators in Section 2, where the connection between hereditariness and modality is initially less obvious. Weakly hereditary cd-operators that are $\mathcal{Z}$-modal admit a particularly nice characterization in terms of their classes of dense elements of $\mathcal{M}$. This leads us to conjecture that the notion of $\mathcal{Z}$-modality can be used to give a more detailed analysis of the Galois connection (introduced by Cassidy, Hébert, and Kelly [2]) between factorization systems and reflective subcategories, since factorization systems may be viewed as idempotent weakly hereditary cd-operators on $\boldsymbol{Mor}(\mathcal{X})$.

In Section 3 we return to a situation where $\mathcal{M}$ is part of a factorization structure $\langle \boldsymbol{E}, \mathcal{M} \rangle$ for sinks; here cd-operators and closure operators coincide. Under mild conditions on $\boldsymbol{E}$ we present a single-step construction of $\mathcal{Z}$-modal cores and $\mathcal{Z}$-modal hulls, utilizing various Galois connections in the same spirit as they were used in [4]. These core and hull constructions turn out to be particularly well-behaved with respect to weakly hereditary closure operators.

For idempotent closure operators the situation is different, as we show in Section 4. Although idempotent $\mathcal{Z}$-modal hulls always exist, their construction can be complicated. If $\mathcal{Z}$ consists of cartesian $\mathcal{M}$-morphisms and has a certain cancellation property, idempotent $\mathcal{Z}$-modal hulls may be formed in a two-step process. This distinguishes modal closure operators from hereditary ones; in the latter case the relevant class $\mathcal{Z}$ need not have the required cancellation property. We illustrate the potential complications with finite examples.

Finally, in Section 5 we present an iterative approach to $\mathcal{Z}$-modal hulls and to idempotent $\mathcal{Z}$-modal hulls in cases where $\mathcal{X}$ is $\mathcal{M}$-well-powered. This technique does not rely on $\boldsymbol{E}$ being well-behaved.

1. Orthogonality, factorization systems, and cd-operators

Our main tool will be a notion of orthogonality that generalizes the one introduced by Tholen (cf. [13]), and encompasses part of the defining properties of factorization structures for morphisms, for sinks and for sources as well as one of the essential features of closure operators (cf. Definition 1.04).

1.00. DEFINITION. (cf. [11]) In a category $\mathcal{X}$, a pair $\langle \boldsymbol{a}, a' \rangle$ consisting of a sink $\boldsymbol{a} = \langle A_i \xrightarrow{a_i} A' \rangle_I$ and a morphism $A' \xrightarrow{a'} A''$ is called **left orthogonal to** a pair $\langle b, \boldsymbol{b}' \rangle$ consisting of a morphism $B \xrightarrow{b} B'$ and a source $\boldsymbol{b}' = \langle B' \xrightarrow{b'_j} B''_j \rangle_J$, written as $\langle \boldsymbol{a}, a' \rangle \perp \langle b, \boldsymbol{b}' \rangle$, iff for any sink $\boldsymbol{f} = \langle A_i \xrightarrow{f_i} B \rangle_I$ and any source $\boldsymbol{f}'' = \langle A'' \xrightarrow{f''_j} B''_j \rangle_J$ with the property that for each $i \in I$ and each $j \in J$ the outer square of following diagram commutes

(1-00)

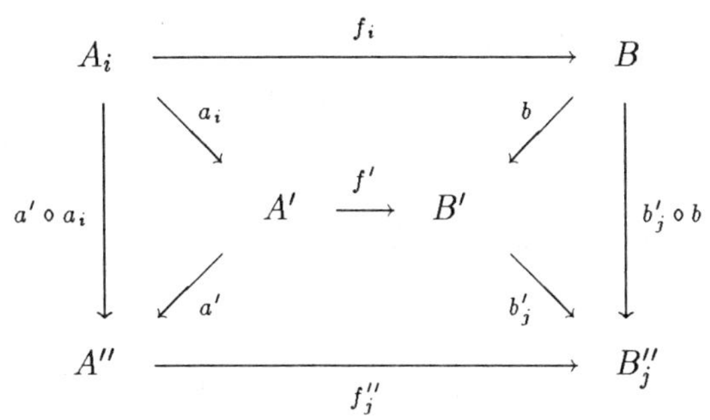

there exists a unique $f' \in \mathcal{X}\langle A', B' \rangle$ such that all inner trapezoids commute. In this case the pair $\langle b, \boldsymbol{b}' \rangle$ is called **right orthogonal to** $\langle \boldsymbol{a}, a' \rangle$. $\perp$ induces relations $P \subseteq \boldsymbol{Mor}(\mathcal{X}) \times \boldsymbol{Mor}(\mathcal{X})$ and $S \subseteq \boldsymbol{Mor}(\mathcal{X}) \times \boldsymbol{Ob}(\mathcal{X})$ via

$$\langle e, m \rangle \in P \iff \langle e, id \rangle \perp \langle id, m \rangle \qquad \text{and} \qquad \langle e, X \rangle \in S \iff \langle e, id \rangle \perp \langle id_X, \emptyset \rangle$$

These relations induce special Galois connections, so-called polarities, from the collection $\boldsymbol{P}(\boldsymbol{Mor}(\mathcal{X}))$ of subclasses of $\boldsymbol{Mor}(\mathcal{X})$ ordered by inclusion to $\boldsymbol{P}^{\mathrm{op}}(\boldsymbol{Mor}(\mathcal{X}))$ and $\boldsymbol{P}^{\mathrm{op}}(\boldsymbol{Ob}(\mathcal{X}))$ that we denote by $\pi = \langle \pi_*, \pi^* \rangle$ and $\sigma = \langle \sigma_*, \sigma^* \rangle$, respectively [1]. The coadjoints π_* and σ_* map a class $\mathcal{E}$ of $\mathcal{X}$-morphisms to the classes $\{ m \in \boldsymbol{Mor}(\mathcal{X}) \mid e \perp m \text{ for every } e \in \mathcal{E} \}$ and to $\{ X \in \boldsymbol{Ob}(\mathcal{X}) \mid e \perp \langle X, \emptyset \rangle \text{ for every } e \in \mathcal{E} \}$, respectively; the adjoints π^* and σ^* are defined analogously.

---

1 $\mathcal{A} \xrightarrow{\rho} \mathcal{B}$ denotes a Galois connection between pre-ordered classes $\mathcal{A}$ and $\mathcal{B}$, i.e., a pair of order-preserving functions $\mathcal{A} \xrightarrow{\rho_*} \mathcal{B}$ and $\mathcal{A} \xleftarrow{\rho^*} \mathcal{B}$ that satisfy $\rho_* \dashv \rho^*$. For more details, see [9].

1.01. DEFINITION. (0) A **pre-factorization system** for $\mathcal{X}$ is a Galois fixed point of π, i.e., a pair $\langle \mathcal{E}, \mathcal{M} \rangle$ of subclasses of $\boldsymbol{Mor}(\mathcal{X})$ with $\pi_*(\mathcal{E}) = \mathcal{M}$ and $\pi^*(\mathcal{M}) = \mathcal{E}$.

(1) A pre-factorization system $\langle \mathcal{E}, \mathcal{M} \rangle$ is called a **factorization system**, if every $\mathcal{X}$-morphism f admits an $\langle \mathcal{E}, \mathcal{M} \rangle$-factorization $f = m \circ e$ with $e \in \mathcal{E}$ and $m \in \mathcal{M}$.

(2) A **factorization system for sinks** on $\mathcal{X}$ is a pair $\langle \boldsymbol{E}, \mathcal{M} \rangle$ consisting of a collection $\boldsymbol{E}$ of sinks and a class $\mathcal{M}$ of $\mathcal{X}$-morphisms such that
- $\mathcal{M} = \{ m \in \boldsymbol{Mor}(\mathcal{X}) \mid \langle \boldsymbol{e}, id \rangle \perp \langle id, m \rangle \text{ for all } \boldsymbol{e} \in \boldsymbol{E} \}$;
- $\boldsymbol{E} = \{ \boldsymbol{e} \in \boldsymbol{Snk}(\mathcal{X}) \mid \langle \boldsymbol{e}, id \rangle \perp \langle id, m \rangle \text{ for all } m \in \mathcal{M} \}$;
- every sink $\boldsymbol{f}$ in $\mathcal{X}$ factors as $\boldsymbol{f} = m \circ \boldsymbol{e}$ with $\boldsymbol{e} \in \boldsymbol{E}$ and $m \in \mathcal{M}$.

Our notion of orthogonality is widely applicable, as the following result shows.

1.02. PROPOSITION. (0) A sink $\boldsymbol{a}$ and an object X are separated in the sense of Pumplün and Röhrl (cf. [12]) iff $\langle \boldsymbol{a}, id \rangle \perp \langle X, \langle id_X, id_X \rangle \rangle$.

(1) A sink $\boldsymbol{a}$ is an epi-sink iff $\boldsymbol{a}$ and every $\mathcal{X}$-object are separated in the sense of (0).

(2) Every Galois fixed point $\mathcal{Q} \subseteq \boldsymbol{Mor}(\mathcal{X})$ of $\sigma^* \circ \sigma_*$ satisfies the following cancellation property

$$r \circ q \in \mathcal{Q} \text{ and } r \in \mathcal{Q} \text{ implies } q \in \mathcal{Q} \tag{1-01}$$

(3) If $\mathcal{A}$ is a reflective subcategory of $\mathcal{X}$ with reflector R, then

$$\sigma^*(\boldsymbol{Ob}(\mathcal{A})) = \{ f \in \boldsymbol{Mor}(\mathcal{X}) \mid R(f) \text{ is iso} \} = \pi^*(\boldsymbol{Mor}(\mathcal{A}))$$

(4) If $\mathcal{X}$ has a terminal object T and if $\langle \mathcal{E}, \mathcal{M} \rangle$ is a pre-factorization system such that $X \xrightarrow{!_X} T$ for each X admits an $\langle \mathcal{E}, \mathcal{M} \rangle$-factorization, then

$$\sigma_*(\mathcal{E}) = \{ X \in \boldsymbol{Ob}(\mathcal{X}) \mid !_X \in \mathcal{M} \}$$

and this is the object class of a reflective subcategory $\mathcal{A}$ of $\mathcal{X}$. Furthermore, if $\mathcal{E}$ has property (1-01) then $\mathcal{E}$ is a Galois fixed point of $\sigma^* \circ \sigma_*$.

Proof. (0) and (1) are straightforward.

(2) It suffices to consider a fixed point Q of the form $Q = \sigma^*\{X\}$ for an object X. Suppose $A \xrightarrow{m} B$ belongs to Q and factors as $m = p \circ n$ with $p \in Q$. For every $A \xrightarrow{f} X$ because of $m \in Q$ there exists a unique g with $g \circ p \circ n = g \circ m = f$. Suppose $h \circ n = f$. Since $p \in Q$ there exists a unique k with $k \circ p = h$. Now $k \circ m = k \circ p \circ n = h \circ n = f$ implies $k = g$.

(3) and (4) are contained in [2]. □

1.03. REMARK. As Walter Tholen has observed in a private communication, factorization systems for $\mathcal{X}$ can be identified as the Eilenberg-Moore algebras (on $\mathcal{X}$) of the following monad on $\boldsymbol{CAT}$, the quasi-category of all categories: the endofunctor on $\boldsymbol{CAT}$ maps $\mathcal{X}$ to the arrow category $\mathcal{X}/\mathcal{X} = [\mathbf{2}, \mathcal{X}]$ (where $\mathbf{2}$ denotes the 2-element chain), the unit is the family of canonical embeddings from $\mathcal{X}$ into $[\mathbf{2}, \mathcal{X}]$ that takes objects to identity morphisms, and the multiplication is the family of functors from $[\mathbf{2}, [\mathbf{2}, \mathcal{X}]] \cong [\mathbf{2} \times \mathbf{2}, \mathcal{X}]$ to $[\mathbf{2}, \mathcal{X}]$ that maps a commutative square to the composition of either path.

The topologically motivated notion of a closure operator for $\mathcal{X}$ is defined relative to a class $\mathcal{M} \subseteq \boldsymbol{Mor}(\mathcal{X})$ (corresponding to the embeddings in $\boldsymbol{Top}$). We regard $\mathcal{M}$ as a full subcategory of the arrow category $\mathcal{X}/\mathcal{X} \cong [\mathbf{2}, \mathcal{X}]$ of $\mathcal{X}$. By an **$\mathcal{M}$-morphism** $\langle f, g\rangle$ from $m \in \mathcal{M}$ to $n \in \mathcal{M}$ we mean a pair of $\mathcal{X}$-morphisms that satisfy $g \circ m = n \circ f$. The **domain functor** $\mathcal{M} \xrightarrow{U} \mathcal{X}$ maps $\langle f, g\rangle$ to f, while the **codomain functor** V maps $\langle f, g\rangle$ to g. An $\mathcal{M}$-morphism $m \xrightarrow{\langle f,g\rangle} n$ is called **cartesian** iff it is **V-initial** (cf. [0]). This is equivalent to $m \xrightarrow{\langle f,g\rangle} n$ constituting a pullback square in $\mathcal{X}$.

1.04. DEFINITION. A **closure** (resp. **density**) **operator** $F = \langle (\,)_F, (\,)^F\rangle$ on $\mathcal{M}$ maps each $m \in \mathcal{M}$ to a pair $\langle m_F, m^F\rangle$ with $m = m^F \circ m_F$ such that m^F (resp. m_F) belongs to $\mathcal{M}$, and $F(m) \perp F(n)$ for all $n \in \mathcal{M}$. We denote the unique fill-in morphism induced by $m \xrightarrow{\langle f,g\rangle} n$ by $F\langle f, g\rangle$. If F is both a closure operator and a density operator, we call it a **cd-operator**.

A **morphism** $F \xrightarrow{r} G$ of closure operators or density operators on $\mathcal{M}$ is a family of $\mathcal{X}$-morphisms $\langle r_m \mid m \in \mathcal{M}\rangle$ subject to $r_n \circ F\langle f, g\rangle = G\langle f, g\rangle \circ r_m$ whenever $m \xrightarrow{\langle f,g\rangle} n$ is an $\mathcal{M}$-morphism.

If F is a closure operator or a density operator, $m \in \mathcal{M}$ is called **F-closed** (resp. **F-dense**) if m_F (resp. m^F) is an isomorphism. $\nabla_*(F)$ and $\Delta^*(F)$ denote the classes of F-closed and of F-dense members of $\mathcal{M}$, respectively.

1.05. DEFINITION. A cd-operator F is called

(0) **idempotent** iff m^F is F-closed for every $m \in \mathcal{M}$;

(1) **weakly hereditary** iff m_F is F-dense for every $m \in \mathcal{M}$;

(2) **hereditary** iff n^F is a pullback of m^F along g whenever $m = g \circ n$ and $g \in \mathcal{M}$;

(3) **modal** iff m^F is a pullback of n^F along g whenever $m \xrightarrow{\langle f,g\rangle} n$ is a cartesian $\mathcal{M}$-morphism.

$\boldsymbol{CD}(\mathcal{M})$ denotes the category of all cd-operators on $\mathcal{M}$, and we write $\boldsymbol{iCD}(\mathcal{M})$, $\boldsymbol{wCD}(\mathcal{M})$, $\boldsymbol{hCD}(\mathcal{M})$, $\boldsymbol{mCD}(\mathcal{M})$, and $\boldsymbol{iwCD}(\mathcal{M})$ for the subcategories of idempotent, weakly hereditary, hereditary, modal, and idempotent weakly hereditary cd-operators, respectively. For the corresponding preordered classes, defined by $F \sqsubseteq G$ iff there exists a morphism from F to G, we use the notation $\boldsymbol{CL}$ instead of $\boldsymbol{CD}$.

1.06. REMARKS. (0) Every hereditary cd-operator F is weakly hereditary and satisfies $m^F \in \boldsymbol{Mono}(\mathcal{X})$ for each $m \in \mathcal{M}$, cf. [8]. Moreover, if $\mathcal{M}$ consists of monos, then every modal cd-operator on $\mathcal{M}$ is hereditary.

(1) In this paper we are mainly concerned with the symmetric notion of cd-operator. If $\mathcal{M}$ satisfies the cancellation property (1-01) of Proposition 1.02(2), then every closure operator on $\mathcal{M}$ is also a density operator. Furthermore, if $\mathcal{M}$ consists of monos, the uniqueness part of the orthogonality condition for a closure operator on $\mathcal{M}$ is automatically satisfied. Both these conditions hold when $\mathcal{M}$ is part of a factorization structure $\langle \boldsymbol{E}, \mathcal{M}\rangle$ for sinks. In this setting Dikranjan and Giuli [7] originally proposed their succinct non-symmetric categorical formulation of the concept of closure operator on $\mathcal{M}$ that views $(\,)^F$ as an endofunctor $\mathcal{M} \xrightarrow{(\,)^F} \mathcal{M}$ that satisfies $V(\,)^F = V$, and views $(\,)_F$ as the domain-part of a natural transformation $id_{\mathcal{M}} \xrightarrow{\delta} (\,)^F$ that satisfies $id_V\delta = id_V$. (The uniqueness part of the orthogonality condition then says that $\langle \delta, (\,)^F\rangle$ is a pre-reflection in the sense of Börger [1], cf. also [14].) Under the given hypotheses on $\mathcal{M}$ this is equivalent to our definition of cd-operator, which in [8] has been called "factorization system on $\mathcal{X}$ with respect to $\mathcal{M}$". In order to avoid confusion with the standard notion of factorization system as given in Definition 1.01(1), we use a name closer to the well-established term "closure operator" here. However, since the original definition of closure operator was non-symmetric, we decided to use the name "cd-operator". Factorization systems correspond to cd-operators on $\boldsymbol{Mor}\,(\mathcal{X})$ that are both idempotent and weakly hereditary.

The pullback of the domain and codomain functors U and V yields the category $\mathcal{M} \diamond \mathcal{M}$ of **composable pairs** from $\mathcal{M}$. Henceforth we assume $\mathcal{M}$ to be closed under composition; $\mathcal{M} \diamond \mathcal{M} \xrightarrow{W} \mathcal{M}$ denotes the **composition functor**. The idea to restrict $\perp$ to $\mathcal{M} \diamond \mathcal{M}$ and to exploit this relation in order to characterize closure operators was first used in the topological context in [4]. Just as factorization systems are special Galois fixed points of the polarity π, cd-operators can be viewed as special Galois fixed points of a suitable polarity.

1.07. Definition. If F is a cd-operator, we call $\langle n,p\rangle \in \boldsymbol{Ob}(\mathcal{M}\diamond\mathcal{M})$ **relatively F-dense** (resp. **relatively F-closed**), if each $m \in \mathcal{M}$ satisfies $\langle n,p\rangle \perp F(m)$ (resp. $F(m) \perp \langle n,p\rangle$), and write $\dot{\omega}^{\sharp}(F)$ (resp. $\ddot{\omega}_{\sharp}(F)$) for the corresponding class of relatively F-dense (resp. relatively F-closed) pairs.

$\boldsymbol{P}(\mathcal{M}\diamond\mathcal{M}) \xrightarrow{\omega} \boldsymbol{P}^{\mathrm{op}}(\mathcal{M}\diamond\mathcal{M})$ denotes the polarity induced by restricting the orthogonality relation $\perp$ to $\mathcal{M}\diamond\mathcal{M}$.

In general, the functions $\dot{\omega}^{\sharp}$ and $\ddot{\omega}_{\sharp}$ need not preserve order from $\boldsymbol{CL}(\mathcal{M})$ to $\boldsymbol{P}(\mathcal{M}\diamond\mathcal{M})$ and $\boldsymbol{P}^{\mathrm{op}}(\mathcal{M}\diamond\mathcal{M})$, respectively. Still, they are useful for recovering cd-operators.

Whenever $C \subseteq \mathcal{M}\diamond\mathcal{M}$ and $m \in \mathcal{M}$ the intersection $C \cap W/m \cap \omega_*(C)$ can have at most one element (up to isomorphism). If all these intersections are non-empty, say $F(m) \in C \cap W/m \cap \omega_*(C)$ for each $m \in \mathcal{M}$, then F is a cd-operator on $\mathcal{M}$. Dually, one obtains a cd-operator from $D \subseteq \mathcal{M}\diamond\mathcal{M}$ if each intersection of the form $\omega^*(D) \cap W/m \cap D$ is non-empty.

1.08. Proposition. Let F be a cd-operator on $\mathcal{M}$. Any class $C \subseteq \mathcal{M}\diamond\mathcal{M}$ with $\dot{\omega}^{\sharp}(F) \subseteq C \subseteq \omega^*\ddot{\omega}_{\sharp}(F)$ satisfies $C \cap W/m \cap \omega_*(C) \neq \emptyset$ for each $m \in \mathcal{M}$, and induces a cd-operator isomorphic to F.

Proof. Clearly, $F(m) \in \dot{\omega}^{\sharp}(F) \cap W/m \cap \ddot{\omega}_{\sharp}(F)$ for each $m \in \mathcal{M}$. Since $C \subseteq \omega^*\ddot{\omega}_{\sharp}(F)$ implies $\ddot{\omega}_{\sharp}(F) \subseteq \omega_*(C)$, we have $F(m) \in C \cap W/m \cap \omega_*(C)$. □

Together with its dual this result shows that the Galois fixed points $\dot{\omega}^{\sharp}(F)$ and $\ddot{\omega}_{\sharp}(F)$ of ω determine intervals of $\boldsymbol{P}(\mathcal{M}\diamond\mathcal{M})$ and $\boldsymbol{P}^{\mathrm{op}}(\mathcal{M}\diamond\mathcal{M})$ each element of which can be used to recover F (up to isomorphism). If F is weakly hereditary or idempotent, smaller classes (possibly outside these intervals) suffice for this recovery. The following characterizations extend results first obtained in [11] for the case that $\mathcal{M}$ is part of a factorization system for sinks.

1.09. Lemma. A cd-operator F is

(0) weakly hereditary iff the class $C = \{\langle n,p\rangle \in \mathcal{M}\diamond\mathcal{M} \mid n \in \Delta^*(F)\}$ satisfies $C \cap W/m \cap \omega_*(C) \neq \emptyset$ for every $m \in \mathcal{M}$;

(1) idempotent iff the class $D = \{\langle n,p\rangle \in \mathcal{M}\diamond\mathcal{M} \mid p \in \nabla_*(F)\}$ satisfies $\omega^*(D) \cap W/m \cap D \neq \emptyset$ for every $m \in \mathcal{M}$. □

In contrast to $\dot{\omega}^{\sharp}$ and $\ddot{\omega}_{\sharp}$, the functions Δ^* and ∇_* from $\boldsymbol{CL}(\mathcal{M})$ to $\boldsymbol{P}(\mathcal{M})$ and to $\boldsymbol{P}^{\mathrm{op}}(\mathcal{M})$, respectively, do preserve order. Lemma 1.09 indicates how Galois connections can be obtained by restricting these functions suitably.

2. $\mathcal{Z}$-modal cd-operators

We now require that $\mathcal{M}$ contains all isos and that pullbacks of $\mathcal{M}$-elements exist and to belong to $\mathcal{M}$. Assume that a choice has been made for each such pullback (such a choice need not be canonical in any sense). This makes both the codomain functor V and the composition functor W into fibrations. More precisely, for an $\mathcal{X}$-morphism $X \xrightarrow{f} Y$ the V-**inverse image functor** $V/Y \xrightarrow{f^{\leftarrow}} V/X$ for f maps $n \in \boldsymbol{Ob}(V/Y)$ to its chosen pullback along f. And if $m \xrightarrow{\langle f,g\rangle} n$ is an $\mathcal{M}$-morphism, the W-**inverse image functor** $W/n \xrightarrow{\langle f,g\rangle^{\leftarrow}} W/m$ maps $\langle s,t\rangle \in \boldsymbol{Ob}(W/n)$ to the unique $\langle q,r\rangle \in \boldsymbol{Ob}(W/m)$ whose second component is the chosen pullback of t along g.

The inverse image function $\boldsymbol{P}(\mathcal{M}) \xrightarrow{p^{\leftarrow}} \boldsymbol{P}(\mathcal{M} \diamond \mathcal{M})$ induced by the projection onto the first factor $\mathcal{M} \diamond \mathcal{M} \xrightarrow{p} \mathcal{M}$ (surjectivity follows from $\boldsymbol{Iso}(\mathcal{X}) \subseteq \mathcal{M}$) is well-known to have a coadjoint, the direct image function, as well as an adjoint that maps $C \subseteq \mathcal{M} \diamond \mathcal{M}$ to the collection $\{ m \in \mathcal{M} \mid \forall_{p \in \mathcal{M}} \ \langle n,p\rangle \in C \}$ (cf. [9]). The resulting Galois connections $\boldsymbol{P}(\mathcal{M}) \xrightarrow{\gamma} \boldsymbol{P}(\mathcal{M} \diamond \mathcal{M}) \xrightarrow{\lambda} \boldsymbol{P}(\mathcal{M})$ defined by $\gamma_* = p^{\leftarrow} = \lambda^*$ satisfy $\lambda \circ \gamma = id_{\boldsymbol{P}(\mathcal{M})}$.

In the following $\mathcal{Z}$ will be a class of $\mathcal{M}$-morphisms that is stable under pullbacks in $\mathcal{M}$. Recall that such pullbacks are formed componentwise, i.e.,

$$\begin{array}{ccc} m' & \xrightarrow{\langle \bar{f},\bar{g}\rangle} & n' \\ {\scriptstyle \langle \bar{a},\bar{b}\rangle}\downarrow & & \downarrow{\scriptstyle \langle a,b\rangle} \\ m & \xrightarrow[\langle f,g\rangle]{} & n \end{array} \tag{2-00}$$

is a pullback in $\mathcal{M}$ iff $\bar{f}$ is a pullback of f along a and $\bar{g}$ is a pullback of g along b.

2.00. Definition. (0) A cd-operator F is called $\mathcal{Z}$-**modal**, if F commutes with W-inverse images along members of $\mathcal{Z}$, i.e., if $m \xrightarrow{\langle f,g\rangle} n$ belongs to $\mathcal{Z}$, then $F(m) \cong \langle f,g\rangle^{\leftarrow}(F(n))$. We write $\boldsymbol{CD}\langle \mathcal{M},\mathcal{Z}\rangle$ for the collection of all $\mathcal{Z}$-modal cd-operators on $\mathcal{M}$.

(1) $C \subseteq \mathcal{M} \diamond \mathcal{M}$ is called $\mathcal{Z}$-**stable**, if whenever $m \xrightarrow{\langle f,g\rangle} n$ belongs to $\mathcal{Z}$, and $\langle s,t\rangle \in C \cap W/n$, then the W-inverse image of $\langle s,t\rangle$ along $\langle f,g\rangle$ belongs to C. We call $A \subseteq \mathcal{M}$ $\mathcal{Z}$-**stable**, if $\gamma_*(A)$ has this property.

Since $\mathcal{Z} \subseteq \boldsymbol{Mor}(\mathcal{M})$ is stable under pullbacks, $\mathcal{Z}$-stability of $A \subseteq \mathcal{M}$ is equivalent to the requirement that whenever $m \xrightarrow{\langle f,g\rangle} n$ belongs to $\mathcal{Z}$ and $n \in A$, then $m \in A$.

2.01. Proposition. (0) Let $\mathcal{C}$ be the class of all cartesian $\mathcal{M}$-morphisms. A cd-operator is modal iff it is $\mathcal{C}$-modal, i.e., $\boldsymbol{mCD}(\mathcal{M}) = \boldsymbol{CD}\langle\mathcal{M},\mathcal{C}\rangle$.

(1) Let $\mathcal{H} = \{\langle f,g\rangle \in \boldsymbol{Mor}(\mathcal{M}) \mid f$ is iso and $g \in \mathcal{M}\}$. A cd-operator is hereditary iff it is $\mathcal{H}$-modal, i.e., $\boldsymbol{hCD}(\mathcal{M}) = \boldsymbol{CD}\langle\mathcal{M},\mathcal{H}\rangle$. □

The next result provides a characterization of weakly hereditary $\mathcal{Z}$-modal cd-operators as well as a partial characterization of $\mathcal{Z}$-modal cd-operators.

2.02. Proposition. (0) For every $\mathcal{Z}$-modal cd-operator F the class $\Delta^*(F)$ is $\mathcal{Z}$-stable, and if $\mathcal{M}$ consists of monos, so is the class $\dot{\omega}^\sharp(F)$.

(1) If $\dot{\omega}^\sharp(F)$ is $\mathcal{Z}$-stable, then F is $\mathcal{Z}$-modal.

(2) If $\Delta^*(F)$ is $\mathcal{Z}$-stable and F is weakly hereditary, then F is $\mathcal{Z}$-modal.

Proof. For $m \xrightarrow{\langle f,g\rangle} n$ in $\mathcal{Z}$ let $\langle q,r\rangle$ be the W-inverse image along $\langle f,g\rangle$ of $\langle s,t\rangle \in W/n$, and let d be the corresponding pullback of g along t.

(2-01)

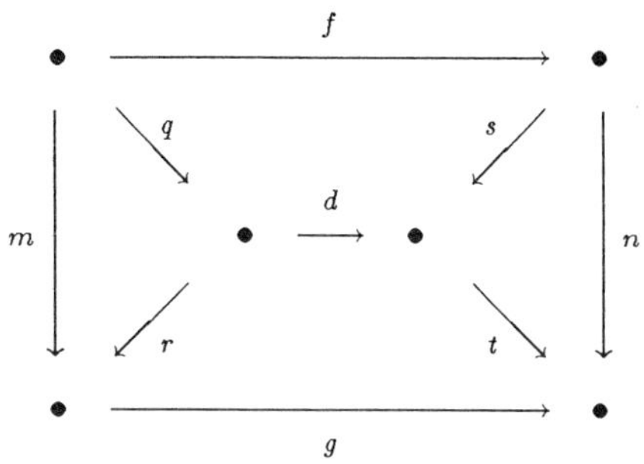

(0) Let F be $\mathcal{Z}$-modal. If n is F-dense, take $\langle s,t\rangle = F(n)$. Then r as a pullback of the iso t is iso as well, and so by $\mathcal{Z}$-modality q is F-dense. Suppose that $\mathcal{M}$ consists of monos, and let $\langle s,t\rangle$ be relatively F-dense. Then there exists a unique W/n-morphism $\langle s,t\rangle \xrightarrow{k} F(n)$. Since $F(m)$ is isomorphic to the W-inverse image of $F(n)$ along $\langle f,g\rangle$, we have that $m^F \xrightarrow{\langle F\langle f,g\rangle, g\rangle} n^F$ is cartesian. Thus we obtain a W/m-morphism $\langle q,r\rangle \xrightarrow{l} F(m)$. Since m^F is mono, there is only one such W/m-morphism, which implies that $\langle q,r\rangle$ is relatively F-dense.

(1) Suppose that $\dot{\omega}^\sharp(F)$ is $\mathcal{Z}$-stable, and take $F(n)$ for $\langle s,t\rangle$. Since $\langle q,r\rangle$ is relatively F-dense, there exists a unique $\langle q,r\rangle \xrightarrow{a} F(m)$ in W/m. On the other hand, the universal property of the pullback induces a unique W/m-morphism $F(m) \xrightarrow{b} \langle q,r\rangle$ that inverts a. Hence $\langle q,r\rangle \cong F(m)$.

(2) If F is weakly hereditary and $\Delta^*(F)$ is $\mathcal{Z}$-stable, again take $F(n)$ for $\langle s,t\rangle$. The stability of $\mathcal{Z}$ under pullbacks implies that $q \xrightarrow{\langle f,d\rangle} n_F$ belongs to $\mathcal{Z}$, and hence that $q \in \Delta^*(F)$. Now the argument proceeds as in (1). □

$\mathcal{S} = \{\langle f,g\rangle \in \mathcal{C} \mid g \text{ mono}\}$ and $\mathcal{S}' = \{\langle f,g\rangle \in \mathcal{S} \mid f \text{ iso}\}$ also satisfy the requirement for $\mathcal{Z}$ to be stable under pullbacks in $\mathcal{M}$. If finite products exist in $\mathcal{X}$ (and hence in $\mathcal{M}$), so does the class $\mathcal{P}$ of all projections of the form $id_X \times n \longrightarrow n$ in $\mathcal{M}$. Notice that each cartesian $\mathcal{M}$-morphism $\langle f,g\rangle$ admits an $\langle \mathcal{S},\mathcal{P}\rangle$-factorization, since $X \xrightarrow{g} Y$ factors through the projection $X \times Y \longrightarrow Y$ by means of its graph.

2.03. COROLLARY. A weakly hereditary cd-operator F on $\mathcal{M}$ is

(0) hereditary iff $\Delta^*(F)$ has the following cancellation property

$$r \circ q \in \Delta^*(F) \text{ and } r \in \mathcal{M} \text{ implies } q \in \Delta^*(F) \tag{2-02}$$

(1) modal iff $\Delta^*(F)$ is stable under pullbacks;
(2) $\mathcal{S}$-modal iff $\Delta^*(F)$ is stable under pullbacks along monos;
(3) $\mathcal{P}$-modal iff $\Delta^*(F)$ is stable under pullbacks along projections, which for idempotent F is equivalent to $\Delta^*(F)$ being stable under products. □

2.04. COROLLARY. Suppose $\mathcal{M}$ has cancellation property (1-01). Then each hereditary cd-operator F satisfies $\dot{\omega}^\sharp(F) = \gamma_*\Delta^*(F)$. A cd-operator F that satisfies this equation is hereditary if m^F is mono for every $m \in \mathcal{M}$.

Proof. Let F be hereditary. $\Delta^* = \gamma^* \circ \dot{\omega}^\sharp$ and the adjunction $\gamma_* \dashv \gamma^*$ imply $\gamma_*\Delta^*(F) \subseteq \dot{\omega}^\sharp(F)$. Since $\lambda_* \dashv \lambda^* = \gamma_*$, to establish the reverse inclusion, it suffices to show $\lambda_*\dot{\omega}^\sharp(F) \subseteq \Delta^*(F)$. Let $\langle q,r\rangle \in W/m$ be relatively F-dense. There exists a unique $\mathcal{X}$-morphism d that satisfies $m^F \circ d = r$ and $d \circ q = m_F$. By (1-01) we have $d \in \mathcal{M}$. Since F is weakly hereditary we have $m_F \in \Delta^*(F)$, and so by (2-02) we conclude $q \in \Delta^*(F)$ as desired.

Conversely, if $\dot{\omega}^\sharp(F) = \gamma_*\Delta^*(F)$, then by Lemma 1.09(0) F is weakly hereditary. Let $m \in \Delta^*(F)$ and $m = r \circ q$ with $r \in \mathcal{M}$. By (1-01) we have $\langle q,r\rangle \in W/m$. For every $n \in \mathcal{M}$, since $\langle m, id\rangle \perp F(n)$ and n^F is mono, it follows immediately that $\langle q,r\rangle \perp F(n)$, and hence that $\langle q,r\rangle \in \dot{\omega}^\sharp(F) = \gamma_*\Delta^*(F)$, i.e., $n \in \Delta^*(F)$. By Corollary 2.03(0) F is hereditary. □

For a finitely complete category $\mathcal{X}$ that has arbitrary intersections of strong monos, Cassidy, Hébert and Kelly [2] have constructed a Galois connection φ between the class of pre-factorization systems on $\mathcal{X}$ for which all morphisms into the terminal object T admit a factorization, ordered by inclusion of the $\mathcal{E}$-part, and the class of reflective subcategories of $\mathcal{X}$, ordered by reverse inclusion. (These orders are chosen to be compatible with the polarities π and σ of Definition 1.00.) Factorization systems $\langle \mathcal{E},\mathcal{M}\rangle$ may be viewed as idempotent weakly hereditary cd-operators on $\boldsymbol{Mor}(\mathcal{X})$; the corresponding cd-operator F then satisfies $\mathcal{E} = \Delta^*(F)$ and $\mathcal{M} = \nabla_*(F)$. A reflective subcategory $\mathcal{A}$ of $\mathcal{X}$ induces a so-called **reflective** pre-factorization system $\varphi^*(\mathcal{A}) = \langle \sigma^*(\boldsymbol{Ob}(\mathcal{A})), \pi_*\sigma^*(\boldsymbol{Ob}(\mathcal{A}))\rangle$ (cf. Proposition 1.02(3)) that under the given hypothesis is a factorization system (cf. [2], Corollary 3.2). In fact, by Proposition 1.02(4) a reflective pre-factorization system $\langle \mathcal{E}, \mathcal{M}\rangle$ is characterized by the fact that $\mathcal{E}$ satisfies the cancellation property (1-01). $\varphi^* \circ \varphi_*$ maps a pre-factorization system for which the objects of $\mathcal{X}/T$ admit factorizations to its **reflective hull**, while $\varphi_* \circ \varphi^*$ is the identity function.

Call a factorization system $\mathcal{Z}$-modal iff the corresponding idempotent weakly hereditary cd-operator on $\boldsymbol{Mor}(\mathcal{X})$ has this property. Theorem 4.7 of [2] shows that under φ localizations of $\mathcal{X}$, i.e., reflective subcategories with a reflector that preserves finite limits, bijectively correspond to modal reflective factorization systems.

2.05. THEOREM. Let $\mathcal{X}$ be finitely complete and have arbitrary intersections of strong monos. Consider a reflective subcategory $\mathcal{A}$ of $\mathcal{X}$ with reflector R, and a factorization system $\langle \mathcal{E}, \mathcal{M}\rangle$.

(0) The induced factorization system $\varphi^*(\mathcal{A})$ is
- hereditary, if R maps all morphisms to monos;
- $\mathcal{S}'$-modal, if R preserves monos;
- $\mathcal{S}$-modal, if R preserves pullbacks of monos;
- $\mathcal{P}$-modal, if R preserves finite products;
- modal, if R preserves pullbacks.

(1) The reflector of the induced reflective subcategory $\varphi_*\langle \mathcal{E}, \mathcal{M}\rangle$
- maps all morphisms to elements of $\mathcal{M}$, if $\mathcal{M} \subseteq \boldsymbol{Mono}(\mathcal{X})$;
- preserves finite products, if $\langle \mathcal{E}, \mathcal{M}\rangle$ is $\mathcal{P}$-modal.

Proof. (0) Suppose the R maps all morphisms to monos, and consider $n = g \circ m$ with n dense. Then $R(n) = R(g) \circ R(m)$ is iso, hence $R(g)$ is a retraction. Since $R(g)$ by hypothesis also is mono, it is iso, which forces $R(m)$ to be iso as well. Hence m is dense.

Suppose that R preserves monos, and consider $n = g \circ m$ with n dense and g mono. The same argument as above applies.

Suppose that R preserves pullbacks of monos, and consider a cartesian $\mathcal{M}$-morphism $m \xrightarrow{\langle f, g\rangle} n$ with n dense and g mono. Since R preserves this pullback and $R(n)$ is iso, $R(m)$ is iso as well. Thus m is dense.

Suppose that R preserves finite products. If n is dense, then $R(n)$ is iso, hence $R(id_X \times n) = id_{R(X)} \times R(n)$ is iso as well. By Corollary 2.03(3) the induced factorization system is $\mathcal{P}$-modal.

Suppose that R preserves pullbacks. As a coadjoint R preserves the terminal object, and therefore preserves finite limits, in particular it preserves finite products. Since every cartesian $\mathcal{M}$-morphism $\langle f, g\rangle$ admits an $\langle \mathcal{S}, \mathcal{P} \rangle$-factorization by factoring g through its graph, the result follows from the previous two results.

(1) Let G be the reflector, and u be the unit of the reflection.
If $\mathcal{M}$ consists of monos, all objects of $\varphi^* \langle \mathcal{E}, \mathcal{M} \rangle$ are $\mathcal{M}$-subobjects of the terminal object T, hence all morphisms of $\varphi^* \langle \mathcal{E}, \mathcal{M} \rangle$ belong to $\mathcal{M}$. Suppose $\langle \mathcal{E}, \mathcal{M} \rangle$ is $\mathcal{P}$-modal, and consider $\mathcal{X}$-objects A and B. Clearly, $u_{A \times B} \in \mathcal{E}$. Since F is $\mathcal{P}$-modal and $\mathcal{E}$ is closed under composition, $u_{G(A) \times G(B)} \circ (u_A \times u_B) \in \mathcal{E}$ as well. If p_A and $p_{G(A)}$ denote projections, we have

$$G\left(p_{G(A)}\right) \circ u_{G(A) \times G(B)} \circ (u_A \times u_B) = G(p_A) \circ u_{A \times B}$$

Since all morphisms in $\varphi_* \langle \mathcal{E}, \mathcal{M} \rangle$ belong to $\mathcal{M}$, orthogonality forces $G(G(A) \times G(B))$ and $G(A \times B)$ to be isomorphic. Hence G preserves finite products. $\square$

While part (0) of Theorem 2.05 slightly refines a result of [2], part (1) seems to be new. Unfortunately we were not able to prove or disprove the corresponding assertions in the cases that $\langle \mathcal{E}, \mathcal{M} \rangle$ is $\mathcal{S}'$-modal, $\mathcal{S}$-modal, or modal.

2.06. Corollary. Let $\mathcal{X}$ be finitely complete (with terminal object T) and have intersections of strong monos.

(0) reflective subcategories of subobjects of T bijectively correspond under φ to hereditary factorization systems.
(1) reflective subcategories with product-preserving reflector bijectively correspond under φ to $\mathcal{P}$-modal reflective factorization systems.

Proof. (0) Observe that in a hereditary factorization system $\langle \mathcal{E}, \mathcal{M} \rangle$ the class $\mathcal{M}$ consists of monos, cf. Remark 1.06(0). Moreover, the cancellation condition (2-02) translates into $p \circ n \in \mathcal{E}$ implies $n \in \mathcal{E}$, which subsumes condition (1-01), hence $\langle \mathcal{E}, \mathcal{M} \rangle$ is already reflective. Now Theorem 2.05 applies.

(1) This follows directly from Theorem 2.05. $\square$

For a different approach to relations between morphisms and objects that is capable of handling morphism classes that are not closed under composition the reader should consult [10].

3. $\mathcal{Z}$-modal closure operators with respect to factorization systems for sinks

In order to have analogues to the complete subspace lattices of topological spaces, for the remainder of this paper we assume that $\mathcal{X}$ has a factorization structure $\langle \boldsymbol{E}, \mathcal{M} \rangle$ for sinks. This insures that $\mathcal{X}$ is sufficiently nice to support certain constructions (cf. [4]). Consequently, in this setting, $\mathcal{M}$ then consists of monos and has the cancellation property (1-01) so that the notions of closure operator and of cd-operator coincide. Thus, in the remainder of this paper we can simplify our exposition by dealing only with the pre-ordered class $\boldsymbol{CL}(\mathcal{M})$ of closure operators and its subclasses as introduced in [4]. Moreover, the functions $\dot{\omega}^{\sharp}$ and $\ddot{\omega}_{\sharp}$ now preserve order, and are in fact parts of Galois connections, hence we can use the notation $\dot{\omega}^{*}$ and $\ddot{\omega}_{*}$ for them.

The V-fibres and the W-fibres introduced in Section 1 now carry complete pre-orders $\leq$ and $\ll$ respectively. To minimize problems resulting from the fact that these pre-orders need not be antisymmetric, we assume that for every sink s in $\mathcal{X}$ a specific $\langle \boldsymbol{E}, \mathcal{M} \rangle$-factorization has been chosen. This choice need not be canonical in any sense, but it does induce a choice of pullbacks of members of $\mathcal{M}$. Now we can speak about *the* infimum and *the* supremum of a sink consisting of $\mathcal{M}$-elements, as well as of *the* pullback of an $\mathcal{M}$-element. We continue to use the term **lattice** for a pre-ordered class that is finitely complete and finitely cocomplete. A lattice is called **complete** if every subclass has an infimum or, equivalently, every subclass has a supremum.

Both V and W now become bifibrations. For $X \xrightarrow{f} Y$ the V-inverse functor $f^{\leftarrow}$ has a coadjoint, the **V-direct image functor** $V/X \xrightarrow{f_{\exists}} V/Y$, that maps an $m \in V/X$ to the $\mathcal{M}$-component of the chosen $\langle \boldsymbol{E}, \mathcal{M} \rangle$-factorization of $f \circ m$. And if $m \xrightarrow{\langle f,g \rangle} n$ is an $\mathcal{M}$-morphism, then the W-inverse image functor $\langle f,g \rangle^{\leftarrow}$ has a coadjoint as well, the **W-direct image functor** $W/m \xrightarrow{\langle f,g \rangle_{\exists}} W/m'$, that maps $\langle n,p \rangle \in W/m$ to the unique $\langle n',p' \rangle \in W/m'$ for which there exists an $\mathcal{X}$-morphism d such that $\langle \langle d,n' \rangle, p' \rangle$ is the chosen $\langle \boldsymbol{E}, \mathcal{M} \rangle$-factorization of the 2-sink $\langle g \circ p, m' \rangle$. Details can be found in [4]. Given a class $C \subseteq \mathcal{M} \diamond \mathcal{M}$, the smallest class containing C that is stable under the formation of W-direct images is denoted by C^{di}.

Under these assumptions arbitrary suprema and infima of closure operators exist. They are formed pointwise in the fibers. In particular, $\boldsymbol{wCL}(\mathcal{M})$ is stable under the formation of suprema and $\boldsymbol{iCL}(\mathcal{M})$ is stable under the formation of infima in $\boldsymbol{CL}(\mathcal{M})$. Thus every closure operator F has an **idempotent hull** (i.e., reflection) $F^{\mathrm{i}} \in \boldsymbol{iCL}(\mathcal{M})$ as well as a **weakly hereditary core** (i.e., coreflection) $F_{\mathrm{w}} \in \boldsymbol{wCL}(\mathcal{M})$. The first construction of these hulls and cores that did not rely on smallness properties of $\mathcal{X}$ with respect to $\mathcal{M}$ can be found in [11], Theorem 1.12, cf. also [4], Lemma 1.10. For additional background on closure operators see, e.g., [3], [6], and [8].

3.00. DEFINITION. $V_\exists(\mathcal{Z})$ denotes the class of $\mathcal{X}$-morphisms g such that there exists $f \in \boldsymbol{Mor}(\mathcal{X})$ with $\langle f, g\rangle \in \mathcal{Z}$.

3.01. PROPOSITION. (0) $\boldsymbol{CL}\langle\mathcal{M},\mathcal{Z}\rangle$ is closed under the formation of infima in $\boldsymbol{CL}(\mathcal{M})$. Consequently, every closure operator F has a **$\mathcal{Z}$-modal hull** $F^{\mathcal{Z}}$ and an **idempotent $\mathcal{Z}$-modal hull** $F^{(\mathrm{i}\mathcal{Z})}$, obtained by taking the infimum of all $\mathcal{Z}$-modal, respectively all simultaneously idempotent and $\mathcal{Z}$-modal, closure operators G that satisfy $F \sqsubseteq G$.

(1) If all V-inverse image functors $g^{\leftarrow}$ for $g \in V_\exists(\mathcal{Z})$ are coadjoint, then $\boldsymbol{CL}\langle\mathcal{M},\mathcal{Z}\rangle$ is closed under the formation of suprema in $\boldsymbol{CL}(\mathcal{M})$ as well, and each $F \in \boldsymbol{CL}(\mathcal{M})$ has a **$\mathcal{Z}$-modal core** $F_{\mathcal{Z}}$. □

If all the V-inverse image functors $m^{\leftarrow}$ for $m \in \mathcal{M}$ are coadjoint, i.e., if all $\mathcal{M}$-subobject lattices are Heyting algebras, then every $F \in \boldsymbol{CL}(\mathcal{M})$ has a **hereditary core** $F_{\mathcal{H}}$. For the **modal core** $F_{\mathcal{C}}$ to exist, we need to require *all* V-inverse image functors $g^{\leftarrow}$ for $g \in \boldsymbol{Mor}(\mathcal{X})$ to be coadjoint. This condition is quite frequently satisfied, e.g., for the class $\mathcal{M}$ of all embeddings in the category $\boldsymbol{Top}$. Thus every closure operator in that context has all $\mathcal{Z}$-modal cores for $\mathcal{H} \subseteq \mathcal{Z} \subseteq \mathcal{C}$ and $\mathcal{Z}$ being pullback-stable.

Recall from [4] that the functions $\dot{\omega}^*$ and Δ^* are the adjoints of Galois connections $\boldsymbol{P}(\mathcal{M}\diamond\mathcal{M}) \xrightarrow{\dot{\omega}} \boldsymbol{CL}(\mathcal{M})$ and $\boldsymbol{P}(\mathcal{M}) \xrightarrow{\Delta} \boldsymbol{CL}(\mathcal{M})$, and that the coadjoint Δ_* factors through $\boldsymbol{wCL}(\mathcal{M})$. More precisely, for $C \subseteq \mathcal{M}\diamond\mathcal{M}$ and $A \subseteq \mathcal{M}$ the closure operators $\dot{\omega}^*(C)$ and $\Delta_*(A)$, respectively, map $m \in \mathcal{M}$ to

$$\text{(3-00)} \qquad \sup_{\ll}(C^{\mathrm{di}} \cap W/m) \qquad \text{and} \qquad \sup_{\ll}((\gamma_*(A))^{\mathrm{di}} \cap W/m)$$

Proposition 2.02 now suggests strategies for constructing the $\mathcal{Z}$-modal hull $F^{\mathcal{Z}}$ (and if it exists, the $\mathcal{Z}$-modal core $F_{\mathcal{Z}}$) of a closure operator F: starting from $\dot{\omega}^*(F)$, produce a suitable $\mathcal{Z}$-stable subclass of $\mathcal{M}\diamond\mathcal{M}$. Notice that whenever $\mathcal{Z}$ contains $\mathcal{H} = \{\langle f, g\rangle \mid f \text{ is iso and } g \in \mathcal{M}\}$, then $\mathcal{Z}$-modal closure operators are hereditary and thus are weakly hereditary.

3.02. PROPOSITION. Let $\boldsymbol{E}$ be stable under $V_\exists(\mathcal{Z})$-pullbacks.

(0) If $C \subseteq \mathcal{M}\diamond\mathcal{M}$ is $\mathcal{Z}$-stable, then the closure operator $\dot{\omega}_*(C)$ is $\mathcal{Z}$-modal.

(1) If $A \subseteq \mathcal{M}$ is $\mathcal{Z}$-stable, then the weakly hereditary closure operator $\Delta_*(A)$ is $\mathcal{Z}$-modal.

Proof. (0) Set $F = \dot{\omega}_* (C)$. The $\mathcal{Z}$-inverse image $\langle x, y\rangle$ of $F(n)$ along $\langle f, g\rangle$ is relatively F-dense. Let z be the pullback of g along n^F. Since $\mathcal{Z}$ is pullback-stable, $\langle f, z\rangle \in \mathcal{Z}$, so that since $z \in V_\exists(\mathcal{Z})$ pullbacks of $\boldsymbol{E}$-sinks along z again belong to $\boldsymbol{E}$. Therefore $\langle x, y\rangle$ is the supremum in W/m of the W-inverse images along $\langle f, g\rangle$ of all $\langle s, t\rangle \in C^{\mathrm{di}} \cap W/n$. It now suffices to show that these W-inverse images belong to $C^{\mathrm{di}} \cap W/m$. Consider $\langle s, t\rangle \in C^{\mathrm{di}} \cap W/n$ with W-inverse image $\langle q, r\rangle$ along $\langle f, g\rangle$, and let d be the corresponding pullback of g along t. The $\mathcal{M}$-morphism $q \xrightarrow{\langle f, d\rangle} s$ is a pullback of $\langle f, g\rangle$ along $\langle id, t\rangle$ and hence belongs to $\mathcal{Z}$. In particular, $d \in V_\exists(\mathcal{Z})$. Since $\langle s, t\rangle \in C^{\mathrm{di}}$, there exist $\bar{n} \in \mathcal{M}$ and $\langle \bar{s}, \bar{t}\rangle \in C \cap W/\bar{n}$ and an $\mathcal{M} \diamond \mathcal{M}$-morphism $\langle \bar{s}, \bar{t}\rangle \xrightarrow{\langle a, b, c\rangle} \langle s, t\rangle$ such that the 2-sink $\langle b, s\rangle$ belongs to $\boldsymbol{E}$. The pullback $\bar{m} \xrightarrow{\langle \bar{f}, \bar{g}\rangle} \bar{n}$ of $\langle f, g\rangle$ along $\langle a, c\rangle$ in $\mathcal{M}$ belongs to $\mathcal{Z}$ by hypothesis, which implies that the W-inverse image $\langle \bar{q}, \bar{r}\rangle$ of $\langle \bar{s}, \bar{t}\rangle$ along $\langle \bar{f}, \bar{g}\rangle$ belongs to C. Moreover, if $\langle a', b', c'\rangle$ is the corresponding pullback of $\langle a, b, c\rangle$ along $\langle f, d, g\rangle$ in $\mathcal{M} \diamond \mathcal{M}$, the 2-sink $\langle b', q\rangle$ as the pointwise pullback of $\langle b, s\rangle$ along $d \in V_\exists(\mathcal{Z})$ belongs to $\boldsymbol{E}$, which establishes $\langle q, r\rangle$ as a W-direct image of $\langle \bar{q}, \bar{r}\rangle$ along $\langle a', c'\rangle$ and hence as an element of C^{di} as desired.

(1) Apply (0) to $C = \gamma_*(A)$. □

Our goal is now to construct the $\mathcal{Z}$-modal hull, and if it exists the $\mathcal{Z}$-modal core, of a closure operator F by expanding (respectively, shrinking) the class of relatively F-dense pairs until it is $\mathcal{Z}$-stable.

The "expanding" and "shrinking" alluded to above are taken care of by another Galois connection. Any $R \subseteq X \times Y$ induces a Galois connection $\boldsymbol{P}(Y) \xrightarrow{\langle R^\exists, R_\forall\rangle} \boldsymbol{P}(X)$ where $R^\exists(B) = \{x \in X \mid \exists_{y \in B}\ \langle x, y\rangle \in R\}$ and $R_\forall(A) = \{y \in Y \mid \forall_{x \in X}\ \langle x, y\rangle \in R \Rightarrow x \in A\}$ (cf. [9]).

3.03. Definition. Let Z^∞ be the transitive closure of the relation Z on $\mathcal{M} \diamond \mathcal{M}$ defined by $\langle\langle n, p\rangle, \langle q, r\rangle\rangle \in Z$ iff $\langle n, p\rangle$ is a W-inverse image of $\langle q, r\rangle$ along some morphism in $\mathcal{Z}$. Similarly, let $\bar{Z}^\infty$ be the transitive closure of the relation $\bar{Z}$ on $\mathcal{M}$ defined by $\langle m, n\rangle \in \bar{Z}$ iff there exists an $\mathcal{M}$-morphism in $\mathcal{Z}$ from m to n. We denote the corresponding Galois connections $\langle (Z^\infty)^\exists, (Z^\infty)_\forall\rangle$ and $\langle (\bar{Z}^\infty)^\exists, (\bar{Z}^\infty)_\forall\rangle$ by $\zeta_{\mathcal{Z}}$ and $\bar{\zeta}_{\mathcal{Z}}$, respectively.

Notice in particular that for $\mathcal{H} = \{\langle f, g\rangle \boldsymbol{Mor}(\mathcal{M}) \mid f \text{ iso and } g \in \mathcal{M}\}$ the Galois connection $\boldsymbol{P}(\mathcal{M} \diamond \mathcal{M}) \xrightarrow{\zeta_{\mathcal{H}}} \boldsymbol{P}(\mathcal{M} \diamond \mathcal{M})$ coincides with $\gamma \circ \lambda$.

3.04. PROPOSITION. (0) For $C \subseteq \mathcal{M} \diamond \mathcal{M}$ among the $\mathcal{Z}$-stable subclasses of $\mathcal{M} \diamond \mathcal{M}$ the smallest superclass of C is $(\zeta_{\mathcal{Z}})_* (C)$ and the largest subclass of C is $(\zeta_{\mathcal{Z}})^* (C)$.

(1) For $A \subseteq \mathcal{M}$ among the $\mathcal{Z}$-stable subclasses of $\mathcal{M}$ the smallest superclass of A is $(\bar{\zeta}_{\mathcal{Z}})_* (A)$ and the largest subclass of A is $(\bar{\zeta}_{\mathcal{Z}})^* (A)$. □

3.05. THEOREM. If $\boldsymbol{E}$ is stable under the formation of $V_\exists (\mathcal{Z})$-pullbacks, then the $\mathcal{Z}$-modal core and hull of $F \in \boldsymbol{CL}(\mathcal{M})$, respectively, are given by

$$F_{\mathcal{Z}} \cong \dot{\omega}_*(\zeta_{\mathcal{Z}})^*\dot{\omega}^* (F) \qquad \text{and} \qquad F^{\mathcal{Z}} \cong \dot{\omega}_*(\zeta_{\mathcal{Z}})_*\dot{\omega}^* (F) \ ,$$

while the weakly hereditary $\mathcal{Z}$-modal core is given by

$$F_{(\mathrm{w}\mathcal{Z})} \cong (F_{\mathcal{Z}})_{\mathrm{w}} \cong ((F_{\mathrm{w}})_{\mathcal{Z}})_{\mathrm{w}} \cong \Delta_*(\bar{\zeta}_{\mathcal{Z}})^*\Delta^* (F) \ .$$

If F is weakly hereditary, then its $\mathcal{Z}$-modal hull in $\boldsymbol{wCL}(\mathcal{M})$ is given by

$$\Delta_*(\bar{\zeta}_{\mathcal{Z}})_*\Delta^* (F) \ .$$

Proof. For any closure operator F all four closure operators $\dot{\omega}_*(\zeta_{\mathcal{Z}})^*\dot{\omega}^* (F)$, $\dot{\omega}_*(\zeta_{\mathcal{Z}})_*\dot{\omega}^* (F)$, $\Delta_*(\bar{\zeta}_{\mathcal{Z}})^*\Delta^* (F)$ and $\Delta_*(\bar{\zeta}_{\mathcal{Z}})_*\Delta^* (F)$ are $\mathcal{Z}$-modal by Propositions 3.02 and 3.04.

Since $(\zeta_{\mathcal{Z}})^*$ is decreasing, we have $\dot{\omega}_*(\zeta_{\mathcal{Z}})^*\dot{\omega}^* (F) \sqsubseteq \dot{\omega}_*\dot{\omega}^* (F) \cong F$.

To show that $\dot{\omega}_*(\zeta_{\mathcal{Z}})^*\dot{\omega}^* (F)$ is the $\mathcal{Z}$-modal core of F suppose that $H \sqsubseteq F$ is $\mathcal{Z}$-modal. By Proposition 2.02 $\dot{\omega}^* (H) = (\zeta_{\mathcal{Z}})^*\dot{\omega}*(H) \subseteq (\zeta_{\mathcal{Z}})^*\dot{\omega}^* (F)$. This implies that $H \cong \dot{\omega}_*\dot{\omega}^* (H) = \dot{\omega}_*(\zeta_{\mathcal{Z}})^*\dot{\omega}^* (H) \sqsubseteq \dot{\omega}_*(\zeta_{\mathcal{Z}})^*\dot{\omega}^* (F)$.

The argument for $F^{\mathcal{Z}}$ is analogous; it uses the fact that $(\zeta_{\mathcal{Z}})_*$ is increasing.

By Propositions 3.02(1) and 3.04(1) weakly hereditary cores of $\mathcal{Z}$-modal closure operators are again $\mathcal{Z}$-modal. Since $F_{(\mathrm{w}\mathcal{Z})} \sqsubseteq F_{\mathrm{w}}$ and $F_{(\mathrm{w}\mathcal{Z})} \sqsubseteq F_{\mathcal{Z}}$, this implies $F_{(\mathrm{w}\mathcal{Z})} \cong (F_{\mathcal{Z}})_{\mathrm{w}} \cong ((F_{\mathrm{w}})_{\mathcal{Z}})_{\mathrm{w}}$. On the other hand, since $(\bar{\zeta}_{\mathcal{Z}})^*$ is decreasing, we have $\Delta_*(\bar{\zeta}_{\mathcal{Z}})^*\Delta^* (F) \sqsubseteq F$. Clearly any weakly hereditary $\mathcal{Z}$-modal closure operator G with $G \sqsubseteq F$ satisfies $G \cong \Delta_*(\bar{\zeta}_{\mathcal{Z}})^*\Delta^* (G) \sqsubseteq \Delta_*(\bar{\zeta}_{\mathcal{Z}})^*\Delta^* (F)$, and so $F_{(\mathrm{w}\mathcal{Z})} \cong \Delta_*(\bar{\zeta}_{\mathcal{Z}})^*\Delta^* (F)$.

Finally, if F is weakly hereditary the fact that $(\bar{\zeta}_{\mathcal{Z}})_*$ is increasing implies $F \sqsubseteq \Delta_*(\bar{\zeta}_{\mathcal{Z}})_*\Delta^* (F)$. Moreover, any weakly hereditary closure operator H with $F \sqsubseteq H$ satisfies $\Delta_*(\bar{\zeta}_{\mathcal{Z}})_*\Delta^* (F) \sqsubseteq \Delta_*(\bar{\zeta}_{\mathcal{Z}})_*\Delta^* (H) \cong H$. □

3.06. COROLLARY. If $\mathcal{X}$ has finite products and $\boldsymbol{E}$ is stable under pullbacks, then for every closure operator F we have

$$F^{\mathcal{C}} = F^{(\mathcal{P}\cup\mathcal{S})} = \left(F^{\mathcal{P}}\right)^{\mathcal{S}} = \left(\left(F^{\mathcal{S}}\right)^{\mathcal{P}}\right)^{\mathcal{S}}$$

$$F_{\mathcal{C}} = F_{(\mathcal{P}\cup\mathcal{S})} = \left(F_{\mathcal{P}}\right)_{\mathcal{S}} = \left(F_{\mathcal{S}}\right)_{\mathcal{P}}$$

The supremum of $F^{\mathcal{P}}$ and $F^{\mathcal{S}}$ is bounded above by $\left(F^{\mathcal{S}}\right)^{\mathcal{P}}$ and hence in general differs from $F^{\mathcal{C}}$. □

3.07. EXAMPLE. Let $\mathcal{M}$ be the class of embeddings in $\mathcal{X} = \boldsymbol{Top}$. The usual topological closure operator T is well-known to be hereditary, and is easily seen to be $\mathcal{P}$-modal.

However, T fails to be $\mathcal{S}$-modal and hence cannot be modal: the set of rational numbers is dense in the set of reals (with the usual Euclidean topology), but its intersection with the set of irrational numbers is empty and hence is not dense in the latter. By Theorem 3.05 $T^{\mathcal{S}}$ is determined by the smallest superclass of all dense embeddings that is stable under pullbacks along monomorphisms. Since every embedding is the pullback of a dense embedding into an indiscrete space along a $\boldsymbol{Set}$-identity, $T^{\mathcal{S}}$ is the indiscrete closure operator. □

3.08. EXAMPLE. Again let $\mathcal{M}$ be the class of embeddings in $\mathcal{X} = \boldsymbol{Top}$. Hausdorff spaces now induce a regular closure operator F as follows: the F-closure of $m \in \mathcal{M}$ is the intersection of all equalizers of those parallel maps $\langle r, s\rangle$ from the codomain of m to a Hausdorff space that satisfy $r \circ m = s \circ m$. It is well-known that on Hausdorff spaces F and the usual closure operator T agree. Still, both operators are actually quite far apart. Clearly, $T \sqsubseteq F$, and it has been shown in [5] that F is not weakly hereditary.

Consider a 3-point space $X = \{a, b, c\}$ with nontrivial open sets $\{a, b\}$, $\{b\}$ and $\{b, c\}$. Every nonempty subset of X is F-dense, so F and F_{w} agree on the subspace lattice of X. But $\{a\}$ is F-closed in $\{a, c\}$, and therefore is also F_{w}-closed. Hence the weakly hereditary core F_{w} of F is not hereditary.

If S is the Sierpinski space on the set $\{0, 1\}$ with $\{0\}$ open, the embedding t of the subspace $\{1\}$ into S is F-dense but not T-dense, i.e., not dense in the usual sense. Since every product $id_X \times t$ is F-dense as well, we have $t \in (\zeta_{\mathcal{P}})^* \Delta^* (F)$ (cf. Theorem 3.05 and Corollary 2.03(2)). Therefore t is $F_{(\mathrm{w}\mathcal{P})}$-dense, and the weakly hereditary $\mathcal{P}$-modal core of F still differs from T. Since t also belongs to $(\zeta_{\mathcal{H}})^* \Delta^* (F)$, the hereditary core $F_{\mathcal{H}}$ differs from T as well. The same is true for the hereditary $\mathcal{P}$-modal core $F_{(\mathcal{H} \cup \mathcal{P})}$. □

4. The relation with idempotent closure operators

We have seen in Theorem 3.05 that the operations of forming the $\mathcal{Z}$-modal core and of forming the $\mathcal{Z}$-modal hull may be restricted to the class $\boldsymbol{wCL}(\mathcal{M})$ of weakly hereditary closure operators. Next we want to address the question of how these operations interact with idempotent closure operators. As indicated earlier, the collections of hereditary closure operators and of modal closure operators behave somewhat differently from each other in this respect.

4.00. THEOREM. If $\boldsymbol{E}$ is stable under the formation of $V_\exists(\mathcal{Z})$-pullbacks, then the following are equivalent:

(a) The idempotent hull of each $\mathcal{Z}$-modal closure operator is $\mathcal{Z}$-modal.

(b) The $\mathcal{Z}$-modal core of each idempotent closure operator is idempotent.

If in addition $\mathcal{Z}$ consists of cartesian $\mathcal{M}$-morphisms and satisfies the following cancellation property relative to the class $\mathcal{C}$ of all cartesian $\mathcal{M}$-morphisms

$$(p \circ n \xrightarrow{\langle a,c\rangle} r \circ q) \in \mathcal{Z} \text{ and } (p \xrightarrow{\langle b,c\rangle} r) \in \mathcal{C} \text{ implies } \langle b,c\rangle \in \mathcal{Z} \tag{4-00}$$

then both (a) and (b) are true. In this case for each closure operator F the idempotent $\mathcal{Z}$-modal hull $F^{(\mathrm{i}\mathcal{Z})}$ is given by $F^{\mathcal{Z}\mathrm{i}} = (F^{\mathcal{Z}})^{\mathrm{i}}$.

Proof. We first establish the equivalence of conditions (a) and (b).

(a)$\Rightarrow$(b). If F is idempotent, we have $F_{\mathcal{Z}} \sqsubseteq F_{\mathcal{Z}}{}^{\mathrm{i}} \sqsubseteq F^{\mathrm{i}} \cong F$. Since $F_{\mathcal{Z}}{}^{\mathrm{i}}$ is $\mathcal{Z}$-modal by hypothesis, it follows that $F_{\mathcal{Z}}{}^{\mathrm{i}} \sqsubseteq F_{\mathcal{Z}}$ and hence that $F_{\mathcal{Z}} \cong F_{\mathcal{Z}}{}^{\mathrm{i}}$.

(b)$\Rightarrow$(a). If F is $\mathcal{Z}$-modal, we have $F \cong F_{\mathcal{Z}} \sqsubseteq F^{\mathrm{i}}{}_{\mathcal{Z}} \sqsubseteq F^{\mathrm{i}}$. Since $F^{\mathrm{i}}{}_{\mathcal{Z}}$ is idempotent by hypothesis, it follows that $F^{\mathrm{i}} \sqsubseteq F^{\mathrm{i}}{}_{\mathcal{Z}}$ and hence that $F^{\mathrm{i}} \cong F^{\mathrm{i}}{}_{\mathcal{Z}}$.

Now assume that $\mathcal{Z} \subseteq \mathcal{C}$ satisfies (4-00). To show (a), let F be $\mathcal{Z}$-modal, and consider $m \xrightarrow{\langle f,g\rangle} n$ in $\mathcal{Z}$. Recall from [11] that

$$F^{\mathrm{i}}(n) = \inf_{\ll}\{\langle q,r\rangle \in W/n \mid r \in \nabla_*(F)\}$$

The class $G = \{\langle f,g\rangle^{\leftarrow}\langle q,r\rangle \mid \langle q,r\rangle \in W/n \text{ and } r \in \nabla_*(F)\}$ is clearly contained in $\{\langle s,t\rangle \in W/m \mid t \in \nabla_*(F)\}$, and the W-inverse image $\langle x,y\rangle$ of $F^{\mathrm{i}}(n)$ along $\langle f,g\rangle$ satisfies $\langle x,y\rangle \cong \inf_{\ll} G$.

In order to show that $\langle x,y\rangle \ll F^{\mathrm{i}}(m)$, consider $\langle s,t\rangle \in W/m$ with $t \in \nabla_*(F)$. Since $g \in V_\exists(\mathcal{Z})$, by hypothesis the g-inverse image functor $g^{\leftarrow}$ preserves suprema and hence has an adjoint $g_\forall$. Form the g-direct image $v = g_\forall(t)$ of t. Since $g^{\leftarrow}(n) \cong m$ and $g_\forall$ preserves order, the adjunction $g^{\leftarrow} \dashv g_\forall$ implies $n \leq g_\forall(m) \leq g_\forall(t) = v$, in particular there exists a unique $v \in \mathcal{M}$ with $v \circ u = n$.

Let $\langle \bar{s},\bar{t}\rangle$ be the W-inverse image of $\langle u,v\rangle$ along $\langle f,g\rangle$, and let d be the corresponding pullback of g along v. By condition (4-00) the $\mathcal{M}$-morphism $\langle d,g\rangle$ belongs to $\mathcal{Z}$. The adjunction $g^{\leftarrow} \dashv g_\forall$ implies that $\bar{t}$ factors through t; we denote the corresponding $\mathcal{X}$-morphism by h. Now apply F to $\bar{t}$ and to v. Since F is $\mathcal{Z}$-modal, $\bar{t}^F$ is a pullback of v^F along g. Moreover, $F(\bar{t}) \ll \langle h,t\rangle$ in $W/\bar{t}$ since t is F-closed, which implies $\bar{t}^F \leq t$. But $v = g_\forall(t)$ is the largest $\mathcal{M}$-subobject of n whose pullback along g is below t, hence v_F is iso, and v is F-closed. This establishes $\langle x,y\rangle \ll F^{\mathrm{i}}(m)$, and hence f^{i} is $\mathcal{Z}$-modal.

Finally, (a) implies that $F^{\mathcal{Z}\mathrm{i}}$ is idempotent and $\mathcal{Z}$-modal for every closure operator F. If G has these properties and satisfies $F \sqsubseteq G$, then $F^{\mathcal{Z}} \sqsubseteq G^{\mathcal{Z}} \cong G$ and hence $F^{\mathcal{Z}\mathrm{i}} \sqsubseteq G^{\mathrm{i}} \cong G$, thus $F^{\mathcal{Z}\mathrm{i}}$ is indeed the idempotent and $\mathcal{Z}$-modal hull $F^{(\mathrm{i}\mathcal{Z})}$ of F. □

4.01. COROLLARY. If $\mathcal{Z} \in \{\mathcal{C}, \mathcal{P}, \mathcal{S}\}$, then $F^{(\mathrm{i}\mathcal{Z})} = F^{\mathcal{Z}\mathrm{i}}$. □

However, even if $\mathcal{Z} \subseteq \mathcal{C}$ satisfies condition (4-00), $F^{\mathrm{i}\mathcal{Z}} = (F^{\mathrm{i}})^{\mathcal{Z}}$ may differ from $F^{(\mathrm{i}\mathcal{Z})}$.

4.02. EXAMPLE. Consider the following partially ordered set $\mathcal{X}$ with $\mathcal{M} = \boldsymbol{Mor}(\mathcal{X}) = \boldsymbol{Mono}(\mathcal{X})$

(4-01)

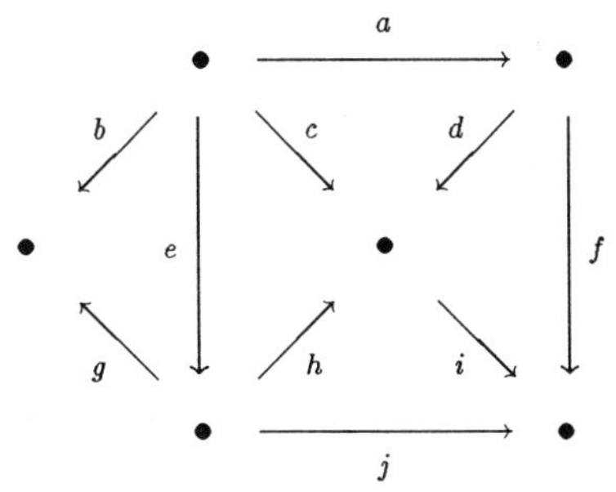

and define k as $f \circ a = i \circ c = j \circ e$. Since each principal ideal with the induced order is a complete lattice, $\mathcal{X}$ is an $\langle \boldsymbol{E}, \mathcal{M}\rangle$-category, where $\boldsymbol{E}$ consists of all supremum-sinks. We obtain a closure operator F by declaring a, b, g, h, i and j to be dense, and d, e and f to be closed. Furthermore we set $F(c) = \langle a, d\rangle$ and $F(k) = \langle a, f\rangle$. Now F is weakly hereditary and idempotent, but not hereditary, since b has a non-dense first factor. To form the $\mathcal{Z}$-modal hull $F^{\mathcal{Z}}$, according to Theorem 3.05 we start with $A = \lambda_* \dot{\omega}^*(F)$. Since $\langle e, g\rangle$ is the only pair in $\dot{\omega}^*(F)$ that does not have an F-dense first component, $A = \Delta^*(F) \cup \{e\}$, which is already stable under pullbacks. Thus $G = \Delta_*(A)$ is the $\mathcal{Z}$-modal hull of F for any admissible $\mathcal{Z}$. Now e and j are G-dense, but k is not, since

$$G(k) = \sup_{\ll}\{\langle a, f\rangle, \langle e, j\rangle, \langle id, k\rangle\} = \langle c, i\rangle .$$

Thus $\Delta^*(G)$ is not closed under composition, and hence G is not idempotent. However, G^{i} is the idempotent $\mathcal{Z}$-modal hull of F, even for $\mathcal{Z} = \mathcal{H}$. □

That idempotent hereditary hulls can be more difficult to form than, say, idempotent modal hulls, is illustrated by the following example. Of course, $\mathcal{H}$ does not satisfy the cancellation condition (4-00).

4.03. EXAMPLE. Consider the following poset $\mathcal{X} = \langle X, \leq \rangle$ with $m = p \circ n = r \circ q$

$$\text{(4-02)} \qquad \begin{array}{ccc} A & \xrightarrow{q} & B \\ {\scriptstyle n}\downarrow & & \downarrow{\scriptstyle r} \\ C & \xrightarrow[p]{} & D \end{array}$$

Define a hereditary closure operator F on $\mathcal{M} = \boldsymbol{Mor}(\mathcal{X})$ by declaring n to be closed, p, q, r to be dense, and $\langle q, r\rangle = F(m)$. Since $r = m^F$ and $id_D = m^{FF}$ are not isomorphic, F is not idempotent. Now n is F^{i}-closed and m is F^{i}-dense, hence F^{i} is not hereditary. □

5. HULLS VIA ITERATION

The stability of $\boldsymbol{E}$ under $V_\exists(\mathcal{Z})$-pullbacks enabled us to construct $F^{\mathcal{Z}}$ directly, and also guaranteed the existence of $F_{\mathcal{Z}}$ (cf. Theorem 3.05). If $\boldsymbol{E}$ is not stable under $V_\exists(\mathcal{Z})$-pullbacks, one can still iterate $\dot\omega_* \circ (\zeta_{\mathcal{Z}})_* \circ \dot\omega^*$ in an attempt to approximate $\mathcal{Z}$-modal hulls "from below". The following proposition shows that if the iteration becomes stationary, then we indeed have found the $\mathcal{Z}$-modal hull:

5.00. PROPOSITION. F is $\mathcal{Z}$-modal iff $F \cong \dot\omega_*(\zeta_{\mathcal{Z}})_*\dot\omega^*(F)$.

Proof. If F is $\mathcal{Z}$-modal, then by Proposition 2.02(0) $\dot\omega^*(F)$ is $\mathcal{Z}$-stable, so $\dot\omega_*(\zeta_{\mathcal{Z}})_*\dot\omega^*(F) = \dot\omega_*\dot\omega^*(F) \cong F$.

Suppose $F \cong \dot\omega_*(\zeta_{\mathcal{Z}})_*\dot\omega^*(F)$. Since $\dot\omega^*(F)$ is a fixed point of the Galois connection $\dot\omega$, it is the largest class $C \subseteq \mathcal{M} \diamond \mathcal{M}$ with the property that $\dot\omega_*(C) \cong F$. Therefore $(\zeta_{\mathcal{Z}})_*\dot\omega^*(F) \subseteq \dot\omega^*(F)$. But $(\zeta_{\mathcal{Z}})_*$ is increasing, so both classes are equal. This implies that $\dot\omega^*(F)$ is $\mathcal{Z}$-stable, and thus by Proposition 2.02 F is $\mathcal{Z}$-modal. □

Recall from Section 1 that $\boldsymbol{CL}(\mathcal{M})$ has suprema and infima of arbitrary subcollections.

5.01. THEOREM. For a closure operator F set

$$\begin{aligned} F_0 &= \dot\omega_*(\zeta_{\mathcal{Z}})_*\dot\omega^*(F) \\ F_{a+1} &= \dot\omega_*(\zeta_{\mathcal{Z}})_*\dot\omega^*(F_a) \quad \text{if } a \text{ is an ordinal} \\ F_b &= \sup\{\, F_a \mid a \in b \,\} \quad \text{if } b \text{ is a limit ordinal} \end{aligned}$$

and let H be the supremum of $\{\, F_a \mid a \in \boldsymbol{Ord} \}$. If $\mathcal{X}$ is $\mathcal{M}$-well-powered, H is the $\mathcal{Z}$-modal hull $F^{\mathcal{Z}}$ of F.

Proof. Since $F \sqsubseteq F^{\mathcal{Z}}$ and $\dot{\omega}_* \circ (\zeta_{\mathcal{Z}})_* \circ \dot{\omega}^*$ preserves order, by Proposition 5.00 we have that $F \sqsubseteq F_a \sqsubseteq F^{\mathcal{Z}}$ for every ordinal a . Hence $H \sqsubseteq F^{\mathcal{Z}}$.

If $\mathcal{X}$ is $\mathcal{M}$-well-powered, for every $\mathcal{X}$-object X there exists an ordinal $c(X)$ such that on the $\mathcal{M}$-subobject lattice of X the closure operators $F_{c(X)}$ and $F_{c(X)+1}$ are isomorphic. By Proposition 5.00 the supremum G of the collection $\{ F_{c(X)} \mid X \in \boldsymbol{Ob}(\mathcal{X}) \}$ is $\mathcal{Z}$-modal. Hence $H \sqsubseteq F^{\mathcal{Z}} \sqsubseteq G \sqsubseteq H$, which shows that $H \cong F^{\mathcal{Z}}$. □

The same strategy as above can also be used to obtain idempotent and $\mathcal{Z}$-modal hulls.

5.02. PROPOSITION. F is idempotent and $\mathcal{Z}$-modal iff $F \cong (\dot{\omega}_*(\zeta_{\mathcal{Z}})_*\dot{\omega}^*(F))^{\mathrm{i}}$.

Proof. If F is $\mathcal{Z}$-modal, Proposition 5.00 implies $F \cong \Delta_*(\zeta_{\mathcal{Z}})_*\lambda_*\dot{\omega}^*(F)$, and since F is idempotent, the rest follows.

Conversely, if $F \cong (\dot{\omega}_*(\zeta_{\mathcal{Z}})_*\dot{\omega}^*(F))^{\mathrm{i}}$, then $\mathcal{Z}$-modality follows from $F \sqsubseteq \dot{\omega}_*(\zeta_{\mathcal{Z}})_*\dot{\omega}^*(F) \sqsubseteq F$, and idempotency is clear. □

The proof of the next result is similar to the proof of Theorem 5.01.

5.03. THEOREM. For a closure operator F set

$$\begin{aligned} G_0 &= (\dot{\omega}_*(\zeta_{\mathcal{Z}})_*\dot{\omega}^*(F))^{\mathrm{i}} \\ G_{a+1} &= (\dot{\omega}_*(\zeta_{\mathcal{Z}})_*\dot{\omega}^*(G_a))^{\mathrm{i}} \quad \text{if } a \text{ is an ordinal} \\ G_b &= \big(\sup\{ G_a \mid a \in b \}\big)^{\mathrm{i}} \quad \text{if } b \text{ is a limit ordinal} \end{aligned}$$

and let K be the supremum of $\{ G_a \mid a \in \boldsymbol{Ord} \}$. If $\mathcal{X}$ is $\mathcal{M}$-well-powered, then K is the idempotent $\mathcal{Z}$-modal hull $F^{(\mathrm{i}\mathcal{Z})}$ of F . □

5.04. REMARK. Similar arguments show that for a weakly hereditary F iterating $\Delta_* \circ (\zeta_{\mathcal{Z}})_* \circ \Delta^*$ and $(\Delta_* \circ (\zeta_{\mathcal{Z}})_* \circ \Delta^*)^{\mathrm{i}}$ produces the $\mathcal{Z}$-modal hull of F in $\boldsymbol{wCL}(\mathcal{M})$ and the idempotent $\mathcal{Z}$-modal hull of F in $\boldsymbol{wCL}(\mathcal{M})$, provided that $\mathcal{X}$ is $\mathcal{M}$-well-powered.

ACKNOWLEDGEMENT. The authors would like to thank the referee for valuable suggestions that led to improvements of this paper.

REFERENCES

[0] ADÁMEK, J., HERRLICH, H., AND STRECKER, G. E. *Abstract and Concrete Categories.* John Wiley, New York, 1990.

[1] BÖRGER, R. *Kategorielle Beschreibungen von Zusammenhangsbegriffen.* PhD thesis, Fernuniversität Hagen, 1981.

[2] CASSIDI, C., HÉBERT, M., AND KELLY, G. M. Reflective subcategories, localizations and factorization systems. *J. Austral. Math. Soc. Ser. A 38* (1985), 287–329.

[3] CASTELLINI, G. Closure operators, monomorphisms and epimorphisms in categories of groups. *Cahiers Topologie Géom. Différentielle Catégoriques 27*, 2 (1986), 151–167.

[4] CASTELLINI, G., KOSLOWSKI, J., AND STRECKER, G. E. Closure operators and polarities. Tech. rep., University of Puerto Rico and Kansas State University, 1991.

[5] CASTELLINI, G., KOSLOWSKI, J., AND STRECKER, G. E. A factorization of the Pumplün-Röhrl connection. *Topology Appl.* (to appear 1992).

[6] DIKRANJAN, D., AND GIULI, E. Closure operators I. *Topology Appl. 27* (1987), 129–143.

[7] DIKRANJAN, D., AND GIULI, E. Factorizations, injectivity and compactness in categories of modules. *Comm. Algebra 19* (1991), 45–83.

[8] DIKRANJAN, D., GIULI, E., AND THOLEN, W. Closure operators II. In *Categorical Topology* (Prague, 1988), J. Adámek and S. MacLane, Eds., World Scientific, Singapore, 1989, pp. 297–335.

[9] ERNÉ, M., KOSLOWSKI, J., MELTON, A., AND STRECKER, G. E. A primer on Galois connections. Tech. rep., Universität Hannover and Kansas State University, 1992.

[10] KOROSTENSKI, M., AND THOLEN, W. On left cancellable classes of morphisms. *Comm. Algebra 14* (1986), 741–766.

[11] KOSLOWSKI, J. Closure operators with prescribed properties. In *Category Theory and its Applications* (Louvain-la-Neuve, 1987), F. Borceux, Ed., no. 1248 in Lecture Notes in Mathematics, Springer-Verlag, Berlin – Heidelberg – New York, 1988, pp. 208–220.

[12] PUMPLÜN, D., AND RÖHRL, H. Separated totally convex spaces. *Manuscripta Math. 50* (1985), 145–183.

[13] THOLEN, W. Factorizations, localizations and the orthogonal subcategory problem. *Math. Nachr. 114* (1983), 63–85.

[14] THOLEN, W. Prereflections and reflections. *Comm. Algebra 14* (1986), 717–740.

Department of Mathematics, University of Puerto Rico, Mayagüez, PR 00681-5000
E-mail address: g_castellini@upr1.upr.clu.edu
Department of Mathematics, Kansas State University, Manhattan, KS 66506-2602
E-mail addresses: koslowj@math.ksu.edu and strecker@math.ksu.edu

Canadian Mathematical Society
Conference Proceeding
Volume **13**, 1992

What is a Gabriel Topology over a Graded Ring?

Hong Andy Chen Po Hsiang Chu *

Abstract

In a non-commutative ring R, there is a one to one correspondence between torsion theories of R-mod and Gabriel topologies over R. We extend this result for any non-commutative group graded ring i.e. $R = \bigoplus_{i\in\mathcal{G}} R_i$ without imposing any restrictions on the torsion theory. This generalization is depended on:1) any group graded ring can be viewed as a small preadditive category and, 2) a characterization of torsion theories over a preadditive category in [Chu,1984].

1 Introduction

Recently there has been some renewed interests in torsion theory in graded rings and modules (see [Nastasescu,1990], [Albu,1991], etc.). Torsion theory in a module category over a ring, and more generally in a Grothendieck category has been studied extensively (see [Stenstrom,1975]). Moreover, every torsion theory in the module category over any ring R can be expressed in terms of a unique Gabriel topology of left ideals i.e. elements of the ring itself ring. Using this correspondence various category theoretical characterizations e.g. torsion free and torsion objects, torsion free divisible objects, etc., can be computed in terms of elements of the ring. However, when R is a graded ring, similar results exist only in special cases e.g. strongly graded, rigid torsion etc. (see [Oystaeyen,1978], [Nastasescu,1990]). A theory of graded rings and modules for the general case is still lacking, even though the category of graded modules for any graded ring and the module category over any ring have essentially the same important properties e.g. complete, cocomplete, Grothendieck, enough injectives etc. (see [Nastasescu, 1982]).

*1991 Mathematics Subject Classification. Primary 13D03, 18K04
This paper is in final form and no version of it will be submitted for publication elsewhere.

Almost twenty years ago, Mitchell [Mitchell, 1972] noticed that many aspects of the graded rings become transparent if they are viewed as small preadditive categories. The aim of this paper is to provide a characterization of the torsion theory in terms of its Gabriel topology for any non-commutative graded ring. In section 2, we review the category theoretical background of the torsion theory. In section 3 we apply Mitchell's observation and Chu's work on small preadditive categories (see [Chu,1984]) to obtain our result.

2 Category Preliminaries

In this section we review various characterizations of torsion theories in an abelian category C.

Definition 2.1 *A torsion theory is a pair (T,F) of classes of objects satisfying:*

1. $\mathrm{Hom}(t, f) = 0$ *for all* t *in* T, f *in* F.
2. *If* c *in* C *such that* $\mathrm{Hom}(c, f) = 0$ *for all* f *in* F, *then* c *is in* T.
3. *If* c *in* C *such that* $\mathrm{Hom}(t, c) = 0$ *for all* t *in* T, *then* c *is in* F.
4. T *is closed under subobjects.*

T is the class of torsion objects and F is the class of torsion free objects.

Definition 2.2 *A class of objects* T *is called a Serre class if* T *is closed under subobjects, quotients, coproducts, and extensions.*

Theorem 2.3 *There is a one to one correspondence between the following:*

1. *Torsion theories in C.*
2. *Serre classes in C.*

Proof. If (T, F) is a torsion theory, then T is a Serre class. On the other hand if T is a Serre class, let F be the class of objects $x \in C$ such that $\mathrm{Hom}(t, x) = 0$.
For more details see [Stenstrom,1975].

□

Suppose R is a ring with unity and C = *R-mod*, the category of left R modules. We can characterize torsion theories by family of left ideals of R.

Definition 2.4 *A family of left ideals* $G = \{J_i | i \in I\}$ *is called a Gabriel topology if it satisfies the following:*

1. $R \in G$.

2. *If* $J \in G$ *and* $r \in R$, $[J : r] \in G$.

3. *If* K *is an ideal and there exists* $J \in G$ *such that* $[K : r] \in G$ *for all* $r \in J$, *then* $K \in G$ *as well.*

Theorem 2.5 *There is a one to one correspondence between:*

1. *Torsion theories on R-mod.*

2. *Serre classes in R-mod.*

3. *Gabriel topologies of R.*

Proof. If T is a Serre class of *R-mod*, let G be the set of left ideals I such that $R/I \in T$.

Before showing the converse, we observe if $M \in$ *R-mod* and $m \in M$, m induces a homomorphism $\tilde{m} : R \longrightarrow M$ by $\tilde{m}(a) = am$. Now suppose G is a Gabriel topology, let T be the class of left R modules M satisfies: for every $m \in M$, the kernel of the map $\tilde{m}$ is a member of G.

For more details see [Stenstrom,1975].

□

3 Main Result

In this this section, we shall first to extend the result of the previous section to an additive functor category $\mathrm{FUN}(C, Ab)$, where C is a small preadditive category, and apply this extension to obtain our general results.

Definition 3.1 *A category* C *is a small preadditive category if* C *satisfies the following:*

1. $|C|$ *is a set.*

2. $\mathrm{Hom}(a, b)$ *is an abelian group for all* $a, b \in |C|$.

3. *The composition* $\mathrm{Hom}(a, b) \times \mathrm{Hom}(b, c) \longrightarrow \mathrm{Hom}(a, c)$ *is bi-homomorphism i.e. homomorphism at each component, for all* $a, b,$ *and* $c \in |C|$.

Proposition 3.2 *If* C *is a small preadditive category, then* $\mathrm{FUN}(C, Ab)$ *is a complete and cocomplete abelian category.*

Proof. See [P. Freyd, 1964].

□

Examples

1. Ring with unity and modules.

If R is a ring with unity, R can be identified as a preadditive category C with $|C| = \{*\}$ and $\mathrm{Hom}(*,*) = R$. There is also a one to one correspondence between left R modules and additive functors from C to Ab: if $M \in R\text{-}mod$, M induces an addtitive functor $\tilde{M} : C \longrightarrow Ab$ by $\tilde{M}(*) = M$ and if $r \in \mathrm{Hom}(*,*)$ and $m \in M$, $\tilde{M}(r)(m) = rm$.

If F is an addtitive functor from C to Ab, let $M = F(*)$, and if $r \in R$ and $m \in M$, $rm = F(r)m$. a natural isomorphism between the category left R modules and $\mathrm{FUN}(C, Ab)$.

This correspondence is a natural isomorphism.

2. Graded ring and graded modules.

Let $\mathcal{G}$ denote any group and $R = \bigoplus_{i\in\mathcal{G}} R_i$ is a graded ring if R satisfies the following:

i. R is a ring with unity 1 and $1 \in R_1$.
ii. For each $i \in \mathcal{G}$, R_i is an abelian group.
iii. $R_iR_j \subseteq R_{ij}$, for all $i, j \in \mathcal{G}$.

$M = \bigoplus_{i\in\mathcal{G}} M_i$ is a graded left R module if M satisfies the following:

i. M is an abelian group.
ii. $1m = m$, for all $m \in M$.
iii. $R_iM_j \subseteq M_{ij}$ for all $i, j \in \mathcal{G}$.

Suppose $R = \bigoplus_{i\in\mathcal{G}} R_i$ is a graded ring, R can be identified as a small preadditive category $\mathcal{C}$ with $|\mathcal{C}| = \mathcal{G}$ such that $\mathrm{Hom}(i,j) = R_{ji^{-1}}$. Moreover, there is also a natural isomorphism between graded left R modules and additive functors from $\mathcal{G}$ to Ab :

If $M = \bigoplus_{i\in\mathcal{G}} M_i$ is a graded left R module, M induces $\tilde{M} : \mathcal{C} \longrightarrow Ab$ given by: $M(i) = M_i$, and if $m \in M_i$ and $r : i \to j$ i.e. $r \in \mathrm{Hom}(i,j) = R_{ji^{-1}}$, then $\tilde{M}(r) = M(r)(m) \in M_j = \tilde{M}(j)$.

If $F : \mathcal{C} \to Ab$ is an additive functor, let $M = \bigoplus_{i\in\mathcal{G}} M_i$ and $M_i = F(i)$ for every $i \in \mathcal{G}$. Suppose $r \in R_i = \mathrm{Hom}(j, ij)$ and $m \in F(j)$, then define $rm = F(r)(m) \in F(ij) = M_{ij}$.

$\square$

Definition 3.3 *Let C be a small preadditive category.*

1. *Let M' be a subfunctor of $M \in \mathrm{FUN}(C, Ab)$ and $m \in M(c)$ for some $c \in |C|$, then the subfunctor $[M' : m]$ of $\mathrm{Hom}(c, -)$ is defined by:*

$$[M' : m](c') = \{x \in \mathrm{Hom}(c, c') | M(x)(m) \in M'(c')\}.$$

2. *An addtitive functor $I : C \longrightarrow Ab$ is called an ideal if I is a subfunctor of $\mathrm{Hom}(c, -)$ for some object $c \in |C|$.*

3. *If $M \in \mathrm{FUN}(C, Ab)$, $c \in |C|$, and $m \in M(c)$, m induces natural transformation $\tilde{m} : \mathrm{Hom}(c, -) \longrightarrow M$ defined by: if $c' \in |C|$ and $f \in \mathrm{Hom}(c, c')$, $\tilde{m}(c')(f) = M(f)(m)$.*

Hence, if $R = \bigoplus_{i \in \mathcal{G}} R_i$ is a graded ring, a left ideal of R corresponds to a graded submodule of $R[i] = \bigoplus_{i \in \mathcal{G}} R_{ji^{-1}}$ for some $i \in \mathcal{G}$.

Now we can extend the notion of Gabriel topology to the functor category $\mathrm{FUN}(C, Ab)$.

Definition 3.4 *Let C be a small preadditive category. A Gabriel topology on C is a set $\{G(c) | c \in |C|\}$ such that*

1. *$G(c)$ is a set of subfunctors of $\mathrm{Hom}(c, -)$, for all $c \in |C|$.*
2. *$\mathrm{Hom}(c, -) \in G(c)$, for all $c \in |C|$.*
3. *If $I \in G(c)$ and $f : c' \longrightarrow c$, then $[I : f] \in G(c')$.*
4. *If I is a subfunctor of $\mathrm{Hom}(c, -)$ and there exists $J \in G(c')$ satisfying the following:*

   For all $c'' \in |C|$ and $f \in J(c'')$, a subset of $\mathrm{Hom}(c', c'')$, then $[I : f] \in G(c')$.

   Then $I \in G(c)$.

Theorem 3.5 *Let C be a small preadditive category, then there is a one to one correspondence between:*

1. *Torsion theories on $\mathrm{FUN}(C, Ab)$.*
2. *Serre classes on $\mathrm{FUN}(C, Ab)$.*
3. *Gabriel topologies on C.*

Proof. If T is a Serre class in FUN(C, Ab), then let $\{G(c)|c \in |C|\}$ be the set satisfies the following properties: for each $c \in |C|$, $G(c)$ is the set of subfunctors I of Hom$(c,-)$ such that Hom$(c,-)/I \in T$. $\{G(c)|c \in |C|\}$ is a Gabriel topology. On the other hand, if $\{G(c)|c \in |C|\}$ is a Gabriel topology, let T be the class of objects $M \in$ FUN(C, Ab) such that for every $c \in |C|$ and $m \in M(c)$, the kernel of the natural transformation $\tilde{m}$: Hom$(c,-) \longrightarrow M \in G(c)$. T is a Serre class of FUN(C, Ab).

For more details see [P. H. Chu,1984].

□

Definition 3.6 *$R = \bigoplus_{i \in \mathcal{G}} R_i$ be a graded ring, a set $\{G(i)|i \in \mathcal{G}\}$ is a Gabriel topology on R if it satisfies:*

1. *$G(i)$ is a set of left ideals of $R[i] = \bigoplus_{i \in \mathcal{G}} R_{ji^{-1}}$, for all $i \in \mathcal{G}$.*
2. *$R[i] \in G(i)$, for all $i \in \mathcal{G}$.*
3. *If $r \in R_{ij^{-1}}$ and $I \in G(i)$, $[I : r] \in G(j)$.*
4. *Suppose I is a left ideal of $R[i]$ and there exists $J \in G(j)$ i.e. J is a left ideal of $R[j]$, such that For all $k \in \mathcal{G}$ and $r \in J(k)$, a subset of $R_{kj^{-1}}$, $[I : r] \in G(j)$. Then $I \in G(i)$.*

Corollary 3.7 *Let $R = \bigoplus_{i \in \mathcal{G}} R_i$ be a graded ring and R-grad be the category of graded left R modules, there is a one to one correspondence between:*

1. *Torsion theories on R-grad.*
2. *Serre classes on R-grad.*
3. *Gabriel topologies on R.*

References

[1] T. Albu, "Pure ideals, quotient categories, and infinite group-graded rings", *Comm. in Algebra*, 18, (839-862), 1990.

[2] P. Freyd, *Abelian categories*, Harper and Row, New York, 1964.

[3] B. Mitchell, *Rings with several objects*, Adv. in Math. 8, (1-161), 1972.

[4] C. Nastasescu and N. Rodino, "Localization on graded modules, relative Maschke's theorem and applications", *Comm. in Algebra*, 18, (811-832), 1990.

[5] C. Nastasescu and F. Van Oystaeyen, *Graded ring theory, Mathematics and its applications*, Reidel Publication Co., 1982.

[6] B. Stenstrom, *Rings of quotients*, Springer-Verlag, Berlin, 1975.

[7] F. Van Oystaeyen, "On graded rings and modules of quotients", *Comm. in Algebra*, 6, (1923-1959), 1978.

Hong Andy Chen
Northeastern Illinois University
5500 N. St. Louis Avenue, Chicago, Illinois, USA 60625
uhchen@uxa.ecn.bgu.edu

Po Hsiang Chu
DePaul University
243 S. Wabash, Chicago, Illinois, USA 60604
chu@ace.bsd.uchicago.edu

Canadian Mathematical Society
Conference Proceeding
Volume **13**, 1992

Strong Categorical Datatypes I

J. ROBIN B. COCKETT AND DWIGHT SPENCER

April, 1991

ABSTRACT. An endofunctor of a cartesian closed category is often called strong if it is enriched over the exponential. Equivalently this strength can be provided as a natural transformation $\theta_{A,X} : F(A) \times X \longrightarrow F(A \times X)$ satisfying some elementary coherence conditions. This latter formulation does not require exponentials, relies only on the presence of an **X**–action over an **X**–strong category, and thereby provides a first–order viewpoint of strength.

The 2-category of **X**–strong categories is not finitely complete. It particularly lacks many standard constructions including the Eilenberg–Moore construction. Thankfully, the suggestion — attributed to Plotkin by Moggi — that strength can also be equivalently framed in terms of fibrations using projections to **X**–objects as display maps can be fully realized. The equivalence can be portrayed as an embedding of **X**–strong categories into the 2–category of **X**–indexed categories or split fibrations over **X**. This embedding can be used to give a coherent declarative implementation of parametrized datatypes-with-strength and their constructors/destructors in the style of Hagino.

The initiality and finality conditions associated with the strong datatype construction in the split fibration setting generate a strongly normalizing weak-head categorical combinator reduction system. The reduction rules permit both "eager" usage of initial datatypes and "lazy" usage of final datatypes to achieve reasonable computing strength. An **X**-action over a strong datatype can be computed in terms of programmed state transformations over its component datatypes. These reduction rules form the abstract engine for computation within the **Charity** categorical programming system.

1991 *Mathematics Subject Classification.* Primary 18A40, 18C20, 18D25, 68P05; Secondary 18A35, 18B15, 18D30, 68Q42.

This paper is in final form and no version of it will be submitted for publication elsewhere.

1. Introduction

In many mathematical instances ordinary functors and natural transformations do not provide sufficient structure to successfully capture a particular phenomenon. A solution usually entails the enrichment of type-forming functors and natural transformations. When the phenomena are programs, we recognize a critical structure-capture requirement: the calculation of a functor should be doable at the same level as the calculation of maps. This has been reinforced by Moggi **[12, 13, 14]** by his demonstration of the direct relation of strong monads to the semantics of computation.

For cartesian closed categories, this causes us to focus on the obvious enrichment over the internal exponentiation. In this setting an endofunctor $F : \mathbf{X} \longrightarrow \mathbf{X}$ is enriched if it is represented as a natural transformation

$$f_{A,B} : A \Rightarrow B \longrightarrow F(A) \Rightarrow F(B)$$

which behaves correctly with respect to the composition as internally given. This natural transformation provides the means for building a "strength" for the ordinary functor via the following equivalence:

$$\frac{X \xrightarrow{\;curry(i_{A\times X})\;} A \Rightarrow (A \times X) \xrightarrow{\;f_{A,A\times X}\;} F(A) \Rightarrow F(A \times X)}{F(A) \times X \xrightarrow{\;\theta_{A,X}\;} F(A \times X)}.$$

Conversely, one may formally define the notion of a functor with strength and show that the notion also implies enrichment by

$$\frac{F(A) \times (A \Rightarrow B) \xrightarrow{\;\theta_{A,A\Rightarrow B}\;} F(A \times (A \Rightarrow B)) \xrightarrow{\;F(ev)\;} F(B)}{A \Rightarrow B \xrightarrow{\;f_{A,B}\;} F(A) \Rightarrow F(B)}.$$

These equivalences make redundant the requirement that exponentiation actually be present. We therefore simply consider functors with a given strength. This paper fully develops the consequences of following such a computational direction. Section 2 provides a definition of $\mathbf{X}$-strong categories which serves, we believe, as the appropriate categorical foundation of strength. The definition departs from previous usage, i.e. as enrichment, by adopting the generality of a non-closed setting while concentrating on the product as the source and effect of a tensor action.

Datatypes are built with limits and colimits, yet the resulting 2-category of $\mathbf{X}$–strong categories disappoints us by lacking arbitrary finite limits. In particular, the category of F-algebras of an endofunctor cannot be constructed. Thus there is no immediate guidance for obtaining the "correct" notion of a free monad or, more generally, a datatype. For example, the "correct" notion of a free monad on an endofunctor F is that it is the monad generated by the left adjoint to the underlying functor from the category of F–algebras.

Section 3 shows how our purview of strength can be moved to a complete setting. There we define and explain an embedding, attributed in **[12, 13]** to

Gordon Plotkin, of the 2-category of **X**-strong categories into the 2-category of split fibrations over **X**. This latter category, by virtue of its completeness, has the algebraic constructions and the guidance we seek.

Section 4 explains the limit constructions of the category of algebras for a strong initial datatype and the category of coalgebras for a strong final datatype within fibrations. The limits are unraveled into the universal initiality and finality diagrams from which all our combinators and computation rules for strong datatypes are derived, as detailed in Section 5. The fibration-based datatype notion we propose is correct in terms of its context parametrization of both constructors and destructors, a feature neglected in other approaches but vital to practical programming practices.

2. X–strength

This section's purpose is to cast a foundational description of strength and then mold it into an elementary description that will be more widely recognized. The molding will be performed quite independently of cartesian closedness.

The starting point in generality is the notion of a **V**–tensored category **Y**. Here **V** is a category with an arbitrary tensor product, as described by Anders Kock [9], and **Y** is a category with a **V**–action on it given by a tensor preserving functor from **V** into *Endo*(**Y**). Our interest narrows to the special case of **V**'s tensor product being the (cartesian) product. We will hereafter designate such a category **V** as **X**. Furthermore, we demand that **Y** have its own tensor, viz. its product, and that the action, now an **X**-action, be representable by this tensor. The term **X**-*strong* will be used to describe such a category **Y**. It should be apparent that **X**-strong categories resemble K-algebras (for K a field) more closely than K-modules.

If **X** is a cartesian category (i.e. has a binary product and a terminal object) an **X**-*strong category* **Y** is a cartesian category **Y** with a bifunctor, called an **X**-*action*,

$$\_\oslash\_ \; : \; \mathbf{Y} \times \mathbf{X} \longrightarrow \mathbf{Y}$$

that is compatible with the coherent natural isomorphisms

- *ass* : $Y \oslash (X_1 \times X_2) \longrightarrow (Y \oslash X_1) \oslash X_2$,
- *rid* : $(Y \oslash 1) \longrightarrow Y$,
- *rep* : $Y \oslash X \longrightarrow Y \times (1 \oslash X)$.

The first two natural isomorphisms are the standard ones and must satisfy the coherence conditions expected for any monoidal action. The last allows the action to be represented by the product of the category **Y** on which **X** acts.

Because these categories are cartesian we can simplify the general commuting requirements to the following four equations:

- $ass\,;rid = id \oslash p_0 \;:\; Y \oslash (X \times 1) \longrightarrow Y \oslash X,$
- $rid = rep\,;p_0 \;:\; (Y \oslash X) \oslash 1 \longrightarrow Y \oslash X,$
- $ass\,;(rid \oslash id) = id \oslash p_1 \;:\; Y \oslash (1 \times X) \longrightarrow Y \oslash X,$
- $rid \oslash id = rep\,;p_1 \;:\; (1 \oslash 1) \oslash X \longrightarrow 1 \oslash X.$

Notice that they are aimed at ensuring that the map *rid* behaves correctly.

PROPOSITION 2.1. *A cartesian category* **Y** *with an* **X***-action is* **X***-strong if and only if there is a cartesian functor*

$$J : \mathbf{X} \longrightarrow \mathbf{Y}$$

and a natural isomorphism $r_{Y,X} : Y \oslash X \longrightarrow Y \times J(X)$.

PROOF. If we are given J and the natural isomorphism r then by setting

$$rid = r_{Y,1}\,;p_0$$

$$ass = r_{Y,X_1\times X_2}\,;\langle\langle p_0, p_1\,;J(p_0)\rangle, p_1\,;J(p_1)\rangle\,;r^{-1}_{Y\times J(X_1),X_2}\,;r^{-1}_{Y,X_1} \oslash id$$

$$rep = r_{Y,X}\,;id \times \langle !, id\rangle\,;id \times r^{-1}_{1,X}$$

the (simplified) coherence requirements are straightforward to verify.

Conversely, given an **X**-action we may choose $J(X) = 1 \oslash X$. Clearly J is a functor and r is definable as *rep*. It remains to show J is cartesian. Because $rid : 1 \oslash 1 \longrightarrow 1$ is an isomorphism it follows that $1 \oslash 1$ is final. To show that products are preserved we consider

$$1 \oslash A \xleftarrow{1 \oslash p_0} 1 \oslash (A \times B) \xrightarrow{1 \oslash p_1} 1 \oslash B$$

and show that it is naturally isomorphic to

$$1 \oslash A \xleftarrow{p_0} 1 \oslash A \times 1 \oslash B \xrightarrow{p_1} 1 \oslash B$$

This is accomplished with the following two diagrams:

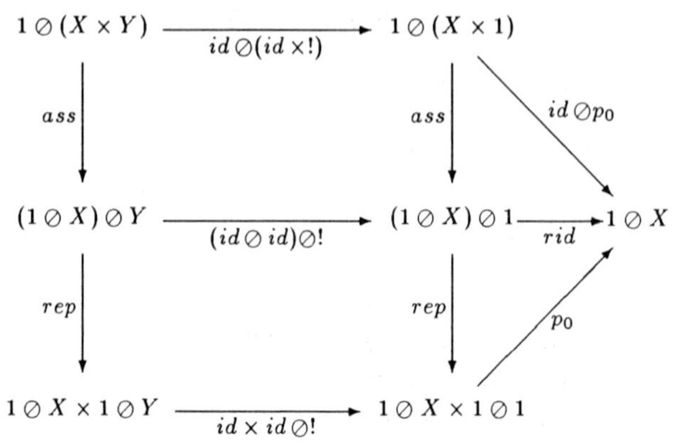

and

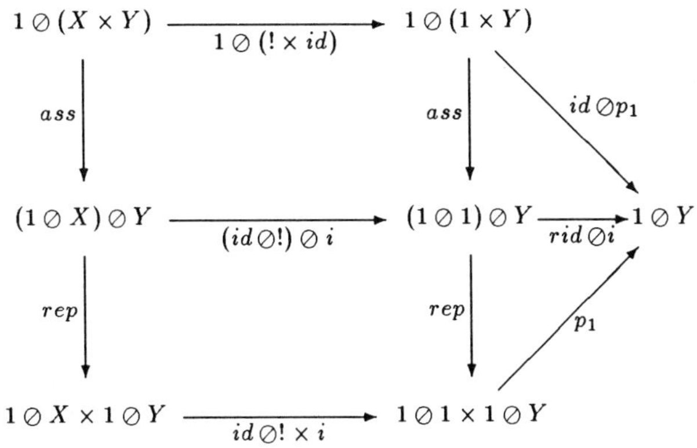

in which the four identities are used in the triangles. □

All the **X**–strong categories of interest can be regarded as having a cartesian embedding J of **X** into it. The **X**–action is, up to isomorphism, simply given by the product with the image of the objects of **X** under this embedding. This observation motivates our dropping the action notation $A \oslash X$ in favor of $A \times J(X)$ or just $A \times X$.

X-*strong functors*, or simply *strong functors*, are functors between **X**–strong categories equipped with a natural transformation

$$\theta^F_{Y,X} : F(Y) \oslash X \longrightarrow F(Y \oslash X),$$

called an **X**-*strength*, or simply a *strength*, such that

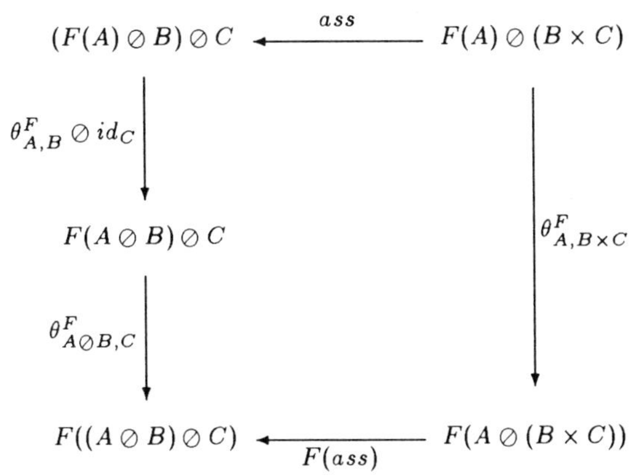

and

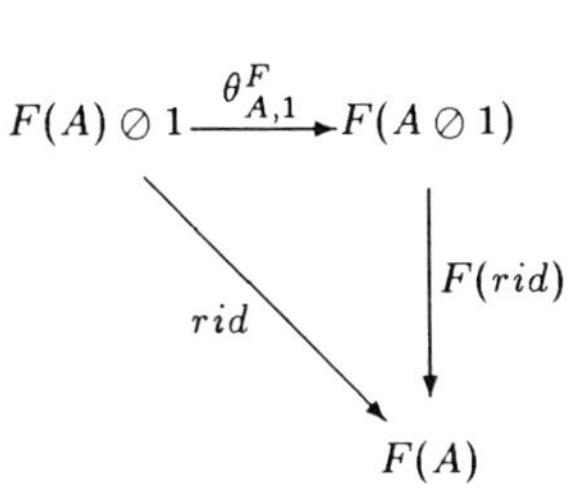

commute.

A strong functor, as used by Moggi for cartesian closed categories, is simply an $\mathbf{X}$-strong endofunctor on $\mathbf{X}$ itself with the natural $\mathbf{X}$–action given by product. It is quite reasonable to discuss "strong functors" for non-cartesian closed categories: the list datatype on a distributive category [**2**] is a strong functor even though its setting is not closed.

An $\mathbf{X}$-*strong natural transformation* between strong functors $F, G : \mathbf{Y}_1 \longrightarrow \mathbf{Y}_2$ is a parameterized natural transformation

$$\alpha_{A,X} : F(A) \oslash X \longrightarrow G(A \oslash X) : \mathbf{Y}_1 \times \mathbf{X} \longrightarrow \mathbf{Y}_2$$

such that

$$\begin{array}{ccc}
(F(A) \oslash X) \oslash Y & \xrightarrow{\theta^F_{A,X} \oslash id_Y} & F(A \oslash X) \oslash Y \\
{\scriptstyle \alpha_{A,X} \oslash id_Y}\downarrow & & \downarrow{\scriptstyle \alpha_{A \oslash X, Y}} \\
G(A \oslash X) \oslash Y & \xrightarrow[\theta^G_{A \oslash X, Y}]{} & G((A \oslash X) \oslash Y)
\end{array}$$

commutes.

Let $\nu : X \xrightarrow{\cong} X_1 \times X_2$ be any decomposition of X into a product. Then the composition of two strong transformations $\alpha_{A,X}$ and $\beta_{A,X}$ is defined by the following sequence of maps:

$$\begin{array}{ccccc}
F(A) \oslash X & \xrightarrow{id \oslash \nu} & F(A) \oslash (X_1 \times X_2) & \xrightarrow{ass} & (F(A) \oslash X_1) \oslash X_2 \\
 & & & & \downarrow{\scriptstyle \alpha_{A,X_1} \oslash id} \\
{\scriptstyle (\alpha.\beta)_{A,X}}\downarrow & & & & G(A \oslash X_1) \oslash X_2 \\
 & & & & \downarrow{\scriptstyle \beta_{A \oslash X_1, X_2}} \\
H(A \oslash X) & \xleftarrow{H(id \oslash \nu^{-1})} & H(A \oslash (X_1 \times X_2)) & \xleftarrow{H(ass^{-1})} & H((A \oslash X_1) \oslash X_2)
\end{array}$$

The composition does not depend on the particular decomposition of X. That is, all decompositions result in the same parameterized composition. Furthermore, this composition straightforwardly commutes with the strengths in the required manner.

By noting that (1) the definition of strength implies $\theta^F_{A,1}$ is an isomorphism and (2) the definition of strong natural transformation trivially implies θ^F is strong and commutative with all other strong transformations, we see θ^F is actually the identity strong transformation on F under this composition. This establishes that **X**–strong categories, **X**–strong functors and **X**–strong transformations form a 2-category **Strong(X)**.

This formulation of an **X**–strong transformation reveals the structural effect of making transformations strong. However, it fails to promote easy manipulation of these transformations. Notice that an ordinary transformation α gives rise to a related strong transformation provided

$$\begin{array}{ccc}
F(A) \oslash X & \xrightarrow{\theta^F_{A,X}} & F(A \oslash X) \\
{\scriptstyle \alpha_A \oslash id_X}\downarrow & & \downarrow{\scriptstyle \alpha_{A \oslash X}} \\
G(A) \oslash X & \xrightarrow[\theta^G_{A,X}]{} & G(A \oslash X)
\end{array}$$

commutes. The associated parameterized transformation is

$$\alpha_{A,X} \;=\; \alpha_A \oslash id_X \,; \theta^G_{A,X} \;=\; \theta^F_{A,X} \,; \alpha_{A \oslash X},$$

which is clearly strong. Conversely, a strong transformation is related to an ordinary natural transformation by

$$F(A) \xrightarrow{rid^{-1}} F(A) \oslash 1 \xrightarrow{\alpha_{A,1}} G(A \oslash 1) \xrightarrow{G(rid)} G(A)$$

and these associations are inverse. Thus we have reorganized our view of **X**–strong categories into a more accessible form:

THEOREM 2.2. *The 2-category* **Strong(X)** *is isomorphic to the 2-category having as*

0-cells: *cartesian functors* $J : \mathbf{X} \longrightarrow \mathbf{Y}$,

1-cells: *pairs* (F, θ^F) *where* F *is a functor* $F : \mathbf{Y} \longrightarrow \mathbf{Y}'$ *and* θ^F *is a strength:*

$$\theta^F_{Y,X} : F(Y) \times J'(X) \longrightarrow F(Y \times J(X)),$$

satisfying

$$\begin{array}{ccc}
(F(A) \times J'(X_1)) \times J'(X_2) & \xleftarrow{ass} & F(A) \times J'(X_1 \times X_2) \\
{\scriptstyle \theta^F_{A,X_1} \times id}\downarrow & & \\
F(A \times J(X_1)) \times J'(X_2) & & \downarrow{\scriptstyle \theta^F_{A,X_1 \times X_2}} \\
{\scriptstyle \theta^F_{A \times J(X_1),X_2}}\downarrow & & \\
F((A \times J(X_1)) \times J(X_2)) & \xleftarrow[F(ass)]{} & F(A \times J(X_1 \times X_2))
\end{array}$$

and

$$\begin{array}{ccc} F(A) \times J'(X) & \xrightarrow{\theta^F_{A,X}} & F(A \times J(X)) \\ & {\scriptstyle p_0} \searrow \quad \swarrow {\scriptstyle F(p_0)} & \\ & F(A) & \end{array}$$

2-cells: *natural transformations* $\alpha : F \longrightarrow G$ *which satisfy :*

$$\begin{array}{ccc} F(A) \times J'(X) & \xrightarrow{\theta^F_{A,X}} & F(A \times J(X)) \\ {\scriptstyle \alpha_A \times id} \downarrow & & \downarrow {\scriptstyle \alpha_{A \times J(X)}} \\ G(A) \times J'(X) & \xrightarrow[\theta^G_{A,X}]{} & G(A \times J(X)) \end{array}$$

Therefore an **X**–strong category may be regarded without loss as a cartesian category with a cartesian functor from **X**, an **X**-strong functor as a functor with a strength, and an **X**–strong transformation as a transformation which commutes with the strengths as shown above. We shall henceforth identify $\oslash$ and $\times$ and avoid the use of the relatively unfamiliar coherence maps for $\_ \oslash \_$.

3. Strength to Fibrations

A strong functor may equivalently be regarded as a morphism of fibrations [**5**] in which **X** is viewed as a category of contexts. Technically this is given by an embedding of the 2–category **Strong(X)** into the 2–category of split fibrations over **X**, **SplFib(X)**. The latter can, of course, be equivalently viewed as **X**–indexed categories. This section derives the embedding; the next section attaches the importance of this embedding to our fabrication methodology for datatypes.

Given an **X**–strong category **Y** we may form the category $P_{\mathbf{X}}(\mathbf{Y})$ to be be thought of as **Y**–maps with context from **X**:

> **Objects:** pairs of objects (Y, X) where $Y \in \mathbf{Y}$ and $X \in \mathbf{X}$,
> **Maps:** pairs of maps $(f, x) : (Y, X) \longrightarrow (Y', X')$ where $f : Y \times X \longrightarrow Y'$ in **Y** and $x : X \longrightarrow X'$ in **X**,
> **Identities:** $(p_0, id) : (Y, X) \longrightarrow (Y, X)$,
> **Composition:** $(f, h) ; (g, k) = (\langle f, p_1 ; h \rangle ; g, h ; k)$.

A map (f, h) uses h to propagate the context in the second component and f to compute new data values from both data values in Y and the available context:

$$(f, h) : (Y, X) \longrightarrow (Y', X') \; ; \; (y, x) \mapsto (f(y, x), h(x)).$$

Composing this map with (g, k) gives:

$$(f, h) ; (g, k) : (Y, X) \longrightarrow (Y'', X'') \; ; \; (y, x) \mapsto (g(f(y, x), h(x)), k(h(x))).$$

Thus g receives a propagated context and the computed value. This can be viewed as the Kleisli composition for the comonad of the $\mathbf{X}$–action (see Remark 3.4).

The objects of this category may be considered as second projections $p_1 : Y \times X \longrightarrow X$ and the maps as naturality squares. The category, seen as arrows, can be viewed as a fibration sitting over $\mathbf{X}$, i.e.

$$\delta_{\mathbf{Y}} : P_{\mathbf{X}}(\mathbf{Y}) \longrightarrow \mathbf{X}$$

where $(f, x) \mapsto x$. The second projections behave as display maps [**21**], i.e. context extractors. The map which carries $\mathbf{Y}$ to the fibration $\delta_{\mathbf{Y}}$ is the object map of a 2–functor from strong categories to split fibrations over $\mathbf{X}$, denoted $P_{\mathbf{X}}(\_)$. We are now at the starting point for discussing the correspondence of strength to fibrations attributed to Plotkin by Moggi. The remainder of this section is dedicated to proving:

THEOREM 3.1. *The 2-functor* $P_{\mathbf{X}}(\_) : \mathbf{Strong(X)} \longrightarrow \mathbf{SplFib(X)}$ *is a full and faithful embedding.*

Let $\mathbf{StrSplFib(X)}$ denote the 2–category of split fibrations of the form $\delta_{\mathbf{Y}}$. The 1-cells are *morphisms of fibrations* (functors $\tilde{F} : P_{\mathbf{X}}(\mathbf{Y}) \longrightarrow P_{\mathbf{X}}(\mathbf{Z})$ that commute with the fibration legs and preserve the splitting on the nose). The 2-cells are *natural transformations of morphisms of fibrations* ($\tilde{\alpha} : \tilde{F}_1 \longrightarrow \tilde{F}_2$ such that $\delta_{\mathbf{Y}} = \delta_{\mathbf{Z}} * \tilde{\alpha}$ (horizontal composite)). We will show that $P_{\mathbf{X}}(\_)$ factored through this full sub–2–category is an isomorphism.

First we assure ourselves that the functors δ are indeed fibrations that individually possess a splitting. The underlying cleavage is defined by

$$(x, (Y, X')) \mapsto (p_0, x) \ .$$

This is illustrated by the diagram

$$\begin{array}{ccc} (Y, X) & \xrightarrow{(p_0, x)} & (Y, X') \\ \vdots\,\delta & & \vdots\,\delta \\ \downarrow & & \downarrow \\ X & \xrightarrow[x]{} & X' \end{array}$$

The fact that this cleavage yields cartesian arrows follows directly from the definitions and parallels the reasoning that

$$\begin{array}{ccc} Y \times X & \xrightarrow{i \times x} & Y \times X' \\ {\scriptstyle p_1}\downarrow & pb & \downarrow{\scriptstyle p_1} \\ X & \xrightarrow[x]{} & X' \end{array}$$

is a pullback. Because the $\mathbf{X}$–action canonically provides the products used in defining the maps in $P_{\mathbf{X}}(\mathbf{Y})$, the cleavage is closed to composition, i.e. it is a splitting.

Suppose $\tilde{F} : P_{\mathbf{X}}(\mathbf{Y}) \longrightarrow P_{\mathbf{X}}(\mathbf{Z})$ is a morphism of fibrations. On the objects it must take the simple form

$$\tilde{F}(A, X) = (F(A), X)$$

for some object function F due jointly to fixing the second coordinate and preserving splitting on the nose, thereby allowing the definition of F to be independent of X. On maps it must take the less precise form

$$\tilde{F}(f, x) = (\tilde{F}_0(f, x), x)$$

since the fibration morphism preserves fibers over morphisms as well.

In the fiber over 1 we may recapture an ordinary functor $F : \mathbf{Y} \longrightarrow \mathbf{Z}$ by setting for $x : Y \longrightarrow Y'$

$$F(x) = \langle id_{F(Y)}, !_{F(Y)} \rangle \,;\, \tilde{F}_0(p_0 \,;\, x \ , \ id_1) \ .$$

To see that this is a functor note first that

$$F(id) = \langle id, ! \rangle \,;\, \tilde{F}_0(p_0, id_1) = \langle id, ! \rangle \,;\, p_0 = id$$

where $\tilde{F}_0(p_0, id_1) = p_0$ since $\tilde{F}$ preserves identities. For composition we exploit the fact

$$\tilde{F}(p_0 \,;\, x \,;\, y \ , \ id_1) = \tilde{F}(p_0 \,;\, x \ , \ id_1) \,;\, \tilde{F}(p_0 \,;\, y \ , \ id_1)$$

so that

$$\begin{aligned} \langle id, ! \rangle \,;\, \tilde{F}_0(p_0 \,;\, x \,;\, y \ , \ id_1) &= \langle id, ! \rangle \,;\, \langle \tilde{F}_0(p_0 \,;\, x \ , \ id_1) \ , \ p_1 \,;\, id_1 \rangle \,;\, \tilde{F}_0(p_0 \,;\, y \ , \ id_1) \\ &= \langle id, ! \rangle \,;\, \tilde{F}_0(p_0 \,;\, x \ , \ id_1) \,;\, \langle id, ! \rangle \,;\, \tilde{F}_0(p_0 \,;\, y \ , \ id_1) \end{aligned}$$

follows, showing composition is preserved.

Next we shall show that this functor has a strength given by $\theta^F = \tilde{F}_0(id, !)$, the first component of

$$\tilde{F}(id, !) \ : \ (F(A), X) \longrightarrow (F(A \times X), 1) \ .$$

θ^F's naturality in each argument derives easily from the commutativity of the following square:

LEMMA 3.2. *The diagram*

$$\begin{array}{ccc} (F(A), X) & \xrightarrow{(\theta^F, !)} & (F(A \times X), 1) \\ {\scriptstyle (p_0 \,;\, F(a) \ , \ x)} \downarrow & & \downarrow {\scriptstyle (p_0 \,;\, F(a \times x) \ , \ id_1)} \\ (F(A'), X') & \xrightarrow[(\theta^F, !)]{} & (F(A' \times X'), 1) \end{array}$$

commutes.

PROOF. The lemma's diagram is the image in $P_{\mathbf{X}}(\mathbf{Z})$ of

$$\begin{array}{ccc}
(A,X) & \xrightarrow{(id,!)} & (A\times X,1) \\
{\scriptstyle (p_0\,;a\ ,\ x)}\Big\downarrow & & \Big\downarrow{\scriptstyle (p_0\,;(a\times x)\ ,\ id_1)} \\
(A',X') & \xrightarrow[(id,!)]{} & (A'\times X',1)
\end{array}$$

in $P_{\mathbf{X}}(\mathbf{Y})$ under $\tilde{F}$. To see this we observe that

$$\tilde{F}(p_0\,;a\ ,\ x) = \tilde{F}(p_0,x)\,;\tilde{F}(p_0\,;a\ ,\ id) = (p_0,x)\,;\tilde{F}(p_0\,;a\ ,\ id)$$

because the splitting over x, (p_0,x), must be taken by $\tilde{F}$ to the specified (i.e. same) splitting over x. Furthermore, any map $\tilde{F}(p_0\,;g\ ,\ id)$ is determined as the unique map over the splitting of $!\ :\ X \longrightarrow 1$ by applying $\tilde{F}$ to

$$\begin{array}{ccc}
(A,X) & \xrightarrow{(p_0,!)} & (A,1) \\
{\scriptstyle (p_0\,;g\ ,\ id_X)}\Big\downarrow & & \Big\downarrow{\scriptstyle (p_0\,;g\ ,\ id_1)} \\
(A',X) & \xrightarrow[(p_0,!)]{} & (A',1)
\end{array}$$

and observing that $(\tilde{F}_0(p_0\,;g\ ,\ id_X)\ ,\ id_X)$ is the unique solution in the resultant image

$$\begin{array}{ccc}
(F(A),X) & \xrightarrow{(p_0,!)} & (F(A),1) \\
\Big\downarrow & & \Big\downarrow{\scriptstyle (\tilde{F}_0(p_0\,;g\ ,\ id_1),id_1)} \\
(F(A'),X) & \xrightarrow[(p_0,!)]{} & (F(A'),1)
\end{array}$$

Tracing the diagram's first coordinate yields $\tilde{F}_0(p_0\,;g\ ,\ id_X) = p_0\,;F(g)$, giving the commutativity of the lemma's diagram (and consequently the naturality of θ^F). □

A more precise form for the $\tilde{F}$-image of an arbitrary map $(a,x)\ :\ (A,X) \longrightarrow (A',X')$. can now be derived:

LEMMA 3.3.

$$\tilde{F}(a,x) = (\theta^F\,;F(a)\ ,\ x).$$

PROOF. A splitting can be factored out by $(a, x) = (a, id)\,;(p_0, x)$, leaving the problem of finding $\tilde{F}(a, id)$. But we note that

$$\begin{array}{ccc} (A, X) & \xrightarrow{(id,!)} & (A \times X, 1) \\ {\scriptstyle (a,id)}\downarrow & & \downarrow{\scriptstyle (p_0\,;a\;,\;id_1)} \\ (A', X) & \xrightarrow[(p_0,!)]{} & (A', 1) \end{array}$$

commutes so $\tilde{F}(a, id)$ is the unique solution (by the familiar splitting argument) in the diagram's $\tilde{F}$-image (created via the proof of Lemma 3.2)

$$\begin{array}{ccc} (F(A), X) & \xrightarrow{(\theta^F,!)} & (F(A \times X), 1) \\ \vdots & & \downarrow{\scriptstyle (p_0\,;F(a)\;,\;id_1)} \\ (F(A'), X) & \xrightarrow[(p_0,!)]{} & (F(A'), 1) \end{array}$$

which is clearly given by $(\theta^F\;; F(a)\;,\;id_X)$. $\square$

This observation is useful in showing how to reconstruct the functor $\tilde{F}$ from (F, θ^F). More immediately we see it provides

$$\begin{array}{ccc} F(A) \times 1 & \xrightarrow{\theta^F} & F(A \times 1) \\ & {\scriptstyle p_0}\searrow \quad \swarrow{\scriptstyle F(p_0)} & \\ & F(A) & \end{array}$$

the first requirement for strength by examining the first coordinate of the image of

$$\begin{array}{ccc} (A, 1) & \xrightarrow{(id_A,!)} & (A \times 1, 1) \\ & {\scriptstyle (p_0,id)}\searrow \quad \swarrow{\scriptstyle (p_0\,;p_0\;,\;id_1)} & \\ & (A, 1) & \end{array}$$

under $\tilde{F}$. For verifying associativity of the strength we can now track the first coordinate in the $\tilde{F}$–image of

$$
\begin{array}{ccccc}
(A, X \times Y) & \xrightarrow{(ass\,;p_0\ ,\ p_1)} & (A \times X, Y) & \xrightarrow{(id, !)} & ((A \times X) \times Y, 1) \\
{\scriptstyle (p_0, id)}\big\downarrow & & & & \big\uparrow{\scriptstyle (p_0\,;ass\ ,\ id_1)} \\
(A, X \times Y) & & \xrightarrow[(id, !)]{\qquad\qquad\qquad} & & (A \times (X \times Y), 1)
\end{array}
$$

This boils down to distilling from the fibration morphism a strong functor. In similar fashion we can extract from a natural transformation $\tilde{\alpha} : \tilde{F}_1 \longrightarrow \tilde{F}_2$ of fibration morphisms a strong transformation α between the respective extracted functors. The (A, X)-component of $\tilde{\alpha}$ has the general form

$$\tilde{\alpha}^{A,X} = (\tilde{\alpha}_0^{A,X}, id_X)$$

due to the horizontal composition requirement. In the fiber over 1 the A-component of α is defined as

$$\alpha_A = \langle id, ! \rangle \,; \tilde{\alpha}_0^{A,1}$$

It is straightforward to verify the required transformational properties of α and to derive the relationship

$$\tilde{\alpha}^{A,X} = (p_0 \,; \alpha_A \ ,\ id_X)$$

directly as the unique comparison over a splitting arrow in the following diagram:

$$
\begin{array}{ccc}
\tilde{F}(A, X) & \xrightarrow{\tilde{F}(p_0, !)} & \tilde{F}(A, 1) \\
\vdots & & \big\downarrow{\scriptstyle (\tilde{\alpha}_0^{A,1}, id_1)} \\
\tilde{G}(A, X) & \xrightarrow[\tilde{G}(p_0, !)]{} & \tilde{G}(A, 1)
\end{array}
$$

This extraction process preserves composition of ordinary functors and transformations as it clearly does so in the fiber over 1. We must also check that it preserves the strong compositions. For functors this fact is given by breaking the defining map for the strength of F into a composition

$$(\theta^F, !) = (id, !) \ ; \ (p_0 \,; \theta_F \ ,\ id_1)$$

Under $\tilde{G}$ this yields in the first coordinate the Kock equation

$$\theta^{G \circ F} = \theta^G \,; G(\theta^F)$$

as desired. For natural transformations of morphisms of fibrations the strengthening of the extracted natural transformations is found in the first coordinate of

$$\begin{array}{ccc}
\tilde{F}(A,X) & \xrightarrow{\tilde{F}(id,!)} & \tilde{F}(A\times X,1) \\
{\scriptstyle \tilde{\alpha}^{A,X}}\downarrow & & \downarrow{\scriptstyle \tilde{\alpha}^{A\times X,1}} \\
\tilde{G}(A,X) & \xrightarrow[\tilde{G}(id,!)]{} & \tilde{G}(A\times X,1)
\end{array}$$

This establishes a 2–functor

$$\mathbf{StrSplFib(X)} \longrightarrow \mathbf{Strong(X)}$$

To show it is an isomorphism it suffices to show bijectivity on objects, functors, and transformations. By construction, bijectivity immediately holds for objects.

Given a strong functor (F, θ^F) we define an opposing lifting map that sends F to $\tilde{F}$ defined by:

$$[(A,X) \xrightarrow{(a,x)} (A',X')] \;\mapsto\; [(F(A),X) \xrightarrow{(\theta^F ; F(a)\ ,\ x)} (F(A'),X')].$$

Note that if $\tilde{F}$ is a morphism of fibrations the strong functor recaptured from the fiber over 1 will be F since

$$\begin{aligned}
\langle id,!\rangle ; \tilde{F}_0(p_0 ; a\ ,\ id_1) &= \langle id,!\rangle ; \theta^F ; F(p_0 ; a) \\
&= \langle id,!\rangle ; p_0 ; F(a) \\
&= F(a).
\end{aligned}$$

So $F \longrightarrow \tilde{F} \longrightarrow F$ will be the identity. Conversely, Lemma 3.3 will tell us that the transition $\tilde{F} \longrightarrow F \longrightarrow \tilde{F}$ is also the identity. This means the 2–functor will be an isomorphism on functors.

$\tilde{F}$ is indeed a morphism of fibrations as it is a functor $P_{\mathbf{X}}(\mathbf{Y}) \longrightarrow P_{\mathbf{X}}(\mathbf{Z})$ by

$$\begin{aligned}
\langle \theta^F ; F(g)\ ,\ x\rangle ; \theta^F ; F(g') &= \langle \theta^F, p_1\rangle ; (F(g)\times x) ; \theta^F ; F(g') \\
&= \langle \theta^F, p_1\rangle ; \theta^F ; F(g\times x) ; F(g') \\
&= \langle id, p_1\rangle ; (\theta^F \times id) ; F(g\times x) ; F(g') \\
&= (id \times \Delta) ; ass ; (\theta^F \times id) ; F(g\times x) ; F(g') \\
&= (id \times \Delta) ; \theta^F ; F(ass) ; F(g\times x) ; F(g') \\
&= \theta^F ; F(id \times \Delta) ; F(ass) ; F(g\times x) ; F(g') \\
&= \theta^F ; F(\langle g\ ,\ p_1 ; x\rangle ; g')
\end{aligned}$$

and it preserves the splitting because $\theta^F ; F(p_0) = p_0$.

Finally, we must show that strong transformations can be lifted back to transformations of fibrations. We now know that a strong transformation is determined by its underlying ordinary transformation. From this fact the process of extracting the strong transformation from a lifted strong transformation is clearly seen to behave as the identity.

If the lifting of the strong transformation extracted from a transformation of fibrations can also be proved to be the identity then we will have established an isomorphism of transformations. To confirm this, we first remark that a strong transformation α may be lifted by using the following diagram as a "naturality square":

$$\begin{array}{ccc} (F(A),X) & \xrightarrow{(\theta^F\,;F(a)\ ,\ x)} & (F(A'),X') \\ {\scriptstyle (p_0\,;\alpha_A\ ,\ id_X)}\Big\downarrow & & \Big\downarrow{\scriptstyle (p_0\,;\alpha_{A'}\ ,\ id_{X'})} \\ (G(A),X) & \xrightarrow[(\theta^G\,;G(a)\ ,\ x)]{} & (G(A'),X') \end{array}$$

Furthermore, we discovered above that transformations of fibrations also have this form. So the lifting of an extracted transformation is truly the identity. This completes the proof of Theorem 3.1.

We shall hereafter refer to the full and faithful embedding of **Strong(X)** into **SplFib(X)** that follows from this development as the *strong–fibration embedding.*

Remark 3.4. A very compact perspective of strength leverages the equivalence between the categories of split fibrations and of indexed categories. Because we can view the **X**–action $\oslash$ of the **X**–strong category **Y** as an ordinary product in **Y**, each functor $\_\oslash X$ for $X \in \mathbf{X}$ is a comonad of **Y**. The unit and multiplier are:

$$\eta : \_\oslash X \xrightarrow{p_0} \_$$

$$\mu : \_\oslash X \xrightarrow{\langle id,p_1\rangle} (\_\oslash X)\oslash X$$

The Kleisli category of this comonad is also Lambek's polynomial category obtained by adjoining an element [**10**].

To each $X \in \mathbf{X}$ we may associate the Kleisli category $\mathbf{Y}_X$. Since any morphism $f : X_1 \to X_2$ in **X** clearly induces a morphism of the respective comonads, then it also induces a functor $f^* : \mathbf{Y}_{X_2} \to \mathbf{Y}_{X_1}$. Our association is consequently an indexed category $\mathcal{F} : \mathbf{X}^{op} \to \mathbf{Cat}$. The morphisms of fibrations in **StrSplFib(X)** restricted to the fiber over 1, i.e. the strong functors, can now be seen to be exactly the morphisms of the Grothendieck construction of the corresponding split fibration for $\mathcal{F}$.

4. Datatypes and Strength

Tatsuya Hagino's thesis [**6**] proposed a categorical programming language that introduced two sorts of datatype declarations: initial datatypes and final datatypes. Gavin Wraith [**25**] somewhat simplified the original system and declarations by assuming explicitly that the underlying setting was a bicartesian closed category. In our treatment we take another direction: assume that the underlying category is a cartesian category and replace the closed requirement by the assumption that the datatypes are strong. Because the strong functors of Section 3 are always covariant, we avoid the well-known nuisance of moving contravariant functors — such as exponentials and hom-functors — to a new venue, e.g. O-categories [**19**], where they can also act covariantly — the limit-colimit coincidence [**17**] — and thereby become eligible for computing ω-colimits as solutions of recursive domain equations for datatypes.

Factorizer versions of the Hagino "right" and "left" forms of datatype declaration done in a style similar to that suggested by Wraith are:

$$
\begin{array}{ll}
\textbf{data} & C \longrightarrow R(A) = \\
\mid & d_1 : C \longrightarrow E_1(A, C) \\
\mid & \ldots \\
\mid & d_n : C \longrightarrow E_n(A, C).
\end{array}
\qquad\qquad
\begin{array}{ll}
\textbf{data} & L(A) \longrightarrow C = \\
\mid & c_1 : E_1(A, C) \longrightarrow C \\
\mid & \ldots \\
\mid & c_n : E_n(A, C) \longrightarrow C.
\end{array}
$$

The first declaration asserts that any map to the new type $R(A)$ from a state type C is to be determined by a set of programmer-chosen maps from C to the types $E_1(A, C), \ldots, E_n(A, C)$. The second definition forms exactly the dual. The labels d_i and c_i name the collection of canonical operations (not to be confused with the programmer-chosen maps) that are universally associated with their respective datatypes and in one-to-one correspondence with the programmer-chosen maps in a particular declaration. Clearly the first definition declares a final datatype, the second an initial datatype.

The chosen parameter type variable A usually ranges over a power of the given category. However, we shall take the interpretation that it simply ranges over some category and assume the typing $E_i : \mathbf{A} \times \mathbf{C} \longrightarrow \mathbf{C}$. The object $R(A)$ comes equipped with canonical maps

$$d_i : R(A) \longrightarrow E_i(A, R(A))$$

which Wraith called **destructors** as they play the dual role to the canonical maps associated with $L(A)$

$$c_i : E_i(A, L(A)) \longrightarrow L(A)$$

which are traditionally called **constructors**. With respect to these maps the initial and final datatypes satisfy the following universal properties:

$$
\begin{array}{ccc}
C & \xrightarrow{\;g_i\;} & E_i(A, C) \\
{\scriptstyle f_R(g)}\,\vdots & & \vdots\,{\scriptstyle E_i(A, f_R(g))} \\
R(A) & \xrightarrow[\;d_i\;]{} & E_i(A, R(A))
\end{array}
\qquad\qquad
\begin{array}{ccc}
E_i(A, L(A)) & \xrightarrow{\;c_i\;} & L(A) \\
{\scriptstyle E_i(A, f_L(h))}\,\vdots & & \vdots\,{\scriptstyle f_L(h)} \\
E_i(A, C) & \xrightarrow[\;h_i\;]{} & C
\end{array}
$$

Here g and h respectively represent the n-tuples of the g_i and h_i morphisms.

We shall denote the datatypes as (R, d) and (L, c), respectively, to emphasize the definitional consolidation of destructors and constructors with their respective datatypes. Additionally, these finality and initiality diagrams often constitute recursive definitions; they shall be referred to as the "recursion" diagrams for their datatypes.

Potentially the category of coalgebras for the final datatypes may be constructed as an **inserter** (see Kelly [**8**]), a 2–categorical limit, if it exists. The inserter resembles the comma category except that the functors giving the domain and codomain originate on the same object. In fact it is a lax equalizer: for general functors $G, H : \mathbf{A} \to \mathbf{B}$, the inserter is the universal pair (U, β) where $U : \mathbf{Insert}(G, H) \to \mathbf{A}$ is a functor and $\beta : G \circ U \to H \circ U$ is a natural transformation.

For example, in the 2-category **Cat** the objects of the inserter category $\mathbf{Insert}(G, H)$ are the pairs (A, b) where A is in $\mathbf{A}$ and $b : G(A) \to H(A)$. A morphism $(A_1, b_1) \to (A_2, b_2)$ corresponds to the existence of a commuting "naturality square" generated by a $\mathbf{A}$-morphism $f : A_1 \to A_2$:

$$\begin{array}{ccc} G(A_1) & \xrightarrow{G(f)} & G(A_2) \\ {\scriptstyle b_1}\downarrow & & \downarrow{\scriptstyle b_2} \\ H(A_1) & \xrightarrow[H(f)]{} & H(A_2) \end{array}$$

The inserter functor U is realized as the underlying (forgetful) functor from $\mathbf{Insert}(G, H)$. The required natural transformation β consequently arises with the components $\beta_{(A,b)} = b$.

In our final datatype situation this means that we seek an underlying functor to the category of origin

$$U_R : \mathbf{Insert}(\Delta_n \circ Pr_1, \langle E_1, ..., E_n \rangle) \longrightarrow \mathbf{A} \times \mathbf{C}$$

and a natural transformation α, regarded as the universal destructor of the final datatype, where

$$\mathbf{Insert}(\Delta_n \circ Pr_1, \langle E_1, ..., E_n \rangle) \underset{\longrightarrow}{\overset{\longrightarrow}{\Downarrow \alpha}} \mathbf{C}^n$$

in which $\alpha : \Delta_n \circ Pr_1 \circ U_R \longrightarrow \langle E_1, ..., E_n \rangle \circ U_R$. Similarly the opposite construction

$$(\mathbf{Insert}(\langle E_1, ..., E_n \rangle, \Delta_n \circ Pr_1), U_L, \alpha')$$

constructs the algebras for the initial datatype and its universal constructor.

To find what the universal properties expressed by the destructor squares translate to in this construction, we must take heed that these datatypes are *parametrized* by objects of $\mathbf{A}$. To naively demand that the underlying functor followed by the first projection, i.e. $Pr_0 \circ U_R$, have a right adjoint for extracting the final coalgebra and the associated destructor squares is inadequate: such an adjunction would permit the parameter to vary within it!

Parametrization is more precisely viewed in fibrational terms: it is elementary to show that $\mathcal{P} = Pr_0 \circ U_R : \mathbf{Insert}(\Delta_n \circ Pr_1, \langle E_1, ..., E_n\rangle) \longrightarrow \mathbf{A}$ is a cofibration where the cocartesian arrows in the inserter category of the cofibration) are the naturality squares of the form $((A, C), \alpha) \to ((A', C), \alpha')$ generated by $\mathbf{A} \times \mathbf{C}$-morphisms $(h_{\mathbf{A}}, h_{\mathbf{C}}) : (A, C) \to (A', C)$ where $h_{\mathbf{C}} = id_C$.

The final coalgebra for each A-object can now be viewed as picking out final objects in each fiber of $\mathcal{P}$. This is formalized by asserting the existence of a *fibered* right adjoint R of $Pr_0 \circ U_R (= \mathcal{P})$ between the identity cofibration $\mathcal{I} : \mathbf{A} \to \mathbf{A}$ and $\mathcal{P}$:

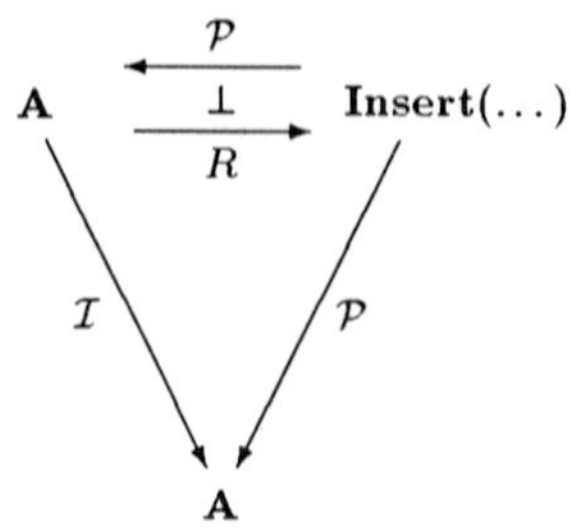

The unit and counit of this adjunction are required by definition to be vertical arrows, thereby allowing the restrictions of the functors $\mathcal{P}$ and R to the respective fibers over any object A to form an adjunction where the parameter is constant. Each fiber-restricted right adjoint of $\mathcal{P}$ does the selection of the final coalgebra for the corresponding base in $\mathbf{A}$ exactly according to the universal destructor square. Furthermore, this definition is slightly redundant: it is easy to show that the unit is vertical if and only if the counit is vertical. Since $\mathcal{I}$-verticals can only be identities, the fundamental requirement becomes having a right adjoint to $\mathcal{P}$ that possesses an identity counit.

Moreover, with R being a fibration morphism, this view of parametrization converts back to the general 2-categorical setting where we essentially insist $\mathbf{A}$ to be a reflective subcategory of $\mathbf{Insert}(\ldots)$ via an inclusion that is the right adjoint to $\mathcal{P}$. Under the expanded requirement, we can now let $\mathcal{P}$ act as a (fiber-wise) underlying functor in order to follow the "correct" notion for building final datatypes.

Likewise, the initial datatype algebra for each A-object can be "correctly" specified by noting that $\mathcal{Q} = Pr_0 \circ U_L : \mathbf{Insert}(\langle E_1, ..., E_n\rangle, \Delta_n \circ Pr_1) \longrightarrow \mathbf{A}$

is a fibration and providing a fibered left adjoint L that has an identity unit as follows:

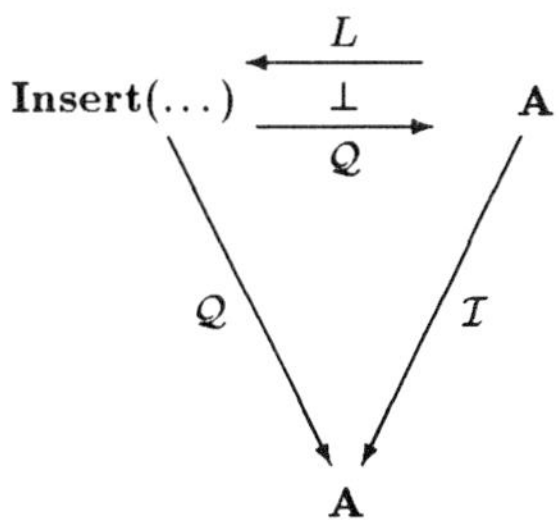

Analogously, here **A** can also be seen as a coreflective subcategory of **Insert**(...) under the inclusion L.

The motivation for laying out such abstract formalization is that once datatype building can be understood in this way we may ship it into any 2–category with finite limits in order to install the correct notion in that setting. The most natural candidate for receiving shipment is, of course, the 2–category **Strong**(**X**). Alas, **Strong**(**X**) is not finitely complete and even fails to have inserters. This situation is rescued by the strong–fibration embedding. With it we can make datatypes using **Strong**(**X**) inside of **SplFib**(**X**) which, by its equivalence to the 2–category of **X**–indexed categories, is finitely complete.

The abstract construction of a strong initial or strong final datatype using the strong–fibration embedding, i.e. respectively the initial object or the final object of an inserter, can be unwound and the computationally unimportant components of morphism pairs then removed. The resulting elementary recursion diagrams expressing the two relevant universal properties are displayed below. We express the collection of functions, $g_1, \ldots, g_n$ as "gs". For a *strong final datatype* the diagram is

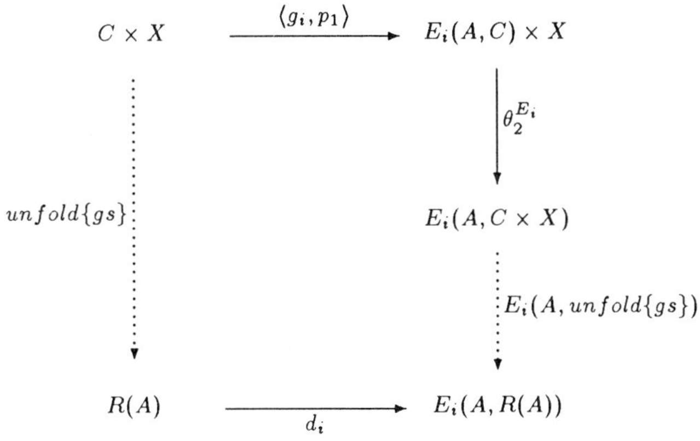

in which $\theta_2^{E_i}$, as $\theta^{E_i}\,; E_i(p_0, id_{C\times X})$ via the naturality of θ^{E_i}, is the strength applied in the second argument only. Similarly, the diagram for a *strong initial datatype* is

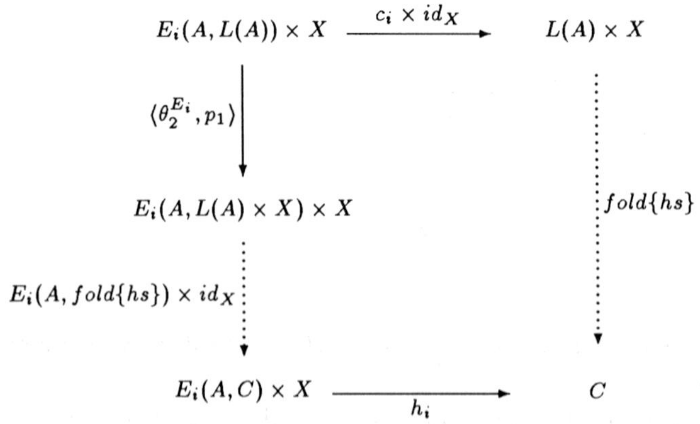

These diagrams are explored further in the next section, but we should note here the serendipitous inclusion of the context parameter X in the domain of each h_i and g_i, the components of actions — *fold* and *unfold*, respectively — that are in turn determined by these components.

It makes moral intuitional sense to allow the context to be processed with the state by the components of an action using that same context. This feature contrasts with other often-used examples, e.g. concerning lists and natural number objects, where this morality has not been kept.

An immediate corollary of the strong-fibration embedding is :

LEMMA 4.1. *In the datatypes (R, d) and (L, c), R and L are strong functors.*

It is well-known that if a category has coproducts and admits the construct of final datatypes then the destructors form a product cone. Dually, if the category has products then the constructors of initial datatypes form coproduct cocones. For strong initial datatypes we have the even stronger result:

LEMMA 4.2. *If (L, c) is an initial $\mathbf{X}$-strong datatype then c forms an $\mathbf{X}$-sum.*

Here we use the term $\mathbf{X}$–*sum* to indicate that the $\mathbf{X}$–action distributes over the coproduct, i.e. the $c_i \times id_X$ morphisms are embeddings.

PROOF. Suppose $h_i : E_i(A, L(A)) \times X \longrightarrow C$ then any comparison map

$$h : L(A) \times X \longrightarrow C$$

with $(c_i \times id_X) ; h = h_i$ satisfies the following recursion diagram

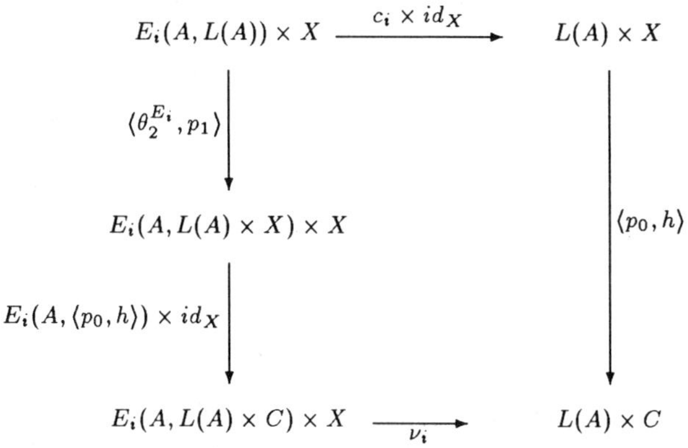

where $\nu_i = \langle p_0\,;E_i(id_A, p_0)\,;c_i\ ,\ (E_i(id_A, p_0) \times id_X)\,;h_i\rangle$. Therefore it exists and is unique. □

Many initial datatypes have been studied in first-order settings. For example, Leopoldo Román proposed various recursion schemes for the natural numbers in [**18**] including the one suggested by the above analysis. The first author studied the list datatype in [**2**] by using a slight variant of the list recursion diagram. For initial datatypes based on functors whose strengths are isomorphisms (these we call *linear functors*) the exact form of the recursion diagrams is not critical. As an example, the recursion suggested in [**2**] is correct — but morally "wrong" — and is shown by

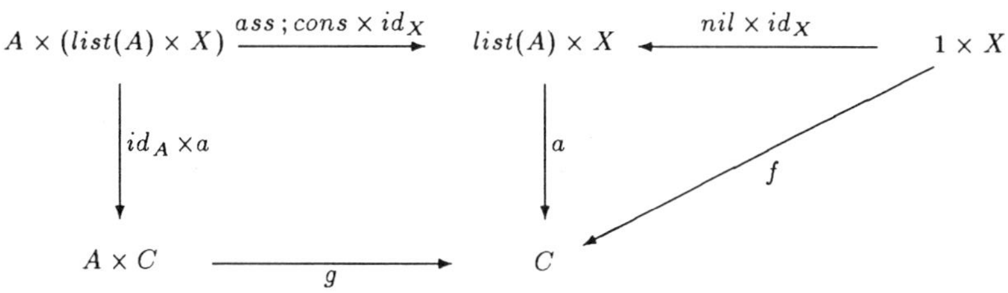

It is quickly seen to be equivalent to the morally correct version

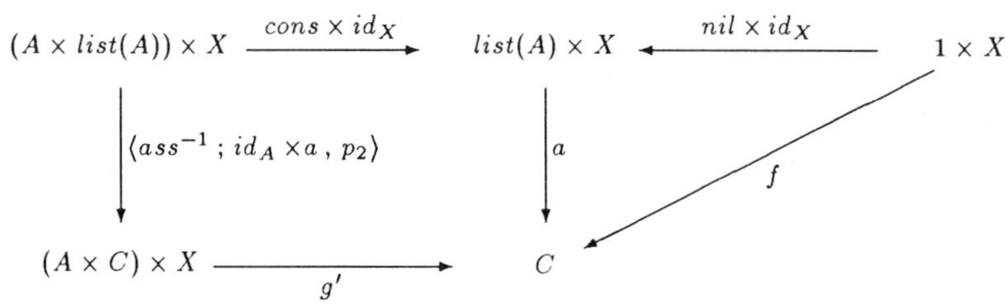

that combines the context X with the state C for processing by the programmer-chosen maps f and g'.

For more complex datatypes based on functors whose strengths are not isomorphisms the equivalence between the datatypes implied by the diagrams above fails.

Choosing the correct diagram in these cases becomes critical, but the proper choice is made evident by our general construction.

A common asserted justification for stronger recursion diagrams are their validity in a cartesian closed category. This justification's weakness is revealed by its not confirming whether we have obtained all the strength it is natural to have.

Yet, admittedly, knowing that our strength-based diagrams are valid in a cartesian closed category certainly provides assurance:

PROPOSITION 4.3. *In a cartesian closed category ordinary datatypes are necessarily strong datatypes.*

PROOF. We shall demonstrate this for the initial datatypes. Suppose (L, c) is an ordinary initial datatype, i.e. it satisfies the simple algebra initiality property firstly-mentioned in this section.

Let $h_i : E_i(A, C) \times X \longrightarrow C$ be a collection of programmer-chosen maps. Define h_i^* as a particular transpose via

$$\frac{E_i(A, X \Rightarrow C) \times X \xrightarrow{\langle \theta_2^{E_i}, p_1 \rangle} E_i(A, (X \Rightarrow C) \times X) \times X \xrightarrow{E_i(id_A, eval) \times id_X \,;\, h_i} C}{E_i(A, X \Rightarrow C) \xrightarrow{h_i^*} X \Rightarrow C}$$

Using the h_i^* as programmer-chosen maps for the ordinary initial datatype L, we construct the diagram below to obtain a unique arrow $a : L(A) \longrightarrow X \Rightarrow C$ which gives commutativity in its top square. The bottom pentagon commutes by adjointness.

By rearranging the left vertical composite, the outermost polygon becomes the strong recursion diagram for $fold\{hs\} = (a \times id_X)\,; eval$. Since any $fold$ map can be factored in the presence of cartesian closure into the form of the right vertical composite, the comparison map is unique. Thus (L, c) is strong as well.

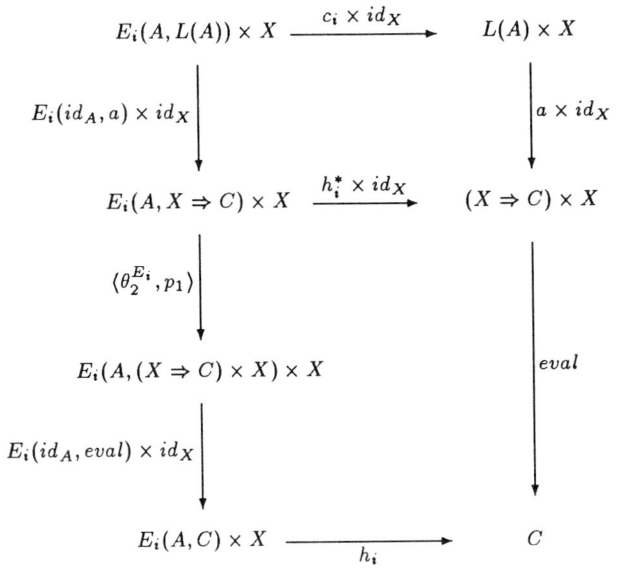

□

5. Datatypes to Combinator Reduction

We now transition from specifying strong datatypes to establishing a categorical combinator reduction system that is directly derivable from our brand of Hagino-Wraith style declarations. The strong initiality and strong finality diagrams of the previous section provide immediately the re-write rules for the $fold\{h_1, \ldots, h_n\}$ and $unfold\{g_1, \ldots, g_n\}$ actions by pasting the diagrams in the direction of decreasing data structure complexity, i.e. towards earlier declaration. These respective actions are uniquely built from a programmer's specification of context-parameterized state transformers h_i coming from components of a datatype and of context-parametrized state de-transformers g_i going to components of a datatype. If components of a datatype are intuitively regarded both as its generalized "subterms" and as "instances" of the domains (codomains) of the transformers (de-transformers), the rewriting rules can be seen to correspond closely to folding and unfolding program transformations as first cataloged by Burstall and Darlington [**1**].

Thus the *fold* rewrite rules for the initial datatype L are easily seen to be:

$$c_i \times id_X \,;\, fold^L\{h_1, \ldots, h_n\} \implies \langle \theta^{E_i}\{p_0, fold^L\{h_1, \ldots, h_n\}\}, p_1\rangle \,;\, h_i$$

and similarly the *unfold* rewrite rules for the final datatype R are:

$$unfold^R\{g_1, \ldots, g_n\} \,;\, d_i \implies \langle g_i, p_1\rangle \,;\, \theta^{E_i}\{p_0, unfold^R\{g_1, \ldots, g_n\}\}$$

where $\theta^F\{f_1, \ldots, f_m\}$ will hereafter abbreviate $\theta^F \,;\, F(f_1, \ldots, f_m)$ for any strong datatype F. We call $fold^L$ the *L fold factorizer* or informally the *fold from L*. Likewise we term $unfold^R$ the *R unfold factorizer* or informally the *unfold to R*.

These two factorizers can form a minimal utilizable set of categorical programming primitives within this setting, but such a severe constraint in expressiveness leads expectedly to over-complicated codes.

This problem is alleviated by defining two specializations, respectively, of both folding and unfolding that correspond to familiar and vital programming tools. The definitions leverage the fact that, although we are working in a cartesian setting, this setting is closed over *all* declarations of strong initial and final datatypes.

First, the respective analogies of case analysis and higher-order transformation for initial datatypes can be specified with the use of the available binary products, which can be considered as the strong final datatype *Prod* of parametric arity 2 where the component functors E_1 and E_2 are respectively $Pr_0 \circ Pr_0$ and $Pr_1 \circ Pr_0$:

(i) *L case factorizer* :

$$\begin{aligned}&case^L\{h_1, \dots, h_n\} \equiv \\ &\quad fold^L\{\dots, \langle p_0 \,;\, E_i(id_A, p_0) \,;\, c_i \;,\; (E_i(id_A, p_0) \times id_X) \,;\, h_i\rangle, \dots\} \,;\, p_1\end{aligned}$$

(ii) *L map factorizer* :

$$map^L\{h_1, \dots, h_n\} \equiv fold^L\{\dots, \theta^{E_i}\{(h_1, \dots, h_n), p_0\} \,;\, c_i, \dots\}$$

Second, the analogies of record formation and higher-order transformation for final datatypes are definable by using the available binary sums, which can be treated as the strong initial datatype *Sum* of arity 2 with the same component functors as *Prod*:

(i) *R record factorizer* :

$$\begin{aligned}&record^R\{g_1, \dots, g_n\} \equiv \\ &\quad b_1 \times id_X \,;\, unfold^R\{\dots, \langle case^{Sum}\{p_0 \,;\, d_i \;,\; g_i\} \,;\, E_i\{id_A, b_0\} \;,\; p_1\rangle, \dots\}\end{aligned}$$

(ii) *R map factorizer* :

$$\begin{aligned}&map^R\{g_1, \dots, g_n\} \equiv \\ &\quad unfold^R\{\dots, d_i \times id_X \,;\, \langle\theta^{E_i}\{(g_1, \dots, g_n), p_0\}, p_1\rangle, \dots\}\end{aligned}$$

Each of the four specializations and their associated rewriting rules are justified below.

The L case factorizer is precisely the comparison map h of the diagram in the proof of Lemma 4.2. The rewrite rule derives directly from the **X**–sum factoring property:

$$c_i \times id_X ; case^L\{h_1, \dots, h_n\} \implies h_i$$

The L map factorizer is defined by the diagram below in which the the collection of passed functions, $h_1, \dots, h_n$ is abbreviated as "hs":

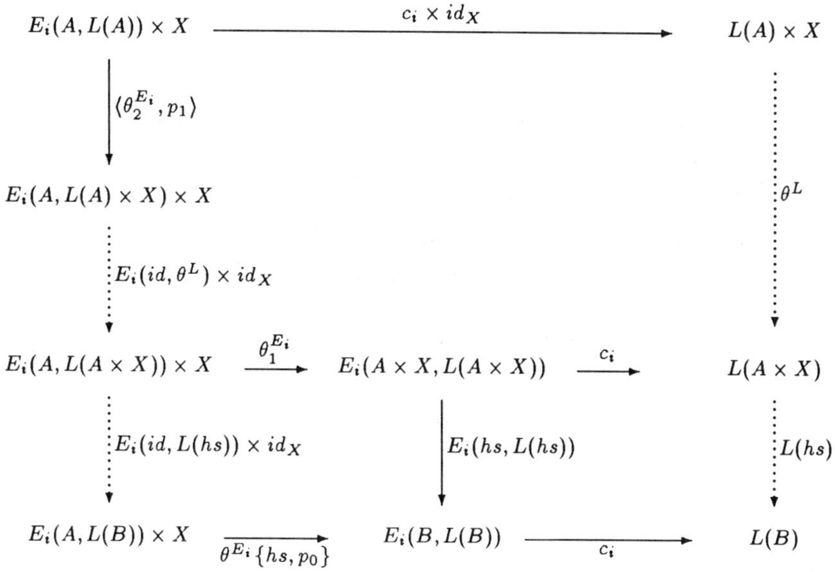

The top hexagon defines L *strength* as a fold operator and the outermost polygon defines the L map factorizer as a fold operator. That is, $map^L\{hs\} = \theta^L\{hs\}$. The diagram's polygons are essentially pasted from upper-right-to-lower-left to derive the following rewrite rule:

$$c_i \times id_X \,; map^L\{h_1, \dots, h_n\} \implies \theta^{E_i}\{(h_1, \dots, h_n), map^L\{h_1, \dots, h_n\}\} \,; c_i$$

The R record factorizer can be defined in a manner dual to the definition of the L case factorizer:

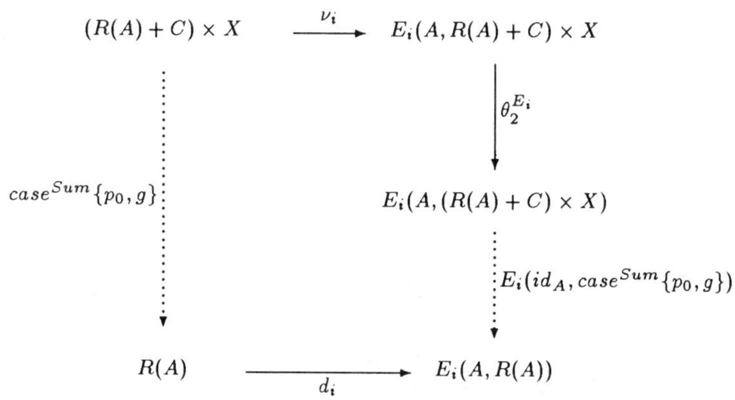

where

$$\nu_i = \langle case^{Sum}\{p_0 \,; d_i \ , \ g_i\} \ , \ p_1 \rangle \,; E_i(id_A, b_0) \times id_X$$

and $g \ : \ C \times X \to R(A)$ is any comparison map such that $g \,; d_i = g_i$. Here $record^R\{g_1, \dots, g_n\}$ is defined to be exactly g which is, in turn, defined by the unique definition of $case^{Sum}\{p_0, g\}$ as an unfold operator. By Lemma 4.2 g is uniquely specified. From the definition's factorization condition we have the rewrite rule:

$$record^R\{g_1, \dots, g_n\}\,; d_i \implies g_i$$

The duality is seen clearly by noting that the L case factorizer definition can now be expressed in terms of record factorizers, viz. by using the relation $record^{Prod}\{f_1, f_2\} = \langle f_1, f_2\rangle$ for f_1 and f_2 having a common domain.

Finally, the R map factorizer is defined by the following diagram in which the collection of passed functions $g_1, \dots, g_n$ is abbreviated as "gs":

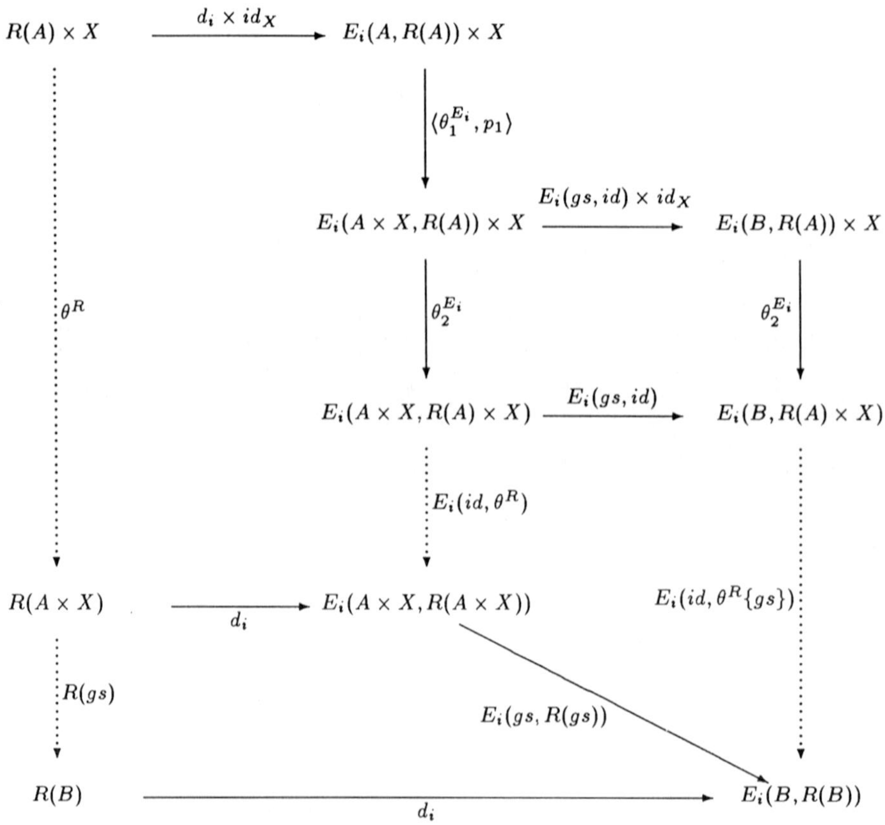

The top left hexagon defines R *strength* as an unfold operator and the outermost polygon defines the R map factorizer as an unfold operator. Again, this means $map^R\{gs\} = \theta^R\{gs\}$. The polygons of the diagram are pasted lower-left-to-upper-right to obtain the rewrite rule:

$$map^R\{g_1, \dots, g_n\}\,; d_i \implies d_i \times id_X\,; \theta^{E_i}\{(g_1, \dots, g_n), map^R\{g_1, \dots, g_n\}\}$$

Conjoining these six rewriting rules with the well-known basic rewriting rules for cartesian categories forms a strongly-normalizing polymorphic combinatory reduction system in which an expansive variety of eager initial and lazy final datatypes can be declared in a cartesian setting and their terms computed. This system forms the abstract machine backbone of the **Charity** categorical programming language implemented by Cockett and Fukushima [**4**].

To balance our discussions of the more familiar initial datatypes, we close this section with two concrete strong final datatype examples: infinite lists and colists, two forms of "lazy lists". Their utility often lies in reducing complexity in list manipulation. Final datatypes are often infinitary and correspond to greatest fixpoints of colimit-preserving functors (e.g. see Manes and Arbib [**11**]). Our first example, infinite lists, are declared with the component functors Pr_0 and Pr_1 and with the destructors that extract respectively the head and the (infinite) tail of an infinite list. Essentially we are solving the domain equation $Inflist(A) = A \times Inflist(A)$ maximally, i.e. finding all truly infinite lists of the parameter type A. The strong infinite list recursion diagram is presented below:

$$
\begin{array}{ccccc}
A \times X & \xleftarrow{\langle f, p_1 \rangle} & C \times X & \xrightarrow{\langle g, p_1 \rangle} & C \times X \\
{\scriptstyle p_0}\big\downarrow & & \big\downarrow{\scriptstyle unfold^{Inflist}\{f,g\}} & & \big\downarrow{\scriptstyle unfold^{Inflist}\{f,g\}} \\
A & \xleftarrow[head]{} & Inflist(A) & \xrightarrow[tail]{} & Inflist(A)
\end{array}
$$

The colist datatype includes the non-empty finite lists as well as the infinite lists. Here Pr_0 and $Sum\{1, C\}$ are the declaration functors and consequently we envision the equation $Colist(A) = A + (A \times Colist(A)) = A \times (1 + Colist(A))$. Recall that the second equality follows from Lemma 4.2. Using like-named destructors, the strong colist recursion diagram reduces to

$$
\begin{array}{ccccc}
A \times X & \xleftarrow{\langle f, p_1 \rangle} & C \times X & \xrightarrow{\langle g, p_1 \rangle} & (1 + C) \times X \\
{\scriptstyle p_0}\big\downarrow & & \big\downarrow{\scriptstyle unfold^{Colist}\{f,g\}} & & \big\downarrow{\scriptstyle \theta^{Sum}\{p_0, unfold^{Colist}\{f,g\}\}} \\
A & \xleftarrow[head]{} & Colist(A) & \xrightarrow[tail]{} & 1 + Colist(A)
\end{array}
$$

The partiality of the colist *tail*, revealed by the appearance of an error type (1) in its codomain, contrasts with the totality of *head*.

6. Conclusions and Future Work

A forthcoming paper will describe a term logic with support for strong datatypes which serves as a categorical programming language. The translation of this term logic into the above datatype combinators will be explained and proven to be consistent and functionally complete.

Such a categorical programming language has been implemented: **Charity** is currently an operational testbed for categorical programming and uses a mixture of eager initial and lazy final datatypes (Cockett and Fukushima [**4**]). As a programming language it is remarkably useable (considering its formal nature and the fact that all programs strongly normalize). The arithmetic strength of the setting is weaker than most comparable systems (as there is no exponentiation) but does allow, for example, the definition of Ackermann's function.

Categorical computation remains relatively inefficient compared even to conventional functional languages, so techniques such as optimizing the translation of the term logic to combinators in order to speed up execution will be important. Recent intriguing results by Barry Jay [**7**] on the efficiency of iteration within categorical computation may help illuminate such optimizations. The discovery of additional specializations of the fold and unfold factorizers to outfit a categorical programmer's "toolkit" in the sense hinted by David Turner [**22**] would further this goal. Also, optimizing via program transformations similar to those developed by Malcolm [**15, 16**] for general Hagino datatypes may be feasible. Finally, the current research surge in functional programming analysis from the perspectives of control, continuations, and partial evaluation might well be worthwhile in this setting.

Cockett's recent investigations [**3**] indicate that attractive abstract data structure specifications, such as those espoused by Walters [**23**], are highly applicable to this programming setting. Also, preliminary investigations have begun towards incorporating dependent types into the system which would yield further improvements in abstract data structure expressiveness.

7. Acknowledgements

Robin Cockett would like to acknowledge the support — financial and intellectual — that the Australian Research Council and the Sydney category seminar provided during the preparation of this work. He is grateful to Richard Wood for explaining his **V**–indexed categories [**24**] amid the Kangaroos and to Barry Jay for his burgeoning interest in this topic. And above all, he would like to acknowledge the constant contributions and support provided by Ross Street throughout his stay in Sydney.

Dwight Spencer wishes to thank the Formal Methods Group of the University of Calgary Computer Science Department for giving warm support and open collaboration to this author's pursuits in categorical computation. He gives particular thanks to Tom Fukushima for his sharp-eyed error-checking of this paper.

References

1. R.M. Burstall and J. Darlington *A transformation system for developing recursive programs.* Journal of the Association for Computing Machinery, 24, No. 1 (1977), 44-67.
2. J.R.B. Cockett, *List-arithmetic distributive categories: locoi.* Journal of Pure and Applied Algebra 66 (1990) 1-29.
3. ______, *Conditional control is not quite categorical control*, IV Higher Order Workshop, Banff 1990, Graham Birtwistle (Ed.) Springer–Verlag, Workshops in Computing (1991) 190-217.

4. J.R.B. Cockett and T. Fukushima, *About Charity*, University of Calgary (1991). Unpublished manuscript.
5. J. W. Gray, *Fibred and cofibred categories.* Proceedings of the Conference on Categorical Algebra, La Jolla, 1965 (Springer, Berlin) 21-84.
6. T. Hagino, *A categorical programming language.* Ph.D. Thesis, University of Edinburgh (1987).
7. C.B. Jay, *Tail recursion via universal invariants*, University of Edinburgh Research Report Preprint (1991).
8. G.M. Kelly, *Elementary observations on 2-categorical limits.* Bulletin of the Australian Mathematical Society 39 (1989) 301-317.
9. A. Kock, *Strong functors and monoidal monads.* Archiv der Mathematik, 23 (1972) 113-120.
10. J. Lambek and P.J. Scott, *Introduction to Higher Order Categorical Logic.* Cambridge University Press (1986).
11. E. G. Manes and M. A. Arbib, *Parametrized datatypes do not need highly constrained parameters.* Information and Control 52 (1982) 139-158.
12. E. Moggi, *Computational lambda-calculus and monads.* Proceedings of the 4th Annual Symposium on Logic in Computer Science (1989) 14-23.
13. ———, *An Abstract View of Programming Languages.* Lecture Notes, University of Edinburgh (June 1989).
14. ———, *Notions of computation and monads.* Information and Control 93 (1991) 55-92.
15. G. Malcolm, *Data Structures and Program Transformation.* Science of Computer Programming 14 (1990) 255-279.
16. ———, *Algebraic Data Types and Program Transformation.* Ph.D. Thesis, University of Groningen (1990).
17. P. Panangaden, P. Mendler and M. Schwartzbach, *Recursively defined types in constructive type theory* in Resolution of Equations in Algebraic Structures. Academic Press (1989).
18. L. Román, *On recursive principles in cartesian categories.* Preprint.
19. M. Smyth and G. Plotkin, *The category theoretic solution of recursive domain equations.* SIAM Journal of Computing 11 (1982) 761-783.
20. D. Spencer, *A survey of categorical computation: fixed points, partiality, combinators, ... control?* Bulletin of the European Association of Theoretical Computer Science 43 (February, 1991) 285-312.
21. P. Taylor, *Recursive domains, indexed category theory and polymorphism*, Ph.D. Thesis, University of Cambridge (1986).
22. D. Turner, *Duality and De Morgan Principles for Lists*, Computer Laboratory Research Report Preprint, University of Kent (1991).
23. R.F.C. Walters, *Data types in distributive categories.* Bulletin of the Australian Mathematical Society 40 (1989) 79-82.
24. R.J. Wood, ***V**-indexed categories.* in Springer Lecture Notes in Mathematics 661 (1978) 126-140.
25. G.C. Wraith, *A note on categorical datatypes* in Springer Lecture Notes in Computer Science 389 (1989) 118-127.

Department of Computer Science, University of Calgary, Calgary, Alberta T2N 1N4

E-mail address: robin@cpsc.ucalgary.ca

Department of Computer Science and Engineering, Oregon Graduate Institute, Beaverton, Oregon 97006-1999

E-mail address: dwights@cse.ogi.edu

Canadian Mathematical Society
Conference Proceeding
Volume **13**, 1992

THE ZARISKI CATEGORY OF GRADED COMMUTATIVE RINGS

Yves DIERS

Abstract. The theory of Zariski Categories provides a general framework for commutative algebra and algebraic geometry.This paper shows how this general theory can be successfully applied to graded commutative algebra, for example to the study of algebraically closed graded fields.

Introduction

The theory of Zariski categories was introduced in the book : "Categories of commutative algebras" [3] in order to provide a general framework for commutative algebra and algebraic geometry. We show that this general theory applies with some success to graded commutative algebra. For example, on algebraically closed graded fields, it gives better results than the classical theory.

We focus on the category $\mathbb{G}$**rad**$\mathbb{C}$**Rng** of graded commutative rings since other categories of graded commutative algebras can be derived directly from it. We prove that $\mathbb{G}$**rad**$\mathbb{C}$**Rng** is a Zariski category which is linked by many morphisms of Zariski categories to the Zariski category $\mathbb{C}$**Rng** of commutative rings. The category $\mathbb{G}$**rad**$\mathbb{C}$**Rng** is isomorphic to the category of coactions of a cogroup in $\mathbb{C}$**Rng**, while the category $\mathbb{C}$**Rng** is equivalent to a coslice category of $\mathbb{G}$**rad**$\mathbb{C}$**Rng**.

The general theory defines the notions of simple objects and algebraically closed simple objects, predicts that any simple object has an algebraically closed simple extension, and states a good Nullstellensatz with respect to any algebraically closed simple object. On the other hand in their book : "Graded ring theory" [8], Năstăsescu and Van Oystaeyen describe graded fields and their graded algebraic closure, and state a weak Nullstellensatz. It turns out that the two notions of simple objects and graded fields are identical, but that the two notions of

1991 Mathematic Subject Classification, Primary 13A02, 18B99
This paper is in final form and no version of it will be submitted for publication elsewhere.

algebraic closeness are different. We prove that our notion of algebraically closed simple object is more adequat for graded algebraic geometry.

Any algebraically closed simple object can be considered as a ground graded field in which we take values. Then any adequatly finitely presentable object A in $\mathbb{G}\mathbf{rad}\mathbb{C}\mathbf{Rng}$ naturally defines the algebraic set $\sum(A)$ of points of A with values in the ground graded field. Any homogeneous ideal J of A naturally defines a zero set $Z(J)$ in $\sum(A)$, and any subset X of $\sum(A)$ naturally defines an homogeneous ideal $I(X)$ in A. Our graded Nullstellensatz asserts that $I(Z(J)) = \mathrm{rad}(J)$. Subsets of $\sum(A)$ of the form $Z(J)$ are called algebraic subsets. They are the closed sets of a topology on $\sum(A)$ naturally called the Zariski topology, and are in one-to-one correspondence with radical homogeneous ideals of A. We prove that the graded prime spectrum of A, $\mathrm{Spec}_{gr}(A)$, is precisely the universal sober space associated to the space $\sum(A)$, and that the projective space, $\mathrm{Proj}(A)$, of the graded ring A is precisely the universal sober space associated to the open subspace $\sum(A)\backslash Z(I_0)$ of $\sum(A)$ where I_0 is the ideal of A generated by all the non zero degree homogeneous elements of A. The scheme structures on $\mathrm{Spec}_{gr}(A)$ and $\mathrm{Proj}(A)$ are investigated by using the geometrical morphisms of Zariski categories between $\mathbb{G}\mathbf{rad}\mathbb{C}\mathbf{Rng}$ and $\mathbb{C}\mathbf{Rng}$.

0. Review of the notion of Zariski Category [3]

Let us be in an arbitrary category $\mathbb{A}$. Let $(g, h) : C \rightrightarrows A$ be a pair of parallel morphisms. The pair (g, h) is *codisjointed* if any morphism $u : A \to X$ which satisfies $ug = uh$ necessarily has, as its codomain, a terminal object. A morphism $f : A \to B$ *codisjoints* the pair (g, h) if the pair (fg, fh) is codisjointed. A *codisjunctor* of (g, h) is a morphism $f : A \to B$ which codisjoints (g, h) and such that any morphism $u : A \to X$ which codisjoints (g, h) factors in a unique way through f. The pair (g, h) is *codisjunctable* if it admits a codisjunctor. An object A is *codisjunctable* if the coproduct $A \amalg A$ exists and the pair of inductions $(\alpha, \beta) : A \rightrightarrows A \amalg A$ is codisjunctable. It is *flatly codisjunctable* if the codisjunctor $f : A \to B$ of (α, β) is a flat morphism i.e. the pushout functor along f preserves monomorphisms. The pair (g, h) is a *congruence* if it is a kernel pair of some morphism $f : A \to B$.

A category $\mathbb{A}$ is a *Zariski Category* if it satisfies the following axioms.

i) $\mathbb{A}$ is cocomplete i.e. $\mathbb{A}$ has all small colimits.
ii) $\mathbb{A}$ has a proper, i.e. strong, generating set whose objects are finitely presentable and flatly codisjunctable.
iii) Regular epimorphisms are universal i.e. are stable under pullbacks.
iv) The terminal object of $\mathbb{A}$ is finitely presentable and has non proper subobject.
v) Binary products of objects are co-universal i.e. stable under pushouts.
vi) For any finite sequence of codisjunctable congruences $r_1, \ldots, r_n$ on any object A, with respective codisjunctors $d_1, \ldots, d_n$, we have :

$$r_1 \overset{c}{\bigvee} \cdots \overset{c}{\bigvee} r_n = 1_{A\times A} \Rightarrow d_1 \bigvee \cdots \bigvee d_n = 1_A,$$

where $\overset{c}{\bigvee}$ denotes the joint in the lattice of congruences on A, while $\bigvee$ denotes the co-union of quotient objects of A.

A morphism of Zariski categories is a functor $U : \mathbb{A} \to \mathbb{B}$

(i) whose domain $\mathbb{A}$ and codomain $\mathbb{B}$ are Zariski categories,
(ii) which preserves filtered colimits,
(iii) which has a left adjoint F, and
(iv) which defines an adjunction $\varphi : \mathrm{Hom}_{\mathbb{A}}(F(\cdot), -) \to \mathrm{Hom}_{\mathbb{B}}(\cdot, U(-))$ which preserves and reflects codisjointed pairs of morphisms.

1. The Zariski category $\mathbb{G}$rad$\mathbb{C}$Rng

1.0. *The category* **$\mathbb{G}$rad$\mathbb{C}$Rng** *of graded commutative rings* has the $\mathbb{Z}$-graded commutative rings with units for objects, and the morphisms of rings preserving units and gradations for morphisms. The *forgetful functor* U : **$\mathbb{G}$rad$\mathbb{C}$Rng** $\to$ **$\mathbb{C}$Rng** to the category of commutative rings with units forgets gradations.

1.1. Proposition. *The functor U has a right adjoint G which assigns to a ring A, the graded ring $G(A) = A[T, T^{-1}]$ where T is an homogeneous indeterminate of degree 1.*

Proof. There is an adjunction $\varphi : \mathrm{Hom}(U(\cdot), -) \to \mathrm{Hom}(\cdot, G(-))$ whose value at $A \in$ **$\mathbb{G}$rad$\mathbb{C}$Rng** and $B \in$ **$\mathbb{C}$Rng** is the map $\varphi_{A,B} : \mathrm{Hom}(U(A), B) \to \mathrm{Hom}(A, G(B))$ which assigns to a ring morphism $f : U(A) \to B$, the morphism of graded rings $\varphi_{A,B}(f) : A \to G(B)$ defined by $\varphi_{A,B}(f)(x) = f(x)T^n$ where x is an homogeneous element of degree n in A. □

1.2. Proposition. *The functor U is comonadic and gives rise to the comonad defined by the comultiplicative cogroup $\mathbb{Z}[T, T^{-1}]$ in* **$\mathbb{C}$Rng**.

Proof. The functor U creates equalizers. Therefore it is comonadic according to Beck's theorem [7, p 147, Theorem 1]. The endofunctor UG is identical to the endofunctor $(-) \amalg \mathbb{Z}[T, T^{-1}]$ obtained by taking the coproduct with the object $\mathbb{Z}[T, T^{-1}]$ in **$\mathbb{C}$Rng**. It is straightforward to prove that the comonad structure on UG is precisely that one defined [7, p 137] by the comultiplicative cogroup structure on the object $\mathbb{Z}[T, T^{-1}]$ in **$\mathbb{C}$Rng**, this latter being such that $\mathrm{Hom}(\mathbb{Z}[T, T^{-1}], A)$ is the multiplicative group of invertible elements in the ring A. □

1.3. Corollary. *The category* **$\mathbb{G}$rad$\mathbb{C}$Rng** *is isomorphic to the category of coactions of the comultiplicative cogroup $\mathbb{Z}[T, T^{-1}]$ on objects in* **$\mathbb{C}$Rng** [cf. 7, p 137, p 170].

1.4. Proposition. *The functor G induces an equivalence of coslice categories* ${}^{\mathbb{Z}}G$: **$\mathbb{C}$Rng** $= \mathbb{Z}/$**$\mathbb{C}$Rng** $\to \mathbb{Z}[T, T^{-1}]/$**$\mathbb{G}rad\mathbb{C}$Rng**.

Proof. For any object A in **$\mathbb{C}$Rng** we denote by $j_A : \mathbb{Z} \to A$ the unique morphism. The functor ${}^{\mathbb{Z}}G$ assigns to (A, j_A) the object $(G(A), G(j_A))$ and to a morphism f the morphism $G(f)$. This functor is easily seen to be full and faithful. Let (A, α) be an object in $\mathbb{Z}[T, T^{-1}]/$**$\mathbb{G}$rad$\mathbb{C}$Rng**. The element $t = \alpha(T)$

is invertible and homogeneous of degree 1 in A. Therefore it is transcendental over A_0. Moreover the pair of elements $\{t, t^{-1}\}$ generates the A_0-algebra A because, for any homogeneous element x of degree n in A, the element $a = xt^{-n}$ belongs to A_0 and $x = at^n$. It follows that $A = A_0[t, t^{-1}] \simeq A_0[T, T^{-1}]$ and $(A, \alpha) \simeq (G(A_0), G(j_{A_0}))$. As a result, the functor ${}^{\mathbb{Z}}G$ is essentially surjective. $\square$

1.5. Corollary. *The functor G is the composite of the equivalence of categories ${}^{\mathbb{Z}}G : \mathbb{C}\mathbf{Rng} \to \mathbb{Z}[T, T^{-1}]/\mathbb{G}\mathbf{rad}\mathbb{C}\mathbf{Rng}$ followed by the projection functor $P : \mathbb{Z}[T, T^{-1}]/\mathbb{G}\mathbf{rad}\mathbb{C}\mathbf{Rng} \to \mathbb{G}\mathbf{rad}\mathbb{C}\mathbf{Rng}$.*

1.6. Proposition. *The adjunction defined by (U, G) preserves and reflects codisjointed pairs of morphisms.*

Proof. It is easy to see that for any category with finite coproducts $\mathbb{A}$ and any object A in $\mathbb{A}$, the adjunction defined by the projection functor $A/\mathbb{A} \to \mathbb{A}$ preserves and reflects codisjointed pairs of morphisms [cf. 2]. Then the result follows from the factorization $G = P{}^{\mathbb{Z}}G$. $\square$

1.7. Proposition. *The functors U and G preserve and reflect codisjunctable pairs of morphisms and their codisjunctors.*

Proof. According to [2, Proposition 4.3.], U preserves codisjunctors, and thus it preserves also codisjunctable pairs of morphisms. Let $(f, g) : A \rightrightarrows B$ be a pair of morphisms in $\mathbb{G}\mathbf{rad}\mathbb{C}\mathbf{Rng}$ whose image by U has a codisjunctor $d : U(B) \to D$. According to Proposition 1.6., the pair of morphisms $(\varphi_{B,D}(d)f, \varphi_{B,D}(d)g) = (\varphi_{A,D}(dU(f)), \varphi_{A,D}(dU(g)))$ is codisjointed. It follows that $\varphi_{B,D}(d)$ codisjoints the pair (f, g) and that $U(\varphi_{B,D}(d))$ codisjoints the pair $(U(f), U(g))$. Therefore there exists a unique morphism $\delta : D \to U(G(D))$ such that $\delta d = U(\varphi_{B,D}(d))$. It is straightforward to prove that (D, δ) is a coalgebra for the comonad defined by U, so that D is equipped with a gradation (Proposition 1.2.). Moreover d becomes a morphism of graded rings, easily seen to be the codisjunctor of (f, g) in $\mathbb{G}\mathbf{rad}\mathbb{C}\mathbf{Rng}$. As a result, the functor U creates codisjunctors and therefore it reflects codisjunctable pairs of morphisms and their codisjunctors. According to [2, Propositions 4.0. and 6.2.], the projection functor $P : \mathbb{Z}[T, T^{-1}]/\mathbb{G}\mathbf{rad}\mathbb{C}\mathbf{Rng} \to \mathbb{G}\mathbf{rad}\mathbb{C}\mathbf{Rng}$ preserves and reflects codisjunctable pairs of morphisms and their codisjunctors. Therefore the functor $G = P{}^{\mathbb{Z}}G$ does it also. $\square$

1.8. Theorem. *The category $\mathbb{G}\mathbf{rad}\mathbb{C}\mathbf{Rng}$ is a Zariski category and the functor $G : \mathbb{C}\mathbf{Rng} \to \mathbb{G}\mathbf{rad}\mathbb{C}\mathbf{Rng}$ a morphism of Zariski categories.*

Proof. $\mathbb{G}\mathbf{rad}\mathbb{C}\mathbf{Rng}$ is an algebraic category in the sense of Gabriel-Ulmer [4]. Therefore it is a cocomplete category in which regular epimorphisms are universal. The terminal object in $\mathbb{G}\mathbf{rad}\mathbb{C}\mathbf{Rng}$ is the zero graded ring $\{0\}$ which is finitely presentable and has no proper subobject. The product of a pair of graded rings $\{A, B\}$ is the graded ring $A \times B$ whose gradation is defined by $(A \times B)_n = A_n \times B_n$. This product is co-universal. For any $n \in \mathbb{Z}$, let us denote by $\mathbb{Z}[X^n]$ the graded polynomial ring $\mathbb{Z}[X]$ where X is an homogeneous element of degree n. This object is finitely presentable in $\mathbb{G}\mathbf{rad}\mathbb{C}\mathbf{Rng}$. Moreover it is flatly codisjunctable in $\mathbb{G}\mathbf{rad}\mathbb{C}\mathbf{Rng}$ because its image by U is flatly codisjunctable in $\mathbb{C}\mathbf{Rng}$, and U preserves and reflects codisjunctable objects (Proposition 1.7.) pushouts and monomorphisms. The set of objects $\{\mathbb{Z}[X^n] : n \in \mathbb{Z}\}$ is a proper generating set in $\mathbb{G}$**rad**$\mathbb{C}$**Rng** because we have $\text{Hom}(\mathbb{Z}[X^n], A) \simeq A_n$ for any object A in $\mathbb{G}$**rad**$\mathbb{C}$**Rng**. Finally, De Morgan's law for codisjunctors [Axiom (vi)] holds in $\mathbb{G}$**rad**$\mathbb{C}$**Rng** because it holds in $\mathbb{C}$**Rng**, and the functor U preserves and reflects codisjunctors and monomorphisms. As a result, $\mathbb{G}$**rad**$\mathbb{C}$**Rng** is a Zariski category [3, Definition 1.2.]. The functor G is a morphism of Zariski categories [3, Definition 7.3.1.] according to Proposition 1.6. □

1.9. The functors J_p, U_p, Q_p are defined for any $p \in \mathbb{Z}$ in the following way on objects and in a canonical way on morphisms.

$J_0 : \mathbb{C}\mathbf{Rng} \to \mathbb{G}\mathbf{rad}\mathbb{C}\mathbf{Rng}$ assigns to the ring A the graded trivial ring A in which any element is homogeneous of degree 0.

$J_p : \mathbb{G}\mathbf{rad}\mathbb{C}\mathbf{Rng} \to \mathbb{G}\mathbf{rad}\mathbb{C}\mathbf{Rng}$, for $p \in \mathbb{Z} \setminus \{0\}$, assigns to the graded ring A the same ring A equipped with the new gradation whose set of homogeneous elements of degree n is $\{0\}$ whenever $n \notin p\mathbb{Z}$, and is A_m whenever $n = mp$.

$U_0 : \mathbb{G}\mathbf{rad}\mathbb{C}\mathbf{Rng} \to \mathbb{C}\mathbf{Rng}$ assigns to A the subring A_0.

$U_p : \mathbb{G}\mathbf{rad}\mathbb{C}\mathbf{Rng} \to \mathbb{G}\mathbf{rad}\mathbb{C}\mathbf{Rng}$, for $p \in \mathbb{Z} \setminus \{0\}$, assigns to A the subring of A whose elements are those whose homogeneous components have degree in $p\mathbb{Z}$, and equipped with the gradation whose set of homogeneous elements of degree n is A_{np}.

$Q_0 : \mathbb{G}\mathbf{rad}\mathbb{C}\mathbf{Rng} \to \mathbb{C}\mathbf{Rng}$ assigns to A the quotient ring A/I_0 where I_0 is the ideal generated by the homogeneous elements of degree $n \neq 0$.

$Q_p : \mathbb{G}\mathbf{rad}\mathbb{C}\mathbf{Rng} \to \mathbb{G}\mathbf{rad}\mathbb{C}\mathbf{Rng}$, for $p \in \mathbb{Z} \setminus \{0\}$, assigns to A the quotient ring A/I_p where I_p is the ideal generated by the homogeneous elements of degree $n \notin p\mathbb{Z}$, and equipped with the gradation whose set of homogeneous elements of degree n is $A_{np}/(I_p \cap A_{np})$.

1.10. Theorem. *The functors J_p and U_p are morphisms of Zariski categories.*

Proof. The functor Q_p is left adjoint to J_p, the adjunction $\varphi : \text{Hom}(Q_p(\cdot), -) \to \text{Hom}(\cdot, J_p(-))$ assigning to a morphism $f : Q_p(A) \to B$ the morphism $\varphi_{A,B}(f) = fq_p : A \to J_p(B)$ where $q_p : A \to A/I_p$ is the canonical projection. This assignment preserves and reflects codisjointed pairs of morphisms, and therefore J_p is a morphism of Zariski categories. The functor J_p is left adjoint to the functor U_p, the adjunction $\psi : \text{Hom}(J_p(\cdot), -) \to \text{Hom}(\cdot, U_p(-))$ assigning to a morphism $f : J_p(A) \to B$ the morphism $\psi_{A,B}(f) : A \to U_p(B)$ induced by f. This assignment also preserves and reflects codisjointed pairs of morphisms, and therefore U_p is a morphism of Zariski categories. □

2. Algebraically closed simple objects

2.0. Proposition. *Congruences in* $\mathbb{G}$**rad**$\mathbb{C}$**Rng** *are precisely congruences modulo homogeneous ideals* [8, p 8].

Proof. Let A be an object in $\mathbb{G}$**rad**$\mathbb{C}$**Rng**. The kernel pair of a morphism $f : A \to B$ in $\mathbb{G}$**rad**$\mathbb{C}$**Rng** is the congruence on A modulo the homogeneous ideal $I = \ker(f)$. Conversely if I is an homogeneous ideal of A, the congruence on A modulo I is the kernel pair of the quotient morphism $q_I : A \to A/I$ in $\mathbb{G}$**rad**$\mathbb{C}$**Rng**. □

2.1. Identification. *In the forthcoming, we identify congruences in* **GradCRng** *with homogeneous ideals.*

2.2. Proposition. *Simple objects in* **GradCRng** *are precisely the so-called graded fields [8, A, 1.4.]. They are precisely of the form K or $K[T,T^{-1}]$ where K is a trivially graded field and T is an homogeneous indeterminate of degree $n \in \mathbb{Z} \setminus \{0\}$.*

Proof. The first statement follows from [8, A, Lemma 1.4.4.]. A field K is a simple object in **CRng**. Because morphisms of Zariski categories preserve simple objects [3, Proposition 7.3.2.], $J_0(K) = K, G(K) = K[T,T^{-1}]$ where $\deg(T) = 1$, and $J_n(G(K)) = K[T,T^{-1}]$ where $\deg(T) = n \in \mathbb{Z} \setminus \{0\}$, are simple objects in **GradCRng**. The converse follows from [8, A, Corollary 1.4.3.]. □

2.3. Lemma. *The functor G preserves and reflects finitely presentable morphisms.*

Proof. Let $f : A \to B$ be a morphism in **CRng** and $G(f) : A[T,T^{-1}] \to B[T,T^{-1}]$ be its image by G. For a family of homogeneous elements $(b_i T^{n_i})_{i \in I}$ in $B[T,T^{-1}]$ we have the following equivalences : $(b_i T^{n_i})_{i \in I}$ generates the graded $A[T,T^{-1}]$-algebra $B[T,T^{-1}] \Leftrightarrow (b_i)_{i \in I}$ generates the graded $A[T,T^{-1}]$-algebra $B[T,T^{-1}] \Leftrightarrow (b_i)_{i \in I}$ generates the A-algebra B. It follows that the graded $A[T,T^{-1}]$-algebra $B[T,T^{-1}]$ is finitely presentable if and only if the A-algebra B is finitely presentable. That is to say $G(f)$ is finitely presentable in **GradCRng** if and only if f is finitely presentable in **CRng**. □

2.4. Theorem. *The functor G preserves and reflects algebraically closed simple objects.*

Proof. The functor G preserves simple objects and monomorphisms and thus it preserves simple extensions [3, p 148]. Let us prove that it lifts simple extensions. Let K be a field. According to Proposition 2.2., any simple extension of $G(K)$ in **GradCRng** is of the form $f : K[T,T^{-1}] \to L[T,T^{-1}]$ where L is a field. The element $f(T)$ is of the form λT where $\lambda \in L^*$. If g is the automorphism of the graded L-algebra $L[T,T^{-1}]$ defined by $g(T) = \lambda^{-1}T$, then the morphism $gf : K[T,T^{-1}] \to L[T,T^{-1}]$ is of the form $G(h)$ where $h : K \to L$ is a simple extension. Consequently f is isomorphic to $G(h)$. According to the preveous Lemma 2.3., f is finitely presentable if and only if $G(h)$ is finitely presentable. It follows that : any finitely presentable simple extension of K is an isomorphism if and only if : any finitely presentable simple extension of $G(K)$ is an isomorphism. Because K is a transnœtherian object in **CRng** and $G(K)$ is a transnœtherian object in **GradCRng** [3, Definition 6.1.4.], the object K is algebraically closed in **CRng** if and only if the object $G(K)$ is algebraically closed in **GradCRng**. □

2.5. Corollary. *Algebraically closed simple objects in* **GradCRng** *are precisely objects of the form $K[T,T^{-1}]$ where K is an algebraically closed field and T is an homogeneous indeterminate of degree 1.*

Proof. A trivially graded field K is not algebraically closed since $K[T,T^{-1}]$ is a finitely presentable simple extension of it. A simple object $K[T,T^{-1}]$ with $\deg(T) \neq 1$ is not algebraically closed, since $K[T,T^{-1}]$ where $\deg(T) = 1$, is a finitely presentable simple extension of it. Consequently, and according to Proposition 2.2., algebraically closed simple objects need to be of the form $G(K)$. □

3. Nullstellensatz in $\mathbb{G}$rad$\mathbb{C}$Rng

3.0. Proposition. *The radical of an object A in* **$\mathbb{G}$rad$\mathbb{C}$Rng** *is identical to the radical of the ring $U(A)$.*

Proof. Let us denote by $\mathrm{rad}_{gr}(A)$ the radical of an object A in **$\mathbb{G}$rad$\mathbb{C}$Rng**, and by $\mathrm{rad}(A)$ the radical of the ring $U(A)$. Let $a \in \mathrm{rad}(A)$. For any simple object L in **$\mathbb{G}$rad$\mathbb{C}$Rng** and any morphism $f : A \to L$ we have $f(a) = 0$. Therefore $a \in \mathrm{rad}_{gr}(A)$. Let $a \notin \mathrm{rad}(A)$. Then a has an homogeneous component $a_i \notin \mathrm{rad}(A)$. The non null graded ring $A[a_i^{-1}]$ has a simple quotient $A[a_i^{-1}] \to L$ in **$\mathbb{G}$rad$\mathbb{C}$Rng**. Therefore there exists some morphism $f : A \to L$ with a simple codomain L, such that $f(a) \neq 0$. Consequently $a \notin \mathrm{rad}_{gr}(A)$. □

3.1. Corollary. *Reduced objects in* **$\mathbb{G}$rad$\mathbb{C}$Rng** *are precisely graded reduced rings.*

3.2. Algebraic sets with values in $K[T, T^{-1}]$.
The following notions are defined with respect to a given algebraically closed simple object $K[T, T^{-1}]$ in **$\mathbb{G}$rad$\mathbb{C}$Rng**. Let A be an object in **$\mathbb{G}$rad$\mathbb{C}$Rng** which is a finitely generated k-algebra for some subfield k of K. A *point* of A is a morphisms $x : A \to K[T, T^{-1}]$. The *algebraic set defined* by A is the $\sum(A)$ of points of A. *The function defined by an element* $a \in A$ is the map $\varphi_a : \sum(A) \to K[T, T^{-1}]$ defined by $\varphi_a(x) = x(a)$. The *zero set* of $a \in A$ is the set $Z(a) = \{x \in \sum(A) : \varphi_a(x) = 0\}$. The *zero set of an homogeneous ideal* J of A is the set $Z(J) = \bigcap_{a \in J} Z(a)$. *An algebraic subset* of $\sum(A)$ is a subset of the form $Z(J)$ for some homogeneous ideal J of A. *The ideal of a subset* $X \subset \sum(A)$ is $I(X) = \{a \in A : \varphi_a(X) = \{0\}\}$.

3.3. Theorem (Nullstellensatz in $\mathbb{G}$rad$\mathbb{C}$Rng). *For any homogeneous ideal J in $A, I(Z(J)) = \mathrm{rad}(J)$.*

Proof. Let a be an arbitrary homogeneous element in A. Let $a \in \mathrm{rad}(J)$. Then $a^n \in J$ for some $n \in \mathbb{N}$. For any $x \in Z(J)$, we have $(x(a))^n = x(a^n) = 0$, and thus $x(a) = 0$. It follows that $a \in I(Z(J))$. Consequently, $\mathrm{rad}(J) \subset I(Z(J))$. Let $a \notin \mathrm{rad}(J)$. Then the non null graded ring $A/J[\overline{a}^{-1}]$ has a simple quotient L in **$\mathbb{G}$rad$\mathbb{C}$Rng** [3, Corollary 2.1.9.]. Then we get a finitely presentable simple extension $k \to L$ in **$\mathbb{G}$rad$\mathbb{C}$Rng**. According to [3, Proposition 6.1.3.], there exists some morphism $f : L \to K[T, T^{-1}]$, and therefore some morphism $x : A \to K[T, T^{-1}]$ such that $x(J) = \{0\}$ and $x(a) \neq 0$. Then $x \in Z(J)$ and $a \notin I(Z(J))$. Consequently $I(Z(J)) \subset \mathrm{rad}(J)$. □

3.4. Corollary. *The assignments $Z(-)$ and $I(-)$ establish a one-to-one Galois correspondence between the ordered set of algebraic subsets of $\sum(A)$ and the ordered set of homogeneous radical ideals of A.*

3.5. Corollary. *The set of algebraic subsets of $\sum(A)$ is the set of closed subsets of a topology on $\sum(A)$, called the Zariski topology.*

4. Graded prime spectra

4.0. Proposition. *Integral objects in* **GradCRng** *are precisely graded integral domains.*

Proof. According to Proposition 2.2., any simple object in **GradCRng** is an integral domain. Any integral object in **GradCRng** being a subobject of a simple one, is an integral domain. Conversely let A be a graded integral domain. The graded ring of fractions $A[S^{-1}]$ where S is the set of non null homogeneous elements of A, is a simple object in **GradCRng**. Then A is an integral object in **GradCRng** as being a subobject of $A[S^{-1}]$. □

4.1. Corollary. *Prime congruences in* **GradCRng** *are precisely congruences modulo homogeneous prime ideals.*

According to the Identification 2.1. prime congruences will be identified with homogeneous prime ideals. We denote by $\mathrm{Spec}_{gr}(A)$ the prime spectrum of A in **GradCRng**, while $\mathrm{Spec}(A)$ denotes the prime spectrum of the ring $U(A)$ in **CRng**.

4.2. Proposition. *$\mathrm{Spec}_{gr}(A)$ is a subspace of $\mathrm{Spec}(A)$.*

Proof. $\mathrm{Spec}_{gr}(A)$ is a subset of $\mathrm{Spec}(A)$ according to Corollary 4.1.. For any ideal J in A, let us denote by $H(J)$ the homogeneous ideal of A generated by J. Then the open set $D(J)$ of $\mathrm{Spec}(A)$ is such that $D(J) \cap \mathrm{Spec}_{gr}(A) = D(H(J)) \cap \mathrm{Spec}_{gr}(A)$ is the open set $D(H(J))$ of $\mathrm{Spec}_{gr}(A)$. Consequently the Zariski topology on $\mathrm{Spec}(A)$ induces the Zariski topology on $\mathrm{Spec}_{gr}(A)$. □

4.3. Theorem. *$\mathrm{Spec}_{gr}(A)$ is the universal sober space associated to $\sum(A)$.*

Proof. Let $\gamma_A : \sum(A) \to \mathrm{Spec}_{gr}(A)$ be the map defined by $\gamma_A(x) = \ker(x)$. According to [3, Proposition 3.1.7.] the closed sets of $\mathrm{Spec}_{gr}(A)$ are of the form $V(J)$ where J is some homogeneous ideal of A. For any $x \in \sum(A)$, we have : $\gamma_A(x) \in V(J) \Leftrightarrow J \subset \ker(x) \Leftrightarrow x \in Z(J)$. It follows that $\gamma_A^{-1}(V(J)) = Z(J)$ and that γ_A is continuous. Futhermore, if J_1, J_2 are homogeneous ideals of A such that $\gamma_A^{-1}(V(J_1)) = \gamma_A^{-1}(V(J_2))$, then $Z(J_1) = Z(J_2), \mathrm{rad}(J_1) = I(Z(J_1)) = I(Z(J_2)) = \mathrm{rad}(J_2)$ according to the Nullstellensatz, and thus $V(J_1) = V(\mathrm{rad}(J_1)) = V(\mathrm{rad}(J_2)) = V(J_2)$. It follows that the continuous map γ_A induces a bijection between the set of closed sets of $\mathrm{Spec}_{gr}(A)$ and the set of closed sets of $\sum(A)$. That is to say, γ_A is a quasi-homeomorphism [5, Chapitre 0, Definition 2.7.2.]. Because $\mathrm{Spec}_{gr}(A)$ is a sober space [3, Theorem 3.1.9.], $\gamma_A : \sum(A) \to \mathrm{Spec}_{gr}(A)$ is the universal sober space associated to $\sum(A)$. □

Let us recall that I_0 denotes the ideal of A generated by all non zero degree homogeneous elements of A (cf. 1.9.). According to [6, p 76] the open set $D(I_0)$ of $\mathrm{Spec}_{gr}(A)$ is *the projective*, $\mathrm{Proj}_{gr}(A)$, *of* the graded ring A. Let us denote by $\sum_*(A)$ the open subspace $\sum(A) \setminus Z(I_0)$ of $\sum(A)$.

4.4. Theorem. *$Proj_{gr}(A)$ is the universal sober space associated to $\sum_*(A)$.*

Proof. $\sum_*(A)$ is the inverse image of $\mathrm{Proj}_{gr}(A)$ by the map $\gamma_A : \sum(A) \to \mathrm{Spec}_{gr}(A)$. Then the result follows from Theorem 4.3. and the fact that $\mathrm{Proj}_{gr}(A)$ is a sober space [3, Proposition 4.3.6.]. □

5. Graded schemes

A scheme on the Zariski category $\mathbb{G}$**rad**$\mathbb{C}$**Rng** [3, Definition 4.3.4.] will be called a *graded scheme*. For example, $\mathrm{Spec}_{gr}(A)$ is canonically an affine graded scheme, and $\mathrm{Proj}_{gr}(A)$ is canonically an open graded subscheme of $\mathrm{Spec}_{gr}(A)$. It is not the case that any morphism of Zariski categories extends naturally to categories of schemes, but it is true for geometrical morphisms [3, Proposition 7.3.13].

5.0. Proposition. *G and J_p are geometrical morphisms of Zariski categories.*

Proof. The proof is straightforward using [3, Proposition 7.3.13.]. □

5.1. Corollary. *For any ring A, we have the following isomorphisms of affine graded schemes :*

$$\left(Spec_{gr}(J_0(A)), \widetilde{J_0(A)}\right) \simeq \left(Spec(A), J_0\widetilde{A}\right)$$

$$(Spec_{gr}A[T,T^{-1}])\ ,\ \widetilde{A[T,T^{-1}]} \simeq \left(Spec(A), G\widetilde{A}\right).$$

The notion of geometrical morphisms of Zariski categories can be localized in the following way. A morphism of Zariski categories $U : \mathbb{A} \to \mathbb{B}$ is said to be *geometrical at an object* A in $\mathbb{A}$ if U preserves codisjunctable congruences on A and lifts prime congruences on $U(A)$.

5.2. Theorem. *The morphism of Zariski categories U_0 is geometrical at any object in* $\mathbb{G}$**rad**$\mathbb{C}$**Rng** *having an invertible homogeneous element of non zero degree.*

Proof. Let A be an object in $\mathbb{G}$**rad**$\mathbb{C}$**Rng** having an invertible homogeneous element a of degree $p \neq 0$.

a) Let us prove that U_0 induces an isomorphism of ordered sets $\mathrm{CongRad}_{gr}(A) \simeq \mathrm{CongRad}(A_0)$ assigning to an homogeneous radical ideal I of A, the ideal $I_0 = I \cap A_0$ of A_0. Let $I, J \in \mathrm{CongRad}_{gr}(A)$ such that $I_0 = J_0$. Let $x \in I_n$. The element $y = x^p a^{-n}$ is homogeneous of degree 0. Then $y \in I_0$. Thus $y \in J_0, x^p = ya^n \in J$, and $x \in J$. Consequently $I \subset J$. Similarly we have $J \subset I$, and thus $I = J$. Let $I \in \mathrm{CongRad}(A_0)$. Let $J = \mathrm{rad}(IA)$. Then $I \subset J_0$. Let $x \in J_0$. There exist some $n \in \mathbb{N}, y_1, \ldots, y_n \in I$ and $z_1, \ldots, z_n \in A_0$ such that $x^n = y_1z_1 + \cdots + y_nz_n$. Then $x^n \in I$ and $x \in I$. It follows that $I = J_0$. As a result the assignment $I \longmapsto I_0$ is one-to-one.

b) Let us prove that U_0 lifts prime congruences on $U_0(A)$. According to a) it is enough to prove that an ideal $I \in \mathrm{CongRad}_{gr}(A)$ is prime whenever

I_0 is prime in A_0. Let $x \in A_n$ and $y \in A_m$ be such that $xy \in I$. Then $(x^p x^{-n})(y^p a^{-m}) = (xy)^p a^{-(n+m)} \in I_0$. If I_0 is prime we have : $x^p a^{-n} \in I_0$ or $y^p a^{-n} \in I_0$, and thus : $x^p \in I$ or $y^p \in I$, and therefore : $x \in I$ or $y \in I$. Consequently I is prime.

c) Let us prove that U_0 preserves finitely generated n-projective ideals of A [1, Definition 2.4.]. Let I be such an ideal of A. According to [1, Proposition 2.5.], there exist a sequence $e_1, \dots, e_k$ of elements of I and a sequence $\varphi_1, \dots, \varphi_k$ of A-linear maps $I^n \to A$ such that : $\forall x \in I^n, x = \varphi_1(x)e_1 + \cdots + \varphi_k(x)e_k$. We can assume that $e_1, \cdots, e_k$ are homogeneous elements of degree $r_1, \dots, r_k$ respectively. There exist some $m \in \mathbb{N}^*$ and a sequence of A-linear maps $\psi_1, \dots, \psi_k : I^m \to A$ such that : $\forall x \in I^m, x = \psi_1(x)e_1^p + \cdots + \psi_k(x)e_k^p$. Then : $\forall x \in I^m, x = \psi_1(x)a^{r_1}(e_1^p a^{-r_1}) + \cdots + \psi_k(x)a^{r_k}(e_k^p a^{-r_k})$. The elements $b_1 = e_1^p a^{-r_1}, \cdots, b_k = e_k^p a^{-r_k}$ are homogeneous of degree 0 i.e. they belongs to I_0, and there exists a sequence of A_0-linear maps : $\psi_{1o}, \dots, \psi_{ko} : I_0^m \to A_0$ such that : $\forall x \in I_0^m, x = \psi_{1o}(x)b_1 + \cdots + \psi_{ko}b_k$. Consequently I_0 is a n-projective ideal of A_0.

d) Let us prove that U_0 preserves codisjunctors of finitely generated n-projective ideals of A. Let I be such an ideal of A. Then I is codisjunctable in $\mathbb{C}\mathbf{Rng}$ [1, Theorem 2.6.] and also in $\mathbb{G}\mathbf{rad}\mathbb{C}\mathbf{Rng}$ (Proposition 1.7.). Let $d : A \to D$ be the codisjunctor of I in $\mathbb{G}\mathbf{rad}\mathbb{C}\mathbf{Rng}$. According to c), I_0 is a finitely generated n-projective ideal of A_0. Let $d_0 : A_0 \to D_0$ be the codisjunctor of I_0 in $\mathbb{C}\mathbf{Rng}$. Since codisjunctors are couniversal, and since $\mathrm{rad}(I) = \mathrm{rad}(AI_0)$ according to a), the morphism d is the pushout along the inclusion morphism $A_0 \to A$ of the morphism d_0. This pushout can be taking in $\mathbb{G}\mathbf{rad}\mathbb{C}\mathbf{Rng}$ since the functor U creates pushouts. Then $D = D_0 \otimes_{A_0} A$. Any homogeneous element of degree zero in D is of the form $x = a_1 \otimes x_1 + \cdots + a_n \otimes x_n$ where $a_1, \dots, a_n \in D_0$ and $x_1, \dots, x_n \in A$ are such that $\deg(a_i) + \deg(x_i) = 0$ which imply that $x_1, \dots, x_n \in A_0$. Then $x = x_1(a_1 \otimes 1) + \cdots + x_n(a_n \otimes 1) \in D_0$. Consequently D_0 is precisely the ring of homogeneous elements of degree zero in D. That is to say $U_0(d) = d_0$ is the codisjunctor of I_0.

e) Let us prove that U_0 preserves codisjunctors of ideals on A. Let I be a codisjunctable ideal of A with codisjunctor $d : A \to D$. According to [1, Theorem 3.0.], I contains a finitely generated codisjunctable ideal J such that $\mathrm{rad}(J) = \mathrm{rad}(I)$. According to [1, Theorem 4.3.] the dense ideal $\overline{J}$ associated to J in the quotient ring $\overline{A} = A / \bigcup_{n \in \mathbb{N}} \mathrm{Ann}(J^n)$ is n-projective for some $n \in \mathbb{N}^*$, and the codisjunctor d factorizes in the form $d = \overline{d}q$ where $q : A \to \overline{A}$ is the quotient morphism and $\overline{d}$ is the monomorphic codisjunctor of $\overline{J}$. According to d), $\overline{d}_0 : \overline{A}_0 \to D_0$ is the monomorphic codisjunctor of $(\overline{J})_0$. According to [1, Proposition 4.0.], $(\overline{J})_0$ is a dense ideal in $\overline{A}_0$, and it is indeed the dense ideal associated to J_0. According to [1, Theorem 4.3.], J_0 is codisjunctable with codisjunctor $d_0 = \overline{d}_0 q_0 : A_0 \to D_0$. The relation $\mathrm{rad}(I_0) = \mathrm{rad}(J_0)$ implies that I_0 is codisjunctable with codisjunctor d_0. □

5.3. Corollary. *The direct image by U_0 of the graded scheme $Proj_{gr}(A)$ is a scheme. It is the scheme $Proj(A)$.*

Proof. According to the proof of [3, Proposition 7.3.13], the functor U_0^{sp} : $\mathbb{S}\mathrm{ch}\mathbb{G}\mathbf{rad}\mathbb{C}\mathbf{Rng} \to \mathbb{S}\mathrm{ch}\mathbb{C}\mathbf{Rng}$ preserves affine schemes of objects of the form

$A[a^{-1}]$ where a is an homogeneous element of A of non zero degree. It follows that U_0^{sp} preserves schemes which are locally of the preveous form. But this is precisely the case for the graded scheme $\mathrm{Proj}_{gr}(A)$. Therefore $U_0^{sp}(\mathrm{Proj}_{gr}(A))$ is a scheme. This is precisely the scheme $\mathrm{Proj}(A)$ according to its Definition [6, p 76]. □

REFERENCES

1. Y. Diers, Codisjunctors and singular epimorphisms in the category of commutative rings, J. Pure. Appl. Algebra **53** (1988), 39-57.
2. Y. Diers, Codisjoncteurs dans les catégories d'algèbres commutatives, Cahiers de topologie et géométrie différentielle catégoriques **XXVIII-1** (1987), 5-28.
3. Y. Diers, *Categories of commutative algebras*, Oxford University Press, New York, 1992.
4. P. Gabriel and F. Ulmer, *Local Präsentierbare Kategorien*, Lecture Notes in Math, 211, Springer Verlag, Berlin-Heidelberg-New York, 1971.
5. A. Grothendieck and J.A. Dieudonné, *Eléments de géométrie algébrique I*, Springer Verlag, Berlin-Heidelberg-New York, 1971.
6. R. Hartshorne, *Algebraic geometry*, Springer Verlag, Berlin-Heidelberg-New York, 1977.
7. S. Mac Lane, *Categories for the Working Mathematician*, Springer Verlag, Berlin-Heidelberg-New York, 1971.
8. C. Năstăsescu and F. Van Oystaeyen, *Graded ring theory*, North-Holland publishing Company, Amsterdam-New York-Oxford, 1982.

Département de Mathématiques
Université de Valenciennes
B.P. 311
F-59304 VALENCIENNES - FRANCE

Canadian Mathematical Society
Conference Proceeding
Volume 13, 1992

Completeness in Continuity Spaces

ROBERT C. FLAGG

ABSTRACT. We apply enriched category theory to study Cauchy completeness in continuity spaces. Our main result is the equivalence in continuity spaces of the category theoretic and the uniform notions of Cauchy completeness. This theorem, which generalizes a result of Lawvere for quasi-metric spaces, makes a natural connection between the category-theoretic and topological aspects of continuity spaces, and provides a simple construction of the Cauchy completion of a continuity space.

The theory of continuity spaces, developed by Kopperman [5], generalizes the theory of metric spaces by replacing the nonnegative real number system by other structures with the appropriate abstract properties to allow their elements to play the role of distances. Since all topologies are "metrizable" in this general setting, the elegant techniques and elementary intuition of the theory of metric spaces are thus extended to arbitrary topological spaces.

In [2], we reformulate Kopperman's theory so that a continuity space can be defined simply as a category enriched in a value quantale. In this way the entire machinery of enriched category theory (Kelly [4]) is immediately available for the study of continuity spaces. The goal of the present paper is to apply some of this machinery to study Cauchy completeness in continuity spaces. Our main result is the equivalence in continuity spaces of the category theoretic and the uniform notions of Cauchy completeness. This theorem, which generalizes a result of Lawvere [6] for quasi-metric spaces, makes a natural connection between the category-theoretic and topological aspects of continuity spaces, and provides a simple construction of the Cauchy completion of a continuity space.

1991 Mathematics Subject Classification. Primary 06B30, 06F20, 54E99, 54F05.
This paper is in final form and no version of it will be submitted for publication elsewhere.

A value quantale is a symmetric monoidal closed category, which is also a value distributive lattice. Since in a poset "all diagrams commute", the specialization of enriched category theory to continuity spaces admits a very elementary formulation. This fact makes possible a self-contained presentation of the few standard results from enriched category theory we need in our treatment of Cauchy completeness for continuity spaces.

1. Value Quantales

Value quantales are the abstract structures whose elements play the role of distances in the theory of continuity spaces. In this section we recall the definition of a value quantale and state a few basic properties needed below. A detailed presentation of this material can be found in [2].

Assume L is a complete lattice. We use the following notation below. The bottom element of L is denoted by 0 and the top element by ∞. For A a subset of L, the least upper bound of A is denoted by $\sup A$ and the greatest lower bound of A by $\inf A$.

1.1. Definition. Assume L is a complete lattice and $x, y \in L$. Then x is *well above* y, denoted by $x \succ y$, iff for any subset $A \subseteq L$, if $y \geq \inf A$, then for some $a \in A$, $x \geq a$.

1.2. Definition. Assume L is a complete lattice. Then L is *completely distributive* iff for all $y \in L$,

$$y = \inf \{x \in X \mid x \succ y \}.$$

1.3. Definition. A *value distributive lattice* is a completely distributive lattice L satisfying the following two conditions:

1) $\infty \succ 0$.
2) If $p \succ 0$ and $q \succ 0$, then $p \wedge q \succ 0$.

The notion of value distributive lattice captures what we regard as the essential *lattice theoretic* properties of the nonnegative real number system. To capture the essential properties of the operation of addition, we use the notion of a quantale.

1.4. Definition. A *quantale* $\mathcal{V} =< V, \leq, + >$ consists of a complete lattice $< V, \leq>$ and an associative and commutative binary operation $+$ on V satisfying:

(q1) the bottom element of V is a unit for $+$: for all $p \in V$,

$$p + 0 = p;$$

(q2) for all $p \in V$ and all families $\{q_i\}_{i \in I}$ of elements of V,

$$p + \inf_{i \in I} q_i = \inf_{i \in I} (p + q_i).$$

Assume $\mathcal{V} =< V, \leq, + >$ is a quantale. For any $p, q \in V$, let

$$q \dot{-} p = \inf \{r \in V | p + r \geq q\}.$$

Then the map $\_ \dot{-} p : V \to V$ is *left adjoint* to the map $\_ + p : V \to V$: for any $q, r \in V$,

$$p + r \geq q \iff r \geq q \dot{-} p.$$

1.5. Lemma. *Assume $\mathcal{V} =< V, \leq, + >$ is a quantale. Then for all $p, q, r \in V$,*

1. $p \dot{-} p = 0$;
2. $(q \dot{-} r) + (r \dot{-} p) \geq (q \dot{-} p)$.

1.6. Lemma. *Assume $\mathcal{V} =< V, \leq, + >$ is a quantale. Then for $p \in V$ and $\{q_i\}_{i \in I}$ a family of elements of V,*

$$p \dot{-} \inf_{i \in I} q_i = \sup_{i \in I} (p \dot{-} q_i) \quad \textit{and} \quad (\sup_{i \in I} q_i) \dot{-} p = \sup_{i \in I} (q_i \dot{-} p).$$

1.7. Definition. A *value quantale* is a quantale $\mathcal{V} =< V, \leq, + >$ such that $< V, \leq >$ is a value distributive lattice.

1.8. Example. (*TRUTH VALUES.*) Let **2** be the two element set $\{0, \infty\}$, where $\infty \geq 0$. Then **2** is a value distributive lattice. Moreover, the binary operation of join, $\vee$, makes **2** a value quantale.

1.9. Example. (*DISTANCES.*) Let $\mathcal{D} = [0, \infty]$ be the extended non-negative reals with the usual ordering. Then $\mathcal{D}$ is a value distributive lattice. The ordinary operation of addition makes $\mathcal{D}$ a value quantale.

1.10. Example. (*DISTANCE DISTRIBUTION FUNCTIONS.*) Let $\mathcal{M}$ be the set of monotone maps from $[0, \infty)$ to $[0, 1]$, ordered pointwise. Then $\mathcal{M}$ is a complete poset with sups and infs computed pointwise. We call $F \in \mathcal{M}$ a *distance distribution function* (d.d.f.) iff F is *left-continuous*:

$$\text{for all } x \in [0, \infty), \ \sup_{y < x} F(y) = F(x).$$

Let Δ be the collection of all d.d.f.'s with the opposite of the pointwise ordering. Then Δ is a value distributive lattice. An associative and commutative binary function $\otimes : \Delta \times \Delta \to \Delta$ which satisfies conditions (q1) and is monotone in each variable is called a *triangle function* ([7], p. 97). Many of the triangle functions discussed in Schweizer and Sklar [7] also satisfy condition (q2). We therefore obtain many examples of value quantales consisting of distance distribution functions.

1.11. Example. (*FREE LOCALES.*) Let R be a set. We write $F \subseteq_f R$ to indicate that F is a finite subset of R. Let $\mathcal{F}(R)$ be the collection of finite subsets of R. A subset p of $\mathcal{F}(R)$ is called a *lower set* iff for all $F, G \in \mathcal{F}(R)$, if $F \in p$ and $G \subseteq F$, then $G \in p$. Let $\Omega(R)$ denote the collection of lower sets of finite subsets of R, ordered by reverse inclusion. Then $\Omega(R)$ is a value distributive lattice. Moreover, the binary operation $\cap$ makes $\Omega(R)$ a value quantale.

1.12. Theorem. *Assume $\mathcal{V} =< V, \leq, + >$ is a value quantale and $\epsilon \in V$. If $\epsilon \succ 0$, then for some δ, $\delta \succ 0$ and $\epsilon \succ 2\delta$.*

1.13. Theorem. *Assume $\mathcal{V} =< V, \leq, + >$ is a value quantale and $p \in V$. Then*

$$p = \inf \{p + \epsilon \mid \epsilon \succ 0\}.$$

2. Continuity Spaces

We assume for the remainder of this paper that $\mathcal{V} =< V, \leq, + >$ is a value quantale.

2.1. Definition. A *$\mathcal{V}$-continuity space* is a pair $\mathcal{X} =< X, d >$ consisting of a set X and a function $d : X \times X \to V$ which is *reflexive*:

$$d(x, x) = 0, \quad \text{for all } x \in X,$$

and *transitive*:

$$d(x, y) \leq d(x, z) + d(z, y), \quad \text{for all } x, y, z \in X.$$

2.2. Definition. Assume $\mathcal{X} =< X, d_X >$ and $\mathcal{Y} =< Y, d_Y >$ are $\mathcal{V}$-continuity spaces, and $f : X \to Y$.

1) f is a *$\mathcal{V}$-functor* iff for all $x_1, x_2 \in X$,

$$d_X(x_1, x_2) \geq d_Y(f(x_1), f(x_2)).$$

2) f is a *full and faithful $\mathcal{V}$-functor* (or an *isometry*) iff for all $x_1, x_2 \in X$,

$$d_X(x_1, x_2) = d_Y(f(x_1), f(x_2)).$$

2.3. Example. $\mathcal{V}$ can itself be made a $\mathcal{V}$-continuity space by defining $d_V : V \times V \to V$ as

$$d_V(p, q) = (q \dot{-} p), \quad \text{for all } p, q \in V.$$

2.4. Definition. Assume $\mathcal{X} = < X, d >$ is an $\mathcal{V}$-continuity space.

1. The *dual* of $\mathcal{X}$ is the pair $\mathcal{X}^* = (X, d^*)$, where for all $x, y \in X$,

$$d^*(x, y) = d(y, x).$$

2. The *symmetrization* of $\mathcal{X}$ is the pair $\mathcal{X}^s = (X, d^s)$, where for all $x, y \in X$

$$d^s(x, y) = d(x, y) + d(y, x).$$

It easily follows that $\mathcal{X}^*$ and $\mathcal{X}^s$ are $\mathcal{V}$-continuity spaces.

There is a natural topology on any continuity space, which is defined in a way completely analogous to the definition of the metric topology on a metric space.

2.5. Definition. Assume $\mathcal{X} = (X, d)$ is a $\mathcal{V}$-continuity space.

1) Assume $x \in X$, and $\epsilon \succ 0$. Then the *open ball of radius ϵ about x* is the set

$$B_\epsilon(x) = \{y \in X \mid \epsilon \succ d(x, y) \}.$$

2) Assume $U \subseteq X$. Then U is *open* in $\mathcal{X}$ iff for all $x \in U$ there is an $\epsilon \succ 0$ such that $B_\epsilon(x) \subseteq U$.

We write $\tau(\mathcal{X})$ for the collection of all open subsets of the $\mathcal{V}$-continuity space $\mathcal{X} = (X, d)$. $\tau(\mathcal{X})$ is a topology on X, which we call the *induced topology* on $\mathcal{X}$.

The following characterization of closed sets in the induced topology is needed below.

2.6. Definition. Assume $\mathcal{X} = (X, d)$ is a $\mathcal{V}$-continuity space, $x \in X$, and $A \subseteq X$. Then the *distance from x to A*, denoted by $d(x, A)$, is given by

$$d(x, A) = \inf_{a \in A} d(x, a).$$

2.7. Lemma. *Assume $\mathcal{X} = (X, d)$ is a $\mathcal{V}$-continuity space and $A \subseteq X$. Then A is closed in $\mathcal{X}$ iff for all $x \in X$, if $d(x, A) = 0$, then $x \in A$.*

There are two other natural topologies on a continuity space, determined by the induced topologies on its dual and symmetrization.

2.8. Definition. Assume $\mathcal{X} = (X, d)$ is a $\mathcal{V}$-continuity space. We call the induced topology on $\mathcal{X}^*$ the *dual topology* on $\mathcal{X}$ and the induced topology on $\mathcal{X}^s$ the *symmetric topology* on $\mathcal{X}$.

2.9. Theorem. *Assume $\mathcal{X} = (X, d)$ is a $\mathcal{V}$-continuity space. Then the symmetric topology on $\mathcal{X}$ is the join of the induced and the dual topologies on $\mathcal{X}$.*

2.10. Example. (*TRUTH VALUES.*) A **2**-continuity space $\mathcal{X} = (X, d)$, where **2** is the value quantale of truth values, is a *preordered set*. The induced topology on $\mathcal{X} = (X, d)$ is the Alexandrov topology on $\mathcal{X}$.

2.11. Example. (*DISTANCES.*) A $\mathcal{D}$-continuity space, where $\mathcal{D}$ is the value quantale of distances, is a quasi-metric space. The induced topology on (X, d) is the standard metric topology on (X, d).

2.12. Example. (*DISTANCE DISTRIBUTION FUNCTIONS.*) Let $\mathcal{V} = (\Delta, \geq, \otimes)$ be a value quantale of distance distribution functions. A $\mathcal{V}$-continuity space $\mathcal{X} = (X, d)$ is a probabilistic metric space. The induced topology on $\mathcal{X}$ is the *strong topology* [7] on the probabilistic metric space $\mathcal{X}$.

2.13. Example. (*FREE LOCALES.*) Generalizing the definition of Henriksen and Kopperman [3], we define a *structure space* to be an $\Omega(R)$-continuity space for some set R. Assume R is a commutative ring with identity and let X be the set of prime ideals on R. Define $d_X : X \times X \to \Omega(R)$ by

$$d_{\Omega(R)}(P, Q) = \{F | F \subseteq_f R \text{ and } F \cap Q \subseteq P\}, \quad \text{for all } P, Q \in X.$$

Then (X, d_X) is an $\Omega(R)$-continuity space. The induced topology on the structure space $\mathcal{X}$ is the hull-kernel topology on X, the dual topology on $\mathcal{X}$ is the dual hull-kernel topology. By Theorem 2.9, the symmetric topology on $\mathcal{X}$ is the patch topology on X. These results can be generalized considerably (cf. Henriksen and Kopperman [3]).

2.14. Theorem. (Kopperman [5], Flagg [2]). *Assume (X, τ) is a topological space. Then there is a value quantale $\mathcal{V}_\tau$ and a $\mathcal{V}_\tau$-continuity space $\mathcal{X} = (X, d_\tau)$ such that τ is the induced topology on $\mathcal{X}$.*

3. Completeness

The theory of completeness in continuity spaces can be seen as a very direct generalization of the corresponding theory for ordinary posets. This generalization is made more transparent if we interpret the elements of a value quantale as *truth values* (Lawvere [6]). We then have the following correspondence for (X, d) a $\mathcal{V}$-continuity space, $x, y \in X$, $p, q \in \mathcal{V}$, $\{p_x\}_{x \in X}$ a family of elements of $\mathcal{V}$, and $\phi : X \to \mathcal{V}$:

$$\begin{aligned}
d(x,y) &\longleftrightarrow x \le y;\\
\phi(x) &\longleftrightarrow x \in \phi;\\
0 &\longleftrightarrow \text{true};\\
(p+q) &\longleftrightarrow (p \mathbin{\&} q);\\
(q \dot{-} p) &\longleftrightarrow (p \Rightarrow q);\\
\sup_{x\in X} p_x &\longleftrightarrow \forall x \in X.\, p_x;\\
\inf_{x\in X} p_x &\longleftrightarrow \exists x \in X.\, p_x;\\
p \ge q &\longleftrightarrow q \vdash p;\\
p = q &\longleftrightarrow q \dashv\vdash p.
\end{aligned}$$

Assume X is a $\mathcal{V}$-continuity space. By an *order ideal* on X, we understand a $\mathcal{V}$-functor $\phi : X^* \to \mathcal{V}$. Thus $\phi : X \to V$ is an order ideal on X iff for all $x, y \in X$,

$$\phi(x) + d(y,x) \ge \phi(y),$$

or in the notation above,

$$x \in \phi \mathbin{\&} y \le x \vdash y \in \phi.$$

Similar translations can be given for the other definitions and results below.

By an *order filter* on X, we understand a $\mathcal{V}$-functor $\psi : X \to \mathcal{V}$.

3.1. Definition. Assume X is a $\mathcal{V}$-continuity space, ϕ is an order ideal on X, and $a \in X$. Then a is the *supremum* of ϕ, denoted by $a = \bigvee \phi$, iff for all $b \in X$,

$$d(a,b) = \sup_{x\in X} (d(x,b) \dot{-} \phi(x)).$$

3.2. Definition. Assume X is a $\mathcal{V}$-continuity space, ψ is an order filter on X, and $a \in X$. Then a is the *infimum* of ψ, denoted by $a = \bigwedge \psi$, iff for all $b \in X$,

$$d(b,a) = \sup_{x\in X} (d(b,x) \dot{-} \psi(x)).$$

3.3. Remark. *The supremum of an order ideal on a continuity space is a special case of the notion of indexed co-limit (Kelly [4], p. 73). Indeed, if X is a $\mathcal{V}$-continuity space and ϕ is an order ideal on X, then $\bigvee \phi$ is just the indexed co-limit $\phi * 1_X$. Moreover, given a $\mathcal{V}$-continuity space K and a pair of $\mathcal{V}$-functors $f : K^* \to V$ and $g : K \to X$, let $\phi : X^* \to V$ be defined by*

$$\phi(x) = \inf_{k\in K} (f(k) + d(x, g(k))).$$

Then ϕ is an order ideal on X, $\bigvee \phi$ exists iff $f * g$ exists, and in this case $\bigvee \phi = (f * g)$. Similar remarks apply to infima and indexed limits.

3.4. Theorem. *Assume X is a $\mathcal{V}$-continuity space. Then the following two conditions are equivalent.*

1. *For every order ideal ϕ on X, $\bigvee \phi$ exists.*
2. *For every order filter ψ on X, $\bigwedge \psi$ exists.*

PROOF.

Assume (1). Let ψ be an order filter on X. Define $\phi : X \to V$ by the condition:

$$(\alpha) \qquad \phi(y) = \sup_{x \in X} (d(y,x) \dot{-} \psi(x)), \qquad \text{for all } y \in X.$$

Then ϕ is an order ideal on X. Let $a = \bigvee \phi$. Then for all $b \in X$,

$$(\beta) \qquad d(a,b) = \sup_{y \in X} (d(y,b) \dot{-} \phi(y)).$$

Since

$$\sup_{y \in X} (\sup_{x \in X} (d(y,x) \dot{-} \psi(x)) \dot{-} \phi(y)) = \sup_{x \in X} (\sup_{y \in X} (d(y,x) \dot{-} \phi(y)) \dot{-} \psi(x)),$$

from (α) and (β), we get

$$0 = \sup_{x \in X} (d(a,x) \dot{-} \psi(x));$$

that is, for all $x \in X$,

$$\psi(x) \geq d(a,x).$$

Thus for all $x, b \in X$,

$$d(b,a) + \psi(x) \geq d(b,a) + d(a,x) \geq d(b,x).$$

Hence for all $b \in X$,

$$d(b,a) \geq \sup_{x \in X} (d(b,x) \dot{-} \psi(x)).$$

Moreover for all $b \in X$,

$$\sup_{x \in X} (d(b,x) \dot{-} \psi(x)) = \phi(b) \geq d(b,a).$$

Evidently then for all $b \in X$,

$$d(b,a) = \sup_{x \in X} (d(b,x) \dot{-} \psi(x)).$$

Hence $a = \bigwedge \psi$.

The proof of the converse is dual.

Q.E.D.

We call a $\mathcal{V}$-continuity space X *complete* iff it satisfies the equivalent conditions of Theorem 3.4. By Remark 3.3, this definition agrees with the standard notion of completeness in enriched category theory.

3.5. Remark. *As we noted in Example 2.3, $\mathcal{V}$ itself is a $\mathcal{V}$-continuity space. An elementary calculation, shows that for ϕ an order ideal on $\mathcal{V}$, $\bigvee \phi$ exists and*

$$\bigvee \phi = \inf_{q \in V} (q + \phi(q)).$$

Similarly, for ψ an order filter on $\mathcal{V}$, $\bigwedge \phi$ exists and

$$\bigwedge \psi = \sup_{q \in V} (q \dot{-} \psi(q)).$$

3.6. Theorem. *$\mathcal{V}$ is a complete $\mathcal{V}$-continuity space.*

For our construction below of the Cauchy completion of a continuity space, we need the standard result from enriched category theory that any $\mathcal{V}$-continuity space can be embedded in a complete one.

Assume X is an $\mathcal{V}$-continuity space. Let $\hat{X}$ denote the set of all order ideals on X. $\hat{X}$ can be made into a $\mathcal{V}$-continuity space by defining

$$d_{\hat{X}}(\phi_1, \phi_2) = \sup_{x \in X} (\phi_2(x) - \phi_1(x)), \quad \phi_1, \phi_2 \in \hat{X}.$$

3.7. Theorem. *Assume X is a $\mathcal{V}$-continuity space. Then $\hat{X}$ is complete.*

PROOF.

Assume Φ is an order ideal on $\hat{X}$. Define $\phi^{\#} : X \to V$ by

$$\phi^{\#}(x) = \inf_{\phi \in \hat{X}} (\phi(x) + \Phi(\phi)), \quad x \in X.$$

Then $\phi^{\#} \in \hat{X}$ and for $\phi_1 \in \hat{X}$,

$$\begin{aligned}
d(\phi^{\#}, \phi_1) &= \sup_x (\phi_1(x) \dot{-} \phi^{\#}(x)), \\
&= \sup_x (\phi_1(x) \dot{-} \inf_{\phi \in \hat{X}} (\phi(x) + \Phi(\phi))). \\
&= \sup_x \sup_{\phi \in \hat{X}} (\phi_1(x) \dot{-} (\phi(x) + \Phi(\phi))). \\
&= \sup_{\phi \in \hat{X}} (\sup_x (\phi_1(x) \dot{-} \phi(x)) \dot{-} \Phi(\phi)). \\
&= \sup_{\phi \in \hat{X}} (d(\phi, \phi_1) \dot{-} \Phi(\phi)).
\end{aligned}$$

Hence $\phi^{\#} = \bigvee \Phi$ and $\hat{X}$ is complete.

Q.E.D.

Each element $a \in X$ gives rise to an order ideal $\downarrow(a)$ defined by

$$\downarrow(a)(x) = d(x,a), \quad \text{for } x \in X.$$

3.8. Yoneda Lemma. *Assume X is a $\mathcal{V}$-continuity space, ϕ is an order ideal on X, and $a \in X$. Then*

$$d_{\hat{X}}(\downarrow(a), \phi) = \phi(a).$$

PROOF.

Simply note that

$$\begin{aligned} d_{\hat{X}}(\downarrow(a),\phi) &= \sup_{x\in X}(\phi(x) \dot{-} \downarrow(a)(x)), \\ &= \sup_{x\in X}(\phi(x) \dot{-} d(x,a)), \\ &= \phi(a). \end{aligned}$$

Q.E.D.

An immediate consequence of the Yoneda Lemma is that the mapping $\mathcal{Y} : X \to \hat{X}$ defined by

$$\mathcal{Y}(x) = \downarrow(x), \quad \text{for all } x \in X.$$

is actually a full and faithful $\mathcal{V}$-functor; this is called the *Yoneda Embedding*.

4. Cauchy Completeness in Continuity Spaces

Assume X and Y are $\mathcal{V}$-continuity spaces. A *distributor* from X to Y (Lawvere [6], Bénabou [1]) is a function $\phi : Y \times X \to V$ satisfying the following two conditions:

$$\phi(y,x) + d_Y(y',y) \geq \phi(y',x), \quad \text{for all } x \in X,\ y,y' \in Y;$$

and

$$\phi(y,x) + d_X(x,x') \geq \phi(y,x'), \quad \text{for all } x,x' \in X,\ y \in Y.$$

These two conditions are equivalent to requiring that ϕ be a $\mathcal{V}$-functor from $Y^* \otimes X$ to $\mathcal{V}$, where $Y^* \otimes X$ is the $\mathcal{V}$-continuity space with underlying set $Y \times X$ and distance function defined by

$$d_{Y^*\otimes X}(<y,x>,<y',x'>) = d_Y(y',y) + d_X(x,x'), \quad x,x' \in X,\ y,y' \in Y.$$

We write $\phi : X \mapsto Y$ to indicate that ϕ is a distributor from X to Y.

Any $\mathcal{V}$-functor $f : X \to Y$ gives rise to a distributor $f_* : X \mapsto Y$ defined by

$$f_*(y,x) = d_Y(y, f(x)), \quad x \in X, y \in Y,$$

and a distributor $f^* : Y \mapsto X$ defined by

$$f^*(x, y) = d_X(f(x), y), \quad x \in X, y \in Y.$$

The essential connection between the distributors f_* and f^* is an adjointness relationship, which is expressed in the following definition.

4.1. Definition. Assume X and Y are $\mathcal{V}$-continuity space, $\phi : X \mapsto Y$, and $\psi : Y \mapsto X$. Then ϕ is *left adjoint* to ψ iff the following two conditions are satisfied:

$$d(x, x') \geq \inf_{y \in Y} [\psi(x, y) + \phi(y, x')], \quad \text{for all } x, x' \in X;$$

and

$$\inf_{x \in X} [\phi(y, x) + \psi(x, y')] \geq d(y, y'), \quad \text{for all } y, y' \in Y.$$

Lawvere [6] observed that for a quasimetric space (X, d), the metric space (X, d^s) is Cauchy complete iff for every pair of distributors $\phi : X \mapsto Y$ and $\psi : Y \mapsto X$ such that ϕ is left adjoint to ψ, there is a $\mathcal{D}$-functor $f : X \to Y$ such that $\phi = f_*$ and $\psi = f^*$. This observation motivates the following definition.

4.2. Definition. Assume Y is a $\mathcal{V}$-continuity space. Then Y is *Cauchy complete* iff for every pair of distributors $\phi : X \mapsto Y$ and $\psi : Y \mapsto X$ such that ϕ is left adjoint to ψ, there is a $\mathcal{V}$-functor $f : X \to Y$ such that $\phi = f_*$ and $\psi = f^*$.

In the arguments below, it is convenient to work with distributors with domain or codomain the terminal continuity space $\mathbf{1}$. These can clearly be identified with the corresponding order ideals or order filters. If ϕ is an order ideal on X and ψ is an order filter on X, then ϕ is left adjoint to ψ iff the following two conditions are satisfied:

$$(a1) \qquad 0 = \inf_{x \in X} (\psi(x) + \phi(x)).$$

$$(a2) \qquad \phi(x) + \psi(y) \geq d(x, y), \quad \text{for all } x, y \in X.$$

We call such a pair a *cut* on X.

Each element $a \in X$ gives rise to a cut $(\downarrow(a), \uparrow(a))$ on X, where $\downarrow(a)$ is the order ideal defined by

$$\downarrow(a)(x) = d(x, a), \quad \text{for } x \in X,$$

and $\uparrow(a)$ is the order filter defined by

$$\uparrow(a)(x) = d(a, x), \quad \text{for } x \in X.$$

4.3. Lemma. *Assume X is a $\mathcal{V}$-continuity space. Then X is Cauchy complete iff for every cut (ϕ, ψ) on X there is an element $a \in X$ such that $\phi = \downarrow(a)$ and $\psi = \uparrow(a)$.*

4.4. Lemma. *Assume X is a $\mathcal{V}$-continuity space and (ϕ, ψ) is a cut on X. Then for all $x \in X$,*

$$\phi(x) = \sup_{y \in X} (d(x,y) \dot{-} \psi(y)),$$

and

$$\psi(x) = \sup_{y \in X} (d(y,x) \dot{-} \phi(y)).$$

PROOF.

Assume (ϕ, ψ) is a cut on X. Let $x \in X$. Then from (a2), for any $y \in X$,

$$\phi(x) + \psi(y) \geq d(x,y).$$

Thus

$$\phi(x) \geq d(x,y) \dot{-} \psi(y).$$

Since y is arbitrary,

$$\phi(x) \geq \sup_{y \in X} (d(x,y) \dot{-} \psi(y)).$$

Moreover from (a1),

$$\begin{aligned}
\sup_{y \in X} (d(x,y) \dot{-} \psi(y)) &= \sup_{y \in X} (d(x,y) \dot{-} \psi(y)) + \inf_{a \in X} (\psi(a) + \phi(a)), \\
&= \inf_{a \in X} (\sup_{y \in X} (d(x,y) \dot{-} \psi(y)) + \psi(a) + \phi(a)), \\
&\geq \inf_{a \in X} ((d(x,a) \dot{-} \psi(a)) + \psi(a) + \phi(a)), \\
&\geq \inf_{a \in X} (d(x,a) + \phi(a)), \\
&\geq \inf_{a \in X} \phi(x), \\
&= \phi(x).
\end{aligned}$$

Hence

$$\phi(x) = \sup_{y \in X} (d(x,y) \dot{-} \psi(y)).$$

A dual argument shows

$$\psi(x) = \sup_{y \in X} (d(y,x) \dot{-} \phi(y)).$$

Q.E.D.

4.5. Remark. *Assume (ϕ, ψ) is a cut on the $\mathcal{V}$-continuity space X. From Lemma (4.4), ψ is uniquely determined by ϕ and conversely. Moreover, an element a of X represents (ϕ, ψ) iff $\downarrow(a) = \phi$ or $\uparrow(a) = \psi$.*

The next result is a direct application of the Yoneda Lemma, the definition of $d_{\hat{X}}$, Lemma 4.4, and (a1).

4.6. Corollary. *Assume ϕ is an order ideal on the $\mathcal{V}$-continuity space X. Then there is an order filter ψ on X such that (ϕ, ψ) is a cut on X iff*

$$0 = \inf_{x \in X} (d_{\hat{X}}(\downarrow (x), \phi) + d_{\hat{X}}(\phi, \downarrow (x)));$$

that is, iff ϕ is in the closure of $\{\downarrow (x) \mid x \in X\}$ in the symmetric topology on $\hat{X}$.

4.7. Theorem. *Assume X is a $\mathcal{V}$-continuity space. If X is complete, then X is Cauchy complete.*

PROOF.

Assume X is complete and (ϕ, ψ) is a cut in X. Let

$$a = \bigvee \phi.$$

Then for all $x \in X$,

$$d(a, x) = \sup_{y \in X} (d(y, x) \dot{-} \phi(y)) = \psi(x);$$

that is, $\uparrow (a) = \psi$. By Remark 4.4, a represents the cut (ϕ, ψ). It follows that X is Cauchy complete.

Q.E.D.

4.8. Lemma. *Assume X is a $\mathcal{V}$-continuity space and $A \subseteq X$. If X is Cauchy complete and A is closed in the symmetric topology, then A is Cauchy complete.*

PROOF.

Assume X is Cauchy complete and A is closed in the symmetric topology. Let (ϕ, ψ) is a cut in A. Define $\bar{\phi} : X \to V$ and $\bar{\psi} : X \to V$ by

$$\bar{\phi}(x) = \inf_{a \in A} (\phi(a) + d(x, a)), \qquad \text{for all } x \in X,$$

and

$$\bar{\psi}(x) = \inf_{a \in A} (\psi(a) + d(a, x)), \qquad \text{for all } x \in X.$$

Then $\bar{\phi}$ is an order ideal on X and $\bar{\psi}$ is an order filter on X. A simple calculation shows that $\bar{\phi}$ is left adjoint to $\bar{\psi}$. Let $\bar{x}$ represent the cut $(\bar{\phi}, \bar{\psi})$ in X. It will suffice to show that $\bar{x}$ is in A. But

$$\begin{aligned}
0 &= \bar{\phi}(\bar{x}) + \bar{\psi}(\bar{x}), \\
&= \inf_{a, a' \in A} (\phi(a) + \psi(a') + d(\bar{x}, a) + d(a', \bar{x})), \\
&\geq \inf_{a, a' \in A} (d(a, a') + d(\bar{x}, a) + d(a', \bar{x})), \\
&\geq \inf_{a \in A} (d(a, \bar{x}) + d(\bar{x}, a)).
\end{aligned}$$

Since A is closed in the symmetric topology, it follows that $\bar{x}$ is in A.

Q.E.D.

4.9. Definition. Assume X and Y are $\mathcal{V}$-continuity spaces. Then Y is a *Cauchy completion* of X iff Y is Cauchy complete and X is isometric to a dense subset of Y.

4.10. Theorem. *Assume $\mathcal{X} = (X, d)$ is a $\mathcal{V}$-continuity space. Then there is a $\mathcal{V}$-continuity space $\bar{X}$ such that $\bar{X}$ is a Cauchy completion of X.*

PROOF.

Let

$$\bar{X} = \{\phi \mid \text{ for some order filter } \psi \text{ on } X,\ (\phi, \psi) \text{ is a cut on } X\ \},$$

reqarded as a subspace of $\hat{X}$. By the Yoneda Lemma and Corollary 4.6, the Yoneda embedding $\mathcal{Y} : X \to \bar{X}$ is an isometric embedding of X onto a dense subset of $\bar{X}$. Moreover, since $\bar{X}$ is a closed subset of $\hat{X}$, by (4.7) and (4.8), $\bar{X}$ is Cauchy complete.

Q.E.D.

We close by making a connection between the definition of Cauchy completeness given here and the standard definition.

Assume $\mathcal{X} = (X, d)$ is a $\mathcal{V}$-continuity space. For each $\epsilon \in V$ such that $\epsilon \succ 0$, let

$$D_\epsilon = \{< x, y >\in X \times X \mid \epsilon \succ d^s(x, y)\ \}.$$

One easily shows that the family $\{D_\epsilon\}_{\epsilon \succ 0}$ is a base for a uniformity, $\mathcal{U}^d$, on X and that the uniform topology generated by $\mathcal{U}^d$ is the symmetric topology on $\mathcal{X}$.

4.11. Theorem. *Assume $\mathcal{V} =< V, \leq, + >$ is a value quantale and $\mathcal{X} = (X, d)$ is a $\mathcal{V}$-continuity space. Then $\mathcal{X}$ is Cauchy complete iff the uniform space $< X, \mathcal{U}^d >$ is Cauchy complete.*

PROOF.

Assume every cut on X is representable. Let $(x_\lambda)_{\lambda \in \Lambda}$ be a Cauchy net in X. Define $\phi, \psi : X \to V$ by

$$\phi(x) = \inf_{\lambda_0} \sup_{\lambda \geq \lambda_0} d(x, x_\lambda), \quad \text{for all } x \in X;$$

and

$$\psi(x) = \inf_{\lambda_0} \sup_{\lambda \geq \lambda_0} d(x_\lambda, x), \quad \text{for all } x \in X.$$

For all $x, y \in X$,

$$\begin{aligned}
\phi(y) + d(x,y) &= \inf_{\lambda_0} \sup_{\lambda \geq \lambda_0} d(y, x_\lambda) + d(x,y), \\
&= \inf_{\lambda_0} (\sup_{\lambda \geq \lambda_0} d(y, x_\lambda) + d(x,y)), \\
&\geq \inf_{\lambda_0} \sup_{\lambda \geq \lambda_0} (d(y, x_\lambda) + d(x,y)), \\
&\geq \inf_{\lambda_0} \sup_{\lambda \geq \lambda_0} d(x, x_\lambda), \\
&\geq \phi(x).
\end{aligned}$$

Hence ϕ is an ideal on X. Similarly, ψ is a filter on X.

Given $\epsilon \succ 0$, choose $\epsilon' \succ 0$ such that $\epsilon \geq 2\epsilon'$. Choose $\bar{\lambda}$ so that for all $\lambda \geq \bar{\lambda}$,

$$\epsilon' \geq d(x_{\bar{\lambda}}, x_\lambda), \quad \text{and} \quad \epsilon' \geq d(x_\lambda, x_{\bar{\lambda}}).$$

Then

$$\begin{aligned}
\epsilon &\geq 2\epsilon', \\
&\geq \sup_{\lambda \geq \bar{\lambda}} d(x_{\bar{\lambda}}, x_\lambda) + \sup_{\lambda \geq \bar{\lambda}} d(x_\lambda, x_{\bar{\lambda}}), \\
&\geq \inf_{x \in X} (\sup_{\lambda \geq \bar{\lambda}} d(x, x_\lambda) + \sup_{\lambda \geq \bar{\lambda}} d(x_\lambda, x)), \\
&\geq \inf_{x \in X} (\inf_{\lambda_0} \sup_{\lambda \geq \lambda_0} d(x, x_\lambda) + \inf_{\lambda_0} \sup_{\lambda \geq \lambda_0} d(x_\lambda, x)), \\
&\geq \inf_{x \in X} (\phi(x) + \psi(x)).
\end{aligned}$$

Hence

$$0 = \inf_{x \in X} (\psi(x) + \phi(x)).$$

Using the fact that Λ is directed, we obtain for all $x, y \in X$,

$$\begin{aligned}
\phi(x) + \psi(y) &= \inf_{\lambda_0} \sup_{\lambda \geq \lambda_0} d(x, x_\lambda) + \inf_{\lambda_0} \sup_{\lambda \geq \lambda_0} d(x_\lambda, y), \\
&= \inf_{\lambda_0, \lambda_1} (\sup_{\lambda \geq \lambda_0} d(x, x_\lambda) + \sup_{\lambda \geq \lambda_1} d(x_\lambda, y)), \\
&\geq \inf_{\lambda_0, \lambda_1} \sup_{\lambda \geq \lambda_0, \lambda' \geq \lambda_1} (d(x, x_\lambda) + d(x_{\lambda'}, y)), \\
&\geq \inf_{\lambda_0, \lambda_1} d(x, y), \\
&= d(x, y).
\end{aligned}$$

Thus (ϕ, ψ) is a cut on X.

Let $a \in X$ represent (ϕ, ψ). Then $0 = \phi(a)$ and $0 = \psi(a)$. Given $\epsilon \succ 0$, choose $\epsilon' \succ 0$ such that $\epsilon \geq 2\epsilon'$. Choose λ_0 and λ'_0 so that for all $\lambda \geq \lambda_0$ and $\lambda' \geq \lambda'_0$,

$$\epsilon' \geq d(a, x_\lambda), \quad \text{and} \quad \epsilon' \geq d(x_{\lambda'}, a).$$

Let $\lambda_1 \geq \lambda_0, \lambda_0'$. Then for all $\lambda \geq \lambda_1$,

$$\epsilon \succ 2\epsilon' \geq d^s(x_\lambda, a).$$

Hence $a = \lim_\lambda x_\lambda$. It follows that $< X, \mathcal{U}^d >$ is Cauchy complete.

Now assume $< X, \mathcal{U}^d >$ is Cauchy complete. Let (ϕ, ψ) be a cut on X. For each $\epsilon \succ 0$ choose x_ϵ so that

$$\epsilon \succ \phi(x_\epsilon) + \psi(x_\epsilon).$$

Then $(x_\epsilon)_{\epsilon \succ 0}$ is a net in X. Given $\epsilon \succ 0$, choose $\epsilon' \succ 0$ such that $\epsilon \geq 2\epsilon'$. Then for all $\epsilon_1, \epsilon_2 \leq \epsilon'$,

$$\begin{aligned}\epsilon &\succ 2\epsilon', \\ &\geq \epsilon_1 + \epsilon_2, \\ &\geq \phi(x_{\epsilon_1}) + \psi(x_{\epsilon_1}) + \phi(x_{\epsilon_2}) + \psi(x_{\epsilon_2}), \\ &\geq d(x_{\epsilon_1}, x_{\epsilon_2}) + d(x_{\epsilon_2}, x_{\epsilon_1}), \\ &= d^s(x_{\epsilon_1}, x_{\epsilon_2}).\end{aligned}$$

Hence $(x_\epsilon)_{\epsilon \succ 0}$ is a Cauchy net in X. Let $a = \lim_\epsilon x_\epsilon$. Since ϕ is continuous for the symmetric topologies,

$$\phi(a) = \phi(\lim_\epsilon x_\epsilon) = \lim_\epsilon \phi(x_\epsilon).$$

In particular, for all $\delta \succ 0$ there is an $\epsilon \succ 0$ such that for $\epsilon' \leq \epsilon$,

$$\delta \succ d_V(\phi(x_{\epsilon'}), \phi(a)) = \phi(a) \dot{-} \phi(x_{\epsilon'}).$$

It follows that

$$\begin{aligned}0 &\geq \inf_\epsilon \sup_{\epsilon' \leq \epsilon} (\phi(a) \dot{-} \phi(x_{\epsilon'})), \\ &\geq \inf_\epsilon \sup_{\epsilon' \leq \epsilon} (\phi(a) \dot{-} \epsilon'), \\ &= \inf_\epsilon (\phi(a) \dot{-} \inf_{\epsilon' \leq \epsilon} \epsilon'), \\ &= \inf_\epsilon (\phi(a) \dot{-} 0), \\ &= \inf_\epsilon \phi(a), \\ &= \phi(a).\end{aligned}$$

Thus $\phi(a) = 0$. Similarly $\psi(a) = 0$. Hence for all $x \in X$,

$$d(x, a) = d(x, a) + \phi(a) \geq \phi(x),$$

and

$$\phi(x) = \phi(x) + \psi(a) \geq d(x, a).$$

So a represents the cut (ϕ, ψ).

Q.E.D.

REFERENCES

1. J. Bénabou, Les Distributeurs, Univ. Cath. de Louvain, Séminaires de Math. Pure, Rapport No. **256**, 1973.

2. R. C. Flagg, Quantales and Continuity Spaces, submitted to Algebra Universalis.

3. M. Henriksen and R. Kopperman, A General Theory of Structure Spaces with Applications to Spaces of Prime Ideals, Algebra Universalis, **28**, 1991, pp. 349–376.

4. G. M. Kelly, *Basic Concepts of Enriched Category Theory*, Cambridge University Press, Cambridge, 1982.

5. R. Kopperman, All Topologies Come From Generalized Metrics, American Mathematical Monthly, **95**, 1988, pp. 89–97.

6. F. W. Lawvere, Metric Spaces, Generalized Logic, and Closed Categories, Rend. Sem. Mat. e. Fisico di Milano, **43** (1973), pp. 135–166.

7. B. Schweizer and A. Sklar, *Probabilistic Metric Spaces*, North Holland, Amsterdam, 1983.

DEPARTMENT OF MATHEMATICS AND STATISTICS, University of Southern Maine, Portland, Me. 04103

E-mail address: flagg@maine.maine.edu

Canadian Mathematical Society
Conference Proceeding
Volume **13**, 1992

Variations on a Theme of Dold

K.A. HARDIE AND K.H. KAMPS

ABSTRACT. Using the techniques of a certain type of coherent homotopy theory under a fixed space A and over a fixed space B we present a new access to comparison (bridging) theorems under A and over B generalizing classical results of Dold.

0. Introduction

In homotopy theory under a fixed space A and over a fixed space B we have the following comparison theorem[1] generalizing classical results of Dold ([4], 6.1, [5], 3.6).

THEOREM. *Let*

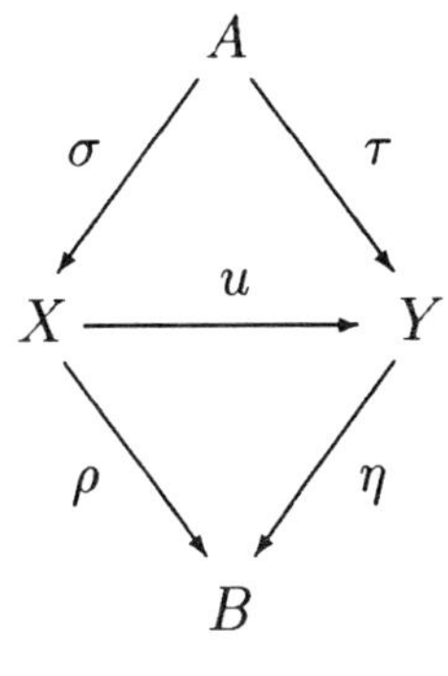

be a commutative diagram of topological spaces where σ, τ *are closed cofibrations and* ρ, η *are fibrations. Then, if* u *is a homotopy equivalence,* u *is a homotopy equivalence under* A *and over* B.

1991 Mathematics Subject Classification. Primary 55P10.
Secondary 55P99, 55P05, 55R05, 55U35.
This paper is in final form and no version of it will be submitted for publication elsewhere.

[1] In [9], Introduction, the terminology bridging theorem has been used.

In this note we present a new proof of a generalization of the theorem separating arguments involving homotopy equivalences from arguments involving fibrations and cofibrations in a transparent way. We exploit the techniques of what we call partially coherent homotopy theory under A and over B.

Note that an alternative approach to the theorem via operations on homotopy sets has been given in [2].

The results of this paper are of formal character and hold in a suitably structured category (see for example [2]). For convenience all the formulations are given for topological spaces.

We point out that the results on partially coherent homotopy theory are of independent interest and will enable better understanding of homotopy theory under A and over B in general (see also [6]).[2] This sort of topic will be dealt with elsewhere.

For the basic notions and notations of homotopy theory we refer to [3], §0.

We wish to thank the referee for valuable comments and suggestions.

1. Partially coherent homotopy under A and over B

Let $b : A \longrightarrow B$ be a continuous map of topological spaces.

We define a coherent homotopy category $\underset{\sim}{\mathcal{H}}(b)$ which is isomorphic to the track homotopy category over b in the sense of [7] associated to the category $\boldsymbol{Top}^A$ of topological spaces under A.

An object X of $\underset{\sim}{\mathcal{H}}(b)$ is a factorization (σ, ρ) of b, i.e. a diagram

$$(1.1) \qquad A \xrightarrow{\sigma} X \xrightarrow{\rho} B$$

where σ, ρ are maps of spaces such that $\rho\sigma = b$.

Let $X = A \xrightarrow{\sigma} X \xrightarrow{\rho} B$ and $Y = A \xrightarrow{\tau} Y \xrightarrow{\eta} B$ be objects of $\underset{\sim}{\mathcal{H}}(b)$. Note $\rho\sigma = \eta\tau = b$. An arrow from X to Y in $\underset{\sim}{\mathcal{H}}(b)$ is an equivalence class $\{u, \rho_t\}$ of a diagram of the form

[2] We understand that in the special case of ex-spaces (see [8]) related subjects have been studied independently by Almusawe [1].

(1.2)

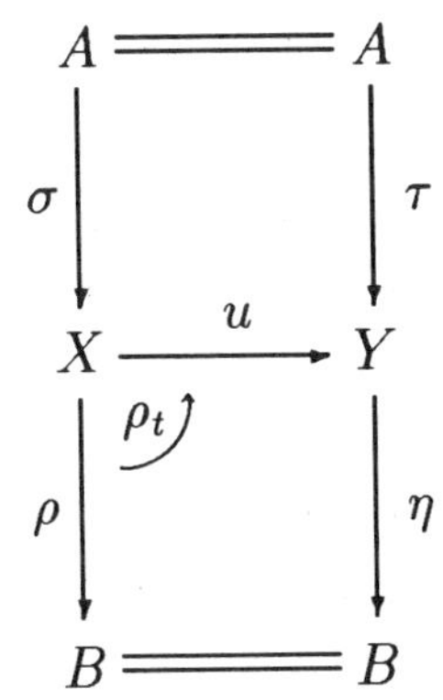

where the upper square commutes (i.e. u is a map under A from σ to τ) and $\rho_t : \rho \overset{A}{\simeq} \eta u$ is a homotopy under A from σ to b. Each diagram (1.2) determines a pair $(u, \{\rho_t\}^A)$ where $\{\rho_t\}^A$ is the track under A (relative homotopy class under A) of ρ_t the latter notion being defined in the obvious way using homotopies under A of homotopies under A. The equivalence relation is given by the formula

$$(1.3) \qquad (u_1, \{\rho_t + \eta u_t\}^A) \sim (u, \{\rho_t\}^A)$$

whenever u_t is a homotopy under A from σ to τ with $u_0 = u$. In (1.3) the $+$ refers to the usual track addition of homotopies.

Denoting the set of equivalence classes by $\underset{\sim}{\pi}(A/X, Y/B)$ we obtain a category $\underset{\sim}{\mathcal{H}}(b)$ with composition induced by juxtaposition of diagrams in the obvious way.

Dually, we have a category $\tilde{\mathcal{H}}(b)$ whose objects are the same as those of $\underset{\sim}{\mathcal{H}}(b)$. The set of morphisms $\tilde{\pi}(A/X, Y/B)$ in $\tilde{\mathcal{H}}(b)$ is obtained in the obvious way from the set of diagrams of the form

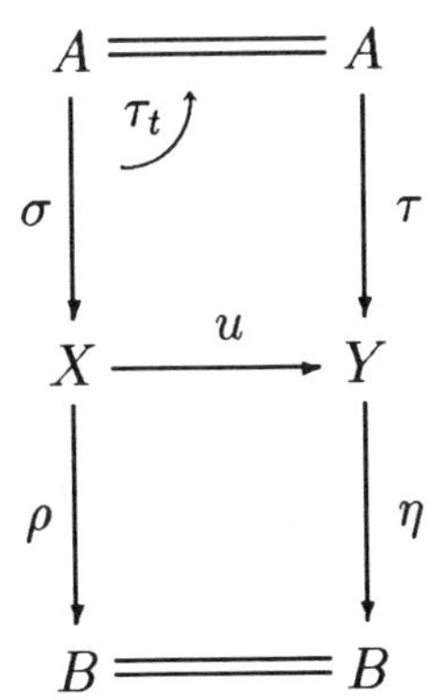

where the lower square commutes and τ_t is a homotopy over B from $u\sigma$ to τ.

Since [7], Theorem 1.3 remains valid when the category $\boldsymbol{\mathcal{T}op}$ of topological spaces is replaced by the category $\boldsymbol{\mathcal{T}op}^A$ of topological spaces under A we record:

THEOREM 1.4. *The arrow* $\{u, \rho_t\}$ *is an isomorphism in* $\underset{\sim}{\mathcal{H}}(b)$ *if and only if* u *viewed as a map under* A, $u : \sigma \longrightarrow \tau$, *is a homotopy equivalence under* A.

By a theorem of Dold ([5], 3.6) we obtain

COROLLARY 1.5. *If in diagram* (1.2) σ *and* τ *are* h-*cofibrations* (*see* [3], (2.2)) *the arrow* $\{u, \rho_t\}$ *is an isomorphism in* $\underset{\sim}{\mathcal{H}}(b)$ *if and only if* u *viewed as a map* $u : X \longrightarrow Y$ *is an ordinary homotopy equivalence.*

2. Comparison theorems

If $b : A \longrightarrow B$ is a map and $A \xrightarrow{\sigma} X \xrightarrow{\rho} B$ and $A \xrightarrow{\tau} Y \xrightarrow{\eta} B$ are objects of $\underset{\sim}{\mathcal{H}}(b)$ (so that $\rho\sigma = \eta\tau = b$) then we have a canonical map

$$\underset{\sim}{\Theta} : [(\sigma,\rho),(\tau,\eta)]^A_B \longrightarrow \underset{\sim}{\pi}(A/X, Y/B)$$

where $[(\sigma,\rho),(\tau,\eta)]^A_B$ denotes the classical homotopy set under A and over B (see [3], (0.26)). The map $\underset{\sim}{\Theta}$ is induced by the assignment which views a commutative diagram in $\boldsymbol{\mathcal{T}op}$

(2.1)

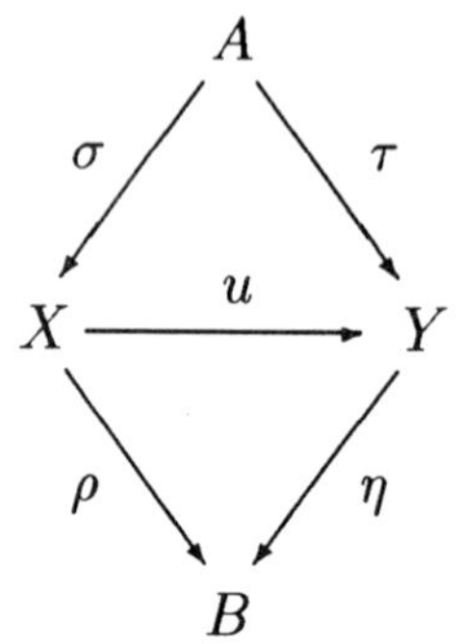

as a homotopy coherent diagram

$$\begin{array}{ccc}
A & = & A \\
\sigma\downarrow & & \downarrow\tau \\
X & \xrightarrow{u} & Y \\
\rho\downarrow & \overset{\rho}{\curvearrowright} & \downarrow\eta \\
B & = & B
\end{array}$$

with the track under A of the constant homotopy ρ under A. Thus we have

$$\underset{\sim}{\Theta}[u]_B^A = \{u, \rho\}$$

where $[u]_B^A$ denotes the homotopy class under A and over B of u.

Dually, we have a canonical map

$$\widetilde{\Theta} : [(\sigma, \rho), (\tau, \eta)]_B^A \longrightarrow \tilde{\pi}(A/X, Y/B)$$

which views a commutative diagram (2.1) in ***Top*** as a homotopy coherent diagram

$$\begin{array}{ccc}
A & = & A \\
\sigma\downarrow & \overset{\tau}{\curvearrowright} & \downarrow\tau \\
X & \xrightarrow{u} & Y \\
\rho\downarrow & & \downarrow\eta \\
B & = & B
\end{array}$$

with the constant homotopy τ over B.

Recall that a pair of maps (σ, η), $\sigma : A \longrightarrow X$, $\eta : Y \longrightarrow B$, is said to have the *covering homotopy extension property* (CHEP) if for any map $f : X \longrightarrow Y$ and any pair of homotopies (φ_t, ψ_t) such that

$$\varphi_t : f\sigma \simeq \varphi_1,\ \psi_t : \eta f \simeq \psi_1 \text{ and } \eta\varphi_t = \psi_t\sigma$$

there exists a homotopy $\phi_t : f \simeq \phi_1$ such that $\phi_t\sigma = \varphi_t$ and $\eta\phi_t = \psi_t$.

Note that by a result of Strøm ([11], Theorem 4) a pair (σ, η) has the CHEP if σ is a closed cofibration and η is a fibration. Further examples are listed in [2], 2.2.

PROPOSITION 2.2. *If* (σ, η) *and* $(\sigma \times id_I, \eta)$ *have the* CHEP *then*

$$\underset{\sim}{\Theta} : [(\sigma, \rho), (\tau, \eta)]_B^A \longrightarrow \underset{\sim}{\pi}(A/X, Y/B)$$

is a bijection.

Proof. Let $\{u, \rho_t\} \in \underset{\sim}{\pi}(A/X, Y/B)$ be represented by diagram (1.2). Then since (σ, η) is assumed to satisfy the CHEP there exists a homotopy $\phi_t : u \simeq \phi_1$ such that $\phi_t\sigma = \tau$ and $\eta\phi_t = \rho_{1-t}$. Hence $\{u, \rho_t\}$ is also represented by

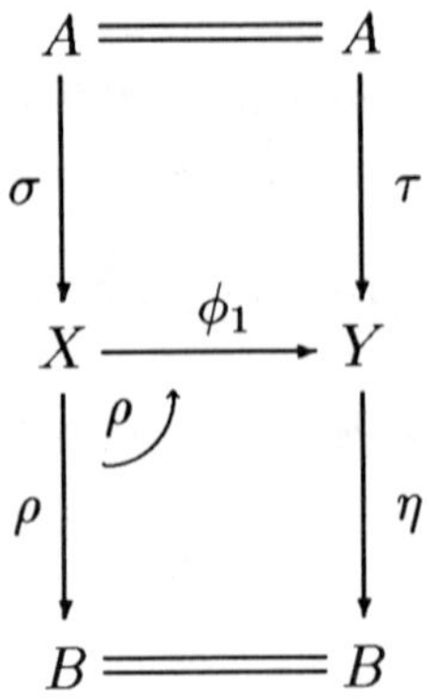

so $\{u, \rho_t\}$ is indeed in the image of $\underset{\sim}{\Theta}$. This proves that $\underset{\sim}{\Theta}$ is surjective.

To see that $\underset{\sim}{\Theta}$ is injective, let $[u]_B^A$, $[u']_B^A \in [(\sigma, \rho), (\tau, \eta)]_B^A$ and suppose $\underset{\sim}{\Theta}[u]_B^A = \underset{\sim}{\Theta}[u']_B^A$. Then by definition of $\sim$ there exists a homotopy under A $\phi_t : u \overset{A}{\simeq} u'$ such that $\{\eta\phi_t\}^A = \{\rho\}^A$. Since $(\sigma \times id_I, \eta)$ is assumed to have the CHEP we can apply [10], 2.1 and obtain $[u]_B^A = [u']_B^A$. $\square$

We note that a completely formalized version of Proposition 2.2 can be proved under the assumption that the pair of morphisms (σ, η) satisfies the properties CHEP (1) and CHEP (2) in the sense of [2], 2.1. These conditions being selfdual we obtain by dualization and specialization to topological spaces

PROPOSITON 2.3. *If* (σ, η) *and* $(\sigma \times id_I, \eta)$ *have the* CHEP *then*

$$\tilde{\Theta} : [(\sigma, \rho), (\tau, \eta)]_B^A \longrightarrow \tilde{\pi}\,(A/X, Y/B)$$

is a bijection.

COROLLARY 2.4. *If σ is a closed cofibration and η is a fibration then we have bijections $\underset{\sim}{\Theta}$, $\tilde{\Theta}$:*

$$\underset{\sim}{\pi}(A/X, Y/B) \xleftarrow{\underset{\sim}{\Theta}} [(\sigma,\rho),(\tau,\eta)]^A_B \xrightarrow{\tilde{\Theta}} \tilde{\pi}(A/X, Y/B).$$

We are now in a position to apply partially coherent homotopy theory under A and over B to comparison theorems under A and over B (cf. [2], 4.5 (1),(2)).

THEOREM 2.5. *Let (2.1) be a commutative diagram of topological spaces. Assume that*

(1) *σ, τ are h-cofibrations,*

(2) *the pairs*

$$(\sigma,\rho),\ (\sigma \times id_I, \rho),\ (\sigma,\eta),\ (\sigma \times id_I, \eta),$$
$$(\tau,\rho),\ (\tau \times id_I, \rho),\ (\tau,\eta),\ (\tau \times id_I, \eta)$$

satisfy the CHEP.

Then, if u is a homotopy equivalence, u is a homotopy equivalence under A and over B.

Proof. We define $b = \rho\sigma = \eta\tau$.

The assertion of the theorem is equivalent to the fact that the set maps

(2.6) $u_* : [(\sigma,\rho),(\sigma,\rho)]^A_B \longrightarrow [(\sigma,\rho),(\tau,\eta)]^A_B$

(2.7) $u_* : [(\tau,\eta),(\sigma,\rho)]^A_B \longrightarrow [(\tau,\eta),(\tau,\eta)]^A_B$

induced by composition of maps under A and over B are bijective.

In order to prove the bijectivity of the first map we consider the following commutative diagram of sets

$$(2.8)\qquad \begin{array}{ccc} [(\sigma,\rho),(\sigma,\rho)]^A_B & \xrightarrow{u_*} & [(\sigma,\rho),(\tau,\eta)]^A_B \\ \Big\downarrow{\scriptstyle \underset{\sim}{\Theta}} & & \Big\downarrow{\scriptstyle \underset{\sim}{\Theta}} \\ \underset{\sim}{\pi}(A/X, X/B) & \xrightarrow[u_\square]{} & \underset{\sim}{\pi}(A/X, Y/B) \end{array}$$

where the arrow $u_\square$ is induced by composition in $\underset{\sim}{\mathcal{H}}(b)$. Since σ and τ are h-cofibrations and u is a homotopy equivalence, the map $u_\square$ is a bijection by Corollary 1.5. Since (σ, ρ), $(\sigma \times id_I, \rho)$, (σ, η), $(\sigma \times id_I, \eta)$ have the CHEP, the vertical arrows in (2.8) are bijective by Proposition 2.2. Hence the map u_* in (2.6) which appears as the upper arrow of (2.8) is bijective.

In a similar way we can prove that the map u_* in (2.7) is bijective. This completes the proof of the theorem. □

EXAMPLE 2.9. In particular Theorem 2.5 holds if either σ, τ are closed cofibrations and ρ, η are fibrations or σ, τ are inclusions of relative CW–complexes and ρ, η are Serre-fibrations.

On the other hand, Theorem 2.5 is not valid in general if σ, τ are closed cofibrations, η is a fibration, but ρ is merely assumed to be an h–fibration. (For a counterexample let $A = \{*\}$ be a singleton, $B = Y = [0,1]$ the unit interval, $X = [0,1] \times \{0\} \cup \{0\} \times [0,1]$, $u = \rho$ the projection onto the first factor, η the identity and define σ, τ by $\sigma(*) = (0,1)$, $\tau(*) = 0$.)

We leave it to the reader to dualize Theorem 2.5 (cf. [2],4.5(1),(2*)).

REFERENCES

1. S. Almusawe, The non-fibrewise category $T(B)$, Thesis, University of Birmingham 1991.
2. A.J. Berrick, K.H. Kamps, Comparison theorems in homotopy theory via operations on homotopy sets, Math. Nachr. **121** (1985), 25-31.
3. T. tom Dieck, K.H. Kamps and D. Puppe, *Homotopietheorie*, Lecture Notes in Math. **157**, Springer, Berlin–Heidelberg– New York, 1970.
4. A. Dold, Partitions of unity in the theory of fibrations, Ann. of Math. **78** (1963), 223-255.
5. A. Dold, *Halbexakte Homotopiefunktoren*, Lecture Notes in Math. **12**, Springer, Berlin–Heidelberg–New York, 1966.
6. K.A. Hardie, K.H. Kamps, Homotopy over B and under A, Cahiers Top. et Géom. Diff. **28** (1987), 183-196.

7. K.A. Hardie, K.H. Kamps, Track homotopy over a fixed space, Glas. Mat. **24 (44)** (1989), 161-179.

8. I.M. James, Ex-homotopy theory I, Ill. J. Math. **15** (1971), 324-337.

9. I.M. James, *General Topology and Homotopy Theory*, Springer, New York–Berlin–Heidelberg–Tokyo, 1984.

10. K.H. Kamps, Operationen auf Homotopiemengen, Math. Nachr. **81** (1978), 233-241.

11. A. Strøm, Note on cofibrations, Math. Scand. **19** (1966), 11-14.

Department of Mathematics
University of Cape Town
Rondebosch 7700
South Africa

Fachbereich Mathematik
Fernuniversität
Postfach 940
D-W-5800 Hagen
Fed. Rep. of Germany

Canadian Mathematical Society
Conference Proceeding
Volume **13**, 1992

Dualities for Accessible Categories

HONGDE HU

ABSTRACT. In this work we prove dualities for Diers categories and Barr categories. The latter duality solves a problem of M. Makkai. Further, we give conceptual completeness statements for the 2-category of Diers categories, the 2-category of Barr categories and the 2-category of locally presentable categories.

1 Introduction

Before the explicit introduction of accessible categories by M. Makkai and R. Paré in [19], various subclasses of accessible categories had been investigated for many years, beginning with Gabriel and Ulmer in 1971 and Artin, Grothendieck and Verdier in 1972, and continuing through the work of Makkai and Reyes in 1977, Diers in 1980, and Guitart and Lair in 1980-1981, and so on. The $\aleph_0$-accessible categories are the categories that are determined, in a quite precise way (free completion by filtered colimits), by some small subcategory. The idea of looking at objects which are filtered colimits of finitely presentable objects has been used in algebra. An early example is Lazare's theorem for flat modules, which says that a flat module is a filtered colimit of finitely generated free modules. The category of Hilbert spaces is another important example for an accessible category, in which all objects are $\aleph_1$-filtered colimits of $\aleph_1$-presentable ones.

Locally presentable categories, introduced by Gabriel and Ulmer, is a subclass of the class of accessible categories. As proved in Makkai and Paré's book, a category is locally presentable if and only if it is accessible and complete, or equivalently cocomplete. The classic work of Gabriel and Ulmer gives a characterization of locally κ-presentable categories which, up to equivalence, can be written as $L_\kappa(\mathbf{C},\mathbf{S})$, the category of the functors preserving κ-limits; here $\mathbf{C}$ is a small category with κ-limits, $\mathbf{S}$ is the category of small sets. The result was rephrased as a duality theorem by Makkai and Pitts in 1987. The duality theorem for accessible categories as given by Makkai and Paré, contains the statement that κ-accessible categories are exactly the categories of the form $L_\kappa Cocts((\mathbf{C},\mathbf{S}),\mathbf{S})$, the category of the functors preserving κ-limits and colimits; here $\mathbf{C}$ is a small category.

1991 Mathematics Subject Classification. Primary 03G30, 18B25, 18E10, 18G05.

This paper is in final form and no version of it will be submitted for publication elsewhere.

We continue the study of dualities for accessible categories: (κ-)Diers categories ((κ)-accessible with connected limits, introduced by Diers) and (κ-)Barr categories ((κ-)accessible with products). A category **C** is said to be coproduct-accessible, if it has small coproducts, and it has a small subcategory **B** consisting of coproduct-presentable objects (the functors represented by objects in **B** preserve small coproducts) such that every object of **C** is the small coproduct of objects in **B**. We have proved the duality for (κ-)Diers categories. The duality theorem for (κ-)Diers categories has the following consequences:
(i) κ-Diers categories are exactly the categories of the form $\coprod L_\kappa(\mathbf{C}, \mathbf{S})$, the category of the functors preserving coproducts and κ-limits, where **C** is a coproduct-accessible category with κ-limits.
(ii) The coproduct-accessible categories with κ-limits are exactly the categories of the form $CoF_\kappa(\mathbf{A}, \mathbf{S})$, the category of the functors preserving κ-filtered colimits and connected limits, with **A** a κ-Diers category.

A Barr-exact category **C** is said to be Barr-exact accessible if it has a small subcategory **B** consisting of regular epi projective objects such that for every object C of **C** there is a regular epimorphism from B into C with B in **B**. We have proved the duality for (κ-)Barr categories. Such a duality theorem has the following consequences:
(i) κ-Barr categories are exactly the categories of the form $\kappa - Reg(\mathbf{C}, \mathbf{S})$, the category of the κ-regular functors, where **C** is a κ-Barr-exact accessible category.
(ii) The κ-Barr exact accessible categories are exactly the categories of the form $\prod F_\kappa(\mathbf{A}, \mathbf{S})$, the category of the functors preserving κ-filtered colimits and products, with **A** a κ-Barr category.

Both duality theorems are Stone-type dualities with **S**, the category of small sets, as the dualizing object. It is interesting to note that a Barr-exact accessible category is necessary a small category.

A strong conceptual completeness (SCC) statement for a logical doctrine means an assertion that any theory of the doctrine can be recovered from an appropriate structure formed by the models of the theory. In classical propositional logic, strong conceptual completeness takes the form of the Stone duality theorem. We give SCC statements for various doctrines: the 2-category of locally presentable categories, the 2-category of Diers categories and the 2-category of Barr categories. Similar results for other accessible doctrines were obtained by Makkai in [15].

The notion of an exact category was introduced by M. Barr in [3], we call it Barr-exact category. The definition of Barr-exact category consists of a combination of finite completeness and exactness conditions. More precisely, it has finite limits and stable quotients of equivalence relations. A functor between Barr-exact categories is regular if it preserves finite limits and quotients of equivalence relations. An important result on this subject is Barr's full embedded on exact categories, that is, a small exact category has a full regular embedded into a set-valued functor category.

A one-sided duality theorem for small Barr-exact categories was found by Makkai in [16]. After showing duality for Barr categories, the relation between Barr categories and Barr-exact categories becomes clearer, although the results of this paper do not contain the results of [16]. Let us mention that, Adámek and

Rosický have recently shown (using a very different method) that small injectivity classes of locally presentable categories are exactly the Barr categories (see [1]). A characterization of the duals of Barr-exact categories, which gives a completion of the one-sided duality of [16] to a perfect duality, will be given in a forthcoming paper of mine (see [11]).

The paper proceeds as follows.

In Section 2 we will summarize a certain amount of material on accessible categories and κ-Barr-exact categories we use later. We make a detailed study of coproduct-accessible categories in Section 3, we have a duality theorem on coproduct-accessible categories, this duality is analogue to the duality for accessible categories. In Section 4 we first make the relation between the categories with κ-multicolimits and coproduct-accessible category with κ-limits, then we show the duality for Diers categories. Section 5 contains the duality for Barr categories. A crucial lemma is that, given a Barr category $\mathbf{A}$, $\prod Acc(\mathbf{A}, \mathbf{S})$ denotes the category of all accessible functors preserving products. For each $F \in \prod Acc(\mathbf{A}, \mathbf{S})$, there is a regular epi $\mathbf{A}(A, -) \to F$ with some $A \in \mathbf{A}$. As corollaries of Gabriel-Ulmer duality and dualities for Diers categories and Barr categories, we give strong conceptual completeness statements in Section 6.1, then we introduce the concept of Barr-exact weak-accessible category. One example of this concept is the opposite category of the category of R-modules, for R being an associative ring. We prove that given two rings R_1 and R_2, if they have an equivalence between the categories of injective modules(left or right), then they are Morita equivalent.

I wish to express my sincere thanks to my Ph.D. supervisor, Professor Micheal Makkai, for his kind encouragement and inspiring motivation as well as financial support. Also, I would like to thank Professor Michael Barr for stimulating conversations and his constant kindness.

2 Preliminaries

Let κ be an infinite regular cardinal. Recall that a category $\mathbf{A}$ is κ-filtered if for any graph $\mathbf{G}$ of cardinality less than κ, any diagram $D : \mathbf{G} \to \mathbf{A}$ has cocone. $\mathbf{A}$ has κ-filtered colimits, if $\mathbf{A}$ has colimits of all diagrams whose domain is a κ-filtered category. Another concept is κ-limit; it refers to the limit of a diagram whose domain category is of size less than κ. An object A of $\mathbf{A}$ is said to be κ-presentable if the representable functor $\mathbf{A}(A, -) : \mathbf{A} \to \mathbf{SET}$ preserves κ-filtered colimits existing in $\mathbf{A}$. Here $\mathbf{SET}$ is the category of sets. When $\mathbf{A}$ is locally small, $\mathbf{SET}$ is replced by $\mathbf{S}$, the category of small sets. The full subcategory of $\mathbf{A}$ whose objects are the κ-presentable ones is denoted by $\mathbf{A}_\kappa$. The following definition is given in [19].

Definition 2.1 *A category* $\mathbf{A}$ *is* κ*-accessible if:*

(i) $\mathbf{A}$ *has* κ*-filtered colimits;*

(ii) There is a small full subcategory $\mathbf{B}$ *of* $\mathbf{A}_\kappa$ *so that every object of* $\mathbf{A}$ *is a* κ*-filtered colimit of a diagram of objects in* $\mathbf{B}$.

A category is accessible if it is κ*-accessible for some infinite regular cardinal* κ.

Let $\mathbf{C}$ be a small category. A functor $F : \mathbf{C} \to \mathbf{S}$ is said to be κ-flat if F is a κ-filtered colimit of representable functors. The category of all κ-flat functors from $\mathbf{C}$ to $\mathbf{S}$, a full subcategory of $(\mathbf{C}, \mathbf{S})$, is denoted as $\kappa - Flat(\mathbf{C})$. The category $\kappa - Flat(\mathbf{C})$ has the universal property of being the free completion of $\mathbf{C}^{op}$ with κ-filtered colimits (Proposition 1.2.4 (ii) in [19]): κ-Flat($\mathbf{C}$) has κ-filtered colimits, and for any category $\mathbf{B}$ with κ-filtered colimits, the functor

$$Z^* : F_\kappa(\kappa - Flat(\mathbf{C}), \mathbf{B}) \to (\mathbf{C}^{op}, \mathbf{B})$$

is an equivalence of categories. Here Z^* is defined by composition with the canonical functor $Z : \mathbf{C}^{op} \to \kappa - Flat(\mathbf{C})$. As proved in [19], a category $\mathbf{A}$ is κ-accessible if and only if it is equivalent to $\kappa - Flat(\mathbf{C})$ for some small category $\mathbf{C}$. Let $\kappa - \mathbf{Acc}$ be the 2-category of all κ-accessible categories as objects with κ-accessible functors(i.e. the functors preserving κ-filtered colimits) as 1-arrows, and all natural transformations as 2-arrows. $\mathbf{P}_\kappa$ denotes the 2-category of all categories which are equivalent to one of the form $(\mathbf{C}, \mathbf{S})$ for a small category $\mathbf{C}$, whose 1-arrows are the functors preserving κ-limits and colimits, and whose 2-arrows are all natural transformations between the latter. The following duality theorem is given in [19] (For the notations, see the list at the end of this section).

Theorem 2.2 *(i) For each* $\mathbf{A}$ *in* $\kappa - \mathbf{Acc}$*, the evaluation functor*

$$\epsilon_{\mathbf{A}} : \mathbf{A} \to L_\kappa Cocts(F_\kappa(\mathbf{A}, \mathbf{S}), \mathbf{S})$$

is an equivalence of categories;

(ii) For each $\mathbf{B}$ *in* $\mathbf{P}_\kappa$*, the evaluation functor*

$$\eta_{\mathbf{B}} : \mathbf{B} \to F_\kappa(L_\kappa Cocts(\mathbf{B}, \mathbf{S}), \mathbf{S})$$

is an equivalence of categories.

The class of accessible categories has a number of subclasses, determined by additional structure on the categories. As proved in [19], A category $\mathbf{A}$ is locally κ-presentable if and only if it is complete κ-accessible, or equivalently cocomplete κ-accessible.

Let $\mathbf{A}$ be an arbitrary category. A set $\mathbf{C}$ of objects of $\mathbf{A}$ is a strong generator of $\mathbf{A}$ if the family of functors represented by the objects in $\mathbf{C}$ are jointly conservative (jointly reflect isomorphisms): for any $f : A \to B$ in $\mathbf{A}$, f is an isomorphism if and only if for all $C \in \mathbf{C}$, $\mathbf{A}(C, f) : \mathbf{A}(C, A) \to \mathbf{A}(C, B)$ is a bijection. Also, recall from [9] that A diagram $D : \mathbf{G} \to \mathbf{A}$ has a multicolimit if the functor

$$D - cocone : \mathbf{A} \to \mathbf{S}$$

assigning to every object A the set of cocones on D, with A as vertex, is isomorphic to a small coproduct of representable functors. The multicolimit of D is then the family of objects of $\mathbf{A}$ representing the functor in the coproduct. A category $\mathbf{A}$ is κ-multicocomplete, if it has the multicolimits of diagrams of size$< \kappa$. The following definition was given in [9].

Definition 2.3 *Let* $\mathbf{A}$ *be a locally small category.* $\mathbf{A}$ *is called locally* κ*-multipresentable if*

(*i*) $\mathbf{A}$ *has* κ*-filtered colimits;*
(*ii*) $\mathbf{A}$ *is* κ*-multicocomplete;*
(*iii*) $\mathbf{A}$ *has a small strong generator consisting of* κ*-presentable objects.*

Theorem 2.4 *Let* $\mathbf{A}$ *be a category. The following are equivalent.*

(*i*) $\mathbf{A}$ *is* κ*-accessible with connected limits;*
(*ii*) $\mathbf{A}$ *is locally* κ*-multipresentable;*
(*iii*) $\mathbf{A}$ *is* κ*-accessible and multicocomplete.*

The above theorem can be found in [9] and [19]. $\mathbf{A}$ in Theorem 2.4 is called κ-Diers.

Recall from [8] that a functor $F : \mathbf{A} \to \mathbf{B}$ has a left multiadjoint if for each object $B \in \mathbf{B}$ there is a small family of morphisms $< g_i : B \to F(A_i) >_{i \in I}$ such that every morphism $g : B \to F(A)$ with $A \in \mathbf{A}$ can be uniquely written as a composite $g = F(f) \circ g_i$ for some g_i and some morphism $f : A_i \to A$. As proved in [8], if a functor $F : \mathbf{A} \to \mathbf{B}$ has a left multiadjoint, then F preserves small connected limits existing in $\mathbf{A}$. The following multiadjoint functor theorem is given by Y. Diers (see Theorem 3.6.1 in [8]).

Theorem 2.5 *Let* $\mathbf{A}$ *be a category with small connected limits, and assume that* $F : \mathbf{A} \to \mathbf{B}$ *preserves all small connected limits. If* F *satisfies the solution-set condition, i.e. for each object* $B \in \mathbf{B}$ *there is a small family of morphisms* $< g_i : B \to F(A_i) >_{i \in I}$ *such that every morphism* $g : B \to F(A)$ *with* $A \in \mathbf{A}$ *can be written as a composite* $g = F(f) \circ g_i$ *for some* g_i *and some morphism* $f : A_i \to A$*. Then* F *has a left multiadjoint.*

Remark 2.6 *Let* $\mathbf{A}$ *be a* κ*-Diers category,* $F : \mathbf{A} \to \mathbf{S}$ *is an* κ*-accessible functor preserving small connected limits. By Proposition 6.1.2 in [19], any accessible functor satisfies the solution-set condition. Therefore* F *has a left multiadjoint. Similar to the Representability Theorem (see Theorem 5.6.3 in [7]), we have that* F *is multi-representable, i.e. it is a sum of representable functors* $\mathbf{A}(A_i, -), i \in I$*. Since* F *preserves* κ*-filtered colimits, it is easily seen that each* $\mathbf{A}(A_i, -)$ *preserves* κ*-filtered colimits, i.e.* $A_i \in \mathbf{A}_\kappa$*.*

Let us recall the notion of regular and Barr-exact categories (see [3] , [4]). A morphism in a category is said to be a regular epimorphism if it is a coequalizer of some pair of morphisms. A category is regular if it has finite limits, coequalizers of kernel-pairs, and in which any pullback of a regular epimorphism is again a regular epi. A functor $F : \mathbf{C} \to \mathbf{D}$ between regular categories is regular if it preserves finite limits and regular epis. $Reg(\mathbf{C}, \mathbf{D})$ denotes the category of regular functors from $\mathbf{C}$ into $\mathbf{D}$, it is a full subcategory of the functor category $(\mathbf{C}, \mathbf{D})$.

A diagram

$$A \underset{g}{\overset{f}{\rightrightarrows}} B$$

in $\mathbf{S}$ is an equivalence relation if the map $a \mapsto <f(a), g(a)> : A \to B \times B$ is one-to-one and its image is an equivalence relation on B in the ordinary sense. Let $\mathbf{C}$ be a category. A diagram as above in $\mathbf{C}$ is an equivalence relation if for each $C \in \mathbf{C}$, the induced diagram

$$\mathbf{C}(C, A) \underset{\mathbf{C}(C,g)}{\overset{\mathbf{C}(C,f)}{\rightrightarrows}} \mathbf{C}(C, B)$$

is an equivalence relation in $\mathbf{S}$. A Barr-exact category is a regular category in which every equivalence relation is a kernel-pair.

The notions of κ-regular category and κ-regular functor are introduced in [16], for κ any infinite regular cardinal. They are a natural generalization of the notions of regular category and regular functor.

Definition 2.7 *A category $\mathbf{C}$ is κ-regular if it is regular, has κ-limits, and satisfies the principle of $< \kappa$ dependent choices (DC_κ): for any ordinal α less than κ, and $\Gamma =< A_\beta, f_{\beta,\gamma} : A_\beta \to A_\gamma >_{\gamma\leq\beta<\alpha}$ is an inverse diagram of type α in $\mathbf{C}$ such that*

(i) $f_{\beta+1,\beta}$ is a regular epi, for every β with $\beta + 1 < \alpha$; and

(ii) the restriction $\Gamma \mid \beta$ of Γ to the domain consisting of all ordinals $\gamma \leq \beta$ is a limit diagram: C_β is a limit of $\Gamma \mid \leq \beta$ (Γ restricted to ordinals $< \beta$) with limit projections $f_{\beta,\gamma} : A_\beta \to A_\gamma (\gamma < \beta)$, for every limit ordinal $\beta < \alpha$.
Then every $f_{\beta,\gamma}$ is a regular epi, for all $\gamma \leq \beta < \alpha$.

A κ-Barr-exact category is a κ-regular category which is Barr-exact. A functor is κ-regular if it is between κ-regular categories, and it preserves all regular epis and all κ-limits.

The following theorem can be found in [2] or [17](Lemma 1.4.9).

Theorem 2.8 *If a regular functor $F : \mathbf{C} \to \mathbf{D}$ is full, and conservative, i.e. it reflects isomorphisms, and $\mathbf{C}$ is Barr-exact. Suppose that for every object D in $\mathbf{D}$ there is an object $C \in \mathbf{C}$ and a regular epi $e : F(C) \to D$. Then F is an equivalence of categories.*

We set up some notations used (before and) later:

κ: an infinite regular cardinal;

Each of the following categories is a full subcategory of the functor category.

$F_\kappa(\mathbf{A}, \mathbf{B})$: the category of functors preserving κ-filtered colimits;

$LF_\kappa(\mathbf{A}, \mathbf{B})$: the category of functors preserving limits and κ-filtered colimits;

$\prod F_\kappa(\mathbf{A}, \mathbf{B})$: the category of functors preserving products and κ-filtered colimits;

$CoF_\kappa(\mathbf{A}, \mathbf{B})$: the category of functors preserving connected limits and κ-filtered colimits;

$CoCocts(\mathbf{A}, \mathbf{B})$: the category of functors preserving connected limits and colimits;

$L_\kappa(\mathbf{A}, \mathbf{B})$: the category of functors preserving κ-limits;

$L_\kappa Cocts(\mathbf{A}, \mathbf{B})$: the category of functors preserving κ-limits and colimits;

$\coprod(\mathbf{A},\mathbf{B})$: the category of functors preserving coproducts;
$L_\kappa\coprod(\mathbf{A},\mathbf{B})$: the category of functors preserving κ-limits and coproducts;
$\kappa - Reg(\mathbf{A},\mathbf{B})$: the category of κ-regular functors;
$LAcc(\mathbf{A},\mathbf{B})$: the category of accessible functors preserving limits;
$CoAcc(\mathbf{A},\mathbf{B})$: the category of accessible functors preserving connected limits;
$\prod Acc(\mathbf{A},\mathbf{B})$: the category of accessible functors preserving products;
$L(\mathbf{A},\mathbf{B})$: the category of functors preserving limits;
$L\coprod(\mathbf{A},\mathbf{B})$: the category of functors preserving limits and coproducts;
$LR(\mathbf{A},\mathbf{B})$: the category of regular functors preserving limits.

3 Coproduct-accessible categories

3.1 Coproduct-flat functors

Definition 3.1 *Let* $\mathbf{C}$ *be a small category, a functor* $F : \mathbf{C} \to \mathbf{S}$ *is called coproduct-flat if* F *is a small coproduct of representable functors in* $(\mathbf{C},\mathbf{S})$.

Let $\mathbf{A}$ be an arbitrary category. By the full closure under coproducts a subcategory $\mathbf{B}$ of $\mathbf{A}$ we means the smallest full subcategory $\mathbf{B}'$ of $\mathbf{A}$ containing $\mathbf{B}$ and closed under coproducts taken in $\mathbf{A}$. Obviously, F is coproduct-flat if and only if F belongs to the full closure under small coproducts of the representable functors in $(\mathbf{C}, \mathbf{S})$.

Let $F_!$ is the left Kan extension of F along the Yoneda embedding, $Y : \mathbf{C} \to (\mathbf{C}^{op},\mathbf{S})$, and assume that F is coproduct-flat. Note that small connected limits commute with coproducts in $\mathbf{S}$, and the left Kan extension $()_!$ is cocontinuous; also for $F = \mathbf{C}(C,-)$, $F_!$ is representable (represented by $\mathbf{C}(-,C)$); it follows that $F_!$ preserves small connected limits.

Theorem 3.2 *Let* $\mathbf{C}$ *be a small category with split idempotents. A functor* $F : \mathbf{C} \to \mathbf{S}$ *is coproduct-flat if and only if* $F_!$ *preserves small connected limits.*

Proof: We only need to show that if $F_!$ preserves small connected limits, then F is coproduct-flat. By Proposition 1.2.4(i) in [19], we have that $F_! \in CoCocts((\mathbf{C}^{op},\mathbf{S}),\mathbf{S})$, and hence $F_! \in CoF_{\aleph_0}((\mathbf{C}^{op},\mathbf{S}),\mathbf{S})$. Since the functor category $(\mathbf{C}^{op},\mathbf{S})$ is an $\aleph_0$-Diers category (it is locally $\aleph_0$-presentable), Remark 2.6 gives that $F_!$ is multipresentable, i.e. $F_! = \coprod_{i\in I} Nat(M_i,-)$, with $M_i \in (\mathbf{C}^{op},\mathbf{S})$. It follows that $Nat(M_i,-)$ is cocontinuous from $F_!$ cocontinuous, i.e. M_i is 0-presentable in $(\mathbf{C}^{op},\mathbf{S})$. Let $M_i = colim\mathbf{C}(-,{C_k}^i)$ be the canonical colimit, then the isomorphism $M_i \to colim\mathbf{C}(-,{C_k}^i)$ factors through a colimit projection:

$$M_i \to \mathbf{C}(-,{C_k}^i) \to colim\mathbf{C}(-,{C_k}^i)$$

so M_i is a retract of $\mathbf{C}(-,{C_k}^i)$, but $\mathbf{C}$ has split idempotents, thus M_i is representable. We conclude that F is coproduct-flat.

The following example shows that the condition on $\mathbf{C}$ in Theorem 3.2 cannot be deleted.

Example 3.3 *Let* $\mathbf{C}$ *be the category of one object* C *and having only one non-trivial morphism* e *which is an idempotent.* $M : \mathbf{C} \to \mathbf{S}$ *is a subfunctor of* $\mathbf{C}(C, -)$ *which is defined by* $M(C) = \{e\}$ *and* $M(e)(-) = (-) \circ e$. *Then* M *is a retract of* $\mathbf{C}(C, -)$. *We have that* $M_!$ *preserves small connected limits. Clearly,* M *isn't coproduct-flat.*

The full subcategory of all coproduct-flat functors of $(\mathbf{C},\mathbf{S})$, is denoted as c-Flat($\mathbf{C}$). Note that the opposite Yoneda embedding Y' factors through $c - Flat(\mathbf{C})$, giving rise to the functor $Z : \mathbf{C}^{op} \to c - Flat(\mathbf{C})$.

Proposition 3.4 *Let* $\mathbf{C}$ *be a small category. Then we have*

(*i*) *The functor* $Z : \mathbf{C}^{op} \to c - Flat(\mathbf{C})$ *has the universal property of being the free completion of* $\mathbf{C}^{op}$ *with small coproducts, i.e. c-Flat(*$\mathbf{C}$*) has small coproducts, and for any category* $\mathbf{A}$ *with small coproducts, the functor*

$$Z^* : \coprod(c - Flat(\mathbf{C}), \mathbf{A}) \to (\mathbf{C}^{op}, \mathbf{A})$$

is an equivalence of categories. Here Z^* *is defined by composition with* Z.

(*ii*) *The quasi inverse of of the equivalence* Z^* *in (i) takes any functor* $F : \mathbf{C}^{op} \to \mathbf{A}$ *to its left Kan extension* $F_!$ *along* Z.

Proof: By Theorem 5.35 in [12].

Corollary 3.5 *For any small category* $\mathbf{C}$ *with split idempotents, we have an equivalence of categories*

$$Y^* : CoCocts((\mathbf{C}^{op}, \mathbf{S}), \mathbf{S}) \to c - Flat(\mathbf{C})$$

defined by composition with the Yoneda embedding Y.

Proof: By Proposition 1.2.4(i) in [19] and Theorem 3.2.

3.2 Coproduct-accessible categories

Let $\mathbf{A}$ be an abitrary category. An object A of $\mathbf{A}$ is *coproduct – presentable*, if the representable functor $\mathbf{A}(A, -) : \mathbf{A} \to \mathbf{SET}$ preserves small coproducts existing in $\mathbf{A}$. As we know, in $\mathbf{SET}$, coproducts are disjoint unions, hence A is a coproduct presentable object if every morphism $A \to \coprod A_i$ into a coproduct $\coprod A_i$ factors uniquely through a coproduct coprojection.

The full subcategory of $\mathbf{A}$ whose objects are the coproduct presentable ones is denoted by $\mathbf{A}_c$.

Proposition 3.6 *In any category, a colimit of a small connected diagram in which the objects are all coproduct presentable is coproduct presentable itself.*

Above proposition is a consequence of the fact that small coproducts commute with small connected limits in $\mathbf{SET}$.

Definition 3.7 *A category* **A** *is called coproduct-accessible if*

(*i*) **A** *has small coproducts, and*

(*ii*) *there is a small full subcategory* **B** *of* $\mathbf{A}_c$ *so that every object of* **A** *is a small coproduct of objects in* **B**.

Example 3.8 **Rel**, *the category of small sets and relations (a morphism from* X *into* Y *is a subset of* $X \times Y$*), is a coproduct accessible category. Indeed,* **Rel** *has all sums (disjoint unions), and one -point set* $*$ *is a coproduct presentable object. For each set* X *, the comma category* $* \downarrow X$ *is just the set of elements of* X*, each regarded as a morphism* $x : * \longrightarrow X$*, hence* X *is the coproduct* $\coprod x$ *of its elements.* **Rel** *has all products, but* **Rel** *does not have all equalizers (see [5]).*

Proposition 3.9 *For every small category* **C**, *c-Flat(***C***) is coproduct-accessible.*

Proof: That small coproducts in $c - Flat(\mathbf{C})$ are computed pointwise follows from the facts that the inclusion functor $c - Flat(\mathbf{C}) \longrightarrow (\mathbf{C}, \mathbf{S})$ preserves the small coproducts, hence representable functors are coproduct presentable objects in c-Flat(**C**).

Recall that a functor $F : I \longrightarrow J$ is final if for each $j \in J$ the comma category j/F is non-empty and connected. We have

Proposition 3.10 *Let* **A** *be a coproduct-presentable category. For every object* A *of* **A**, *let* $A = \coprod_{i \in I} A_i$ *with* A_i *in* $\mathbf{A}_c$, *and the functor*

$$F : I \longrightarrow \mathbf{A}_c/A$$

is defined by $F(i) = p_i$; *here* p_i *is the coprojection. Then* F *is final.*

Proof: If $(f : B \longrightarrow A) \in \mathbf{A}_c/A$, then, since B is coproduct presentable, there is $i \in I$ with $f = p_i \circ f'$; here $f' : B \longrightarrow A_i$ is some morphism, and p_i is the coprojection. This shows that f/F is non-empty. Let $g_i : f \longrightarrow p_i$ and $g_j : f \longrightarrow p_j$ be two morphisms in f/F, that is, $f = p_i \circ g_i$ and $f = p_j \circ g_j$. By the coproduct-presentability of B, we have $i = j$ and $g_i = g_j$. Thus f/F is connected.

Corollary 3.11 *Let* **A** *be a coproduct-accessible category, then*

(*i*) $\mathbf{A}_c$ *is essentially small, and an object of* **A** *is coproduct presentable if and only if it is a retract of some object in* **B**, *the category* **B** *referred to in Definition 3.7.*

(*ii*) $\mathbf{A}_c$ *is dense in* **A**, *i.e. for every object* A *of* **A**, *the canonical cocone*

$$t_A : (\mathbf{A}_c/A)^+ \longrightarrow \mathbf{A}$$

with vertex A *is colimiting.*

Proof: The proof of (i) is essentially same as Proposition 2.1.5(i) in [19]; (ii) follows from Theorem IX.3.1 in [7] and Proposition 3.10.

Proposition 3.12 *Let* $\mathbf{A}$ *be a coproduct-accessible category,* $\mathbf{C}^{op} = \mathbf{A}_c$*, and let* $i : \mathbf{C}^{op} \to \mathbf{A}$ *be the inclusion. Then the functor*

$$\Sigma : \mathbf{A} \to (\mathbf{C}, \mathbf{S})$$
$$A \longmapsto \mathbf{A}(i(-), A)$$

is full and faithful, and its essential image consists of the coproduct-flat functors from $\mathbf{C}$ *to* $\mathbf{S}$*, i.e. we have an equivalence*

$$\mathbf{A} \simeq c - Flat(\mathbf{C}).$$

Also, the diagram

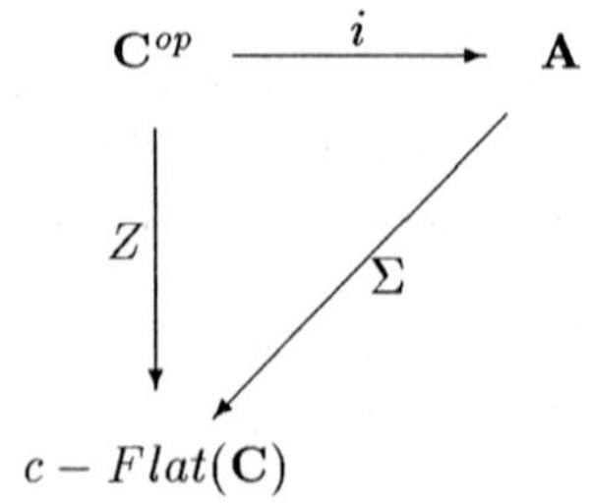

commutes(for Z*,see Proposition 3.4).*

Proof: For any C in $\mathbf{C}$, by the coproduct presentability of C, the functor $\Sigma(-)(C) : \mathbf{A} \to \mathbf{S}$ preserves coproducts. That Σ preserves coproducts follows from the fact that colimits are computed pointwise in $(\mathbf{C}, \mathbf{S})$. The density of $\mathbf{A}_c$ in $\mathbf{A}$ is equivalent to saying that Σ is full and faithful. Note that $\Sigma(C) \cong \mathbf{C}(C, -)$ for every C in $\mathbf{C}$, and c-Flat($\mathbf{C}$) consists of those objects which are coproducts of objects of the form $\Sigma(C)$. Since $\mathbf{A}$ has small coproducts and Σ is full and faithful functor preserving them, we have that the essential image of Σ is c-Flat($\mathbf{C}$).

The last assertion of the proposition is immediate from the definitions.

Corollary 3.13 *A category is coproduct-accessible if and only if it is equivalent to c-Flat(*$\mathbf{C}$*) for some small category* $\mathbf{C}$*.*

Corollary 3.14 *In a coproduct-accessible category, we have*

(i) Small coproducts are stable under pullback, and

(ii) Small coproducts commute with small connected limits to the extent that the latter exist.

Proof: Note that c-Flat($\mathbf{C}$) is closed under the small coproducts in the functor category ($\mathbf{C}$,$\mathbf{S}$), and in the latter, small coproducts are stable under pullback. As for (ii), the full and faithful functor $\Sigma : \mathbf{A} \to (\mathbf{C}, \mathbf{S})$ of Proposition 3.12 preserves small coproducts, as well as all limits existing in $\mathbf{A}$, the reason is that representable functors preserve all existing limits. Also note that small coproducts commute with small connected limits in the functor category.

3.3 Duality of coproduct accessible categories

Let $\mathcal{A}$ denote the 2-category of all categories with small connected limits and small colimits, with functors preserving such limits and small colimits as 1-arrows, and all natural transformations as 2-arrows. Let $\mathcal{B}$ be the 2-category of categories with small coproducts and functors that preserve small coproducts, whose 2-arrows are all natural transformations between them. $\mathbf{S}$, the category of small sets, is an object living in both $\mathcal{A}$ and $\mathcal{B}$ such that the two structures on $\mathbf{S}$ commute with each other, i.e. small coproducts commute with small connected limits and all colimits in $\mathbf{S}$. Such a state of affairs gives rise to a pair of adjoint functors. We obtain the 2-adjunction

$$\mathcal{A}^{op} \underset{G}{\overset{F}{\rightleftarrows}} \mathcal{B}$$

$$\eta : id_{\mathcal{B}} \longrightarrow G \circ F(unit)$$

$$\varepsilon : F \circ G \longrightarrow id_{\mathcal{A}^{op}}(counit)$$

given by

$$G = CoCocts(-, \mathbf{S}), F = \coprod(-, \mathbf{S});$$

both unit and counit are defined as the following evaluation functors.

$$\eta_{\mathbf{B}} : \mathbf{B} \longrightarrow CoCocts(\coprod(\mathbf{B}, \mathbf{S}), \mathbf{S})$$

$$\varepsilon_{\mathbf{A}} : \mathbf{A} \longrightarrow \coprod(CoCocts(\mathbf{A}, \mathbf{S}), \mathbf{S})$$

We consider the full sub-2-category $\mathcal{C}$ of $\mathcal{A}$ on the objects that are categories equivalent to one of the form $(\mathbf{C}, \mathbf{S})$ for a small category $\mathbf{C}$ with split idempotents, and the full sub-2-category $\mathcal{D}$ of $\mathcal{B}$ whose objects are coproduct-accessible categories $\mathbf{A}$ so that $\mathbf{A}_c$ has split idempotents. By Proposition 3.4, Corollary 3.5, and Corollary 3.13, we have the 2-adjunction

$$\mathcal{C}^{op} \underset{G}{\overset{F}{\rightleftarrows}} \mathcal{D}$$

with the restricted unit and counit

$$\eta : id_{\mathcal{D}} \longrightarrow G \circ F, \varepsilon : F \circ G \longrightarrow id_{\mathcal{C}^{op}}.$$

Theorem 3.15 *The 2-adjunction $(F, G, \eta, \varepsilon)$ between $\mathcal{C}$ and $\mathcal{D}$ is a biequivalence, in other words,*

(i) if $\mathbf{C}$ is in $\mathcal{C}$, then $CoCocts(\mathbf{C}, \mathbf{S}) \in \mathcal{D}$, and the evaluation functor

$$\varepsilon_{\mathbf{C}} : \mathbf{C} \longrightarrow \coprod(CoCocts(\mathbf{C}, \mathbf{S}), \mathbf{S})$$

is an equivalence;

(*ii*) *if* **A** *is in* $\mathcal{D}$, *then* $\coprod(\mathbf{A}, \mathbf{S}) \in \mathcal{C}$, *and the evaluation functor*

$$\eta_{\mathbf{A}} : \mathbf{A} \longrightarrow CoCocts(\coprod(\mathbf{A}, \mathbf{S}), \mathbf{S})$$

is an equivalence.

Proof (i) follows from Proposition 3.4 and Corollary 3.5. (ii) can be obtained from Corollary 3.4 and Corollary 3.13.

4 Duality for Diers categories

4.1 Coproduct-accessible categories with κ-limits

Proposition 4.1 *Let* **C** *be a coproduct-accessible category. For every object* A *of* **C**, *the slice category* $\mathbf{C} \downarrow A$ *is coproduct-accessible.*

Proof: Consider the forgetful functor $U : \mathbf{C} \downarrow A \to \mathbf{C}$. U creates colimits, hence $\mathbf{C} \downarrow A$ has small coproducts.

If C is a coproduct presentable object in **C**, then any morphism $k : C \to A$ is coproduct presentable one in $\mathbf{C} \downarrow A$. In fact, given a set of arrows $f_i : C_i \to A$, consider coproduct $\coprod C_i$ with coprojections q_i. By the universal property of coproduct, there is a unique morphism $f : \coprod C_i \to A$ such that $f_i = f \circ q_i$ for all i. Such a morphism f is the coproduct of f_i in $\mathbf{C} \downarrow A$. If $h : k \to f$, then, since $c \in \mathbf{C}_c$, we have a unique i and a unique $l : C \to C_i$ such that

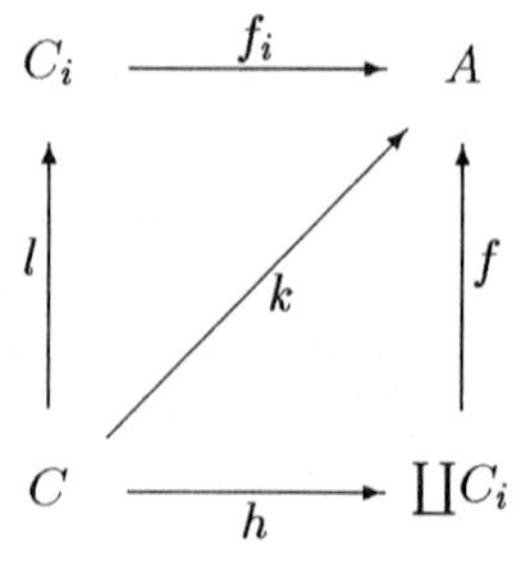

commutes; this means $h : k \to \coprod f_i$ factors uniquely through a unique f_i, showing that k is coproduct presentable.

For any object $g : B \to A$ of $\mathbf{C} \downarrow A$, let $B = \coprod C_i$ with coprojections $q_i : C_i \to B$, we already have that $g = \coprod g \circ q_i$, and $g \circ q_i$ are coproduct presentable objects in $\mathbf{C} \downarrow A$.

Remark 4.2 *Note that if* **C** *has finite limits, then the slice category* $\mathbf{C} \downarrow A$ *has finite limits too.*

Let **X** be any category. Recall ([19], p.115) that the category of families in **X**, Fam(**X**), has as objects pairs (I, X) where I is a set and X is an I-indexed family of objects X_i of $\mathbf{X}$. A morphism $(I, X) \to (J, Y)$ in Fam($\mathbf{X}$) is a pair (f, x), here $f : I \to J$ is a function and x is a family of morphisms $x_i : X_i \to Y_{f(i)}$ indexed by I. Composition of morphisms is given by $(g, y) \circ (f, x) = (g \circ f, f^* y \cdot x)$ where $(f^* y \cdot x)_i = y_{f(i)} \circ x_i$.

Proposition 4.3 *Let $\mathbf{C}$ be an arbitrary coproduct-accessible category, and $i : \mathbf{C}_c \to \mathbf{C}$ be the inclusion functor. Then the functor*

$$\Sigma : Fam(\mathbf{C}_c) \to \mathbf{C}$$

$$(I, < C_i >) \longmapsto \coprod C_i$$

is an equivalence of categories.

Proof: The functoriality of Σ is directly verified by the universal property of coproduct. To show that Σ is full and faithful, let $(I, < C_i >)$ and $(J, < D_j >)$ be any two objects in $Fam(\mathbf{C}_c)$, and $f : \coprod C_i \to \coprod D_j$ is a morphism. Let $p_i : C_i \to \coprod C_i$ be the coprojections of $\coprod C_i$, $q_j : D_j \to \coprod D_j$ are the coprojections of $\coprod D_j$, for each $i \in I$, by the coproduct presentability of C_i, we have $f \circ p_i = q_{i'} \circ f_i$ for a unique morphism $f_i : C_i \to D_{i'}$. Thus we define a morphism $(k, < f_i >)$ between $(I, < C_i >)$ and $(J, < D_j >)$ as follows.

$$k : I \to J \ (i \mapsto i')$$

Then $\Sigma((k, < f_i >)) = f$. This proves the fullness of Σ, the faithfulness of Σ follows from the uniqueness of f_i. Finally , Σ is surjective on objects, since every object of $\mathbf{C}$ is a coproduct of coproduct presentable objects.

Remark 4.4 *As shown in [19], if category $\mathbf{X}$ is accessible, then Fam($\mathbf{X}$) is an accessible category. Given a coproduct-accessible category $\mathbf{C}$, $\mathbf{C}_c$ is a small category. If $\mathbf{C}_c$ has split idempotents, by Theorem 2.2.2 in [19], $\mathbf{C}_c$ is accessible, in that case , $\mathbf{C}$ is accessible.*

Recall that a category $\mathbf{A}$ is κ-multicocomplete, if it has the multicolimits of diagrams of size$< \kappa$. The notion of κ-multicomplete is dual to that of κ-multicocomplete.

Proposition 4.5 *Let $\mathbf{C}$ be a coproduct-accessible category with κ-limits. We have*

(i) $\mathbf{C}_c$ is κ-multicomplete; and

(ii) $\mathbf{C}_c$ has split idempotents.

Proof: (i) Let $\mathbf{I}$ be a graph of size$< \kappa$. For any diagram $G : \mathbf{I} \to \mathbf{C}_c$, let $C = limG$ in $\mathbf{C}$, with limit projections $< p_i : C \to C_i >$. For any cone $< f_i : D \to C_i >$ in $\mathbf{C}_c$, there is a unique morphism $f : D \to C$ such that $f_i = p_i \circ f$. Let $C = \coprod A_j$ with all A_j in $\mathbf{C}_c$, the coproduct presentability of D gives that f factors uniquely through a coproduct coprojection q_j, i.e. there is a unique morphism $g : D \to A_j$ so that $f = q_j \circ g$. We have

$$f_i = p_i \circ (q_j \circ g) = (p_i \circ q_j) \circ g.$$

This shows that every cone on $\mathbf{I}$ factors uniquely through a unique cone $< p_i \circ q_j : A_j \to C_i >$.

(ii) Given an idempotent $e : A \to A$ in $\mathbf{C}_c$, we need to show that e factors as $e = f \circ g$ with $g \circ f = id$ in $\mathbf{C}_c$. Consider the equalizer of (id_A, e), say that $f : D \to A$ is the equalizer of them. Since $e = e \circ e$, there is a unique morphism $h : A \to D$ such that $e = f \circ h$. Let $D = \coprod D_i$ with the coprojections $g_i : D_i \to D$, where all D_i in $\mathbf{C}_c$. The coproduct presentability of A gives that h factors uniquely through a coprojection g_i, thus $e = f \circ (g_i \circ h') = (f \circ g_i) \circ h'$. But $f = e \circ f$, hence $f \circ g_i = e \circ (f \circ g_i) = (f \circ g_i) \circ h' \circ (f \circ g_i)$. Note that the coproduct coprojections are monomorphisms in coproduct-accessible category, also equalizer f is monomorphism , therefore $h' \circ (f \circ g_i) = id$.

Theorem 4.6 *Every coproduct-accessible category with κ-limits is accessible.*

Proof: By Proposition 2.2.2 in [19], every small category with split idempotents is accessible, given a coproduct-accessible category $\mathbf{C}$ with κ-limits, hence $\mathbf{C}_c$ is accessible by proposition 4.5 (ii). We conclude that $\mathbf{C}$ is accessible from Remark 4.4.

Proposition 4.7 *Let $\mathbf{A}$ be an arbitrary κ-Diers category. Then $CoF_\kappa(\mathbf{A}, \mathbf{S})$ is a coproduct-accessible category having κ-limits, and the full subcategory of coproduct presentable objects of $CoF_\kappa(\mathbf{A}, \mathbf{S})$ is equivalent to $\mathbf{A}_\kappa^{op}$, the opposite category of the full subcategory of κ-presentable objects of $\mathbf{A}$.*

Proof: Note that κ-limits and coproducts commute with connected limits and κ-filtered colimits in $\mathbf{S}$, hence the category $CoF_\kappa(\mathbf{A}, \mathbf{S})$ is closed under the κ-limits and coproducts in $(\mathbf{A}, \mathbf{S})$. For any κ-presentable object A in $\mathbf{A}$, the functor $\mathbf{A}(A, \text{-})$ is in $CoF_\kappa(\mathbf{A}, \mathbf{S})$, using the Yoneda Lemma and the fact that small colimits in $(\mathbf{A}, \mathbf{S})$ are computed pointwise, it is easy to see that $\mathbf{A}(A, \text{-})$ is a coproduct presentable object in $CoF_\kappa(\mathbf{A}, \mathbf{S})$. By Remark 2.6, every functor F in $CoF_\kappa(\mathbf{A}, \mathbf{S})$ is multi-representable, i.e. it is a coproduct of representable functors $\mathbf{A}(A_i, -)$; here A_i is κ-presentable. Thus $CoF_\kappa(\mathbf{A}, \mathbf{S})$ is a coproduct-accessible category with κ-limits.

Corollary 4.8 *Let $\mathbf{A}$ be an arbitrary κ-Diers category. Then $\mathbf{A}_\kappa$ is κ-multicocomplete.*

Proof: By proposition 4.5(i) and Proposition 4.7.

Corollary 4.9 *A category is a coproduct-accessible with κ-limits if and only if it is equivalent to the category of the form $c - Flat(\mathbf{C})$; here $\mathbf{C}$ is a small category with κ-multicolimits.*

Proof: By Corollary 3.13, a category is coproduct-accessible if and only if it is equivalent to $c - Flat(\mathbf{C})$ with $\mathbf{C}$ small. If $\mathbf{C}$ is a small category with κ-multicolimits, by Theorem 3.0 in [9], then $\kappa - Flat(\mathbf{C}^{op})$ is κ-Diers. Note that $c - Flat(\mathbf{C})$ is equivalent to the category $CoF_\kappa(\kappa - Flat(\mathbf{C}^{op}), \mathbf{S})$ in Proposition 4.7, hence $c - Flat(\mathbf{C})$ has κ-limits. Conversely, if $c - Flat(\mathbf{C})$ has κ-limits, by Proposition 4.5 (i), then $\mathbf{C}$ is κ-multicocomplete.

4.2 Duality for Diers categories

Proposition 4.10 *Let* $\mathbf{C}$ *be a coproduct-accessible category with* κ*-limits, and* $i : \mathbf{C}_c \to \mathbf{C}$ *be the inclusion functor. Then the functor*

$$\sum : L_\kappa \coprod(\mathbf{C}, \mathbf{S}) \to (\mathbf{C}_c, \mathbf{S})$$

$$F \longmapsto F \circ i$$

is full and faithful, and its essential image consists of the κ*-flat functors from* $\mathbf{C}_c$ *to* $\mathbf{S}$*, i.e. we have an equivalence*

$$L_\kappa \coprod(\mathbf{C}, \mathbf{S}) \simeq \kappa - Flat(\mathbf{C}_c).$$

Proof: First of all, for every functor M in $L_\kappa \coprod(\mathbf{C}, \mathbf{S})$, we have that $M \circ i \in \kappa - Flat(\mathbf{C}_c)$. Indeed, by Theorem 1.2.2 in [19], it is equivalent to say that the category $el(M \circ i)$, the category of elements of $M \circ i$, is κ-filtered. Let I be a graph of size less than κ, and a diagram

$$D : I \to el(M \circ i)$$

$$k \longmapsto x_k \in M \circ i(C_k)$$

Since $\mathbf{C}$ has κ-limits, let $(\pi_k : C \to C_k)_{k \in I}$ be the limit cone in $\mathbf{C}$. So $(M(\pi_k) : M(C) \to M(C_k))_{k \in I}$ be the limit cone in $\mathbf{S}$. Assume that $C = \coprod C_j$ with $C_j \in \mathbf{C}_c$, then $M(C)$ is the disjoint union of $M(C_j)$. It follows that there is some C_j and $x \in M(C_j)$ so that $M(\pi_k)(x) = x_k$ for each $k \in I$.

The fact that $\sum$ is full and faithful follows from Proposition 3.4. Also, note that $\kappa - Flat(\mathbf{C}_c)$ consists exactly of those objects which are κ-filtered colimits of the form $\sum(\mathbf{C}(C, -))$ with $C \in \mathbf{C}_c$. We conclude that the essential image of $\sum$ is $\kappa - Flat(\mathbf{C}_c)$.

Corollary 4.11 *For any coproduct-accessible category* $\mathbf{C}$ *with* κ*-limits, the category* $L_\kappa \coprod(\mathbf{C}, \mathbf{S})$ *is a* κ*-Diers category, moreover, the full subcategory of* κ*-presentable objects in* $L_\kappa \coprod(\mathbf{C}, \mathbf{S})$ *is equivalent to the opposite category of* $\mathbf{C}_c$*.*

Proof: That the category $L_\kappa \coprod(\mathbf{C}, \mathbf{S})$ has connected limits and κ-filtered colimits follows from the fact that connected limits and κ-filtered colimits commute with κ-limits and coproducts in $\mathbf{S}$. So, by Proposition 4.10 and Theorem 2.4, we conclude that $L_\kappa \coprod(\mathbf{C}, \mathbf{S})$ is a κ-Diers category.

Let $\mathbf{A}$ be a κ-Diers category. We have the evaluation functor

$$e_{\mathbf{A}} : \mathbf{A} \to L_\kappa \coprod(CoF_\kappa(\mathbf{A}, \mathbf{S}), \mathbf{S})$$

defined by

$$A \longmapsto [h \longrightarrow h_A]$$

$$f \longmapsto [M \longrightarrow M(f)].$$

By Proposition 4.7, $CoF_\kappa(\mathbf{A}, \mathbf{S})$ is a coproduct-accessible category having κ-limits, and the full subcategory of coproduct presentable objects in $CoF_\kappa(\mathbf{A}, \mathbf{S})$ is equivalent to the opposite category of $\mathbf{A}_\kappa$. Proposition 4.10 gives an equivalence

$$\sum : L_\kappa \coprod (CoF_\kappa(\mathbf{A}, \mathbf{S}), \mathbf{S}) \to \kappa - Flat(\mathbf{A}_\kappa^{op}).$$

Let E be the functor

$$E : \mathbf{A} \to \kappa - Flat(\mathbf{A}_\kappa^{op})$$

$$A \longmapsto \mathbf{A}(i(-), A)$$

By Proposition 2.1.8 in [19], such a functor E is an equivalence. Clearly , $E = \sum \circ e_{\mathbf{A}}$. Thus we have

Theorem 4.12 *For any κ-Diers category $\mathbf{A}$, the canonical functor $e_{\mathbf{A}}$ is an equivalence of categories.*

Let $\mathcal{A}$ be the 2-category of all categories with κ-limits and coproducts as objects and functors preserving κ-limits and coproducts as arrows, and all natural transformations between them as 2-arrows. $\mathcal{B}$ the 2-category of all categories with connected limits and κ-filtered colimits and functors preserving κ-filtered colimits and connected limits. $\mathbf{S}$ is an object living in both $\mathcal{A}$ and $\mathcal{B}$ such that the two structures on $\mathbf{S}$ commute with each other . We obtain the 2-adjunction

$$\mathcal{A}^{op} \underset{G}{\overset{F}{\rightleftarrows}} \mathcal{B}$$

Here $F = CoF_\kappa(-, \mathbf{S}), G = L_\kappa \coprod(-, \mathbf{S})$; both unit and counit are defined, at any object of the respective kind, as the evaluation functor. We consider the full sub-2-category $L_\kappa \coprod$ of $\mathcal{A}$ whose objects are coproduct-accessible with κ-limits, and the full sub-2-category $\kappa - \mathbf{Diers}$ of $\mathcal{B}$ whose objects are κ-Diers categories. By Proposition 4.7 , 4.10, we have the 2-adjunction

$$L_\kappa \coprod^{op} \underset{G}{\overset{F}{\rightleftarrows}} \kappa - \mathbf{Diers}$$

with restricted unit and counit, i.e. the component $e_{\mathbf{A}}$ at $\mathbf{A}$ in $\kappa - \mathbf{Diers}$ is the functor in Theorem 4.12, the counit ε at $\mathbf{C}$ in $L_\kappa \coprod$ is defined as the evaluation functor

$$\varepsilon_{\mathbf{C}} : \mathbf{C} \longrightarrow CoF_\kappa(L_\kappa \coprod(\mathbf{C}, \mathbf{S}), \mathbf{S}).$$

Theorem 4.13 *The pair of adjoint 2-functors obtained is a biequivalence. In other words,*

(i) if $\mathbf{A}$ *in* $\kappa-$ **Diers**, *then* $CoF_\kappa(\mathbf{A},\mathbf{S})$ *in* $L_\kappa\coprod$, *and* e_A *is an equivalence;*

(ii) if $\mathbf{C}$ *in* $L_\kappa\coprod$, *then* $L_\kappa\coprod(\mathbf{C},\mathbf{S})$ *in* $\kappa-$ **Diers**, *and* $\varepsilon_{\mathbf{C}}$ *is an equivalence.*

Proof: By Proposition 4.7, 4.11, and Theorem 4.12, we only need to show that $\varepsilon_{\mathbf{C}}$ is an equivalence of categories, for any coproduct-accessible category with κ-limits. First of all, we show that $\varepsilon_{\mathbf{C}}$ is full and faithful functor. Proposition 1.2.4 (ii) in [19] and Proposition 4.10 give an equivalence $Y^* : (\mathbf{C}_c{}^{op},\mathbf{S}) \to F_\kappa(L_\kappa\coprod(\mathbf{C},\mathbf{S}),\mathbf{S})$. By observation, the following diagram

$$\begin{array}{ccc}
\mathbf{C} & \xrightarrow{\varepsilon_{\mathbf{C}}} & F_\kappa(L_\kappa\coprod(\mathbf{C},\mathbf{S}),\mathbf{S}) \\
{\scriptstyle\Sigma}\downarrow & \nearrow {\scriptstyle Y^*} & \\
(\mathbf{C}_c{}^{op},\mathbf{S}) & &
\end{array}$$

commutes. We obtain that $\varepsilon_{\mathbf{C}}$ is full and faithful from Proposition 3.12.

Secondly, $\varepsilon_{\mathbf{C}}$ in Theorem 4.13 (i) is essentially surjective on objects. Indeed, for any functor M in $CoF_\kappa(L_\kappa\coprod(\mathbf{C},\mathbf{S}),\mathbf{S})$, by what we know, M is a pointwise coproduct of $Nat(\mathbf{C}(C_i,-),-)$, i.e. $\varepsilon_{\mathbf{C}}(C_i)$; here all C_i in $\mathbf{C}_c$. But the evaluation functor $\varepsilon_{\mathbf{C}}$ preserves coproducts, we have $M = \varepsilon_{\mathbf{C}}(\coprod C_i)$.

The following conceptual completeness theorem is an immediate consequence of Theorem 4.13.

Proposition 4.14 *(i) If* $F : \mathbf{C} \to \mathbf{D}$ *is a functor between coproduct-accessible categories having* κ*-limits that preserves coproducts and* κ*-limits, if the induced functor*

$$F^* : L_\kappa\coprod(\mathbf{D},\mathbf{S}) \to L_\kappa\coprod(\mathbf{C},\mathbf{S})$$

is an equivalence of categories, then F *itself is an equivalence;*

(ii) If $M : \mathbf{A} \to \mathbf{B}$ *is a functor between* κ*-Diers categories preserving connected limits and* κ*-filtered colimits, and the induced functor*

$$M^* : CoF_\kappa(\mathbf{B},\mathbf{S}) \to CoF_\kappa(\mathbf{A},\mathbf{S})$$

is an equivalence of categories, then M *is an equivalence .*

The following result also proved in [9](Corollary 6.2).

Corollary 4.15 *If* $\mathbf{C}$ *and* $\mathbf{D}$ *are small* κ*-multicocomplete categories, and the categories* $Mul_\kappa(\mathbf{C}^{op},\mathbf{S})$ *and* $Mul_\kappa(\mathbf{D}^{op},\mathbf{S})$ *are equivalent, then* $\mathbf{C}$ *and* $\mathbf{D}$ *are equivalent. Here* $F \in Mul_\kappa(\mathbf{C}^{op},\mathbf{S})$ *is* κ*-multicontinuous functor, i.e.* F *makes* κ*-multicolimits in* $\mathbf{C}$ *into the correspondence* κ*-limits in* $\mathbf{S}$.

Proof: For any small κ-multicocomplete category $\mathbf{C}$, as shown in [9], the category $Mul_\kappa(\mathbf{C}^{op}, \mathbf{S})$ is a κ-Diers category.

Here is a set of conditions ensuring that an accessible category is a κ-Diers category (also see Proposition 6.1.8 in [19]).

Proposition 4.16 *Suppose* $\mathbf{B}$ *is a* κ*-Diers category,* $\mathbf{A}$ *is an accessible category with* κ*-filtered colimits and small connected limits,* $F : \mathbf{A} \to \mathbf{B}$ *preserves them, and* F *is conservative, i.e. it reflects isomorphisms. Then* $\mathbf{A}$ *is* κ*-Diers.*

Proof: F is an accessible functor, by Proposition 6.1.2. in [19], so F satisfies the solution set condition. It follows that F has a left multiadjoint, i.e. for each object $B \in \mathbf{B}$, there exists a small cone $(f_i : B \to F(A_i))(i \in I)$, such that every morphism $f : B \to F(A), A \in \mathbf{A}$, factors uniquely as $f = F(f') \circ f_i$ for some $i \in I$ and $f' : A_i \to A$. Consider now the object $B \in \mathbf{B}_\kappa$, we show that every object A_i is a κ-presentable one of $\mathbf{A}$. Indeed, Let A is a κ-filtered colimit of C_j in $\mathbf{A}$ with colimit coprojections $p_j : C_j \to A$, since F preserves filtered colimits, then $F(A)$ is the κ-filtered colimit of $F(C_j)$ in $\mathbf{B}$ with coprojections $F(p_j)$. For any morphism $g : A_i \to A$, by the presentability of B, the morphism $F(g) \circ f_i$ factors through some coprojection $F(p_j)$, says that $F(g) \circ f_i = F(p_j) \circ g'$; here $g' : B \to F(C_j)$. It is easy to see that g' factors uniquely as $g' = F(f') \circ f_i$ with some morphism $f' : A_i \to C_j$, thus $F(g) \circ f_i = F(p_j) \circ (F(f') \circ f_i)$. It follows that $g = p_j \circ f'$.

Next, we prove that the set $(A_i, B \in \mathbf{B}_\kappa)$ is a strong generator for $\mathbf{A}$. Given any morphism $f : A \to C$ in $\mathbf{A}$, if for each A_i ,

$$f_{A_i} : \mathbf{A}(A_i, A) \to \mathbf{A}(A_i, C)$$

is a bijection. Note that for $B \in \mathbf{B}_\kappa$,

$$\mathbf{B}(B, F(A)) \cong \coprod \mathbf{A}(A_i, F(A))$$

for every $A \in \mathbf{A}$. Thus we have

$$F(f)_B : \mathbf{B}(B, F(A)) \cong \mathbf{B}(B, F(C))$$

for every $B \in \mathbf{B}_\kappa$. Since $\mathbf{B}_\kappa$ is a strong generator in $\mathbf{B}$, it follows that $F(f)$ is an isomorphism. By F reflecting isomorphism, we obtain that f is an isomorphism. So, by Definition 2.3, $\mathbf{A}$ is a κ-Diers category.

5 Duality for Barr categories

5.1 Barr-exact accessible categories

Let $\mathbf{C}$ be an arbitrary category. An object C of $\mathbf{C}$ is called projective presentable (regular epi projective) if the representable functor $\mathbf{C}(C, -) : \mathbf{C} \to \mathbf{SET}$ takes regular epimorphisms into surjective morphisms.

Remark 5.1 *C is regular epi projective means that C is a projective object with respect to the class of regular epimorphisms of $\mathbf{C}$, i.e. if $e : A \to B$ is a regular epimorphism in $\mathbf{C}$, then every morphism from C into B factors through e.*

The full subcategory of $\mathbf{C}$ whose objects are the regular epi projective ones is denoted by $\mathbf{C}_p$.

Definition 5.2 *A category $\mathbf{C}$ is called regular epi projective-accessible if*
(i) $\mathbf{C}$ has coequalizers of kernel-pairs;
(ii) $\mathbf{C}_p$ is essentially small, and for any object C of $\mathbf{C}$ there is a regular epimorphism from B into C with B in $\mathbf{C}_p$.

Let κ be any infinite regular cardinal. A regular epi projective-accessible category is called κ-Barr-exact accessible if $\mathbf{C}$ is a κ-Barr exact category. $\mathbf{C}$ is Barr-exact accessible if it is κ-Barr-exact accessible for some κ. The following Proposition can be found in some form in [4].

Proposition 5.3 *Let $\mathbf{C}$ be a Barr-exact category, and $\mathbf{D}$ a full subcategory of $\mathbf{C}$. If for every object C of $\mathbf{C}$ there is a regular epimorphism from D into C with D in $\mathbf{D}$, then $\mathbf{D}$ is dense in $\mathbf{C}$.*

Proof: We have to show that for any $C \in \mathbf{C}$, the canonical cocone y_C : $(\mathbf{D}/C)^+ \longrightarrow \mathbf{C}$ with vertex C is colimiting. Let C be a fixed object of $\mathbf{C}$, take an arbitrary cocone

$$(f : D \to C) \longmapsto (f^* : D \to C')$$

on $y_C|(\mathbf{D}/C)$; for any morphism g in $\mathbf{D}$, if g and f are composable, we have $(f \circ g)^* = f^* \circ g$. We will prove that there is a unique morphism $h : C \to C'$ such that $f^* = h \circ f$ for every $f : D \to C$.

Take a regular epimorphism $e : A \to C$ with $A \in \mathbf{D}$. Let e be the coequalizer of the kernel pair $(u, v : A' \to A)$, and $a : B \to A'$ be a regular epimorphism with $B \in \mathbf{D}$. Then e is coequalizer of the pair of $(u \circ a, v \circ a)$. Writing $u' = u \circ a$, $v' = v \circ a$, since A and B are in $\mathbf{C}_p$. we have

$$(e \circ u')^* = e^* \circ u', (e \circ v')^* = e^* \circ v'.$$

By the universal property of coequalizer , there is a unique morphism $h : C \to C'$ such that $e^* = h \circ e$.

It remains to show that for any $f : D \to C$, we have $f^* = h \circ f$. Consider the following diagram

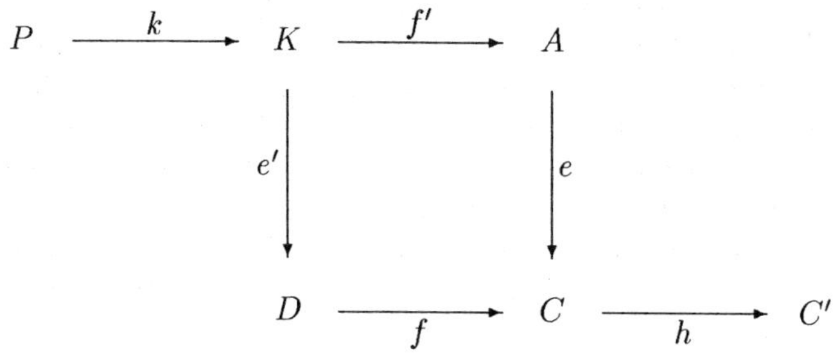

Where K is the pullback of morphisms (e, f), and k is a regular epimorphism with $P \in \mathbf{D}$. Since $\mathbf{C}$ is a Barr-exact category, e' is a regular epimorphism too. We have

$$h \circ e \circ f' \circ k = e^* \circ f' \circ k = (e \circ f' \circ k)^* = (f \circ e' \circ k)^* = f^* \circ e' \circ k.$$

Thus $h \circ f \circ e' \circ k = f^* \circ e' \circ k$. Note that a composite of two regular epimorphisms is a regular epimorphism , thus $e' \circ k$ is a regular epimorphism. It follows that $f^* = h \circ f$ as required.

The uniqueness of h is assured by the equality $e^* = h \circ e$ as we noted above.

Corollary 5.4 *Let $\mathbf{C}$ be a Barr-exact accessible category. Then $\mathbf{C}_p$ is dense in $\mathbf{C}$.*

Proposition 5.5 *Let $\mathbf{A}$ be a κ-accessible category with small products. Then for any functor F in $\prod F_\kappa(\mathbf{A}, \mathbf{S})$, there is a κ-presentable object A in $\mathbf{A}$, and a regular epimorphism $\eta : \mathbf{A}(A, -) \to F$. Moreover, every $F \in \prod F_\kappa(\mathbf{A}, \mathbf{S})$ is the codomain object of a coequalizer of a pair of morphisms between the representable functors.*

Proof: Let $\mathbf{B}$ be a small full subcategory $\mathbf{A}$ consisting of κ-presentable objects so that every object of $\mathbf{A}$ is a κ-filtered colimit of a diagram of objects in $\mathbf{B}$. Given a functor $F \in \prod F_\kappa(\mathbf{A}, \mathbf{S})$, for every $B \in \mathbf{B}$, let us enumerate all elements of $F(B) :< a_i^B >_{i \in J_B}$ here J_B is an ordinal number. Consider the small product $\prod_{B \in \mathbf{B}} B^{J_B}$ in $\mathbf{A}$. The product is the colimit of a κ-filtered diagram $(< B_s >_{s \in S}, < a_{st} : B_s \to B_t >_{s \leq t})$ with the colimit coprojections $< e_s : B_s \to \prod_{B \in \mathbf{B}} B^{J_B} >_{s \in S}$, where each $B_s \in \mathbf{B}$. Let K be the join of all J_B, and by $< a_k >_{k \in K}$ denoting the set of elements of $F(B)$ with all $B \in \mathbf{B}$, $\pi_k : \prod_{B \in \mathbf{B}} B^{J_B} \to B$ be the product projections. Since F preserves products, then $F(\prod_{B \in \mathbf{B}} B^{J_B}) = \prod_{B \in \mathbf{B}} F(B)^{J_B}$, and $F(\pi_k)$ are the product projections in $\mathbf{S}$. We have that there is $a \in F(\prod_{B \in \mathbf{B}} B^{J_B})$ such that $F(\pi_k)(a) = a_k$, for all $k \in K$. Also note that F preserves κ-filtered colimits, the morphisms $F(e_s) : F(B_s) \to F(\prod_{B \in \mathbf{B}} B^{J_B})$ make $F(\prod_{B \in \mathbf{B}} B^{J_B})$ a κ-filtered colimit of the diagram $(< F(B_s) >_{s \in S}, < F(a_{st}) >_{s \leq t})$ in $\mathbf{S}$. Thus there is $s \in S$, and some $c \in F(B_s)$ such that $F(e_s)(c) = a$. It follows that

$$F(\pi_k \circ e_s)(c) = a_k, k \in K$$

We use A for B_s. The Yoneda Lemma gives a natural transformation $\eta : \mathbf{A}(A, -) \longrightarrow F$, is defined by $\eta(id_A) = c$. For every $B \in \mathbf{B}$, we have that η_B is surjective in $\mathbf{S}$. Note that every object of $\mathbf{A}$ is a κ-filtered colimit of objects in $\mathbf{B}$, and A is κ-presentable, it is easy to see that $\eta_{A'}$ is surjective , for all $A' \in \mathbf{A}$. $\prod F_\kappa(\mathbf{A}, \mathbf{S})$ is Barr-exact and the inclusion $\prod F_\kappa(\mathbf{A}, \mathbf{S}) \to (\mathbf{A}, \mathbf{S})$ is regular; in the latter , η being a regular epimorphism means that $\eta_{A'}$ is surjective for all $A' \in \mathbf{A}$. We conclude that η is regular epimorphism .

We have established that every $F \in \prod F_\kappa(\mathbf{A}, \mathbf{S})$ has a regular epimorphism from a representable functor $\mathbf{A}(A, -)$ into F with $A \in \mathbf{B}$. Thus we have a coequalizer diagram

$$G \underset{g}{\overset{f}{\rightrightarrows}} \mathbf{A}(A,-) \xrightarrow{\ \eta\ } F$$

Using again the previous conclusion, there is a regular epimorphism $e : \mathbf{A}(B,-) \to G$ with $B \in \mathbf{B}$. Then η is a coequalizer of the morphisms $(f \circ e, g \circ e)$ between the representable functors $\mathbf{A}(A,-)$ and $\mathbf{A}(B,-)$.

Proposition 5.6 *Let* $\mathbf{A}$ *be an arbitrary* κ*-accessible category having small products.* $\prod F_\kappa(\mathbf{A}, \mathbf{S})$*, the category of the functors preserving* κ*-filtered colimits and small products, is a small* κ*-Barr exact accessible category, and the full subcategory of projective presentable objects of* $\prod F_\kappa(\mathbf{A}, \mathbf{S})$ *is equivalent to* $\mathbf{A}_{\kappa^{op}}$.

Proof: By Proposition 6.9. in [16], $\prod F_\kappa(\mathbf{A}, \mathbf{S})$ is an essentially small κ-Barr exact category. Note that , for any A in $\mathbf{A}_\kappa$, the representable functor $\mathbf{A}(A,-)$ is in $\prod F_\kappa(\mathbf{A}, \mathbf{S})$. Clearly, if $A \in \mathbf{A}$, then $\mathbf{A}(A,-)$ is projective presentable. The assertion of above proposition is immediate from the Proposition 5.5.

Definition 5.7 *A* κ*-accessible category having small products is called a* κ*-Barr category. A category is a Barr category if it is a* κ*-Barr category for some infinite regular cardinal number* κ.

Proposition 5.8 *Let* $\mathbf{C}$ *be a small* κ*-Barr exact accessible category,* $i : \mathbf{C}_p \to \mathbf{C}$ *be the inclusion functor. Then the functor*

$$Z : \kappa - Reg(\mathbf{C}, \mathbf{S}) \longrightarrow (\mathbf{C}_p, \mathbf{S})$$

$$M \longmapsto M \circ i$$

is full and faithful, and its essential image consists of the κ*-flat functors from* $\mathbf{C}_p$ *into* $\mathbf{S}$*, i.e. we have an equivalence*

$$\kappa - Reg(\mathbf{C}, \mathbf{S}) \simeq \kappa - Flat(\mathbf{C}_p).$$

Proof: First of all , for every functor M of $\kappa - Reg(\mathbf{C}, \mathbf{S})$, we verify that $M \circ i \in \kappa - Flat(\mathbf{C}_p)$. As we know , it is sufficient to show that the category $el(M \circ i)$ of elements of the functor $M \circ i$ is κ-filtered . Let I be a graph of size less than κ, and a diagram

$$D : I \longrightarrow el(M \circ i)$$

$$k \longmapsto x_k \in M \circ i(C_k)$$

Since $\mathbf{C}$ has κ-limits, we have the limit cone $(\pi_k : C \to C_k)_{k \in I}$ in $\mathbf{C}$. By Definition 5.2, there is a regular epimorphism $e : D \to C$ with D in $\mathbf{C}_p$. Note that M is a κ-regular functor; thus $(M(\pi_k) : M(C) \to M(C_k))_{k \in I}$ is a limit cone in $\mathbf{S}$, and $M(e) : M(D) \to M(C)$ is surjective. There is a $x \in M(C)$ such that $M(\pi_k)(x) = x_k$ for all $k \in I$. Chosing some $d \in M(D)$ with $M(e)(d) = x$, then $(x_k \in M \circ i(C_k) \to d \in M \circ i(D))_{k \in I}$ is a cocone on D.

Secondly, we are going to show that the functor Z is full and faithful . Let $M, N \in \kappa - Reg(\mathbf{C}, \mathbf{S})$, and $t : M \circ i \to N \circ i$ a natural transformation. We want to find $\eta : M \to N$ such that $Z(\eta) = t$. For any C in $\mathbf{C}$, take a regular epimorphism $e : A \to C$ with A in $\mathbf{C}_p$, and let

$$D \underset{g}{\overset{f}{\rightrightarrows}} A \xrightarrow{\quad e \quad} C$$

be the kernel pair of e; e is the coequalizer of (f, g). Let $d : B \to D$ be a regular epimorphism, then e is a coequalizer of the morphisms $(f \circ d, g \circ d)$. Let $f' = f \circ d$, $g' = g \circ d$. By M and N being κ-regular functors, we have that $M(e)$ is a coequalizer of the morphisms $(M(f'), M(g'))$, and $N(e)$ is a coequalizer of the morphisms $(N(f'), N(g'))$. Naturality of t gives rise to the commutative diagrams as follows.

$$\begin{array}{ccc} M(B) & \xrightarrow{\quad t_B \quad} & N(B) \\ {\scriptstyle M(f')}\Big\downarrow & & \Big\downarrow{\scriptstyle N(f')} \\ M(A) & \xrightarrow[\quad t_A \quad]{} & N(A) \end{array}$$

$$\begin{array}{ccc} M(B) & \xrightarrow{\quad t_B \quad} & N(B) \\ {\scriptstyle M(g')}\Big\downarrow & & \Big\downarrow{\scriptstyle N(g')} \\ M(A) & \xrightarrow[\quad t_A \quad]{} & N(A) \end{array}$$

Since $N(e) \circ N(f) = N(e) \circ N(g)$, we obtain that

$$N(e) \circ t_A \circ M(f) \circ M(d) = N(e) \circ t_A \circ M(g) \circ M(d).$$

But $M(d)$ is surjective, it follows that

$$(N(e) \circ t_A) \circ M(f) = (N(e) \circ t_A) \circ M(g).$$

By the universal property of the coequalizer $M(e)$, there is a unique morphism $t_C : M(C) \to N(C)$ such that the following diagram

$$\begin{array}{ccc} M(A) & \xrightarrow{t_A} & N(A) \\ {\scriptstyle M(e)}\downarrow & & \downarrow{\scriptstyle N(e)} \\ M(C) & \xrightarrow[t_C]{} & N(C) \end{array}$$

commutes.

Claim: Let C and D be arbitrary objects in $\mathbf{C}$, $h : D \to C$ is any morphism, we have the following commutative diagram

$$\begin{array}{ccc} M(D) & \xrightarrow{t_D} & N(D) \\ {\scriptstyle M(h)}\downarrow & & \downarrow{\scriptstyle N(h)} \\ M(C) & \xrightarrow[t_C]{} & N(C) \end{array}$$

Proof of claim: First of all, consider morphism $g : B \to C$ with B in $\mathbf{C}_p$, as before , we let that $e : A \to C$ is regular epimorphism with A in $\mathbf{C}_p$. Taking the pullback of (e, g), say that Q is its pullback with $e' : Q \to B$ and $g' : Q \to A$. For Q , there is a regular epimorphism $a : P \to Q$ with $P \in \mathbf{C}_p$. Note that e' is a regular epimorphism, so is $e' \circ a$. Let $e'' = e' \circ a$ and $g'' = g' \circ a$. Consider the following commutative diagram

$$\begin{array}{ccc} P & \xrightarrow{e''} & B \\ {\scriptstyle g''}\downarrow & & \downarrow{\scriptstyle g} \\ A & \xrightarrow[e]{} & C \end{array}$$

Applying M and N on above the commutative diagram, we have a diagram as follows.

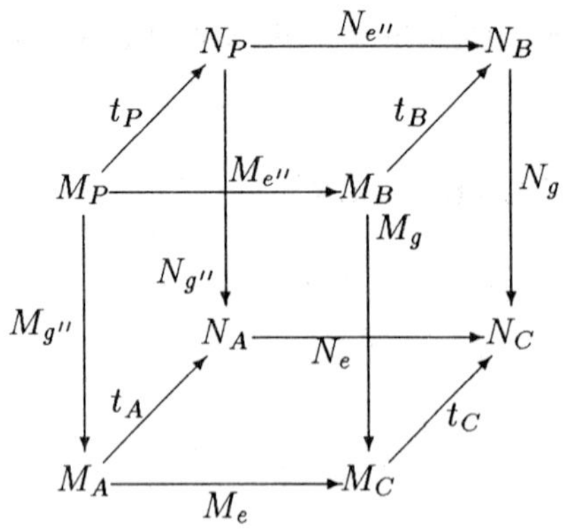

Here M_S, N_S, M_h and N_h denote $M(S)$, $N(S)$, $M(h)$ and $N(h)$, respectively, for any object S and any morphism h in $\mathbf{C}$. It is easily to verify that

$$t_C \circ M(g) \circ M(e' \circ a) = N(g) \circ t_B \circ M(e' \circ a).$$

It follows that $t_C \circ M(g) = N(g) \circ t_B$ from $M(e' \circ a)$ being surjective.

Consider now an arbitrary $h : D \to C$, let $b : P \to D$ be a regular epimorphism with $P \in \mathbf{C}_p$, then

$$t_C \circ M(h \circ b) = N(h \circ b) \circ t_P = N(h) \circ (t_D \circ M(b)).$$

We obtain that $t_C \circ M(h) = N(h) \circ t_D$ from $M(b)$ being surjective.

The claim shows that there is a natural transformation $\eta : M \to N$ such that $Z(\eta) = t$. This proves the fullness of Z, the faithfulness of Z is clear (by t_C uniqueness).

Note that the category $\kappa - Reg(\mathbf{C}, \mathbf{S})$ is closed under the κ-filtered colimits in $(\mathbf{C}, \mathbf{S})$,and colimits are computed pointwise in the latter, thus Z preserves κ-filtered colimits. For any C in $\mathbf{C}_p$, $\mathbf{C}(C, -) \in \kappa - Reg(\mathbf{C}, \mathbf{S})$, $\kappa - Flat(\mathbf{C}_p)$ consists exactly of those objects of the codomain of Z that are κ-filtered colimits of objects of the form $Z(\mathbf{C}(C, -))$. We obtain that the essential image of Z is $\kappa - Flat(\mathbf{C}_p)$.

Remark 5.9 *For any κ-Barr-exact accessible category $\mathbf{C}$, by the above proposition and Proposition 2.1.4 in [19], the category κ-Reg($\mathbf{C}$, $\mathbf{S}$) is a κ-Barr category. Using duality theorem for Barr categories (which will be proved next section), we have that any κ-Barr exact accessible category is essentially small.*

5.2 Duality for Barr categories

Let $\kappa - \mathbf{BARR} - \mathbf{EX}$ be the 2-category of all κ-Barr-exact categories as objects and κ-regular functors as 1-arrows, and all natural transformations between the latter as 2-arrows. $\prod\mathcal{F}_\kappa$ is the 2-category of all categories with κ-filtered colimits and products as objects , all functors preserving κ-filtered colimits and small products as 1-arrow, and all natural transformations between the latter as 2-arrows. $\mathbf{S}$, the category of small sets, is an object in both $\kappa - \mathbf{BARR} - \mathbf{EX}$ and $\prod\mathcal{F}_\kappa$, and each of the $\kappa - \mathbf{BARR} - \mathbf{EX}$-operations commute with each of the

$\prod\mathcal{F}_\kappa$-operations on $\mathbf{S}$(see [16]). Such a statement gives rises to a pair of adjoint 2-functors

$$\kappa - \mathbf{BARR} - \mathbf{EX}^{op} \underset{G}{\overset{F}{\rightleftarrows}} \prod\mathcal{F}_\kappa$$

$$F = \prod F_\kappa(-, \mathbf{S}), G = \kappa - Reg(-, \mathbf{S});$$

both unit and counit are defined , at any object of the respective kind, as the evaluation functor. Consider the full sub-2-category $\kappa - \mathbf{PAcc}$ of $\kappa - \mathbf{BARR} - \mathbf{EX}$ whose objects are κ-Barr exact accessible categories. $\kappa - \mathbf{Barr}$ is the full sub-2-category of $\prod\mathcal{F}_\kappa$ with objects κ-Barr categories. By Proposition 5.6 and Remark 5.9, we have the 2-adjunction

$$\kappa - \mathbf{PAcc}^{op} \underset{G}{\overset{F}{\rightleftarrows}} \kappa - \mathbf{Barr}$$

with restricted unit and counit.

Theorem 5.10 *The pair of adjoint 2-functors as mentioned is biequivalence. In other words,*

(*i*) *If* $\mathbf{C} \in \mathbf{PAcc}$, *then* $\kappa - Reg(\mathbf{C}, \mathbf{S}) \in \kappa - \mathbf{Barr}$, *and the evaluation functor*

$$\varepsilon_{\mathbf{C}} : \mathbf{C} \to \prod F_\kappa(\kappa - Reg(\mathbf{C}, \mathbf{S}), \mathbf{S})$$

is equivalence of categories; and

(*ii*) *if* $\mathbf{A} \in \kappa - \mathbf{Barr}$, *then* $\prod F_\kappa(\mathbf{A}, \mathbf{S}) \in \kappa - \mathbf{PAcc}$, *and the evaluation functor*

$$\eta_{\mathbf{A}} : \mathbf{A} \to \kappa - Reg(\prod F_\kappa(\mathbf{A}, \mathbf{S}), \mathbf{S})$$

is an equivalence of categories.

Proof: (i) For any $\mathbf{C} \in \mathbf{PAcc}$, by Proposition 5.4 , the canonical functor $\sum : \mathbf{C} \to [\mathbf{C}_p{}^{op}, \mathbf{S}]$ is full and faithful. Using Proposition 1.2.4(ii) in [19] and Proposition 5.8, we have an equivalence

$$Z : (\mathbf{C}_p{}^{op}, \mathbf{S}) \to F_\kappa((\kappa - Reg(\mathbf{C}, \mathbf{S}), \mathbf{S}).$$

Also, the diagram

$$\begin{array}{ccc}
\mathbf{C} & \xrightarrow{\varepsilon_{\mathbf{C}}} & F_\kappa(\kappa - Reg(\mathbf{C}, \mathbf{S}), \mathbf{S}) \\
{\scriptstyle\sum}\downarrow & \nearrow{\scriptstyle Z} & \\
(\mathbf{C}_p{}^{op}, \mathbf{S}) & &
\end{array}$$

commutes. Thus $\varepsilon_{\mathbf{C}}$ is full and faithful functor, using the fact that small limits and colimits are computed pointwise, we easily see that $\varepsilon_{\mathbf{C}}$ is a κ-regular functor. Proposition 5.5 gives that every object in $\prod F_\kappa(\kappa - Reg(\mathbf{C},\mathbf{S}),\mathbf{S})$ is a codomain of a regular epimorphism , whose domain is a functor of the form $\varepsilon_{\mathbf{C}}(C)$ with $C \in \mathbf{C}_p$. By Theorem 2.8, we obtain that $\varepsilon_{\mathbf{C}}$ in Theorem 5.10(i) is an equivalence. For (ii), since $\prod F_\kappa(\mathbf{A},\mathbf{S})$ is a κ-Barr-exact accessible category, we can apply Proposition 5.8. Let

$$Y : \mathbf{A}_\kappa^{op} \to \prod F_\kappa(\mathbf{A},\mathbf{S})$$
$$A \mapsto \mathbf{A}(A,-)$$

be the canonical functor, we have the induced functor

$$Y^* : \kappa - Reg(\prod F_\kappa(\mathbf{A},\mathbf{S}),\mathbf{S}) \longrightarrow \kappa - Flat(\mathbf{A}_\kappa^{op})$$

$$M \longmapsto M \circ Y$$

is an equivalence of categories. By Proposition 2.1.8. in [19], the functor

$$\sum : \mathbf{A} \longrightarrow \kappa - Flat(\mathbf{A}_\kappa^{op})$$

$$A \longmapsto \mathbf{A}(i(-),A)$$

is an equivalence, where $i : \mathbf{A}_\kappa \to \mathbf{A}$ be the inclusion functor. It is easy to see that $\sum = \eta_{\mathbf{A}} \circ Y^*$, we conclude that $\eta_{\mathbf{A}}$ is an equivalence of categories.

Remark 5.11 *J. Adámek and R. Rosický have recently shown in [1] that accessible categories with products are exactly the small injectivity classes of locally presentable categories. By Theorem 5.10, a category is a κ-Barr iff it is equivalent to $\kappa - Reg(\mathbf{C},\mathbf{S})$ for some κ-Barr exact accessible category $\mathbf{C}$, such a category $\mathbf{C}$ is essentially small. As we know, the category $\kappa - Reg(\mathbf{C},\mathbf{S})$ is the injectivity class for the small set of regular monomorphisms between representable functors in the category $L_\kappa(\mathbf{C},\mathbf{S})$.*

Corollary 5.12 *Let $\mathbf{A}$ and $\mathbf{B}$ be any two κ-Barr categories. If $F : \mathbf{A} \to \mathbf{B}$ is a κ-accessible functor preserving small products such that the induced functor*

$$F^* : \prod F_\kappa(\mathbf{B},\mathbf{S}) \longrightarrow \prod F_\kappa(\mathbf{A},\mathbf{S})$$

is an equivalence of categories, then F is an equivalence as well.

Corollary 5.13 *A small κ-Barr exact category has a full regular embedding into a Barr-exact accessible category.*

Proof: Let $\mathbf{C}$ be small κ-Barr-exact, then the evaluation functor

$$e_{\mathbf{C}} : \mathbf{C} \longrightarrow \prod F_\kappa(\kappa - Reg(\mathbf{C},\mathbf{S}),\mathbf{S})$$

is an equivalence (see [16]). Let $\kappa - Reg(\mathbf{C},\mathbf{S})$ be λ-accessible, for some regular cardinal $\lambda > \kappa$. Thus we have a full κ-regular functor from $\mathbf{C}$ into $\prod F_\lambda(\kappa - Reg(\mathbf{C},\mathbf{S}),\mathbf{S})$, the latter is Barr-exact accessible from Theorem 5.10.

Let $\mathbf{D}$ be a small category with finite limits. Recall that the Barr-exact hull $\mathbf{C}$ of $\mathbf{D}$, means that $\mathbf{C}$ is Barr-exact , and we have a $F \in Lex(\mathbf{D}, \mathbf{C})$ such that for any Barr-exact category $\mathbf{E}$, the induced functor

$$F^* : Reg(\mathbf{C}, \mathbf{E}) \longrightarrow Lex(\mathbf{D}, \mathbf{E})$$
$$M \mapsto M \circ F$$

is an equivalence of categories(see [16]). We have

Proposition 5.14 *Let* $\mathbf{D}$ *be a small category with finite limits. Then the Barr-exact hull of* $\mathbf{D}$ *is a Barr-exact accessible category, more precisely,* $\mathbf{C}$ *is equivalent to the category* $\prod F_{\aleph_0}(Lex(\mathbf{D}, \mathbf{S}), \mathbf{S})$.

Proof: Note that $Lex(\mathbf{D}, \mathbf{S})$ is an $\aleph_0$-Barr category, so, by Theorem 5.10, we have an equivalence

$$e_{\mathbf{D}}{}^* : Reg(\textstyle\prod F_{\aleph_0}(Lex(\mathbf{D}, \mathbf{S}), \mathbf{S}), \mathbf{S}) \longrightarrow Lex(\mathbf{D}, \mathbf{S})$$
$$M \longmapsto M \circ e_{\mathbf{D}}$$

Where $e_{\mathbf{D}} : \mathbf{D} \to \prod F_{\aleph_0}(Lex(\mathbf{D}, \mathbf{S}), \mathbf{S})$ is the evaluation functor. We conclude that the Barr-exact hull of $\mathbf{D}$ is equivalent to $\prod F_{\aleph_0}(Lex(\mathbf{D}, \mathbf{S}), \mathbf{S})$ from Corollary 6.5. in [16]. Using Theorem 5.10 again, the latter is a Barr-exact accessible category.

6 Some reconstruction results

6.1 SSC on doctrines: LP, Diers and Barr

In this section, we will give strong conceptual completeness results for doctrines: **LP**, **Diers** and **Barr**.

LP is the 2-category of all locally presentable categories, whose 1-arrows are accessible functors preserving small limits, and whose 2-arrows are all natural transformations between the latter.

Diers is the 2-category of Diers categories, its 1-arrows are accessible functors preserving small connected limits, 2-arrows are all natural transformations between them. **Barr** is the 2-category of Barr categories, its 1-arrows are accessible functors preserving products, and 2-arrows are all natural transformations.

Theorem 6.1 *(i) For each* $\mathbf{A}$ *in* **LP**, *the evaluation functor*

$$\epsilon_{\mathbf{A}} : \mathbf{A} \to L(LAcc(\mathbf{A}, \mathbf{S}), \mathbf{S})$$

is an equivalence of categories.

(ii) For each $\mathbf{A}$ *in* **Diers**, *the evalution functor*

$$\epsilon_{\mathbf{A}} : \mathbf{A} \to L\textstyle\coprod(CoAcc(\mathbf{A}, \mathbf{S}), \mathbf{S})$$

is an equivalence of categories.

(iii) For each $\mathbf{A}$ *in* **Barr**, *the evaluation functor*

$$\epsilon_{\mathbf{A}} : \mathbf{A} \to LR(\prod Acc(\mathbf{A}, \mathbf{S}), \mathbf{S})$$

is an equivalence of categories.

Proof: The above results follow from Gabriel-Ulmer duality and duality theorems for (κ)-Diers and (κ)-Barrs, respectively. Since the proofs are essentially same in each case, we only give the proof of (ii). Let $\mathbf{A}$ be any Diers category. We assume that $\mathbf{A}$ is a κ-Diers category, let $i : CoF_\kappa(\mathbf{A}, \mathbf{S}) \to CoAcc(\mathbf{A}, \mathbf{S})$ be the inclusion functor, then the functor induced by i

$$i^* : L\coprod(CoAcc(\mathbf{A}, \mathbf{S}), \mathbf{S}) \longrightarrow L_\kappa\coprod(CoF_\kappa(\mathbf{A}, \mathbf{S}), \mathbf{S})$$

$$M \longmapsto M \circ i$$

is an equivalence of categories.

Firstly, Y^* is full and faithful. Let $M, N \in L\coprod(CoAcc(\mathbf{A}, \mathbf{S}), \mathbf{S})$, and $t : M \circ i \to N \circ i$ be a natural transformation. Any object $A \in \mathbf{A}$ is a κ-filtered colimit of a diagram whose objects are in $\mathbf{A}_\kappa$, say $A = colim_{i \in I} A_i$ with $A_i \in \mathbf{A}_\kappa$. We have $\mathbf{A}(A, -) = lim_{i \in I}\mathbf{A}(A_i, -)$. Note that $\mathbf{A}(A_i, -) \in CoF_\kappa(\mathbf{A}, \mathbf{S})$, and M and N preserve limits, the naturality of t and the universal property of the limit cone $lim_{i \in I} N(\mathbf{A}(A_i, -))$ give a unique morphism $t_{\mathbf{A}(A,-)} : M(\mathbf{A}(A, -)) \to N(\mathbf{A}(A, -))$. It is easy to check that such morphisms fulfil the naturality . Consider now any functor $F \in CoAcc(\mathbf{A}, \mathbf{S})$, as we know, F is a small coproduct of representable functors, note that M and N preserve small coproducts, by above $t_{\mathbf{A}(A,-)}$, we have a unique morphism $t_F : M(F) \to N(F)$.By checking the naturality of t_F (similar to the proof of Proposition 5.8), we obtain that a natural tranformation $\eta : M \to N$ satisfies $i^*(\eta) = t$. This prove the fullness of i^*, the faithfulness of i^* is given by the uniqueness of t_F.

Secondly, i^* is essentially surjective on objects, the reason is that for every object $A \in \mathbf{A}$, $Nat(\mathbf{A}(A, -), -) \in L\coprod(CoAcc(\mathbf{A}, \mathbf{S}), \mathbf{S})$, and $i^*(Nat(\mathbf{A}(A, -), -)) \cong e_{\mathbf{A}}(A)$; here $e_{\mathbf{A}}$ is in Theorem 4.12. Also, note that $e_{\mathbf{A}} = i^* \circ \epsilon_{\mathbf{A}}$. We conclude that $\epsilon_{\mathbf{A}}$ is an equivalence.

Remark 6.2 *For any functor* $F \in LAcc(\mathbf{A}, \mathbf{S})$, *F has left adjoint, hence F is a representable functor. We have that if* $\mathbf{A}$ *is locally presentable, then* $\mathbf{A}$ *is equivalent to the category* $L(\mathbf{A}^{op}, \mathbf{S})$; *if* $\mathbf{B}$ *is locally copresentable (i.e.* $\mathbf{B}^{op}$ *is locally presentable), then* $\mathbf{B}$ *is equivalent to the category* $LAcc(\mathbf{B}^{op}, \mathbf{S})$.

6.2 Barr-exact weak-accessible categories

Let $\mathbf{A}$ be a Barr category. The category $\prod Acc(\mathbf{A}, \mathbf{S})$ is a complete Barr-exact category. We call it Barr-exact weak-accessible category. The meaning of 'weak' is that the full subcategory of projective presentable objects of it isn't necessary small, but whose opposite category is accessible.

Definition 6.3 *A category* $\mathbf{B}$ *is called Barr-exact weak-accessible, if*

(i) $\mathbf{B}$ *is a complete Barr-exact category; and*

(ii) The opposite category of $\mathbf{B}_p$ *is a Barr category, and for every objects* $C \in \mathbf{B}$ *there is a regular epimorphism from* B *into* C *with* B *in* $\mathbf{B}_p$.

Proposition 6.4 *Let* $\mathbf{B}$ *be a Barr-exact weak-accessible category. Then*

(i) $\mathbf{B}$ *is locally small;*

(ii) $\mathbf{B}_p$ *is dense in* $\mathbf{B}$.

Proof: Since $\mathbf{B}_p$ is locally small, the proof of (ii) is same as that of Proposition 2.1.5(i) in [19]. (ii) follows from Proposition 5.3.

Proposition 6.5 *For any Barr category* $\mathbf{A}$, $\prod Acc(\mathbf{A}, \mathbf{S})$ *is a Barr-exact weak-accessible category, and the full subcategory of projective presentable objects of* $\prod Acc(\mathbf{A}, \mathbf{S})$ *is equivalent to the category* $\mathbf{A}^{op}$.

Proof: For every object A of $\mathbf{A}$, note that the representable functor $\mathbf{A}(A, -)$ is projective presentable in $\prod Acc(\mathbf{A}, \mathbf{S})$. Given $F \in \prod Acc(\mathbf{A}, \mathbf{S})$, say that $F \in \prod F_\kappa(\mathbf{A}, \mathbf{S})$, by Propositon 5.5, we have a regular epimorphism from a representable functor into F. The last assertion follows obviously .

Proposition 6.6 *Let* $\mathbf{B}$ *be a Barr-exact weak-accessible category,* $i : \mathbf{B}_p \to \mathbf{B}$ *be the inclusion functor. Then the functor*

$$\sum : \mathbf{B} \longrightarrow ({\mathbf{B}_p}^{op}, \mathbf{S})$$

$$B \longmapsto \mathbf{B}(i(-), B)$$

is full and faithful, and its essential image consists of the accessible functors from $\mathbf{B}_p^{op}$ *to* $\mathbf{S}$ *that preserve products, i.e. we have an equivalence*

$$\mathbf{B} \simeq \prod Acc(\mathbf{B}_p^{op}, \mathbf{S}).$$

Proof: Note that , by the definition of projective presentability, for any $B \in \mathbf{B}_p$, the functor $\sum(-)(B) : \mathbf{B} \to \mathbf{S}$ is a regular functor, since colimits are computed pointwise in $(\mathbf{B}_p^{op}, \mathbf{S})$, then $\sum$ is regular.

That $\sum$ is full and faithful follows from the density of $\mathbf{B}_p$. Given $F \in \prod Acc(\mathbf{B}_p^{op}, \mathbf{S})$, there is a regular epimorphism $\mathbf{B}_p^{op}(B, -) \to F$ with $B \in {\mathbf{B}_p}^{op}$. Note that for $B \in \mathbf{B}_p$, $\sum(B) \cong \mathbf{B}_p^{op}(B, -)$, by Theorem 2.8, we obtain that the essential image of $\sum$ is $\prod Acc(\mathbf{B}_p^{op}, \mathbf{S})$.

Proposition 6.7 *Let* $\mathbf{B}$ *be a Barr-exact weak-accessible category. For every* $B \in \mathbf{B}_p$ *, we have* $\mathbf{B}(B, -) \in LR(\mathbf{B}, \mathbf{S})$*, and the canonical functor*

$$y_{\mathbf{B}} : {\mathbf{B}_p}^{op} \longrightarrow LR(\mathbf{B}, \mathbf{S})$$

$$B \mapsto \mathbf{B}(B, -)$$

is an equivalence of categories. Thus $LR(\mathbf{B}, \mathbf{S})$ *is a Barr category.*

Proof: That $y_{\mathbf{B}}$ is full and faithful follows from the Yoneda lemma. By Proposition 6.6, we have

$$\sum^{*} : LR(\prod Acc(\mathbf{B}_p^{op}, \mathbf{S}), \mathbf{S}) \longrightarrow LR(\mathbf{B}, \mathbf{S})$$

is an equivalence of categories. By Theorem 6.1(iii), we conclude that $y_{\mathbf{B}}$ is an equivalence.

Let **LR** be the 2-category of all Barr-exact weak-accessible categories as objects and regular functors preserving limits as 1-arrows, all natural transformations between the latter as 2-arrows. **Barr** is the 2-category of all Barr categories, whose 1-arrows are accessible functors preserving small products, and whose 2-arrows are all natural transformations between the latter. We have the following duality result.

Theorem 6.8 *(i) If* $\mathbf{B} \in \mathbf{LR}$, *then* $LR(\mathbf{B}, \mathbf{S}) \in \mathbf{Barr}$, *and the evaluation functor*

$$\varepsilon_{\mathbf{B}} : \mathbf{B} \longrightarrow \prod Acc(LR(\mathbf{B}, \mathbf{S}), \mathbf{S})$$

is an equivalence of categories;

(ii) If $\mathbf{A} \in \mathbf{Barr}$, *then* $\prod Acc(\mathbf{A}, \mathbf{S}) \in \mathbf{LR}$, *and the evaluation functor*

$$\eta_{\mathbf{A}} : \mathbf{A} \longrightarrow LR(\prod Acc(\mathbf{A}, \mathbf{S}), \mathbf{S})$$

is an equivalence of categories.

Proof: (i) By Proposition 6.7, the functor induced by $y_{\mathbf{B}}$

$$Y_{\mathbf{B}} : \prod Acc(LR(\mathbf{B}, \mathbf{S}), \mathbf{S}) \longrightarrow \prod Acc(\mathbf{B}_p^{op}, \mathbf{S})$$

$$M \longmapsto M \circ y_{\mathbf{B}}$$

is an equivalence. clearly , we have $\sum = Y_{\mathbf{B}} \circ \varepsilon_{\mathbf{B}}$, here $\sum$ is given in Proposition 6.6. Thus $\varepsilon_{\mathbf{B}}$ is an equivalence of categories.

(ii) is given by Theorem 6.1(iii).

Example 6.9 *Let R be an associative ring with unit. Mod_R denotes the category of right R-modules. Both of Mod_R and ${Mod_R}^{op}$ are Barr-exact categories. Note that Mod_R is locally κ-presentable category, for any infinite regular cardinal κ (see [10]). Thus ${Mod_R}^{op}$ is complete Barr-exact. By Gabriel-Ulmer duality, We have that $Mod_R \simeq L_\kappa(({Mod_R})_\kappa{}^{op}, \mathbf{S})$. Let Inj_R denote the subcategory of Mod_R consisting of the injective modules (projective presentable objects of ${Mod_R}^{op}$). As shown in [16], $Inj_R \simeq \kappa - Reg(({Mod_R})_\kappa{}^{op}, \mathbf{S})$ for some regular cardinal κ. Since each functor in $L_\kappa(({Mod_R})_\kappa{}^{op}, \mathbf{S})$ is the domain of some regular monomorphism whose codomain is a functor in $\kappa - Reg(({Mod_R})_\kappa{}^{op}, \mathbf{S})$ (see [4], [16]). So the category ${Mod_R}^{op}$ is a Barr-exact weak-accessible category. It follows that ${Mod_R}^{op} \simeq \prod Acc(Inj_R, \mathbf{S})$ from Theorem 6.8. Using Theorem 6.8 again, we have*

Proposition 6.10 *Give two rings R_1 and R_2, if they have equivalent categories of injective modules (left or right), then $Mod_{R_1} \simeq Mod_{R_1}$, i.e. they are Morita equivalent.*

This result is also shown in [M3].

References

[1] J.Adámek and J.Rosický, On injectivity in locally presentable categories, to appear in Trans. Amer. Math. Soc.

[2] M.Artin, A.Grothendieck, and J.L.Verdier, Theorie des Topos et Cohomologie Etale des Schemas, Lecture Notes in Math., no. 269, 270, Spring-Verlag, 1972.

[3] M.Barr, Exact categories, Lecture Notes in Math., no.236, Spring-Verlag, 1971, 1-120

[4] M.Barr, Representations of categories, J. Pure Appl. Algebra 41 (1986), 113-137

[5] M.Barr, Accessible categories and models of linear logic, J. Pure Appl. Algebra 69(1990), 219-232

[6] M.Barr and C.Wells, Toposes, Triples and Theories, Grundlehren Math. Wiss. 278, Spring-Verlag, 1985

[7] S.Maclane, Categories for the working mathematician, Spring-Verlag, 1971

[8] Y.Diers, Familles universelles de morphismes, Ann. Soc. Bruxelles, Tome 93, no.3, 1979

[9] Y.Diers, Catégories localement multiprésentables , Arch. Math. 34(1980), 344-356

[10] P.Gabriel and F.Ulmer, Lokal präsentierbare kategorrien, Lecture Notes in Math., no.221, Spring-Verlag, 1971

[11] H.Hu, Duality for Barr-exact categories, Preprint

[12] G.M.Kelly, Basic concepts of enriched category theory, London Math. Lecture Note Ser., no.64, 1982

[13] M.Makkai, Ultraproducts and categorical logic, Lecture Notes in Math., no.1130, Spring-Verlag 1985, 222-309

[14] M.Makkai, Stone duality for first order logic, Adv. in Math. 65 (1987), 97-170

[15] M.Makkai, Strong conceptual completeness for first order logic, Ann. Pure Appl. Logic 40 (1988), 167-215

[16] M.Makkai, A theorem on Barr-exact categories, with an infinitary generalization, Ann. Pure Appl. Logic 47(1990), 225-268

[17] M.Makkai and G.R.Reyes, First order categorical logic, Lecture Notes in Math., no.611, Spring-Verlag, 1977

[18] M.Makkai and A.M.Pitts, Some results on locally finitely presentable categories, Trans. Amer. Math. Soc. 229 (1987), 473-496

[19] M.Makkai and R.Paré, Accessible categories: the foundations of categorical model theory, Contemporary Math. 104 (1990)

[20] R.Paré, Some application of categorical model theory, Contemporary Math. 92(1989), 325-339

Department of Mathematics and Statistics, McGill University, Montreal, PQ, CANADA, H3A 2K6

E-mail address: hu@ gauss.math.mcgill.ca

Canadian Mathematical Society
Conference Proceeding
Volume **13**, 1992

Some Problems in Descriptive Locale Theory

JOHN ISBELL

ABSTRACT. Partial results on the questions: (1) In the real line **R**, does every Borel class of sublocales have representatives nowhere locally Borel of lower class? (Yes, except for the first ambiguous class and possibly other limit ambiguous classes.) (2) Is every absolutely Borel (metric) locale spatial? (3) Is every spatial Borel sublocale of **R** an $O_{\delta\sigma}$? (4) Question 1 for the space **Q** of rationals.

Introduction

I should have liked to call this paper "Test problems ..."; but I do not, like Kaplansky [K], have a general scheme (for "descriptive ... theory"), only some moderately interesting results and some infuriating unsolved problems. The best results are in the preceding paper [I_3], which is really a prerequisite for reading this paper, though some points from it will be recalled below. In particular, the natural inference from the massive unsolved problems offered here, that descriptive locale theory should either not be attempted at all or be postponed until adequate tools for it have been developed -- I think that that inference is refuted by the main theorem of [I_3], which extends a theorem of Hausdorff and Montgomery from completely metrizable spaces to arbitrary metrizable spaces by replacing G_δ (countable *subset* intersection of open sets) by O_δ (countable sublocale meet of open sets: the same thing in the complete case).

The present subject matter fits well under the head of four problems. Problem 1: *Does the real line* **R** *have* proper *representatives of all Borel*

1991 Mathematics Subject Classification. Primary 54H 05; Secondary 03E15, 06B99.
Supported by NSF Grant DMS9003834.
This paper is in final form and no version of it will be submitted for publication elsewhere.

classes of sublocales? Of course the first question is whether there are sublocales exactly of each Borel class: of that class, but not of any lower class. That is not an easy question, but the answer is over eighty years old, for it is easy to get from Borel sets exactly representing the possible classes to sublocales of the same description. I ask, though, for representatives of each Borel class $\mathcal{B}$ which are nonzero and nowhere locally of any lower Borel class (here called *proper* representatives). It is routine to construct — from suitable Borel sets — proper representatives of the unambiguous classes, like $F_{\sigma\delta\sigma\delta\sigma}$. There are no proper representatives of the first ambiguous class: a subspace (or sublocale) of any metrizable space which is F_σ and O_δ has a point at which it is locally closed.

Isolated ambiguous classes after the first have proper representatives -- this is not hard. For ambiguous classes indexed by limit ordinals, it is unsolved.

Problem 2: *Is every absolutely Borel locale spatial*? Background: First recall how 'absolutely Borel' works classically. If a metric space X is a Borel set in its completion (in any metric), of Borel class $\mathcal{B}$, then it is absolutely $\mathcal{B}$, i.e. Borel of class $\mathcal{B}$ in every metric extension space. (That is, past the lowest classes, open, closed, and F_σ.) This holds for O_δ in locales but fails for $F_{\sigma\delta}$ [I3]. In fact we find that it holds for spaces — that is, if a metric space X is a Borel (class $\mathcal{B} \geq O_\delta$) sublocale of a complete space, it is Borel (of that class) in every metric extension locale.

One may think of the O_δ situation as a corollary of that, since every O_δ in a complete space is a subspace. I think that is like deducing the fundamental theorem of algebra from the theory of Banach algebras (which some like to do), for all the O_δ result requires is the classical proof; it works for locales. Anyway, the question is as indicated: are there any absolutely Borel locales besides the spatial ones? It reduces to this: is there a (strictly) pointless absolutely Borel locale? Such a locale must be first category in any metric completion. The most familiar pointless locales are eliminated: (i) the smallest dense sublocale D(X) of a metrizable space X, if nonzero pointless, is not Borel in X; (ii) the largest pointless parts pl(X) of spaces X, when nonzero, are not absolutely Borel.

Do not misunderstand the affirmative result on absolute Borelness to mean that classical absolute Borel sets are absolutely Borel locales. 'Most' of them are Borel sets but not Borel sublocales. Problem 3 begins to show why I referred to infuriating unsolved problems: *Is every spatial Borel sublocale of* **R** *an* $O_{\delta\sigma}$? Or one can say calmly, *What Borel classes are exactly represented by subspaces*? The difference is just between concentrating our ignorance and leaving it room to sprawl. One attempt on the problem produces the peculiar result: *If a complete metric space* E *has* *subspaces which are exactly* $F_{\sigma\delta}$ *sublocales, then it has subspaces which are* $F_{\sigma\delta\sigma}$ *sublocales but not* $F_{\sigma\delta}$ *nor* $O_{\delta\sigma}$, *in fact nowhere locally so.*

Problem 4 is, broadly, *What about the space* Q *of rationals*? Of course Q is not a very suitable space to do descriptive set theory (or locale theory) in. It is, however, an interesting Borel set, the smallest one with any internal complexity to speak of. The descriptive set theory of Q is degenerate; every subspace is F_σ, therefore every subspace is G_δ. By the way, there are only c ($=2^{\aleph_0}$) subspaces. However, in locale theory the question arises: (4a) Does Q *have proper (or at least exact) representatives of all Borel classes of sublocales*? One approach to constructing such Borel sublocales: (4b) *Is there a nice quotient mapping from the pointless part* pl(Q) *to a perfect set*? At least we find in pl(Q) c disjoint closed sublocales -- so pl(Q), and Q, have 2^c nonisomorphic sublocales. But (4b) calls for c neatly arranged disjoint closed sublocales.

1. Borel classes

The Borel sublocales of a metrizable (or other) locale are defined [I_3] like Borel sets, starting with open or closed sublocales and using countable joins and meets in the lattice S(A) of all sublocales (which is anti-isomorphic with the frame of a related locale A_d called [I_1] the *dissolution* of A). In particular, we call countable joins of closed sublocales F_σ *sublocales*; however, a countable meet of open sublocales is called an O_δ *sublocale.* Not "G_δ", because, while in spaces (metrizable or not) F_σ sublocales are precisely F_σ subspaces, the O_δ's need not be spatial. In completely metrizable spaces, O_δ's are G_δ subspaces, and then $O_{\delta\sigma}$'s are $G_{\delta\sigma}$'s (and therefore spatial). Even in the real line, locale $F_{\sigma\delta}$'s are different from classical $F_{\sigma\delta}$'s.

Precisely where are O_δ sublocales the same as G_δ subspaces? It happens that, in regular spaces, this (not very reasonable) question has a simple answer: precisely in those spaces every nonempty closed subset of which is second category in itself. The term for these seems to be (unfortunately, I think) *Baire spaces in the strong sense* (after [F]. [HM] offers the alternative *totally non-meager.*) The property is much stronger than being a *Baire space*: locally of the second category. Let us abbreviate to *strongly Baire.*

It is shown in [I_3, 1.6] that every O_δ sublocale of a strongly Baire regular space is a subspace. The converse doesn't even require regularity:

1.1. *A sober space all of whose* O_δ *sublocales are subspaces is strongly*

Baire.

Proof. Let H be a nonempty closed subset of the space X, supposedly a countable union of sets K_i nowhere dense in H. We may as well have K_i closed. Then the open sublocales $U_i = X\backslash K_i$ meet in an O_δ, by hypothesis a subspace M. Now since $K_i \subset H$ is nowhere dense, U_i contains the smallest dense sublocale D(H); hence $D(H) \leq M$. But M is a subspace, none of whose points p belongs to $H = \cup K_i$ (since $p \in U_i$). So $M \subset X\backslash H$, which is open; $M \wedge D(H) = 0$, a contradiction.

Evidently *a* G_δ *in a strongly Baire space is strongly Baire*; its closed sets K are dense G_δ's in second category K^-, so second category.

As noted in [I_3] just before 1.6, wherever every O_δ is a subspace, every G_δ subspace is an O_δ sublocale. I should guess the converse is false, but have no example.

Recall the individual Borel classes [Ku]: open sublocales form *additive* class 0, closed ones *multiplicative* class 0. Additive class $\alpha+1$ consists of countable joins of sublocales of multiplicative class α, and more generally, additive class β consists of countable joins of sublocales of lower Borel class. (Each additive or multiplicative class contains all previous classes.) Similarly multiplicative class β consists of countable meets of sublocales of lower Borel class. A sublocale is of *ambiguous class* α if it is both of additive and of multiplicative class α.

1.2. *Each Borel class in* S(A) *is a semilattice, closed under binary join.*

Proof. For the open sublocales and the closed sublocales this is familiar. For the higher additive classes it is trivial. So let B and C be of multiplicative class α: $B = \wedge B_i$, $C = \wedge C_j$, with the B_i and C_j of classes $< \alpha$. The point is that in the 'anti-frame', S(A), $(\wedge B_i) \vee (\wedge C_j) = \wedge_j ((\wedge B_i) \vee C_j) = \wedge_j(\wedge_i(B_i \vee C_j))$, so 1.2 follows by induction.

The original Borel classes of Borel *sets* [Ku] still exist and will, not surprisingly, be useful for us. The new ones may be called *localic* Borel classes sometimes, to avoid any danger of confusion. (1.2 illustrates another approach: a Borel class "in S(A)" can scarcely be classical.) The original Borel classes are (similarly) *classical*, or "of Borel sets".

Localic Borel classes are not closed under binary meet. The meet of two "disjoint" countable dense sets (i.e. having no common point) is dense (since the meet of all dense sublocales is dense) and pointless, and certainly not an F_σ. N.B. The reader may wish to check that this does not damage the stratification by additive and multiplicative classes.

Of course the multiplicative classes are closed under meet. For O_δ's something more is true:

1.3. *Two* O_δ *subspaces of a metrizable space which have no common point are disjoint. Indeed, the meet of two* O_δ *subspaces is a subspace.*

Proof. Let $S = \bigwedge U_i$ and $T = \bigwedge V_j$ be O_δ subspaces of a metrizable space A. Suppose the meet $S \wedge T$ is not the subspace $S \cap T$; thus there is (by [I_3, 1.5]) a nonzero closed set H disjoint from $S \cap T$, in which both $H \wedge S$ and $H \wedge T$ are dense.

Closed H is complemented, so $H \wedge S = H \cap S$, $H \wedge T = H \cap T$. Now consider the subspace $S \cup T = B$. The sets $B \setminus U_i$ are closed and their union is T. So T is F_σ and O_δ, and therefore complemented in B. But T is also non-complemented in B, because $H \cap B$ is a nonzero closed set in which $H \cap T$ is dense with empty relative interior. The contradiction proves $S \wedge T =$ $= S \cap T$.

Note on $X \setminus Y$, $X - Y$. For subspaces X, Y of a locale A, $X \setminus Y$ is the subspace on the points of X not in Y. For sublocales X, Y of a locale A, if Y is either complemented in A or contained in X and relatively complemented there, we may write '$X - Y$' for the relative complement of $X \wedge Y$ in X.

The Borel classes are closed under locally finite join, in a metrizable locale, we shall see. They are also local: with a reservation. All classes except "closed" are local, i.e. B belongs to the class if it is contained in a join of open parts U_i of space -- say A -- such that each $U_i \wedge B$ is in the class; we prove this in 1.5 and 1.6 below. Even "closed" is externally local, i.e. if A is covered by open parts U_i such that $U_i \wedge B$ is closed, then B is closed. Proof: (a) "0" is local: a sublocale is empty if it has an open covering (or any other covering) by empty sublocales. (b) "1" is externally local: a sublocale containing every element of an (open) covering of A is A. (c) If $\{U_i\}$ is an open covering of A, and each $U_i \wedge B$ is closed, $A' = \bigvee(U_i - (U_i \wedge B))$ is open since it is a join of open parts, $A \wedge A'$ is 0 by (a), and $A \vee A'$ is 1 by (b).

For the rest of the description of Borel classes, 1.4-1.8, *all locales are metrizable.* "Locally finite" is *externally* locally finite; so closure under locally finite join will follow from 1.2, 1.5 and 1.6. We need a direct proof of

1.4. *Each Borel class in* S(A) *is closed under discrete join.*

Proof. We shall use

(*) *Each Borel class except "closed" is open-hereditary,*

which is obvious. (Technically, it is proved by induction. The frame law $x \wedge (\bigvee y_\lambda) = \bigvee (x \wedge y_\lambda)$, false for general sublocales, is true if x is open or, more generally, complemented [I_1].)

"Open" is closed under arbitrary join. A discrete (indeed, a locally finite) join of closed parts is externally locally closed, so it is closed. A discrete join of Borel sublocales of additive class $\alpha > 0$ is a countable join of discrete joins of Borel sublocales of class β, for the (countably many) $\beta < \alpha$; so if there is a first Borel class not closed under discrete join, it is not an additive class.

Finally consider the join of a discrete family $\{B_\lambda\}$ of Borel sublocales of multiplicative class $\alpha > 0$: each B_λ is $\bigwedge_i C_{\lambda i}$, $C_{\lambda i}$ of Borel class $< \alpha$. Discreteness means we have an open cover $\{U_m\}$ each of whose elements meets at most one B_λ. But in a metrizable locale, the discrete family $\{B_\lambda\}$ has a family of neighborhoods $\{N_\lambda\}$ which is still discrete. Then B_λ is $\bigwedge_i (N_\lambda \wedge C_{\lambda i})$, and the rest is obvious.

1.5. *In a metrizable locale* A, *a sublocale* B *which is locally Borel of additive class* α *is so globally. In fact, every sublocale* B *has a largest open sublocale* C *which is locally (hence globally) Borel of additive class* α *in* A.

Proof. A has a σ-discrete basis $\{U_{i\lambda}\}$, with each $\mathfrak{D}_i = \{U_{i\lambda}\colon \text{all } \lambda\}$ a discrete collection [I_2]. Let $\mathfrak{C}$ be the set of all $U_{i\lambda}$ such that $B \wedge U_{i\lambda}$ is Borel of additive class α, being $\bigvee_k F_{i\lambda k}$, $F_{i\lambda k}$ Borel of class $< \alpha$. (The case $\alpha = 0$ is obvious.) A discrete join $\bigvee_\lambda F_{i\lambda k}$ of Borel sublocales of class $\beta < \alpha$ is Borel of class β (1.4). But $\bigvee_{\mathfrak{C}} (B \wedge U_{i\lambda})$ is a countable join of those. Since the $U_{i\lambda}$ form a basis, this is the largest such open sublocale of B.

1.6. *A locally Borel sublocale of multiplicative class* $\alpha > 0$ *in a metrizable locale is so globally and every sublocale* B *of* A *has a largest open sublocale which is (locally) of multiplicative class* α *in* A.

Proof. The second half is true since those sublocales of B are closed under join. For the first half, A has a σ-discrete basis $\{U_{i\lambda}\}$ as before. For each i, let J_i be the discrete join of all $U_{i\lambda}$ for which $B \wedge U_{i\lambda}$ is of multiplicative class α. Since B is locally of that class, $\bigvee J_i \geq B$. Also each $B \wedge J_i$ is of that class, by 1.4. Thus $B \wedge J_i$ is a meet $\bigwedge_j V_{ij}$ of ω sublocales of class $< \alpha$. The closed complements $A - J_i$ are O_δ: $A - J_i = \bigwedge_k W_{ik}$. Then I claim B is the same as $B' = (\bigvee J_i) \wedge (\bigwedge_i \bigwedge_j \bigwedge_k (V_{ij} \vee W_{ik}))$, which is visibly (in view of (*) and 1.2) of multiplicative class α. Since both are contained in $\bigvee J_i$, it will suffice to prove $B \wedge J_r = B' \wedge J_r$ for each r. $B \wedge J_r$ is $\bigwedge_j V_{rj}$. For each j, $B' \wedge J_r \leq$ $\leq J_r \wedge (\bigwedge_k (V_{rj} \vee W_{rk}))$. Since S(A) is anti-isomorphic with a frame, $V_{rj} \vee (\bigwedge_k W_{rk}) = \bigwedge_k (V_{rj} \vee W_{rk})$); so $B' \wedge J_r \leq J_r \wedge (V_{rj} \vee (\bigwedge_k W_{rk})) =$ $= J_r \wedge (V_{rj} \vee (A - J_r)) = V_{rj}$. This gives $B' \wedge J_r \leq B \wedge J_r$. On the other hand, $V_{ij} \vee W_{ik} \geq (\bigwedge_j V_{ij}) \vee (\bigwedge_k W_{ik}) = (B \wedge J_i) \vee (A - J_i) \geq B$, so $B' \wedge J_r \geq$

$\geq (\bigvee J_i) \wedge B \wedge J_r = B \wedge J_r$, completing the proof.

1.7. Corollary. *Each Borel class in* S(A) *is closed under locally finite join.*

The same reasons show that the Borel classes, except "open" and "closed", are closed under *operation* $\mathfrak{M}$. Instances of operation $\mathfrak{M}$ are performed on a transfinite sequence of sublocales $\{B_\lambda\}$, using a like-indexed sequence of open sublocales $\{U_\lambda\}$. The result of the operation is $\bigvee_\lambda (B_\lambda \wedge (U_\lambda - \bigvee_{\kappa<\lambda} U_\kappa))$. The disjunctified bits $U_\lambda - \bigvee_{\kappa<\lambda} U_\kappa$ are locally closed, and we need to note

(**) *Each Borel class except "open" is closed-hereditary,*
proved like (*).

Then if each B_λ is in a Borel class $\mathcal{C}$ other than "open" or "closed", so is each $B_\lambda \wedge (U_\lambda - \bigvee_{\kappa<\lambda} U_\kappa) = C_\lambda$. As for $S_\lambda = \bigvee_{\kappa<\lambda} C_\kappa$, $S_0 = 0 \in \mathcal{C}$. Inductively, $S_{\lambda+1} = S_\lambda \vee C_\lambda \in \mathcal{C}$ by 1.2. Also for $\kappa < \lambda$, $S_\lambda \wedge U_\kappa = S_{\kappa+1} \wedge U_\kappa$. So at a limit ordinal λ, $S_\lambda \in \mathcal{C}$ by 1.5 or 1.6.

1.8. *Borel classes other than "open" or "closed" are closed under operation* $\mathfrak{M}$.

One concept of descriptive set theory which develops surprising complications in locales is that of *absolutely Borel* objects. "Absolute G_δ" more or less bifurcates; the classical absolute G_δ (metrizable) spaces are precisely the absolute O_δ (metrizable) locales, and their pointless parts are precisely the pointless (-metrizable-) absolute O_δ's [I_3, 2.19]. For F_σ, of course absolute F_σ means σ-compact, while there are no nonzero pointless-absolute F_σ's [I_3, 2.24]. For the higher Borel classes, [I_3, 2.18.b] exhibits an $F_{\sigma\delta}$ sublocale of the real line which is not absolute $F_{\sigma\delta}$. The trouble is in pullbacks; if $f: A \to B$ is a morphism and L a Borel sublocale of B, $f^{-1}(L)$ need not be Borel of the same class [I_3, 2.18.a or b] (L is F_σ but $f^{-1}(L)$ not $F_{\sigma\delta}$ nor $O_{\delta\sigma}$). For Borel sublocales which are spatial, absoluteness nevertheless works, as follows.

1.9. *If* $f: A \to B$ *is a locale morphism and* $L \subset B$ *a Borel sublocale, then* A *has a Borel sublocale of the same Borel class which is contained in* $f^{-1}(L)$ *and contains all the points of* $f^{-1}(L)$.

Proof. If L is open or closed, so is $f^{-1}(L)$. Then we may assume inductively that the result is true for Borel classes $< \alpha$. If L is of multiplicative class α, it is $\bigwedge K_i$ where the K_i are of lower class; so $f^{-1}(L) = \bigwedge f^{-1}(K_i)$, and the result holds for L. Suppose L is additively $\bigvee K_i$, where the K_i are of lower class, so that there are Borel sublocales C_i of A of like Borel class, between $f^{-1}(K_i)$ and its largest subspace. Then $\bigvee C_i$ is of additive class α, contained in $\bigvee f^{-1}(K_i) \leq f^{-1}(L)$, and every point of $f^{-1}(L)$ is in it as required.

1.10. *A subspace of a completely metrizable space which is a Borel sublocale of additive or multiplicative class* $\alpha \geq 2$, *or an* O_δ, *is Borel of that class in every metrizable locale containing it.*

Proof. Let E be complete metric, A a subspace of E which is a Borel sublocale of class ≥ 2, L a metric locale containing A homeomorphically and $\hat{L}$ the completion of L. By Lavrentiev's theorem there is a homeomorphism between G_δ subspaces D of $\hat{L}$, D' of E, extending the identity on A. Since A is of (additive or multiplicative) class α in E, it is so in D' and therefore in homeomorphic D. Hence A is of that class in $\hat{L}$. Since $A \cap L = A$, which is spatial, 1.9 gives that A is Borel of the same class in L.

We call a metrizable locale A *absolutely Borel (of class* $\mathcal{B}$*)* if in every metrizable locale $E \supset A$, A is a Borel sublocale (of class $\mathcal{B}$). By the way, one naturally wonders if every absolutely Borel locale is absolutely Borel of class $\mathcal{B}$ for some $\mathcal{B}$. I do not see a way to separate that question from the central question, what do absolutely Borel locales look like? We do have

1.11. *A closed sublocale of an absolutely Borel locale (of class* $\mathcal{B} \neq$ *'open') is absolutely Borel (of class* $\mathcal{B}$*).*

Proof. If A is absolutely Borel (of class $\mathcal{B}$) and H is closed in A, consider any metrizable extension E of H. The pushout locale $P = A +_H E$ has for frame T(P) just the pullback $T(A) \times_{T(H)} T(E)$. This is regular, as follows. For each of its elements $(u, v) \in T(A) \times T(E)$, consider the common trace w of u and v in T(H). Since E is regular, v is the join of elements $v_\alpha \,\bar{<}\, v$ in T(E) [J], i.e. $v_\alpha^- \subset v$ in E. So w is the join of the $w_\alpha = v_\alpha \cap H$, with $w_\alpha^- \subset w$ (w_α^- being the H-closure). Since A is normal, the w_α can be extended to opens u_α ($u_\alpha \wedge H$ being w_α) with $u_\alpha^- \subset u$; moreover, we can throw in opens $u_\beta^\star$ so chosen that $\bigvee u_\beta^\star = u - H$ and $(u_\beta^\star)^- \subset u$. Put $t_{\alpha\beta} = u_\alpha \vee u_\beta^\star$. Then $p_{\alpha\beta} = (t_{\alpha\beta}, v_\alpha) \in T(P)$, and $p_{\alpha\beta} \,\bar{<}\, (u, v)$. That is, there is $(j, k) \in T(P)$ with $t_{\alpha\beta} \wedge j = 0$, $u \vee j = 1$, $v_\alpha \wedge k = 0$, $v \vee k = 1$; j is $1 - t_{\alpha\beta}^-$ in T(A), and k is $1 - (t_{\alpha\beta}^- \wedge H)$ in T(E).

Also P has a σ-locally finite base, constructed similarly from an arbitrary σ-locally finite base of E and a suitable base of A, with the same trace on H. Hence [I_2] P is metrizable. Therefore the sublocale A is Borel (of class $\mathcal{B}$), so its closed sublocale H is likewise. By construction E is closed in P, so H is Borel (of class $\mathcal{B}$) also in E, as was to be shown.

1.12. Corollary. *If there is a nonspatial absolutely Borel locale then there is a nonzero absolutely Borel locale with no points.*

Proof. Every sublocale of a regular locale A is an intersection of open sublocales [I_1]. So if the spatial part pt(A) is not A, there is an open proper sublocale containing it; thus there is a nonzero closed sublocale H of A without points. By 1.11, H is absolutely Borel if A is.

1.13. Corollary. *An open sublocale of an absolutely Borel locale* A *(of class* $\mathcal{B} > 0$*) is absolutely Borel (of class* $\mathcal{B}$*).*

Proof. Let U be open in absolutely Borel A as described, and let L be a metrizable extension locale of U. Observe that U is covered by ω open sublocales whose closures S_i in A are contained in U; if A is absolutely Borel of fixed class $\mathcal{B}$, so are the S_i, and in any case the S_i must be Borel in L of bounded class, so of some class $\mathcal{B}$ (> 0). Consider the largest open part V of U in L that is Borel of class $\mathcal{B}$. It must contain the interior (relative to U) of each S_i, so it is U as claimed.

We can get some more information on absolutely Borel locales by using Baire category.

2. Category

For sublocales $S \subset A$, as for subspaces, define: S is *first category* (*in* A) if S is contained in a countable join of nowhere dense sublocales of A. N.B. For subspaces, it is the same to say S *is* a countable join of nowhere dense subspaces ($S \cap N_i$); but for sublocales, e.g. the pointless part pl(Q) of the rationals is not equal to a countable join of nowhere dense sublocales (of **R**, or of itself), yet it is first category in **R** because it is contained in Q, a countable join of nowhere dense singletons. *A sublocale of a first category sublocale is first category.*

$S \subset A$ is *second category* (*in* A) if it is not first category.

2.1. *For any sublocale* A *of a metrizable locale* E *there is an open sublocale* $U \subset A^-$ *such that for every nonempty open part* V *of* U, $A \wedge V$ *is second category, while* $A \wedge (E-U)$ *is first category.*

Proof. In a σ-discrete basis $\{U_{i\lambda}\}$ consider the family $\mathcal{F}$ of all $U_{i\lambda}$ whose meet with A is first category; it is a countable union of discrete subfamilies $\mathcal{F}_i$. Then $J_i = \bigvee[U_{i\lambda}\colon U_{i\lambda} \in \mathcal{F}_i]$ meets A in a first category sublocale, and therefore so does $J = \bigvee J_i$. Let U be the interior of $E - J$. Evidently $E - A^- \subset J$, so $U \subset A^-$. By definition of $\mathcal{F}$, all $U_{i\lambda}$ not in $\mathcal{F}$ have second category meet with A; hence so does each open $V \subset U$. $A \wedge (E - U)$ is the join of first category $A \wedge J$ and nowhere dense $E - J - U$, and thus first category as required.

2.1 is of course not impressive (though since A is arbitrary, not much should be expected). We now come to the more considerable result that if A is a Borel sublocale, then $A \wedge U$ is residual in U. *Stating* this correctly turns out to be two-thirds of the problem. First, U is not unique. In 2.1 we constructed a particular open U, and it is easy to verify that that is the largest such U. The most general U for 2.1 is any dense open part of the one we constructed. In 2.4 below, "U from 2.1" is mentioned. The proof uses the constructed U, and one may verify that any dense open part of it works as well.

Now, "residual", and by the way, "meager". *Meager* simply means "first category". The point of the alternative terminology is this: we have first category and second category, which is "not first category"; and then we have meager and residual, which is (classically) "containing a dense G_δ"; equivalently, having a meager complement. The notion is not in use, though, outside Baire spaces (where intersecting ω_0 dense open sets gives a dense G_δ). We define:

Definition. A sublocale A of a metrizable locale E is *residual* in E if (1) E is Baire (locally second category in itself), and (2) there is a dense O_δ sublocale D of E with $D \le A$.

2.2. *A dense* O_δ *sublocale of a Baire locale is Baire.*

Proof. First, a dense $O_\delta \bigwedge U_i$ of second category E is second category in E. For if it were first category it would be under the join of ω nowhere dense K_i's of E. The closed sublocales $H_i = E - U_i$ are nowhere dense; so the join J of all H_i and all K_i is not E. Hence there is nonzero complemented M with $M \wedge J = 0$. But this puts M in all U_i, in $\bigwedge U_i$, but disjoint from all K_i and therefore (since $K_i \le E - M$) disjoint from their join, a contradiction. Now E is Baire if and only if each nonzero open sublocale U is second category in itself, so dense O_δ's of opens U are second category in U, and in E, which means opens of dense O_δ's are second category in E and the O_δ's are Baire.

The idea that "locally" has a fixed, comprehensible meaning is an agreeable one, but it keeps letting you down. A sublocale A of a locale E is called *locally residual* if E is metrizable Baire and A is covered by open sublocales U such that $A \wedge U$ is residual *in* U (not, of course, in E).

2.3. *A dense locally residual sublocale of a metrizable Baire locale is residual.*

Proof. If $A \wedge U$ is residual in U then $A \wedge V$ is residual in V, for any open $V \subset U$. Hence, having a family of such open sublocales U_i, we get a locally finite family $\{V_\lambda\}$ of the same description with the same union — a dense open part of the whole, since it contains A. Residuality of $A \wedge V_\lambda$ in V_λ means containing an O_δ sublocale K_λ dense in V_λ. By 1.7, $\bigvee K_\lambda$ is O_δ. It is dense in $\bigvee V_\lambda$ and thus in the whole locale, as required.

2.4. Theorem. *For a Borel sublocale* A *of a metrizable locale* E, *and the open* U *from* 2.1 *in which* A *is everywhere second category,* $A \wedge U$ *is residual in* U. *So everywhere in* E, *except on the boundary of* U, A *is locally meager or locally residual.*

Proof. By construction A is meager in the open sublocale J, so locally meager there. It remains to show that in the locale U, where $A' = A \wedge U$ is Borel and everywhere second category, A' contains a dense O_δ D. (U is Baire since A' is everywhere second category.)

If A' is of Borel class 0, i.e. open or closed, then it is dense open or U, respectively, and D exists. Inductively, suppose this is true for lower-class Borel sublocales A_i. Then for $A' = \bigwedge A_i$, it holds; for each A_i must be everywhere second category, so it contains a dense O_δ D_i, and A' contains $\bigwedge D_i$.

Now consider the other step $A' = \bigvee A_i$. Since A' is everywhere (in U) second category, the open sublocales U_i associated to A_i must have dense join. (By 2.1, outside the join they are all first category, so A' is too.) In the join, A' is locally residual, so by 2.3 it is residual, completing the proof.

One application of 2.4 is to $pl(Q) \subset pl(\mathbf{R})$, which was shown in [$I_3$] to be not Borel of class 2 -- though it is $F_{\sigma\delta}$ in **R**, and is a pullback of the F_σ $Q \subset \mathbf{R}$. It turns out to be not Borel at all in pl(**R**). Naturally, this generalizes, along the lines "For $A \subset B$ dense, pl(A) is not Borel in pl(B) unless A is most of B". A 'ought to be' residual; but there is no such result even if B = **R**. Why not? Because pl( ) fits too loosely; A need not be residual to contain *all* of pl(B). If pl(A) is Borel in pl(B), it must be residual, and then if A is a classical Borel set in B, it too is residual.

2.5. *If* Y *is a dense-in-itself completely metrizable space and* X *is a dense subspace, then* pl(X) *is second category in* pl(Y); *so if it is Borel, it is residual. If also* X *is a classical Borel set in* Y, *it is residual.*

Proof. If pl(X) were first category it would be in a join of ω_0 (closed) nowhere dense sublocales N_i of pl(Y); these are the traces of closed nowhere dense subspaces L_i of Y. But in the proof of 2.14 in [I_3] we have (‡), saying that in this situation Y has a closed dense-in-itself subset H with each $H \cap L_i$ finite and $H \cap X$ dense in H. Hence the complement Y−H contains $\bigvee N_i$ (since $N_i \subset \mathrm{pl}(L_i)$) but not pl(X), a contradiction.

Now observe that pl(X) must be *everywhere* second category in pl(Y) -- since it is everywhere dense. So if it is Borel, it is residual by 2.4.

For the final assertion we want the classical versions of 2.1, 2.4 [Ku], from which a nonresidual Borel set X in Y is meager in some open set. We may relativize to that open set (for Y); it is completely metrizable with the trace of X dense, and also it is dense in itself. So pl(X) is still everywhere second category in pl(Y). Being Borel, it is residual; there is a dense O_δ D of pl(Y) contained in pl(X). But D is the trace on pl(Y) of an O_δ Δ of pl(Y) [I_3, 1.9], which must be dense since its closure contains pl(Y). Δ is spatial, and $D = \Delta \cap \mathrm{pl}(Y)$ is pl(Δ). We have a dense G_δ Δ with $\mathrm{pl}(\Delta) \subset X$, X meager. By [$I_3$, 2.8], this means every subspace of $\Delta \backslash X$ closed in $\Delta \cup X$ is scattered. But this is absurd; $\Delta\backslash$(meager) is residual in Δ and contains Cantor sets. The contradiction shows that X is not nonresidual in Y.

Now note:

2.6. *A sublocale of a space is first category if and only if it is contained in a first category* F_σ *subspace.*

For a join of certain nowhere dense sublocales is contained in the join of their closures, and closures and their joins (in a space) are spatial. Countable joins are F_σ.

Accordingly

2.7. Theorem. *If* X *is a dense-in-itself metrizable space,* pl(X) *is not absolutely Borel.*

Proof. Suppose the contrary, and consider a completion Y of X. pl(X) is not residual in Y, nor is any part of it residual in a nonempty open set of Y, since G_δ's of Y have points; so by 2.4, pl(X) is first category in Y. By 2.6, it is contained in a first category F_σ subspace M, which is dense in Y since pl(X) is so. Now by 2.5, pl(X) is residual in pl(Y). Therefore $\mathrm{pl}(M) \supset \mathrm{pl}(X)$ is residual in pl(Y). But M is a classical Borel set in Y, so by 2.5 M must be residual in Y, a contradiction.

2.8. *A pointless absolutely Borel locale is first category in any metric completion.*

Proof. The first two sentences of the proof of 2.7 apply.

We now pursue pl(X) further, to show that it is 'normally' not even Borel in X (2.10), and in a complete or near-complete space X, Borel sublocales containing pl(X) are bigger than residual (2.13). First

2.9. *In a regular space* X, pl(X) *is second category if and only if* X *is not the union of two subspaces* Y, Z, *where* Z *is first category and every closed subspace of* X *that is contained in* Y *is scattered.*

Proof. Immediate from 2.6 and [I_3, 2.7]: in a regular space X, pl(X\A) contains pl(X) iff every subspace of A closed in X is scattered.

Now, to apply 2.9. Once we show that it applies to a Cantor set **C**, it extends at once to all possible classical Borel sets (Borel sets in a Polish space); for such a set B either is countable, making pl(B) an $F_{\sigma\delta}$ in any complete space containing it, or it contains a homeomorph of **C**, necessarily closed and meeting pl(B) in pl(**C**) -- so if pl(B) were Borel pl(**C**) would be also. In fact, it extends further; an uncountable classical analytic set contains a copy of **C** [Ku, §32].

2.10. *In an absolute* G_δ *the pointless part is second category; hence in any uncountable classical analytic set* X, pl(X) *is not Borel.*

Proof. In an absolute G_δ A, if $Z \subset A$ is first category it is contained in a first category F_σ set Ω, and $A \setminus \Omega$ is a dense G_δ of A and contains a Cantor set. Thus pl(A) is second category, and as noted above, pl(X) is not Borel.

Second categoriness of pl(X) extends further. The key is

2.11. *A pointless-absolute* O_δ *is second category in its closure in any metrizable locale containing it.*

(There is a chance of confusing the hypothetical pointless absolutely Borel locales with the relative pointless-absolute O_δ's; not much chance, especially at the O_δ level where absolute O_δ's must be spatial.)

Proof. We need only show that the given locale L is second category in its closure in the completion E of any metric container A; for if it were in a countable join of nowhere dense parts N_i of A, they would serve as well in E. Now L is the pointless part pl(X) of a complete metric space X [I_3], and by Lavrentiev's theorem the isomorphism $L \approx$ pl(X) extends to a homeomorphism of containing G_δ's $F \to Y$. Since $X \supset Y \supset$ pl(X), pl(X) = pl(Y). By 2.9 pl(Y) is second category in Y, so L is second category in F. Therefore L is second category in E as required.

The metrizable spaces whose pointless part is a pointless-absolute O_δ have another property or two. Let us call them *approximate* G_δ spaces.

2.12. *The class of approximate* G_δ *spaces is open- and closed-hereditary and* G_δ*-hereditary. Approximate* G_δ *spaces are strongly Baire.*

Proof. If U is open in X, pl(U) is open in pl(X). Here pl(X) is pl(E) for some complete E, so pl(U) is also. Now consider H closed in X. pl(H) = H$\wedge$pl(X) (by (†) in the proof of 2.8 [I_3]), which is O_δ in pl(X). That is O_δ in pl(E) for a complete metric space E; by 1.9 [I_3] this is pl(F) for an O_δ F of E, i.e. a completely metrizable space. Thus H is an approximate G_δ.

Nonempty approximate G_δ spaces H are second category in themselves, as follows. Recall the proof of 2.19 of [I_3]; in a metric completion Y of H we get (by Lavrentiev's theorem) a dense G_δ E with pl(E) = pl(H) as sublocales of Y. We may assume H, Y, and E have no isolated points. So if H were first category, a countable union of nowhere dense closed sets K_i, then the sets $K_i^- \cap E$ would be closed nowhere dense in the absolute G_δ E. The rest of E is a dense-in-itself absolute G_δ. Hence it contains a Cantor set, which is disjoint from $H \subset \cup K_i$. But this conflicts with $H \supset$ pl(E) [I_3, 2.8]; thus H is not first category.

Since the class is closed-hereditary, we now know that all nonempty closed subspaces are second category; the spaces are strongly Baire. O_δ sublocales are then subspaces; thus $G_\delta = O_\delta$ in these spaces, and the class is G_δ-hereditary.

2.13. Theorem. *A Borel sublocale of an approximate* G_δ *space* X *that contains* pl(X) *has a subspace which in every dense-in-itself closed set of* X *is residual.*

Proof. A dense-in-itself closed set H of X is a dense-in-itself approximate G_δ. The Borel sublocale has Borel intersection with H, containing pl(H), which by 2.11 is second category in H; so the result follows from 2.4.

We turn to first category Borel sublocales.

2.14. Lemma. *If a subspace* S *is first category in its closure* S^-, *and* $\{D_i\}$ *is a sequence of dense subspaces of* S^- *having no common point with* S, *then there is a nonzero closed set* $H \subset S^- \setminus S$ *in which all* $D_i \cap H$ *are dense.*

Proof. First we want to enlarge S to a first category F_σ in S^-, say S^*, while keeping each $D_i{}^* = D_i \setminus S^*$ dense in S^- (= S^{*-}). For that, let $\{N_j\}$ be a sequence of closed sets nowhere dense in S^- whose union contains S. Let $\{U_{k\lambda}\}$ be a σ-discrete basis of S^-. For each i, k, and λ, $D_i \cap U_{k\lambda}$ is dense in $U_{k\lambda}$, so not contained in $J_{ik} = \bigcup [N_j\colon j \leq i+k]$; choose a point $d_{ik\lambda}$ of $D_i \cap U_{k\lambda} \setminus J_{ik}$. Then the set $D_i{}^+$ of all $d_{ik\lambda}$ is still dense in S^-. The points $d_{ik\lambda}$ in each N_j form a discrete collection, there being less than j^2 of them for each λ. So $N_j{}^* = N_j \setminus \bigcup D_i{}^+$ is open in N_j, thus F_σ, and $S^* =$
$= \bigcup N_j{}^*$ is a first category F_σ. $D_i{}^* = D_i \setminus S^*$ contains $D_i{}^+$, so the first objective is attained.

Now write S for S^*, $S = \bigcup M_j$ with each M_j closed nowhere dense in S^-; and write D_i for $D_i{}^*$. We describe a sequence of points p_k of $\bigcup D_i$; the closure of $\{p_k\}$ will be the set H. First, choose an ∞-to-1 function $\phi\colon \omega \to \omega$, p_k to be in $D_{\phi(k)}$. Thus we will be choosing a sequence σ_i in D_i for each i. Second, once p_k has been chosen (beginning with any p_1 in $D_{\phi(1)}$), for each i take a subsequence Λ_{ik} of the set of indices $\phi^{-1}(i) \setminus \{1, 2, \ldots, k\}$, disjoint from the previously chosen Λ_{ij}'s and still not exhausting $\phi^{-1}(i)$; for m in Λ_{ik}, p_m is to be within m^{-1} of p_k, but $p_m \neq p_k$. Thus all σ_i will have the same closure H, in which $D_i \cap H$ is dense. Finally, after p_k has been chosen, consider the finite set $I_k =$
$= \{p_1, \ldots, p_k\}$ and the closed set $J_k = M_1 \cup \ldots \cup M_k$. They are disjoint, since $I_k \subset \bigcup D_i$ and $J_k \subset S$. So they are at distance $\varepsilon_k > 0$, and we can hereafter choose all p's (p_m, for $m > k$) outside the $(\varepsilon_k/2)$-neighborhood of J_k. Thus H will be disjoint from S. Now observe that since each D_i is dense, p_{k+1} can always be chosen satisfying the indicated conditions, so the lemma holds.

This gives an awkwardly complicated result which has some simple special cases.

2.15. Theorem. *A subspace* S *of a metrizable space* E *which is first category in itself and Borel in* E, *being a countable meet of spatial Borel sublocales of additive class* $<\alpha$, *has a dense subspace open in* S *and Borel of additive class* $\alpha - 1$, *or* α *if* α *is a limit ordinal, in* S^-.

Proof. We may assume S is dense, i.e. $E = S^-$. We have $S = \bigwedge T_i$, T_i spatial and Borel of additive class $< \alpha$. The result will be precisely that the union U of the sets U_i which are defined in T_i as the interior of S is dense in S; since each U_i is obviously open in S, that implies 2.15. Now if U were not dense in S, there would be a nonempty open set V (of space, S^-) in which all $(T_i \backslash S) \cap V$ are dense ($\Leftrightarrow$ $S \cap V$ has empty interior in $T_i \cap V$). Then 2.14 applies; S is first category in S^-, this relativizes to V, and $D_i = (T_i \backslash S) \cap V)$ is dense in V and has no common point with $S \cap V$. We get a nonzero closed set H of V disjoint from S in which all $D_i \cap H$ are dense. So the dense part $D(H) \leq \bigwedge T_i$, which is incompatible with $H \cap S = 0$ and $H \neq 0$. So U is indeed dense in S.

For $\alpha > 1$, the conclusion that U is Borel of a certain class in S^- simply means it is Borel of that class in E. What 2.15 means for $\alpha = 1$ is

2.15.a. *In a first category metrizable space, every dense spatial* O_δ *sublocale has dense interior.*

That this implies the case $\alpha = 1$ of 2.15 is left as an exercise; the converse implication is immediate. $\alpha = 2, 3$ give us

2.15.b. *In a metrizable space* E, *every spatial* $F_{\sigma\delta}$ *first category in itself has a dense relatively open subset which is* F_σ *in* E, and

2.15.c. *In an approximate* G_δ *space* E, *every spatial* $O_{\delta\sigma\delta}$ *first category in itself has a dense relatively open subset which is* $O_{\delta\sigma}$ *in* E.

The case $\alpha = 3$ says something further about metric spaces which are not approximate G_δ's, but we are not equipped to profit from it.

No spatial $F_{\sigma\delta}$ which is not an $O_{\delta\sigma}$ is known; in fact, as stated in the Introduction, no spatial Borel sublocale of a metrizable space which is not $O_{\delta\sigma}$ is known. Considering just the possible $F_{\sigma\delta}$'s, 2.15.b says that if there is an example there is an example second category in itself (obtained by removing the largest open subspace that is F_σ in E). We can say a bit more about such an example, after some preparation concerning proper Borel class.

We say that ambiguous Borel class α is *lower* than additive or multiplicative class α, and class β is *lower* than class α if $\beta < \alpha$. A sublocale $S \neq 0$ of A is *properly* of a certain class, higher than ambiguous class 1, if it is of that class, and none of its nonzero open sublocales is of lower class.

Remarks. (1) If ambiguous class 1 were not excluded by definition, what happens there would depend on how the definition is worded. Worded as above, $\mathbf{R}^2$ would have subspaces properly of ambiguous class 1, such as an open line segment; but one can easily check that **R** and the Cantor set **C** would not have them. If we said, as in the Introduction, "nowhere locally of any lower Borel class", that is impossible in every metrizable space. It calls for a nonzero complemented subspace S that is nowhere locally closed; but S does not split S^- inseparably, by [I_3, 1.1], which means that relatively in S^-, S has an interior point p; and S is locally closed at p. (2) Since ambiguous class 1 is excluded, the two forms of words are equivalent. Clearly an open sublocale of lower class is somewhere locally of lower class, and the converse follows from open hereditariness for classes higher than "closed".

2.16. *A sublocale of additive or multiplicative class* $\alpha \geq 1$, *and not of ambiguous class* α, *has a closed sublocale properly of its class.*

Proof. Let B be Borel of additive class α in A. By 1.6, B has a largest open sublocale T which is (locally) of multiplicative class α in A. B$-$T is closed in B, and therefore it is Borel of additive class α. If U were a nonzero open part of B$-$T of lower Borel class (i.e. of ambiguous class α), it would be the trace of open V of B. Now V has the open part $V \cap T$, open in T and therefore of multiplicative class α, and the complementary part U of the same Borel class; so (by 1.2) $V = (V \cap T) \vee U$ is of multiplicative class α, impossible because T is the largest such. Thus B$-$T is properly of additive class α, as required. The other case, B of multiplicative class α, goes the same way.

2.17. *A spatial* $F_{\sigma\delta}$ *in a strongly Baire metrizable space* E *which is not* $O_{\delta\sigma}$ *must contain a spatial* $F_{\sigma\delta\sigma}$ *which is nowhere locally* $F_{\sigma\delta}$ *nor* $O_{\delta\sigma}$ *in* E.

Proof. Let A be such an $F_{\sigma\delta}$ in strongly Baire E, and B a closed subspace properly $F_{\sigma\delta}$ in E (from 2.16). By 2.15.b, B is second category in itself, and locally second category in itself. It is so also in B^-, so by 2.4 it contains a dense G_δ D of B^-; B^- is strongly Baire, so G_δ is O_δ in it and D is spatial. D is G_δ in E, so it contains no open subset of B: $B\backslash D = S$ is dense in B. Then it is dense in B^-; but it is first category in B^- (since D is dense G_δ), so S is first category in itself. It is a countable join of closed subsets of B, so it is $F_{\sigma\delta\sigma}$. Can it have an open subset V which is $O_{\delta\sigma}$ in E? Then the open-in-B extension $B-(S\backslash V)^-$ would be $O_{\delta\sigma}$ in E, contrary to the propriety of B. If S has an open subset W which is $F_{\sigma\delta}$ in E, 2.15.b gives an open set V F_σ in E; so V is $O_{\delta\sigma}$ in E and again we have a contradiction. Thus S is the required subspace.

Of course, 2.17 arose in a search for a contradiction. One suspects one has it when S turns out to be a relative F_σ in an $F_{\sigma\delta}$, but not itself $F_{\sigma\delta}$. But that *is* possible, as 2.18 shows. Can it happen in subspaces? We do not even know if the $F_{\sigma\delta}$ (which is not more than an $F_{\sigma\delta}$) can be found in subspaces.

2.18. Example. pl(Q) *(which is* $F_{\sigma\delta}$ *in* **R**) *has an* F_σ *sublocale which is not itself* $F_{\sigma\delta}$ *in* **R**.

The example $B\subset$ pl(Q) was already used as a counterexample in 2.18.c of $[I_3]$: an F_σ of pl(Q)$\subset$Q which is not the trace of an F_σ of Q (equivalently, of **R**). We are now to show that not even a (decreasing) sequence of F_σ's will describe B.

Put $Q\times Q$ for Q; it is homeomorphic. Count $Q = \{q_i\}$, and let the i-th *line* L_i be $\{q_i\}\times Q$. The example B is $\bigvee \mathrm{pl}(L_i)$, F_σ in $\mathrm{pl}(Q\times Q)$. An F_σ of $Q\times Q$ containing B is the same as a subspace H of $Q\times Q$ such that each line L_i is, except for a scattered subset, contained in H. Such a subspace contains the graph of a continuous function $f\colon Q\to Q$; in fact, (†) given any set $S\subset Q\times Q$ with each $S\cap L_i$ scattered, we can build up a continuous function $f\colon Q\to Q$ with graph contained in $H\backslash S$. For this, choose $f(q_1)$ in $(L_1\cap H)\backslash S$, which is all of L_1 with scattered exceptions, and resolve that for all q_j within n^{-1} of q_1 (nonstrictly: $|q_j-q_1|\le n^{-1}$), $f(q_j)$ will be within (say) n^{-1} of $f(q_1)$. Make that strict, $|f(q_j)-f(q_1)| < n^{-1}$. Continue in the same manner; coming to choose $f(q_k)$, there are only finitely many continuity conditions, and scattered values to avoid. There will be further conditions, that for later q_m near q_k, $f(q_m)$ is near $f(q_k)$. Those $f(q_m)$ are also required to be near certain earlier $f(q_j)$. There are finitely many of them; for each, there is a greatest n such that $|q_k - q_j| \le n^{-1}$, and sufficiently near q_k that n does not increase.

So for q_m sufficiently near q_k, all these constraints would be satisfied by $f(q_m) = f(q_k)$. That value may be excluded, but we left a margin ($|f(q_k) - f(q_j)| < \varepsilon(j, k)$), so we can keep the constraints consistent and the recursion runs.

I claim that for each sequence $\{H_i\}$ of F_σ's containing B, $\bigwedge H_i \neq B$. The test for this is [I_3, 1.5] that $Q \times Q$ has a nonzero closed set J disjoint from B with all $H_i \cap J$ dense in J. We may assume each $H_{i+1} \subset H_i$. Then put $H = H_1$, and remove the set S which on each line L_i is $L_i \backslash H_i$. The construction (†) above gives us a continuous function whose graph J (closed and homeomorphic with Q) is contained in each H_i, except for $i-1$ of the nowhere dense sets L_j. Then $H_i \cap J$ is certainly dense, and the open set $(Q \times Q) - J$ contains each $pl(L_i)$ and therefore B.

Returning to proper Borel class, we can use 2.16 to get proper representatives of all nonambiguous classes. We will be working with the long transfinite expressions $\bigvee_i \bigwedge_j \bigvee_k \ldots$ describing the construction of Borel sets, or sublocales. (We will not need a really adequate notation.)

Start with one of the known constructions for a (classical) Borel set B exactly of additive or multiplicative class α. (See e.g. [Ku, §26, XIV].) Perform the same construction in the lattice of sublocales S(C); it gives a sublocale A, certainly of the indicated class because it is so constructed. The space of points of A is B (inductively, since $\bigvee$ and $\bigwedge$ commute with taking the points). The space of points is a functor, call it pt(A). Then whatever (lower) class B is known *not* to belong to (as Borel set), neither does A (as sublocale), for if it did, so would pt(A). Thus A satisfies the hypothesis of 2.16, so also the conclusion.

There remain the ambiguous classes. For ambiguous class $\alpha + 1$, roughly, proper representatives can be had by combining M properly of multiplicative class α, and A, of the additive class, by the formula "$(P \times M) \cup (Q \times A)$" (Q a countable dense set, P its classical complement). Correctly, fix a *counted* dense set $Q = \{q_i\} \subset C$. For sublocales M, A of

C, presented by Borel constructions Γ, Γ' as of multiplicative resp. additive class α (for some a), M finally as $\wedge E_i$ and A as $\vee F_j$, let $\Pi(M, A) \subset \subset \mathbf{C} \times \mathbf{C}$ be the join of $\wedge((\mathbf{C}-\{q_i\}) \times E_i)$ and $\vee(\{q_j\} \times F_j)$.

2.19. *If* M *is properly of multiplicative class* α *in* C *and* A *is properly of additive class* α *then* Π(M, A) *is properly of ambiguous class* $\alpha+1$, *for* $\alpha \geq 1$.

Proof. The factors E_i of M are Borel of additive classes cofinal in the predecessors of α, so their products by the open sets $\mathbf{C}-\{q_i\}$ are likewise (by an obvious induction). The meet of those products is then of multiplicative class α. Similarly $\vee(\{q_j\} \times F_j)$ is of additive class α. So both are of ambiguous class $\alpha+1$, and by 1.2 so is their join Π(M, A). But each nonzero open sublocale U of Π(M, A) has closed parts isomorphic with open parts of M (in $\{p\} \times M$, $p \in P$), so U is not of additive class α, and similarly no open part of Π(M, A) is of multiplicative class α: it is properly ambiguous-($\alpha+1$).

Summarizing:

There are no Borel sublocales properly of ambiguous class 1. *All nonambiguous classes, and all isolated ambiguous classes after class* 1, *have proper representatives in* C. *For each limit ambiguous class it is unknown.*

Returning to 2.14, it can be pushed further; first by putting a sublocale for the subspace S.

2.20. *If a sublocale* L *of a metric space is first category in* L^-, *and* $\{D_i\}$ *is a sequence of dense subspaces of* L^- *having no common point with* L, *then there is a nonzero closed set* H *of* L^- *whose complement contains* L, *in which all* $D_i \cap H$ *are dense.*

Proof. As before, let $\{N_j\}$ be a sequence of closed sets nowhere dense in L^- whose join (the subspace union) contains L. Let $\{U_{k\lambda}\}$ be a σ-discrete basis of L^-. For each i, k, and λ, $D_i \cap U_{k\lambda}$ is dense in $U_{k\lambda}$, so not contained in $J_{ik} = \cup[N_j\colon j \leq i+k]$; choose a point $d_{ik\lambda}$ of $D_i \cap U_{k\lambda} \setminus J_{ik}$. Then the set D_i^+ of all $d_{ik\lambda}$ is still dense in L^-. The points $d_{ik\lambda}$ in each N_j form a discrete collection, there being less than j^2 of them for each λ. So $N_j^* = N_j \setminus \cup D_i^+$ is open in N_j, thus F_σ, and $L^* = \cup N_j^*$ is a first category F_σ subspace containing L. Again $D_i^* = D_i \setminus L^*$ contains D_i^+, so it is dense in L^-. Then by 2.14 there is nonzero closed $H \subset L^- \setminus L^*$ in which all $D_i \cap H$ are dense. The complement contains L^*, so it contains L as required.

This applies e.g. to the sublocale D(X) (which is always first category in a dense-in-itself metric space X, being contained in a σ-discrete dense subspace). To say that the complement of H contains D(X) is to say that it is dense, i.e. the closed set H is nowhere dense. This approaches a proof that D(X) is not Borel, but we need a refinement of 2.20. Call a locale *spacious* if it has no nonzero open pointless Boolean sublocale.

2.21. Lemma. *Given a dense-in-itself metrizable locale* X *and a sequence of dense spacious sublocales* D_i *of* X, *there exists a nonzero dense-in-itself nowhere dense closed part* J *of* X *in which each* $D_i \wedge J$ *is dense.*

Proof. Again we build up J by closing the result of a sequence of choices, $\bigvee \pi_j$. Again choose $\phi:\omega\to\omega$ infinity-to-one; π_j will be a nonzero nowhere dense part of $D_{\phi(j)}$. The D_i's being spacious *and* dense in themselves (since $D_i^- = X$ is dense in itself), every nonzero open part of D_i has a nonzero nowhere dense part. Again, after π_j has been chosen (beginning with any nonzero nowhere dense π_1 in $D_{\phi(1)}$), we take for each i a subsequence Λ_{ij} of $\phi^{-1}(i)\backslash\{1, 2, \ldots, j\}$, disjoint from the previously chosen Λ_{ik}'s and still not exhausting $\phi^{-1}(i)$. For m in Λ_{ij}, π_m is to come within m^{-1} of all of π_j without meeting π_j. This just means that, without meeting π_j, π_m must meet each of a suitable collection of open parts of X. So, as before, all the $\sigma_i = \bigvee[\pi_j: \phi(j) = i]$ will have the same closure J, $D_i \wedge J$ ($\supset \sigma_i$) is dense, and J is dense in itself. Finally (for nowhere-denseness), take a σ-discrete basis $\{U_{n\alpha}\}$. When π_n is being chosen, first choose nonempty open subsets $V_{n\alpha}$ of $U_{n\alpha}$ with $V_{n\alpha}^- \wedge \pi_j = 0$ for $j < n$: all π_k, $k \geq n$, must be disjoint from $V_{n\alpha}^-$. The conditions are consistent, so the construction can be completed.

2.22. Theorem. *The spacious sublocales of a metrizable locale include the subspaces and are closed under join and countable meet. Hence a pointless Boolean sublocale of a metrizable space contains no nonzero Borel sublocale. In particular, the dense part of a dense-in-itself metrizable space contains no nonzero Borel sublocale.*

Proof. The subspaces have no pointless open part. Also the spacious sublocales are closed under join, for assuming that a join of them $\bigvee F_\alpha$ has an open pointless Boolean part U, we get $F_\alpha \wedge U = 0$ since all sublocales of U are open (and pointless Boolean) and thus not nonzero sublocales of any F_α. But in $\bigvee F_\alpha$, open U is complemented, so U = 0.

As for a countable meet M of spacious sublocales S_i of X, suppose it has a nonzero open sublocale U which is pointless Boolean. Then U is $M \wedge V$ for a V open in X; and we can relativize to U^-, $Y = V \cap U^-$. There U is $\bigwedge S_i$ ($\wedge Y$) and is dense; the S_i are dense spacious, and $\bigwedge S_i = U = D(Y)$. U^- is dense in itself since U is atomless, and therefore Y, open in U^-, is dense in itself. So 2.21 applies, giving us nonzero J which is nowhere dense — thus disjoint from D(Y) — and has all $S_i \wedge J$ dense in it. So $\bigwedge S_i$ must contain D(J), a contradiction. Hence there is no such U, as was to be shown.

3. Q

3.1. *Q has* 2^c *nonisomorphic sublocales* ($c = 2^{\aleph_0}$).

Proof. We need only find 2^c distinct sublocales S_θ. For an isomorphism $i: S_0 \to S_\alpha$ will extend, by the localic Lavrentiev theorem [H], to a homeomorphism $I: D \to D'$, where D and D' are G_δ's of **R** containing S_0 and S_α. There are only c choices for D, D', and I; the isomorphic image of S_0 under I is unique; so S_0 is isomorphic with at most c of the S_θ. Hence there must be 2^c equivalence classes.

For the S_θ, we want

3.2. Lemma. *There are* c *closed dense-in-themselves subspaces of* Q, *any two of which have discrete intersection.*

Proof. Q is homeomorphic with $Q \times Q$; and that contains the graphs of c differentiable functions $f_\alpha: Q \to Q$ with different constant derivatives α. For one need only pick an enumeration of $Q = \{q_i\}$ and successively choose $f_\alpha(q_i)$, consistently with restrictions previously adopted, and restrict future choices $f_\alpha(x)$ for x very near q_i, to lie in a narrow band assuring $f_\alpha'(q_i) = \alpha$. Then the intersection of the graphs of two such functions cannot have a limit point (x, y) in itself, since $f_\alpha'(x) \neq f_\beta'(x)$.

I do not know if Q has c closed dense-in-themselves subsets which are almost disjoint (with finite intersections). From the graphs of the f_α in 3.2 we have disjoint nonzero closed H_α in pl(Q). Since the H_α are complemented, each of them is disjoint from the join of all other H_β. Therefore the joins of different subsets of $\{H_\alpha\}$ are all different; and this gives 3.1.

We found not only c disjoint closed sublocales H_α of pl(Q), but a discontinuous partial function from Q×Q to **R**, taking the graph of f_α to α; well, nearly a partial function (since the graphs are nearly disjoint). One thinks of constructing in this way a continuous partial function on pl(Q), i.e. a morphism $f: S \to \mathbf{R}$ where S is a sublocale of pl(Q), which might be a quotient map: no subsets of **R** except the open ones have open inverse image (sublocale). That is much more than I know how to do. It seems one would need a little more control to be able to infer that f^{-1} preserves the class of Borel sublocales. (Lavrentiev has removed one difficulty: S can be assumed O_δ in Q, so that its Borel sublocales are Borel in Q.)

REFERENCES

F. Frolík, Z., Baire spaces and some generalizations of complete metric spaces, Czech. Math. J. **11** (1961), 237-247.

H. Hager, A. W., Isomorphisms of some completions of C(X), Topology Proc. **4**, 1979, 407-435 (1980).

HM. Haworth, R., and McCoy, R., Baire spaces, Diss. Math. **CXLI** (1977), 73 pp.

I_1. Isbell, J., Atomless parts of spaces, Math. Scand. **31** (1972), 5-32.

I_2. Isbell, J., Graduation and dimension in locales, *Aspects of Topology*, London Math. Soc. Lecture Notes 93, Cambridge, 1985, 195-210.

I_3. Isbell, J., First steps in descriptive theory of locales, Trans. Amer. Math. Soc. **327** (1991), 353-371.

J. Johnstone, P. T., *Stone Spaces*, Cambridge Univ. Press, Cambridge, 1982.

K. Kaplansky, I., *Infinite abelian groups*, Univ. of Michigan Press, Ann Arbor, 1954.

Ku. Kuratowski, K., *Topologie*, vol. I, 4th ed., Polish Scientific Publishers, Warsaw, 1958.

DEPARTMENT OF MATHEMATICS, STATE UNIVERSITY OF NEW YORK AT BUFFALO, BUFFALO, NEW YORK 14214

mthisbel@ubvmsb.cc.buffalo.edu

Canadian Mathematical Society
Conference Proceeding
Volume **13**, 1992

Modelling homotopy coherence

J.F. JARDINE

ABSTRACT. This paper contains an alternate definition of homotopy coherent diagrams, which applies easily, for example, to lax functors. A proof of the realization theorem for homotopy coherent diagrams is given for the model discussed here. A homotopy coherent version of Quillen's Theorem B is proved.

1. Introduction

In their seminal paper on homotopy coherence [3], Dwyer and Kan show (Theorem 2.1) that every weak r-equivalence $f : \mathcal{A} \to \mathcal{B}$ of simplicial categories induces an equivalence $f^* : \mathbf{S}^{\mathcal{B}} \to \mathbf{S}^{\mathcal{A}}$ of homotopy theories, and conversely. Here, for example, $\mathbf{S}^{\mathcal{A}}$ is the set of functors of simplicial functors from $\mathcal{A}$ to the category $\mathbf{S}$ of simplicial sets, where $\mathbf{S}$ is endowed with a simplicial structure in the usual way, via function complexes. A consequence of this result is a realization theorem, which asserts that, if the canonical functor

$$\phi : \mathcal{A} \to \pi_0\mathcal{A} = C$$

is a weak equivalence of simplicial categories, then every simplicial functor $\omega : \mathcal{A} \to \mathbf{S}$ can be replaced up to weak equivalence by an C-diagram $\omega_* : C \to \mathbf{S}$.

In particular, *all simplicial functors of the form* ω can be regarded as homotopy coherent C-diagrams in the simplicial set category. This is a rather unorthodox view of homotopy coherence. The usual approach to the theory is to decree that a homotopy coherent C-diagram (or "infinitely homotopy commutative diagram", as in [3] – see also [2]) is a simplicial functor $F_*C \to \mathbf{S}$, where the "standard resolution" F_*C of the category C arises in the canonical way from free category FC which has the same objects as C and which has exactly one generator for

1991 Mathematics Subject Classification. Primary 18G30, 55P65, 55U35

Supported by NSERC.

Typeset with $\mathcal{A}_{\mathcal{M}}\mathcal{S}$-TEX.

This paper is in final form and no version of it will be submitted for publication elsewhere.

every non-identity map of C. But this definition is rather too restrictive for some applications, since it does not allow for identity maps to vary up to homotopy.

The main purpose of this paper is to present a different resolution BC_s of the category C. The simplicial category BC_s is essentially constructed from the simplices of the nerve of the category C, and simplicial functors $BC_s \to \mathbf{S}$ allow the identities to vary up to homotopy in a controlled way. Such simplicial functors also accept lax functors quite naturally as input, so the categorical notion of homotopy coherence fits into this model quite well.

For expositional purposes, I have given a self-contained proof of the realization theorem for the simplicial functor $BC_s \to C$. The key point of this proof is that a homotopy coherent diagram X of simplicial sets which is defined on a category D with terminal object has a homotopy colimit which is weakly equivalent to the simplicial set of sections $X(t)$ of X at the terminal object. This is Theorem of this paper. This result leads immediately to the realization construction (Proposition), as well as an analogue of Quillen's Theorem B for homotopy coherent diagrams of simplicial sets.

The version of homotopy coherence that is given here can be used in any closed simplicial model category $\mathbf{T}$, but proving the analogue of Theorem for simplicial functors $BC_s \to \mathbf{T}$ requires extra structure. The categories of pointed simplicial sets and spectra have this structure, essentially because they are constructed from simplicial sets, and the corresponding realization theorems appear as consequences of Proposition and Corollary below. There are analogous results for chain complexes which are left to the reader.

Simplicial functors of the form $BC_s \to \mathbf{S}$ may give a better picture of homotopy coherence than those defined on the standard resolution F_*C in many circumstances, but it's not clear that this model fits all known examples. Suppose, for example, that $X : C \to \mathbf{S}$ is an ordinary functor taking values in simplicial sets, and form the homotopy colimit

$$\underrightarrow{\text{holim}}_{c \in C} X(c).$$

Then everyone would agree that the maps

$$in_\sigma : X(c_0) \times \Delta^n \to \underrightarrow{\text{holim}}_{c \in C} X(c),$$

which are associated to the various simplices

$$\sigma : \quad c_0 \to c_1 \to \cdots \to c_n$$

of the nerve BC, fit together to give a homotopy coherent cone

$$X \to \underrightarrow{\text{holim}}_{c \in C} X(c). \tag{1}$$

Furthermore, if we define a functor $c \mapsto Y(c)$ by letting the space $Y(c)$ be the diagonal of the bisimplicial set

$$\bigsqcup_{c_0 \to \cdots \to c_n \to c} X(c_0),$$

then the obvious forgetful maps $Y(c) \to \underrightarrow{\text{holim}}_{c \in C} X(c)$ determine an honest cone

$$(2) \qquad Y \to \underrightarrow{\text{holim}}_{c \in C} X(c)$$

in which each space $Y(c)$ is homotopy equivalent to the corresponding space $X(c)$, and the equivalence respects maps in C up to canonical homotopy. Thus, the homotopy coherent cone (1) is weakly equivalent to space level cone (2), and the so-called "lifting problem" is solved.

On the other hand, after a certain amount of hair pulling, one sees that there's no obvious model for this cone within the framework described in this paper, except after the fact. The moral is that there is no such thing as a definition of homotopy coherence as such, but rather there are homotopy coherence phenomena which have to be dealt with on a case by case basis, and usually by extravagantly technical means.

This project was initiated as a result of a series of conversations with Marc Levine about categories of motives, for which I thank him. I would also like to take this opportunity to thank the U.E.R de Mathématique et Informatique, Université Paris VII, for its hospitality and support while much of this research was carried out. I am particularly grateful to Bruno Kahn and Max Karoubi for making that visit possible.

2. Definitions and Results

Let $\mathbf{n}$ denote the ordinal number

$$0 \to 1 \to 2 \to \cdots \to n.$$

Let C be a category, and let a and b be objects of C. There is a category, denoted $C_s(a,b)$, whose objects are the functors of the form $\theta : \mathbf{n} \to C$, with $\theta(0) = a$ and $\theta(n) = b$. The morphisms of $C_s(a,b)$ are commutative diagrams of the form

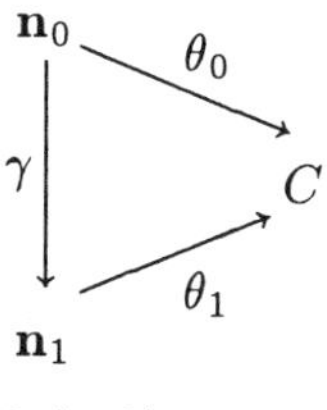

where θ_0 and θ_1 are objects of $C_s(a,b)$ and $\gamma : \mathbf{n}_0 \to \mathbf{n}_1$ is an ordinal number map which is *end point preserving* in the sense that $\gamma(0) = 0$ and $\gamma(n_0) = n_1$.

Suppose that $\theta : \mathbf{n} \to C$ is an object of $C_s(a,b)$ and $\omega : \mathbf{m} \to C$ is an object of $C_s(b,c)$. The poset join $\mathbf{n} * \mathbf{m}$ may be identified up to isomorphism with the ordinal number $\mathbf{n} + \mathbf{m}$, in such a way that the inclusion $\mathbf{n} \hookrightarrow \mathbf{n} * \mathbf{m}$ is identified with the ordinal number map $\mathbf{n} \to \mathbf{n} + \mathbf{m}$ defined by $j \mapsto j + n$ for $0 \le j \le m$. The objects θ and γ together determine a functor $\theta * \gamma : \mathbf{n} + \mathbf{m} \to C$, defined by

$$\theta * \gamma(j) = \begin{cases} \theta(j), & \text{if } 0 \le j \le n, \\ \gamma(j-n), & \text{if } n \le j \le m+n. \end{cases}$$

This is plainly the object level description of a functor

$$C_s(a,b) \times C_s(b,c) \xrightarrow{*} C_s(a,c),$$

which I shall refer to as the *join functor*. Note that the join functor is associative, and has a two-sided identity in each $C_s(a,a)$ given by the object $a : \mathbf{0} \to C$.

We have therefore constructed a category C_s which has the same objects as the original category C, and is enriched in the category **cat** of categories. Applying the nerve construction to each of the categories $C_s(a,b)$ yields simplicial sets of the form $BC_s(a,b)$, which then form the morphism objects for a category BC_s, which has the same objects as C, and which is enriched in the category $\mathbf{S}$ of simplicial sets. These constructions are obviously natural: a functor $f : C \to D$ induces a functor $f : C_s \to D_s$ of categories enriched in **cat**, and hence a functor $f : BC_s \to BD_s$ of categories enriched in $\mathbf{S}$.

Recall that the simplicial set category $\mathbf{S}$ is already enriched in simplicial sets in a standard way (see Chapter II of [8]). I shall continue the notational abuse of the Introduction, and denote this enriched category also by $\mathbf{S}$. The objects of $\mathbf{S}$ are simplicial sets, and the morphism objects are the function complexes $\mathbf{hom}(X,Y)$; explicitly, if X and Y are simplicial sets, then $\mathbf{hom}(X,Y)$ is the simplicial set whose n-simplices are the simplicial set maps $X \times \Delta^n \to Y$.

For the purposes of what follows (but see also the disclaimer in the Introduction), I define a *homotopy coherent diagram* on C to be a functor of enriched categories (or *simplicial functor*) of the form $X : BC_s \to \mathbf{S}$.

Such a structure X captures standard homotopy coherence phenomena. For example, suppose that

$$a_0 \xrightarrow{\alpha_1} a_1 \xrightarrow{\alpha_2} a_2$$

is a composable pair of morphisms of C, and let these morphisms canonically determine a functor $\alpha : \mathbf{2} \to C$, which in turn is a 0-simplex of $BC_s(a_0,a_2)$. Then X associates simplicial set morphisms $(\alpha_1)_* : X(a_0) \to X(a_1)$ and $(\alpha_2)_* : X(a_1) \to X(a_2)$ to the morphisms (aka. 0-simplices) α_1 and α_2 respectively, and associates the composite simplicial set map

$$(\alpha_2)_*(\alpha_1)_* : X(a_0) \to X(a_2)$$

to the simplex α, since X takes joins to composites. But there is also the 1-simplex σ of $BC_s(a_0,a_2)$, defined by the picture

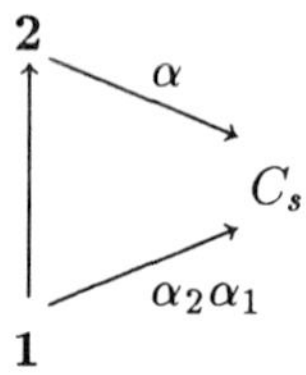

where $\mathbf{1} \to \mathbf{2}$ is the unique endpoint preserving ordinal number map, and this 1-simplex is mapped to a homotopy $X(\sigma) : X(a_0) \times \Delta^1 \to X(a_2)$ from $(\alpha_2\alpha_1)_*$ to the composite $(\alpha_2)_*(\alpha_1)_*$. I shall leave it to the reader to show that the pictures of higher homotopies that people usually draw for strings of three and four arrows (see [2], [9]) can be extracted from this construction.

But a simplicial functor $X : BC_s \to \mathbf{S}$ also contains more information. In particular, if a is an object of C, then the identity map $1_a : a \to a$ defines a 0-simplex $1_a : \mathbf{1} \to C$, and hence gets mapped to the simplicial set map $(1_a)_* : X(a) \to X(a)$, whereas the 0-simplex $a : \mathbf{0} \to C$ is an identity for BC_s, and is therefore sent to the identity $1_{X(a)}$ on $X(a)$. There is a 1-simplex η of $BC_s(a,a)$, of the form

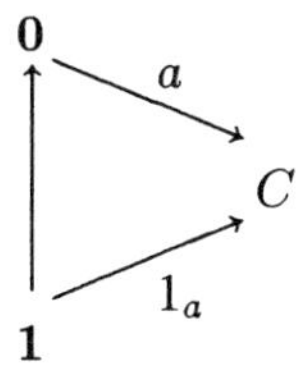

which is then mapped to a homotopy $X(\eta) : X(a) \times \Delta^1 \to X(\alpha)$ from $(1_a)_*$ to $1_{X(a)}$.

Any 1-simplex

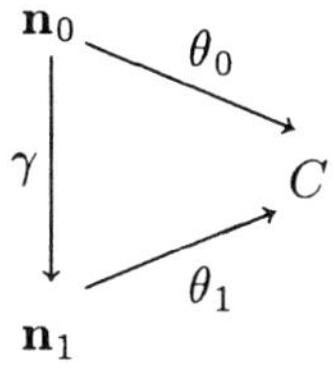

of $BC_s(a,b)$ can be thought of as defining a rearrangement of the arrows of θ_1 by operations involving composition and insertion of identities, giving θ_0. $X(\theta_1)$ is the composite of the morphisms associated to the arrows defining θ_1, and $X(\theta_0)$ is defined similarly. Then $X(\gamma) : X(a) \times \Delta^1 \to X(b)$ is a homotopy from the composite $X(\theta_0)$ to the composite $X(\theta_1)$, which can be "broken up" up to higher homotopy into homotopies arising from compositions of adjacent arrows and insertions of identities, since γ is a composite of cofaces and codegeneracies.

EXAMPLE 1. Note that the category $C_s(a,b)$ has the form

$$C_s(a,b) = \bigsqcup_{\substack{f:a\to b\\ \text{in } C}} C_s(a,b)_f,$$

where $C_s(a,b)_f$ is the subcategory of those strings $\alpha : \mathbf{n} \to C_s$ whose composite is $f : a \to b$. The category $C_s(a,b)_f$ has an initial object, namely the functor $f : \mathbf{1} \to C_s$ which is determined by f, so that, on the simplicial set level, we have a decomposition

$$BC_s(a,b) = \bigsqcup_{\substack{f:a\to b\\ \text{in } C}} BC_s(a,b)_f,$$

of $BC_s(a,b)$ into connected components, each of which is contractible. It follows that the connected component functor π_0 gives rise to a simplicial functor $\pi_0 : BC_s \to C$, where C is identified with a discrete simplicial category, and this simplicial functor is a resolution of C in the sense of [3]. In particular, any simplicial set-valued functor $f : C \to \mathbf{S}$ determines a canonical functor

$$f : C \to \mathbf{S},$$

and hence gives rise to a homotopy coherent diagram of the form

$$BC_s \xrightarrow{\pi_0} C \xrightarrow{f} \mathbf{S}.$$

EXAMPLE 2. Recall [7] that a lax functor $F : C \rightsquigarrow cat$ associates a category $F(a)$ to each object a of C, and associates a functor $\alpha_* : F(a) \to F(b)$ to each morphism $\alpha : a \to b$, in such a way that there are natural transformations

$$\theta(\beta,\alpha) : (\beta\alpha)_* \to \beta_*\alpha_* \quad \text{and} \quad \eta_a : (1_a)_* \to 1_{F(a)},$$

which together satisfy the cocycle conditions

$$\begin{array}{ccc}
(\beta\alpha\gamma)_* & \xrightarrow{\theta(\beta\alpha,\gamma)} & (\beta\alpha)_*\gamma_* \\
{\scriptstyle \theta(\beta,\alpha\gamma)}\downarrow & & \downarrow{\scriptstyle \theta(\beta,\alpha)\gamma_*} \\
\beta_*(\alpha\gamma)_* & \xrightarrow[\beta_*\theta(\alpha,\gamma)]{} & \beta_*\alpha_*\gamma_*
\end{array}$$

$$\begin{array}{ccccccc}
\alpha_* 1_{F(a)} & \xleftarrow{\alpha_*\eta_a} & \alpha_*(1_a)_* & & (1_b)_*\alpha_* & \xrightarrow{\eta_b\alpha_*} & 1_{F(b)}\alpha_* \\
\| & & \uparrow{\scriptstyle \theta(\alpha,1_a)} & & {\scriptstyle \theta(1_b,\alpha)}\uparrow & & \| \\
\alpha_* & = & (\alpha 1_a)_* & & (1_b\alpha)_* & = & \alpha_*.
\end{array}$$

The category **cat** of categories is enriched in categories in the standard way, and I shall use the same notation for the enriched object. Its morphism categories $\mathbf{hom}(A,B)$ have the functors from A to B as objects with natural transformations as morphisms. Recall that one has to be a little careful with the definition of the composition functor

$$\mathbf{hom}(A,B) \times \mathbf{hom}(B,C) \to \mathbf{hom}(A,C)$$

on morphisms: given natural transformations α_1 and α_2 of the form

$$\begin{array}{ccccc}
 & \xrightarrow{F_1} & & \xrightarrow{F_2} & \\
A & \downarrow \alpha_1 & B & \downarrow \alpha_2 & C \\
 & \xrightarrow[G_1]{} & & \xrightarrow[G_2]{} &
\end{array}$$

one defines the composition $\alpha_2\alpha_1$ of α_1 and α_2 to be the diagonal transformation appearing in the commutative diagram

$$\begin{array}{ccc}
F_2F_1 & \xrightarrow{F_2\alpha_1} & F_2G_1 \\
{\scriptstyle \alpha_2F_1}\downarrow & \searrow & \downarrow{\scriptstyle \alpha_2G_1} \\
G_2F_1 & \xrightarrow[G_2\alpha_1]{} & G_2G_1.
\end{array}$$

There is a one to one correspondence between functors

$$C_s \to \mathbf{cat}$$

and lax functors on C. In effect, given a lax functor F as above, associate to each string $\mathbf{n} \to C$ in C of the form

$$a_0 \xrightarrow{\alpha_1} a_1 \xrightarrow{\alpha_2} \dots \xrightarrow{\alpha_n} a_n$$

the composite functor

$$(\alpha_n)_* \dots (\alpha_1)_* : F(a_0) \to F(a_n).$$

The requisite functors

$$C_s(a_0, a_n) \to \mathbf{hom}(F(a_0), F(a_n))$$

are defined on morphisms

$$\begin{array}{ccc} \mathbf{n}_0 & \xrightarrow{\theta_0} & \\ {\scriptstyle\gamma}\downarrow & & C \\ \mathbf{n}_1 & \xrightarrow{\theta_1} & \end{array}$$

of $C_s(a_0, a_n)$ by first making a definition on cofaces and codegeneracies, which are completely determined up to join by the transformations θ and η, and then by showing that the relevant cosimplicial identities are satisfied. The non-trivial cosimplicial identities amount to the cocycle conditions which appear in the definition of the lax functor F.

The object **cat** has associated to it a category B**cat** which is enriched in simplicial sets, in the obvious way: one uses the nerves $B\mathbf{hom}(A, C)$ of the categories of morphisms $\mathbf{hom}(A, C)$ appearing in **cat**. Note that $B\mathbf{hom}(A, C)$ is canonically isomorphic to the function complex $\mathbf{hom}(BA, BC)$. The collection of such canonical isomorphisms respects the composition laws of B**cat** and **S** and hence determine a functor $B\mathbf{cat} \to \mathbf{S}$. It follows in particular that any lax functor $F : C_s \to \mathbf{cat}$ determines a homotopy coherent diagram

$$BC_s \xrightarrow{BF} B\mathbf{cat} \to \mathbf{S}$$

on C.

Every homotopy coherent diagram $X : BC_s \to \mathbf{S}$ has a *homotopy colimit* $\underrightarrow{\mathrm{holim}}\, X$. The homotopy colimit of X is a bisimplicial set (or its associated diagonal) whose vertical simplicial sets have the form

$$\bigsqcup_{\substack{(a_0, \dots, a_n) \\ \in Ob(C)^{n+1}}} X(a_0) \times BC_s(a_0, a_1) \times \dots \times BC_s(a_{n-1}, a_n).$$

Alternatively, $\underrightarrow{\text{holim}}\, X$ can be thought of as the nerve of a simplicial category. For each $n \geq 0$, there is a *translation category at level* n, $Tr_n X$ whose objects consist of all pairs (c, x) with c an object of C and x an n-simplex of $X(c)_n$. A morphism $(c, x) \to (d, y)$ of $Tr_n X$ is an element α of $BC_s(c, d)_n$ such that $\alpha_*(x, \iota_n) = y$ (recall that $\alpha_* : X(c) \times \Delta^n \to X(d)$ is defined to be the image of α under $X : BC_s \to \mathbf{S}$). Then the translation categories $Tr_n X$, $n \geq 0$, plainly form a simplicial category. Furthermore the nerve $BTr_n X$ can be identified with the horizontal simplicial set of $\underrightarrow{\text{holim}}\, X$ in degree n.

In particular, if $* : BC_s \to \mathbf{S}$ is the homotopy coherent diagram which assigns the one point simplicial set $*$ (or Δ^0) to each object c of C_s (see Example 1 above), then the vertical simplicial set of $\underrightarrow{\text{holim}} *$ in degree n may be identified with the simplicial set

$$\bigsqcup_{\substack{(a_0,\ldots,a_n) \\ \in Ob(C)^{n+1}}} BC_s(a_0, a_1) \times \cdots \times BC_s(a_{n-1}, a_n).$$

I shall denote $\underrightarrow{\text{holim}} *$ by $\mathcal{B}C_s$. Observe that there is a canonical map

$$\underrightarrow{\text{holim}}\, X \xrightarrow{\pi_X} \mathcal{B}C_s$$

for each homotopy coherent diagram $X : BC_s \to \mathbf{S}$, which map is defined in each vertical degree as a disjoint union of projections. Note that, once again, $\mathcal{B}C_s$ is a nerve of a certain simplicial category, namely BC_s. From this point of view, it is obvious that the path component functor π_0 induces a morphism of $\pi_0 : \mathcal{B}C_s \to BC$ of bisimplicial sets, once we identify BC with a bisimplicial set which is constant in the vertical direction. I shall write $p_X : \underrightarrow{\text{holim}}\, X \to BC$ for the composite of π_0 with π_X.

Let the functor $\sigma : \mathbf{n} \to C$ be an n-simplex of BC, and recall that σ induces a simplicial functor $\sigma_* : B\mathbf{n}_s \to BC_s$ of enriched categories. For each homotopy coherent diagram $X : BC_s \to \mathbf{S}$, σ therefore induces a map of bisimplicial sets

$$c_\sigma : \underrightarrow{\text{holim}}\, X\sigma_* \to \underrightarrow{\text{holim}}\, X$$

which fits into a commutative diagram of the form

$$\begin{array}{ccccc} & & \underrightarrow{\text{holim}}\, X\sigma_* & \xrightarrow{c_\sigma} & \underrightarrow{\text{holim}}\, X \\ & & \downarrow{\scriptstyle p_{X\sigma_*}} & & \downarrow{\scriptstyle p_X} \\ \Delta^n & = & B\mathbf{n} & \xrightarrow[\sigma]{} & BC \end{array}$$

Suppose that the string

$$\theta(0) \xrightarrow{\theta_1} \theta(1) \xrightarrow{\theta_2} \cdots \xrightarrow{\theta_m} \theta(m)$$

in $\mathbf{n}$ defines an ordinal number map $\theta : \mathbf{m} \to \mathbf{n}$. Then the component of $\underrightarrow{\text{holim}}\, X\sigma_*$ lying over θ in horizontal degree n has the form

$$X\sigma(\theta(0)) \times B\mathbf{n}_s(\theta(0), \theta(1))_{\theta_1} \times B\mathbf{n}_s(\theta(1), \theta(2))_{\theta_2} \times \cdots \times B\mathbf{n}_s(\theta(m-1), \theta(m))_{\theta_m}.$$

The simplicial sets $B\mathbf{n}_s(\theta(i), \theta(i+1))_{\theta_{i+1}}$ are contractible, so such a component is weakly equivalent to $X\sigma(\theta(0))$. On the other hand, the component in horizontal degree n of the pullback

$$B\mathbf{n} \times_{BC} \underrightarrow{\text{holim}}\, X$$

which lies over θ has the form

$$X\sigma(\theta(0)) \times BC_s(\sigma\theta(0), \sigma\theta(1))_{\sigma(\theta_1)} \times \cdots \times BC_s(\sigma\theta(m-1), \sigma\theta(m))_{\sigma(\theta_m)},$$

and each of the simplicial sets

$$BC_s(\sigma\theta(i), \sigma\theta(i+1))_{\sigma(\theta_{i+1})}$$

is contractible. It follows that the canonical map

$$\underrightarrow{\text{holim}}\, X\sigma_* \to B\mathbf{n} \times_{BC} \underrightarrow{\text{holim}}\, X$$

is a weak equivalence for each simplex σ of BC. In view of the fact (see Lemma 2.7 of [5]) that $\underrightarrow{\text{holim}}\, X$ is a homotopy colimit of pullbacks over the simplices of BC, we obtain

LEMMA 3. *Suppose that $X : BC_s \to \mathbf{S}$ is a homotopy coherent diagram on C, and define a simplicial set-valued functor on the simplex category $\mathbf{\Delta} \downarrow BC$ of BC by associating to $\sigma : \mathbf{n} \to C$ the (diagonal) simplicial set $\underrightarrow{\text{holim}}\, X\sigma_*$. Then there is a weak homotopy equivalence of the form*

$$\underrightarrow{\text{holim}}\, X \simeq \underrightarrow{\text{holim}}_{\sigma \in \Delta \downarrow BC} \underrightarrow{\text{holim}}\, X\sigma_*.$$

Suppose that $Z : D \to \mathbf{S}$ is a simplicial set-valued functor which is defined on a category D having a terminal object t. It is well known that, in this case, $\underrightarrow{\text{holim}}\, Z$ is weakly equivalent to the simplicial set $Z(t)$ which is associated to the terminal object t. There is an easy proof of this fact, based on the observation that the horizontal simplicial sets in $\underrightarrow{\text{holim}}\, Z$ are the nerves of the translation categories associated to the set-valued functors $Z_n : D \to Sets$. The translation category TrZ_n associated to Z_n has as objects all pairs (c, x), where $c \in D$ and $x \in Z_n(c)$, and a morphism $(c, x) \to (d, y)$ is a morphism $\alpha : c \to d$ of D such that $\alpha_*(x) = y$. The full subcategory of TrZ_n whose objects consist of all pairs of the form (t, x) may be identified with the discrete category $Z_n(t)$ whose set of objects in $Z_n(t)$. Let $i : Z_n(t) \subset TrZ_n$ denote the inclusion functor. There is a functor $p : TrZ_n \to Z_n(t)$ which associates to the object (c, x) the object $(t, p_*(x))$, where $p : c \to t$ is the unique map. Then pi is the identity on $Z_n(t)$ and there is a natural transformation from the identity on TrZ_n to the composite ip which is given by the maps $p : (c, x) \to (t, p_*(x))$, and so the induced simplicial set inclusion $BZ_n(t) \subset BTrZ_n$ is a weak equivalence. The collection of all such inclusions gives the desired weak equivalence $Z(t) \to \underrightarrow{\text{holim}}\, Z$.

There is a corresponding statement for homotopy coherent diagrams $BD_s \to \mathbf{S}$ defined on categories D with a terminal object. One has to work harder to prove it, essentially because the commutative diagrams implicit in the argument above lose their meaning in this case.

THEOREM 4. *Suppose that D is a category with terminal object t, and let $X : BD_s \to \mathbf{S}$ be a homotopy coherent diagram on D. Then there is a weak equivalence of the form*

$$X(t) \xrightarrow{\omega_t} \underrightarrow{\operatorname{holim}} X.$$

PROOF: The first step is to prove the result in the case where D is an ordinal number. Let $X : B\mathbf{n}_s \to \mathbf{S}$ be a homotopy coherent diagram on $\mathbf{n}$. Identify $\mathbf{n}$ with a discrete simplicial category in the obvious way. There is a simplicial functor of the form $\omega_n : \mathbf{n} \to B\mathbf{n}_s$ that is determined by the functor on the 0-simplex level which is the identity on objects and takes the morphism $i \to i+1$ to the unique element of the form $\mathbf{1} \to \mathbf{n}$ in $\mathbf{n}_s(i, i+1)$. Then the composite

$$\mathbf{n} \xrightarrow{\omega_n} B\mathbf{n}_s \xrightarrow{X} \mathbf{S}$$

may be identified with a functor $X|_{\mathbf{n}} : \mathbf{n} \to \mathbf{S}$ such that $X|_{\mathbf{n}}(i) = X(i)$. Furthermore, the map ω_n is a section of the path component functor $\pi_0 : B\mathbf{n}_s \to \mathbf{n}$, and it follows that the induced map $\underrightarrow{\operatorname{holim}} X|_{\mathbf{n}} \to \underrightarrow{\operatorname{holim}} X$ is a weak equivalence. Suppose that $n : \mathbf{0} \to \mathbf{n}$ is the functor which picks out the (terminal) object n in $\mathbf{n}$. Then there is a commutative diagram of functors enriched in simplicial sets of the form

$$\begin{array}{ccc}
\mathbf{0} & \xrightarrow{\omega_0} & B\mathbf{0}_s \\
{\scriptstyle n}\downarrow & & \downarrow{\scriptstyle n_*} \\
\mathbf{n} & \xrightarrow[\omega_n]{} & B\mathbf{n}_s
\end{array}$$

and so there is an induced commutative diagram of bisimplicial set maps of the form

$$\begin{array}{ccccc}
X(n) & \simeq & \underrightarrow{\operatorname{holim}} X|_{\mathbf{0}} & \xrightarrow{(\omega_0)_*} & \underrightarrow{\operatorname{holim}} X|_{\mathbf{0}_s} \\
 & & {\scriptstyle n_*}\downarrow & & \downarrow{\scriptstyle n_*} \\
 & & \underrightarrow{\operatorname{holim}} X|_{\mathbf{n}} & \xrightarrow[(\omega_n)_*]{} & \underrightarrow{\operatorname{holim}} X
\end{array}$$

in which all of the maps are weak equivalences, and the claim is proved.

To pass to the general case, let $\mathbf{\Delta} \downarrow BD$ denote the simplex category of the simplicial set BD. Its objects (see [5]) may be described as all functors $\mathbf{n} \to D$, $n \geq 0$, and its morphisms are the commutative diagrams of functors of the form

$$\begin{array}{ccc}
\mathbf{n} & & \\
 & \searrow^{\gamma_1} & \\
{\scriptstyle \theta}\downarrow & & D. \\
 & \nearrow_{\gamma_2} & \\
\mathbf{m} & &
\end{array}$$

Suppose, quite generally, that $U : \mathbf{\Delta} \downarrow BD \to Set$ is a set-valued functor on the simplex category for BD, and recall that we are assuming that D has a terminal object t. The simplex category $\mathbf{\Delta} \downarrow D$ has a subcategory E whose objects are those functors $\gamma : \mathbf{n} \to D$ such that $\gamma(n) = t$, and whose morphisms are those commutative triangles as above such that $\theta(n) = m$. Let $i : E \hookrightarrow \mathbf{\Delta} \downarrow BD$ denote the inclusion functor, and consider the corresponding functor $i_* : Tr(Ui) \to TrU$ of translation categories. Each object $\theta : \mathbf{n} \to D$ of the simplex category has associated to it a unique object $\theta_* : \mathbf{n+1} \to D$ of E such that the diagram of functors

$$\begin{array}{ccc} \mathbf{n} & \xrightarrow{d^{n+1}} & \mathbf{n+1} \\ & {\scriptstyle\theta}\searrow \quad \swarrow{\scriptstyle\theta_*} & \\ & D & \end{array}$$

commutes. It follows that there is a functor $r : TrU \to Tr(Ui)$ which is defined by sending the object (θ, x) to the object $(\theta_*, d_*^{n+1}(x))$. Furthermore, there is a natural transformation from the identity functor on TrU to i_*r which is given by the maps

$$d^{n+1} : (\theta, x) \to (\theta_*, d_*^{n+1}(x)).$$

There is a natural transformation from ri_* to the identity functor on the translation category $Tr(Ui)$ which is given by the maps

$$s^n : (\gamma_*, d_*^{n+1}(x)) \to (\gamma, x).$$

Thus, the inclusion $i : E \to \mathbf{\Delta} \downarrow BD$ determines a weak equivalence

$$i_* : BTr(Ui) \to BTrU.$$

This statement immediately bootstraps to a statement about simplicial set-valued functors $V : \mathbf{\Delta} \downarrow BD \to \mathbf{S}$: the induced map $i_* : \underrightarrow{\text{holim}}\, Vi \to \underrightarrow{\text{holim}}\, V$ is a weak equivalence for all such V.

Lemma 3 says that $\underrightarrow{\text{holim}}\, X$ is weakly equivalent to a homotopy colimit over the simplex category $\mathbf{\Delta} \downarrow BD$ of the form

$$\underrightarrow{\text{holim}}_{\theta \in \Delta \downarrow BD}\ \underrightarrow{\text{holim}}\, X\theta_*,$$

and so the argument just given implies that $\underrightarrow{\text{holim}}\, X$ is weakly equivalent to the homotopy colimit

$$\underrightarrow{\text{holim}}_{\gamma \in E}\, \underrightarrow{\text{holim}}\, X\gamma_*.$$

The argument given in the first paragraph of this proof implies that, for each $\gamma : \mathbf{n} \to D$ in E, the unique E-morphism

$$\begin{array}{ccc} \mathbf{0} & & \\ \downarrow & \searrow^{t} & \\ & & D \\ \downarrow & \nearrow_{\gamma} & \\ \mathbf{n} & & \end{array}$$

induces a weak equivalence

$$\underrightarrow{\operatorname{holim}}\, Xt_* \to \underrightarrow{\operatorname{holim}}\, X\gamma_*.$$

It follows that there are weak equivalences of the form

$$BE \times X(t) \simeq BE \times \underrightarrow{\operatorname{holim}}\, Xt_* \simeq \underrightarrow{\operatorname{holim}}_{\gamma \in E}\, \underrightarrow{\operatorname{holim}}\, X\gamma_*,$$

whereas BE is contractible since it has an initial object. It follows that restriction along the composite

$$\mathbf{0} \xrightarrow{\omega_0} B\mathbf{0}_s \xrightarrow{t_*} BD_s$$

induces the desired weak equivalence ω_t. ∎

Suppose that $X : BC_s \to \mathbf{S}$ is an arbitrary homotopy coherent diagram. Any arrow $\alpha : i \to j$ of C canonically determines a functor $\alpha : \mathbf{1} \to C$, and hence induces a composite enriched functor of the form

$$\mathbf{1} \xrightarrow{\omega_1} B\mathbf{1}_s \xrightarrow{\alpha_*} BC_s \xrightarrow{X} \mathbf{S}.$$

Precomposing the resulting map

$$\underrightarrow{\operatorname{holim}}\, X|_{\mathbf{1}} \xrightarrow{\alpha_*} \underrightarrow{\operatorname{holim}}\, X$$

with the cofaces d^0 and d^1 therefore determines a commutative diagram of the form

$$\begin{array}{ccc}
X(i) & & \\
{\scriptstyle d^1}\downarrow & \searrow^{\omega_i} & \\
\underrightarrow{\operatorname{holim}}\, X|_{\mathbf{1}} & \xrightarrow[\alpha_*]{} & \underrightarrow{\operatorname{holim}}\, X, \\
{\scriptstyle d^0}\uparrow & \nearrow_{\omega_j} & \\
X(j) & &
\end{array}$$

where the maps ω_i and ω_j are defined by analogy with the map ω_t which appears in the proof of Theorem 4. Standard nonsense about diagonals of bisimplicial sets (see [5]) implies that the triangle

$$\begin{array}{ccc}
X(i) & \searrow^{d^1} & \\
{\scriptstyle X(\alpha)}\downarrow & & \underrightarrow{\operatorname{holim}}\, X|_{\mathbf{1}} \\
X(j) & \nearrow_{d^0} &
\end{array}$$

commutes up to canonical homotopy. It therefore follows that the triangle

$$\begin{array}{ccc} X(i) & \xrightarrow{\omega_i} & \\ {\scriptstyle X(\alpha)}\downarrow & & \underrightarrow{\operatorname{holim}} X \\ X(j) & \xrightarrow{\omega_j} & \end{array}$$

canonically homotopy commutes.

For each i in the category C, let X_i denote the homotopy coherent functor

$$B(C \downarrow i)_s \to BC_s \xrightarrow{X} \mathbf{S}$$

which arises by precomposing with the functor of enriched categories which is induced by the forgetful functor $Q_i : C \downarrow i \to C$, where $C \downarrow i$ is the category of arrows over i in C. Let $q_i : \underrightarrow{\operatorname{holim}} X_i \to \underrightarrow{\operatorname{holim}} X$ denote the resulting induced map of simplicial sets.

Any morphism $\alpha : i \to j$ of C determines a functor

$$C \downarrow i \xrightarrow{\alpha_*} C \downarrow j$$

which is defined on objects by sending a morphism

$$k \xrightarrow{\beta} i$$

to the composite

$$k \xrightarrow{\beta} i \xrightarrow{\alpha} j.$$

The functor α_* therefore induces a map

$$\alpha_* : \underrightarrow{\operatorname{holim}} X_i \to \underrightarrow{\operatorname{holim}} X_j$$

in the obvious way.

Note that we have succeeded in defining a functor

$$i \mapsto \underrightarrow{\operatorname{holim}} X_i$$

on C. This functor is the *realization* of the homotopy coherent diagram X. It has the basic properties that one might expect from a realization functor, since each category $C \downarrow i$ has a terminal object, namely $1_i : i \to i$, and $X_i(1_i) = X(i)$, so that Theorem 4 implies that there is a weak homotopy equivalence of the form

$$X(i) \xrightarrow{\omega_{1_i}} \underrightarrow{\operatorname{holim}} X_i.$$

Write $\tilde{\omega}_i = \omega_{1_i}$, for the purposes of ease of notation. Then the above discussion can be used to show

PROPOSITION 5. *Suppose that $\alpha : i \to j$ is an arrow of a category C, and that $X : BC_s \to \mathbf{S}$ is a homotopy coherent diagram on C. Then the following diagram of simplicial set maps canonically homotopy commutes:*

$$\begin{array}{ccc} X(i) & \xrightarrow[\simeq]{\tilde{\omega}_i} & \underrightarrow{\text{holim}}\, X_i \\ {\scriptstyle X(\alpha)}\downarrow & & \downarrow{\scriptstyle \alpha_*} \\ X(j) & \xrightarrow[\tilde{\omega}_j]{\simeq} & \underrightarrow{\text{holim}}\, X_j. \end{array} \quad \blacksquare$$

Subject to the notation above, $\underrightarrow{\text{holim}}\, X_i$ may be identified with the diagonal of a bisimplicial set which has vertical simplicial sets of the form

$$\bigsqcup_{\substack{(a_0,\dots,a_n) \\ \in Ob(C)^{n+1}}} X(a_0) \times BC_s(a_0, a_1) \times \cdots \times BC_s(a_{n-1}, a_n) \times C(a_n, i)$$

where $C(a_n, i)$ denotes the (discrete) set of morphisms from a_n to i in C. Thus, the homotopy colimit $\underrightarrow{\text{holim}}\, X_i$ over C is the diagonal of a trisimplicial set having simplicial sets in degrees $(*, n, q)$ of the form

$$\bigsqcup_{\substack{(a_0,\dots,a_n) \\ \in Ob(C)^{n+1}}} X(a_0) \times BC_s(a_0, a_1) \times \cdots \times BC_s(a_{n-1}, a_n)$$
$$\times\, C(a_n, i_0) \times C(i_0, i_1) \times \cdots \times C(i_{q-1}, i_q).$$

Furthermore, the map

$$q : \underrightarrow{\text{holim}}_{i \in C}\, X_i \to \underrightarrow{\text{holim}}\, X$$

which is induced by the maps q_i defined above may be identified with the realization of a bisimplicial set map that is a disjoint union of projection maps of the form

$$\begin{array}{c} X(a_0) \times BC_s(a_0, a_1) \times \cdots \times BC_s(a_{n-1}, a_n) \times B(a_n \downarrow C) \\ \downarrow \\ X(a_0) \times BC_s(a_0, a_1) \times \cdots \times BC_s(a_{n-1}, a_n) \end{array}$$

in each vertical degree. Each of the simplicial sets $B(a_n \downarrow C)$ is contractible, and so we have proved

PROPOSITION 6. *The map*

$$\underrightarrow{\text{holim}}_{i \in C}\, \underrightarrow{\text{holim}}\, X_i \xrightarrow{q} \underrightarrow{\text{holim}}\, X$$

is a weak equivalence for each homotopy coherent diagram $X : BC_s \to \mathbf{S}$.

Proposition 6 leads immediately to the following homotopy coherent version of Quillen's Theorem B (see [5]):

PROPOSITION 7. *Suppose that $X : BC_s \to \mathbf{S}$ is a homotopy coherent diagram such that $X(\alpha) : X(i) \to X(j)$ is a weak equivalence for each morphism $\alpha : i \to j$ of C. Then, for each object i of C, the homotopy fibre of the simplicial set map $p_X : \underrightarrow{\operatorname{holim}} X \to BC$ over the vertex i of BC may be identified with $X(i)$, up to weak homotopy equivalence.*

Suppose that $\mathbf{T}$ is an arbitrary closed simplicial model category in the sense of [8], and let $\mathbf{T}$ also denote the canonically associated simplicial category. A homotopy coherent C-diagram in $\mathbf{T}$ is a simplicial functor $X : BC_s \to \mathbf{T}$. One uses the tensor functors which are part of the structure of $\mathbf{T}$ to define a homotopy colimit for such homotopy coherent diagrams X in $\mathbf{T}$; explicitly, $\underrightarrow{\operatorname{holim}} X$ is defined to be the *simplicial object* in $\mathbf{T}$ whose object of n-simplices has the form

$$\bigsqcup_{\substack{(a_0,\dots,a_n)\\ \in Ob(C)^{n+1}}} X(a_0) \otimes (BC_s(a_0, a_1) \times \cdots \times BC_s(a_{n-1}, a_n)),$$

where $\sqcup$ is the coproduct in $\mathbf{T}$.

One is entitled to construct the homotopy coherent diagrams X_i in $\mathbf{T}$ by restricting along the functors $Q_i : C \downarrow i \to C$, just as before, and the formal nonsense in the proof of Proposition 6 translates into this more general context, giving

PROPOSITION 8. *Suppose that $X : BC_s \to \mathbf{T}$ is a homotopy coherent diagram taking values in a closed simplicial monoidal category $\mathbf{T}$. Then the map*

$$\underrightarrow{\operatorname{holim}}_{i\in C} \underrightarrow{\operatorname{holim}} X_i \xrightarrow{q} \underrightarrow{\operatorname{holim}} X$$

induces a pointwise weak equivalence of simplicial objects in $\mathbf{T}$.

Proposition 8 doesn't mean too much without the existence of a realization functor

$$| \; | : \mathbf{T}^{\Delta^0} \to \mathbf{T}$$

(in a different sense of the term) which associates objects of $\mathbf{T}$ to simplicial objects of $\mathbf{T}$, and takes pointwise weak equivalences to equivalences.

The diagonal functor from bisimplicial sets to simplicial sets is the primary example of such a gadget, and we have been using it extensively here. The diagonal functor restricts to a realization functor from pointed bisimplicial sets to the category $\mathbf{S}_*$ of pointed simplicial sets, which naturally inherits this preservation of weak equivalence property. The pointed diagonal functor can be extended to a realization functor which take simplicial spectra to the category $\mathbf{Spt}$ of spectra: one simply diagonalizes each of the simplicial diagrams of pointed simplicial sets which make up the simplicial spectrum (the word "spectrum" here is in the sense of [1]). The realization functor from simplicial spectra to spectra takes pointwise strict weak equivalences to strict weak equivalences. The various categories of chain complexes also have an adequate notion of realization of simplicial objects, provided by the total complex functor.

We shall concentrate on pointed simplicial sets and spectra henceforth. The homotopy colimit of a homotopy coherent diagram X in either category will be identified with the realization $|\underrightarrow{\text{holim}}\, X|$ of the simplicial object described above. I shall write

$$\underrightarrow{\text{holim}}_* X = |\underrightarrow{\text{holim}}\, X|$$

in both cases.

Suppose that $X : BD_s \to \mathbf{S}_*$ is a homotopy coherent diagram in the category of pointed simplicial sets. Then the homotopy colimit $\underrightarrow{\text{holim}}_* X$ is the diagonal of the bisimplicial set whose vertical simplicial sets have the form

$$\bigvee_{\substack{(a_0,\dots,a_n) \\ \in Ob(D)^{n+1}}} X(a_0) \ltimes (BD_s(a_0,a_1) \times \cdots \times BD_s(a_{n-1},a_n)). \tag{3}$$

This is a wedge of a collection of half-smash products, where, generally, $Y \ltimes K$ is defined for a pointed simplicial set Y and a simplicial set K to be the quotient

$$Y \ltimes K = \frac{Y \times K}{* \times K},$$

where $*$ is the base point of Y. It follows that the pointed simplicial set (3) can be identified with the cofibre of the inclusion

$$\begin{array}{c} \displaystyle\bigsqcup_{\substack{(a_0,\dots,a_n) \\ \in Ob(D)^{n+1}}} * \times (BD_s(a_0,a_1) \times \cdots \times BD_s(a_{n-1},a_n)) \\ \downarrow \\ \displaystyle\bigsqcup_{\substack{(a_0,\dots,a_n) \\ \in Ob(D)^{n+1}}} X(a_0) \times (BD_s(a_0,a_1) \times \cdots \times BD_s(a_{n-1},a_n)) \end{array}$$

which is determined by the base points. These maps form an inclusion of bisimplicial sets, and so there is a cofibre sequence of the form

$$\mathcal{B}D_s \hookrightarrow \underrightarrow{\text{holim}}\, X \twoheadrightarrow \underrightarrow{\text{holim}}_* X,$$

where $\underrightarrow{\text{holim}}\, X$ is the unpointed homotopy colimit. If D has a terminal object t, then $\mathcal{B}D_s$ is contractible, and so the map $\underrightarrow{\text{holim}}\, X \twoheadrightarrow \underrightarrow{\text{holim}}_* X$ is a weak equivalence. The canonical map $X(t) \to \underrightarrow{\text{holim}}\, X$ is a weak equivalence by Theorem 4, and so we have proved

PROPOSITION 9. *Suppose that $X : BD_s \to \mathbf{S}_*$ is a homotopy coherent diagram in the category of pointed simplicial sets, and that D has a terminal object t. Then there is a pointed weak equivalence of the form*

$$X(t) \xrightarrow{\omega_t} \underrightarrow{\text{holim}}_* X.$$

Proposition 9 immediately implies

COROLLARY 10. *Suppose that $X : BD_s \to$ **Spt** is a homotopy coherent diagram in the category of spectra, and that D has a terminal object t. Then there is a strict weak equivalence of the form*

$$X(t) \xrightarrow{\omega_t} \underrightarrow{\text{holim}}_* X.$$

As we saw above for the simplicial set case, results having the form of Proposition 9 and Corollary 10 give rise to realization results for homotopy coherent diagrams. If X is a homotopy coherent diagram defined on BC_s which takes values in either pointed simplicial sets or spectra, then the functor $i \mapsto \underrightarrow{\text{holim}}_* X_i$ is a functor on C such that each $\underrightarrow{\text{holim}}_* X_i$ is canonically weakly equivalent to $X(i)$, and an analogue of Proposition 5 holds. These weak equivalences are strict weak equivalences if one is dealing with spectra.

The reader is invited to establish the corresponding results for the usual categories of chain complexes. The analogue of Theorem 4 may be proved in such cases with a spectral sequence argument.

REFERENCES

1. A.K. Bousfield and E.M. Friedlander, *Homotopy theories of* Γ*-spaces, spectra and bisimplicial sets*, Springer Lecture Notes in Math. **658** (1978), 85–150.
2. J-M. Cordier and T. Porter, *Vogt's theorem on categories of homotopy coherent diagrams*, Math. Proc. Camb. Phil. Soc. **100** (1986), 65–90.
3. W.G. Dwyer and D.M. Kan, *Equivalences between homotopy theories of diagrams*, in "Algebraic Topology and Algebraic K-theory," W. Browder, ed., Annals of Math. Studies, Vol. 113, Princeton University Press, Princeton, 1987, pp. 180–205.
4. W.G. Dwyer, D.M. Kan and J.H. Smith, *Homotopy commutative diagrams and their realizations*, Journal Pure App. Algebra **57** (1989), 5–24.
5. J.F. Jardine, *Homotopical foundations of algebraic K-theory*, Contemp. Math. **83** (1987), 57–82.
6. J.F. Jardine, *Supercoherence*, J. Pure Applied Algebra **75** (1991), 103–194.
7. J.P. May, *Pairings of categories and spectra*, J. Pure Applied Algebra **19** (1980), 299–346.
8. D.G. Quillen, "Homotopical Algebra," Springer Lecture Notes in Math., Vol. 43, Springer, Berlin-Heidelberg-New York, 1967.
9. R.M. Vogt, *Homotopy limits and colimits*, Math. Z. **134** (1973), 11–52.

MATHEMATICS DEPARTMENT
UNIVERSITY OF WESTERN ONTARIO
LONDON, ONTARIO, N6A 5B7
CANADA

E-mail: jardine@hydra.uwo.ca

Canadian Mathematical Society
Conference Proceeding
Volume **13**, 1992

ELEMENTARY REFLECTIONS OF CONCRETE CATEGORIES OF MIXED STRUCTURES

GRZEGORZ JARZEMBSKI

INTRODUCTION

Working with a category $\mathcal{K}$ of concrete categories we often ask the following questions:
- what does the subcategory $\mathcal{EL}(\mathcal{K})$ of categories of models of first order theories which are in $\mathcal{K}$ look like?
- given any category $\mathbf{A}$ in $\mathcal{K}$; does there exist a "best possible" description of $\mathbf{A}$ which can be done using some first order language? More precisely; does there exist a $\mathcal{K}$-functor $H : \mathbf{A} \longrightarrow \mathbf{El}(\mathbf{A})$ with $\mathbf{El}(\mathbf{A})$ in $\mathcal{EL}(\mathcal{K})$ such that any $\mathcal{K}$-functor $G : \mathbf{A} \longrightarrow B$ with codomain $\mathbf{B}$ in $\mathcal{EL}(\mathcal{K})$ factorizes uniquely as $G = G^{el} \cdot H$ for some functor $G^{el} : \mathbf{El}(\mathbf{A}) \longrightarrow \mathbf{B}$ preserving ultraproducts?

If such a category $\mathbf{El}(\mathbf{A})$ exists, we shall call it an *elementary reflection* of $\mathbf{A}$ in $\mathcal{K}$.

properties of the class of categories considered.

In the present paper we answer these questions for $\mathcal{K}$ consisting of *partially monadic categories* over sets ([4]). This is a wide class of concrete categories which are generally recognized as "categories of mixed, topologically–algebraic character". In particular, this class contains all monadic and all topological categories over sets. On the other hand, every concretely complete partially monadic category over sets is topologically–algebraic.

1. PRELIMINARIES

1. Partially monadic categories. We refer on this subject to [4] but for the readers' convenience we recall briefly basic definitions.

Throughout **Set** denotes the category of sets while **Pos** denotes the category of posets and order preserving functions.

A concrete category $(\mathbf{A}\ , U : \mathbf{A} \longrightarrow \mathbf{Set})$ is said to be *partially monadic over* **Set** (*p.m.category* , for short) with respect to a subcategory $\mathbf{D} \subset \mathbf{A}$ provided the following hold true:

1991 Mathematics Subject Classification. Primary 18C10.
This paper is in final form and no version of it will be submitted for publication elsewhere.

I. for every set X we may distinguish an *"ordered spectrum of U over X"* i.e. there exists a poset $S_U X$ endowed with a functor

$\tilde{J}_X : S_U X \longrightarrow \mathbf{A}$

and a natural transformation

$\eta_X : \Delta X \longrightarrow U\tilde{J}_X$ (Δ denotes here the "constant" functor)

subject to the following conditions:

for any pair $(h : X \longrightarrow UA : A \in \mathbf{A})$ there is a unique $i \in S_U X$ together with a unique morphism $\tilde{h} : J_X(i) \longrightarrow A$ in $\mathbf{D}$ such that $U\tilde{h} \cdot \eta_i^X = h$

and, moreover,

whenever $g : \tilde{J}_X(j) \longrightarrow A$ is a morphism such that $Ug \cdot \eta_j^X = h$, then

$j \leq i$ in $S_U X$ and $g = \tilde{h} \cdot \tilde{J}_X(j \leq i)$

II. $U_{|\mathbf{D}} : \mathbf{D} \longrightarrow \mathbf{Set}$ creates coequalizers of U-contractible pairs and the embedding $\mathbf{D} \hookrightarrow \mathbf{A}$ preserves them.

In what follows, $\mathbf{D}$-morphisms are called *perfect* morphisms. The $\mathbf{D}$-morphism $\tilde{h}$ is called a *perfect extension* of a U-morphism (h, A).

If every spectrum $S_U X$ is a one-point set, then $\mathbf{D} = \mathbf{A}$ and a partially monadic functor considered is a monadic one. On the other hand, if the category $\mathbf{A}$ is concretely complete and, moreover, every η_i^X is an identity function, then $(\mathbf{A}, U)$ is a topological category and then $\mathbf{D}$ consists of all U-initial morphisms.

2. *Spectral algebraic theories.* The important feature of p.m. categories which we shall use in the paper, is the existence of their "constructive descriptions" based on the concept of the ordered spectrum presented above.

Namely, given a p.m. category $(\mathbf{A},\ U : \mathbf{A} \longrightarrow \mathbf{Set},\ \mathbf{D} \subset \mathbf{A})$ with the spectrum

$$(S_U = \{S_U X : X \in \mathbf{Set}\}\ ,\ \tilde{J} = \{\tilde{J}_X : X \in \mathbf{Set}\},\quad \eta = \{\eta^X : X \in \mathbf{Set}\})$$

by the *associated spectral algebraic theory* in $\mathbf{Set}$ we mean a 4-tuple

$$\mathbf{S}_U \ = \ (S_U, J, \eta, (\ )^*)$$

such that:

$S_U : Ob\mathbf{Set} \longrightarrow Ob\mathbf{Pos}$, where $S_U(X) = S_U X$ for every set X,

$J = \{ J_X = U\tilde{J}_X :\ S_U X \longrightarrow \mathbf{Set}\ : X \in \mathbf{Set}\}$

$\eta = \{ \eta^X : \Delta X \longrightarrow J_X : X \in \mathbf{Set}\}$

and the operator $(\ )^*$ which assings to every pair $(f : Y \longrightarrow J_X(i), i)$ a pair $(f, i)^* = (f^* = U\tilde{f} : J_Y(j) \longrightarrow J_X(i), j)$, where

$\tilde{f} : \tilde{J}_Y(j) \longrightarrow \tilde{J}_X(i)$ is the strong extension of $(f, \tilde{J}_X(i))$.

It is easily checked that the following conditions are satisfied:

i.) $f^* \cdot \eta_j^Y = f$, for every $(f : Y \longrightarrow J_X(i), i))$

ii.) $(\eta_i^X, i)^* = (id_{J_X(i)}, i)$ for every set X and $i \in S_U X$

iii.) whenever $(f, i)^* = (f^*, j)$ and $(g, j)^* = (g^*, k)$, then

$$(f^* \cdot g, i)^* = (f^* \cdot g^*, k)\,,$$

iv.) whenever $(f, i)^* = (f^*, j)$, $i \leq j$ in $S_U X$ and

$$(J_X(i \leq k) \cdot f,\ k)^* = ((J_X(i \leq k)^*, s)$$

then $j \leq s$ in $S_U Y$ and

$$J_X(i \leq k) \cdot f^* = (J_X(i \leq k) \cdot f)^* \cdot J_Y(j \leq s).$$

It has been proved in [4] that the p.m.category **A** is concretely isomorphic to the category $\mathbf{S}_U$-Alg of *S_U-algebras* defined as follows:
objects ($\mathbf{S}_U$-algebras) are triples $(A,\ i \in S_U A, a : J_A(i) \longrightarrow A)$ such that
- $a \cdot \eta_i^A = id_A$
- for every pair of functions $f, g : X \longrightarrow J_A(i)$ with $(f, i)^* = (f^*, k)$ and $(g, i)^* = (g^*, j)$,

$$\text{whenever } a \cdot f = a \cdot g \quad \text{then} \quad af^* = ag^* \quad \text{and} \quad j = k\ .$$

An *$\mathbf{S}_U$-morphism* from an $\mathbf{S}_U$-algebra (A, i, a) to an $\mathbf{S}_U$-algebra (B, j, b) is a function $h : A \longrightarrow B$ such that if $(\eta_j^B \cdot h)^* : J_A(k) \longrightarrow J_B(j)$, then

$$i \leq k \text{ and } h \cdot a = b \cdot (\eta_j^B \cdot h)^* \cdot J_A(i \leq k)$$

If, moreover, $i = k$ then h is called *perfect*. Perfect S_U-morphisms correspond, via the isomorphism $\mathbf{A} \longrightarrow \mathbf{S}_U$-Alg , to **D**-morphisms.

The "converse" is also true: any 4-tuple $\mathbf{S} = (S, J, \eta, (\ )^*)$ satisfying all the conditions distinguished above for the 4-tuple $\mathbf{S}_U$ is called a *spectral algebraic theory* (*s.a.t.*, for short). Then the category **S**-Alg of **S**-algebras (defined in the same way as the category of $\mathbf{S}_U$-algebras above) considered together with the obvious underlying functor U^S : **S**-Alg $\longrightarrow$ **Set** is a partially monadic category with respect to the subcategory of perfect **S**-morphisms. Moreover, the spectral algebraic theory associated to the triple (**S**-Alg , U^S, *perfect* **S**-*morphisms*) is the original s.a.t. **S** ([4]).

Thus, in what follows, discussing a p.m.category we will always have in mind a category of **S**-algebras for some s.a.t. **S**.

Notice that if **S**-Alg is a monadic category, then the description (A, i, a) of **S**-algebras reduces to the pair (A, a) i.e., **S**-Alg is the category of Eilenberg-Moore algebras of the algebraic theory **S**. If **S**-Alg is topological, then the "triple description" (A, i, a) reduces to the pair (A, i) (since a must be an identity) and we get the well known description of topological categories.

For simplicity, in the rest of the paper we shall assume that p.m.categories considered have at most one object with an empty carrier.

3. *Semiperfect functors.* Let $\mathbf{S} = (S, J, \eta, (\ )^*)$ and $\hat{\mathbf{S}} = (\hat{S}, \hat{J}, \hat{\eta}, (\ )^o)$ be spectral algebraic theories in **Set**.

We call a concrete functor H : **S**-Alg $\longrightarrow$ $\hat{\mathbf{S}}$-Alg *semiperfect* provided it preserves perfect morphisms.

Semiperfect functors have a useful description in terms of the associated s.a.ts: by a *morphism of s.a.ts* from **S** to $\hat{\mathbf{S}}$ we mean a pair

$$(w, \beta) = \{(w_X \in \mathbf{Pos}(SX, \hat{S}X),\ \beta^X : \hat{J}_X w_X \longrightarrow J_X) : X \in Ob\mathbf{Set}\}$$

such that for every set X:
- $\beta^X \cdot \hat{\eta}^X_{w_X} = \eta^X$

- for every $i \in SX$ and $h : Y \longrightarrow \hat{J}_X(w_x(i))$, whenever

$$(\beta_i^X \cdot h,\, i)^* = ((\beta_i^X \cdot h)^*,\, r) \quad , \text{ then}$$

$$(h, w_X(i))^o = (h^o, w_X(r)) \quad \text{and} \quad (\beta_i^X \cdot h,\, i)^* \cdot \beta_r^Y = \beta_i^X \cdot h^o$$

Lemma 1.1. *There is one-to-one contravariant correspondence between semiperfect functors from* **S**-*Alg to* $\hat{\mathbf{S}}$-*Alg and morphisms of s.a.ts from* $\hat{\mathbf{S}}$ *to* **S**.

Proof. Given a functor $H : \mathbf{S}\text{-Alg} \longrightarrow \hat{\mathbf{S}}\text{-Alg}$ and an **S**-algebra A, for every $h : X \longrightarrow U^S A = U^{\hat{S}} HA$ consider perfect extensions $\tilde{h}$, $\hat{h}$ of (h, A) and (h, HA) in **S**-Alg and $\hat{\mathbf{S}}$-Alg , respectively. The following diagram illustrates the connection between $\tilde{h}$, $\hat{h}$ and the morphism of s.a.ts (w, β) associated to the functor H:

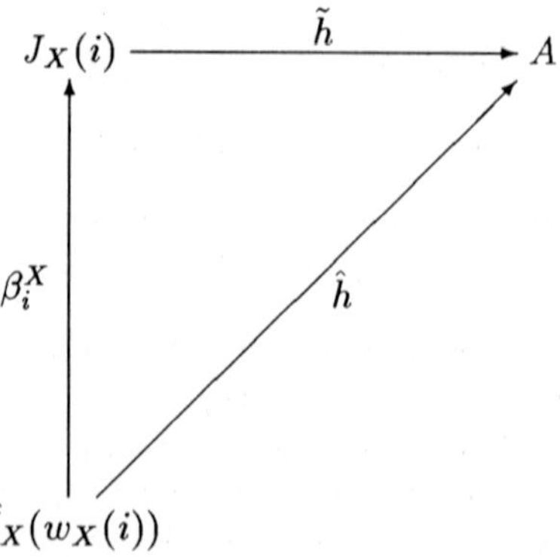

Given a morphism $(v, \gamma) : \hat{\mathbf{S}} \longrightarrow \mathbf{S}$ we define the associated functor $H : \mathbf{S}\text{-Alg} \longrightarrow \hat{\mathbf{S}}\text{-Alg}$ by $H(A, i, a) = (A, v_A(i), a \cdot \gamma_i^A)$ for every **S**-algebra (A, i, a). □

Notice that the composition of morphisms of s.a.ts (defined in the obvious way) correspond to the composition of the associated functors.

4. *Weak varieties of partial algebras.* First we make our concept of an elementary p.m.category more precise: a p.m. category $(\mathbf{A}, U, \mathbf{D} \subset \mathbf{A})$ is *elementary* provided **A** is concretely isomorphic to a class of models of some first order theory in such a way that **D**-morphisms correspond to *strong homomorphisms* ([2]).

Remark. Let $((\mathbf{A}, U), \mathbf{D} \subset \mathbf{A})$ be a p.m.category and let $\tilde{J}_X : SX \longrightarrow A$ be the spectrum of U over X. Then for every $i \in SX$ and $\kappa \in J_X(i)$ there exists an "*X-ary partial operation in* **S**-*Alg*" determined by κ i.e. there is a family of partial functions
$\{\kappa^A : (UA)^X \longrightarrow UA : A \in \mathbf{S}\text{-Alg}\}$ such that for every **S**-morphism $h : A \longrightarrow B$ and $a \in (UA)^X$ (i.e., $a : X \longrightarrow UA$), whenever $\kappa^A(a)$ exists then $\kappa^B(ha)$ exists and

$$h \cdot \kappa^A(a) = \kappa^B(ha)$$

(i.e. h "preserves" the operation κ.)

To prove this remark, recall that every $(a : X \longrightarrow UA, A)$ has a perfect

extension $\tilde{a} : \tilde{J}_A(j) \longrightarrow A$. We put:
$\kappa^A(a)$ is defined iff $i \leq j$ and then

$$\kappa^A(a) = \tilde{a} \cdot J_X(i \leq j)(\kappa).$$

This remark shows a natural relationship between p.m.categories and the theory of partial algebras. We recall briefly some basic facts about partial algebras now.

For a given finitary type $\Omega = \{\Omega_n : n \in N\}$ by $Palg\Omega$ we shall denote the concrete category of *partial* Ω-*algebras* and their homomorphisms. A partial Ω-algebra is a pair $(A, (q^A : q \in \Omega))$ where for every $q \in \Omega_n$, q^A is a *partial n-ary operation* - a partial function from A^n to A.

A homomorphism from $(A, (q^A))$ to $(B, (q^B))$ is a function $h : A \longrightarrow B$ "preserving operations" i.e. for every $q \in \Omega_n$ and $\hat{a}$ in A^n, whenever $q^A(\hat{a})$ is defined, then $q^B(h\hat{a})$ is defined and then $h \cdot q^A(\hat{a}) = q^B(h\hat{a})$.

If, moreover, whenever $q^B(h\hat{a})$ is defined , then $q^A(\hat{a})$ is defined, we call the homomorphism h *strong* ([2]).

A class $\mathbf{V}$ of partial Ω-algebras (as well as the full subcategory of $Palg\Omega$ it determines) is called a *weak variety* iff the following hold ([5]):

- whenever $(f^i : A \longrightarrow A^i : i \in I) \subset Palg\Omega$ is a monosource containing at least one strong homomorphism and every A^i is in $\mathbf{V}$, then $A \in \mathbf{V}$.
- if $e : A \longrightarrow B$ is strong and surjective and $A \in \mathbf{V}$, then $B \in \mathbf{V}$, too.

Every weak variaty is a p.m.category with strong homomorphisms as perfect morphisms ([5]). A description of ordered spectra in weak varieties will be given in section 3 of this paper.

In genaral, weak varieties are not closed under formation of ultraproducts. If it is so, a weak variety is called *elementary.*

For a first order description and for examples of elementary weak varieties we refer the reader to [5].

Now we are able to fromulate the main theorem of the present paper:

MAIN THEOREM. *For every p.m. category* $((\mathbf{A}, U), \mathbf{D} \subset \mathbf{A})$ *there exists an elementary weak variety* $\mathbf{V}(A)$ *together with a semiperfect functor* $H : \mathbf{A} \longrightarrow \mathbf{V}(\mathbf{A})$ *such that any semiperfect functor* $G : \mathbf{A} \longrightarrow \mathbf{V}$ *into an elementary weak variety factorizes uniquely as* $G = G^{el} \cdot H$ *where the functor* $G^{el} : \mathbf{V}(\mathbf{A}) \longrightarrow \mathbf{V}$ *preserves ultraproducts .*

We call the weak variety $\mathbf{V}(\mathbf{A})$ an *elementary reflection* of $\mathbf{A}$ within the class of partially monadic categories.

2. First step - finitary reflections

Notation. By $FinX$ we denote the direct system of all finite subsets of the set X. For a given s.a.t $\mathbf{S} = (S, j, \eta, (\ )^*)$ and $Y \in FinX$ with the embedding $y : Y \longrightarrow X$, by $Sy : SX \longrightarrow SY$ we shall denote the order preserving function such that for every $i \in SX$,

$$(\eta_i^X \cdot y)^* : J_y(Sy(i)) \longrightarrow J_X(i).$$

Definition 2.1. An s.a.t. **S** as well as the associated category **S**-Alg is called finitary provided for every set X,

$$(SX,\ \{Sy : SX \longrightarrow SY : Y \in FinX\}) \quad \text{is a limit cone in } \mathbf{Pos}$$

and for every $i \in SX$,

$$(J_X(i),\ \{(\eta_i^X \cdot y)^* : J_Y(Sy(i)) \longrightarrow J_X(i) : Y \in FinX\})$$
is a colimit cone in **Set**.

Roughly speaking, a finitary s.a.t. is determined by its restriction to finite sets, i.e. spectra over infinite sets may be constructed from those over finite sets in some canonical way. This has also "semantic" counterpart: a s.a.t. **S** is finitary iff the restriction of the functor U^S : **S**-Alg $\longrightarrow$ **Set** to the subcategory **S**-Alg$_p$ of perfect morphisms creates direct colimits and the embedding **S**-Alg$_p$ $\longrightarrow$ **S**-Alg preserves them ([4]).

It is easily checked that for every elementary weak variety **V** the underlying functor U^V : **V** $\longrightarrow$ *Set* satisfies this condition, i.e. **V** is a finitary p.m.category.

In what follows, discussing finitary s.a.ts we shall consider its restrictions to finite sets only.

Lemma 2.1. *Finitary partially monadic categories form a full, reflective subcategory of the category of p.m.categories and semiperfect functors.*

Proof. For a given s.a.t $\mathbf{S} = (S, J, \eta, (\ )^*)$ its restriction to finite sets determnies uniquely a finitary s.a.t $\mathbf{S}^f = (S^f, J^f, \bar{\eta}, (\ )^o)$ in the obvious way. Namely, for every set X we put

$S^f X = lim(SY : Y \in FinX)$

while for every i in $S^f X$,

$J^f_X(i) = colim(J_Y(p_Y(i)); Y \in FinX)$

($p_Y : S^f X \longrightarrow S^f Y = SY$ denotes the limit projection).

We left the details of this easy construction to the reader. It is also clear that there exists an obvious morphism of s.a.ts $(w, \beta) : \mathbf{S}^f \longrightarrow \mathbf{S}$.

Moreover, given a morphism $(v, \gamma) : \mathbf{S}_1 \longrightarrow \mathbf{S}$ from its "restriction to finite sets" we may define a new morphism of s.a.ts using suitable limits and colimits constructions in the obvious way. It is quite clear that the constructed morphism satisfies the equation

$$(w, \beta) \cdot (\bar{v}, \bar{\gamma}) = (v, \gamma) .$$

Now the lemma follows from Lemma 1.1. □

3. Elementary spectral algebraic theories.

Finitary reflections are obvious - we simply "forget" infinitary partial operations in a p.m.category considered. The next step is more complicated. First we are going to show that a characteristic feature distinguishing elementary p.m.categories among finitary ones is the existence of suitable topological structures on their spectra.

We shall use a concept of a *Priestley space* - an ordered, compact (Hausdorff) space $(X, \leq, \mathcal{T})$ which is *totally order disconnected* i.e., for every $x, y \in X$, $x \leq y$ provided for every clopen increasing set F whenever $x \in F$, then $y \in F$, too ([7]).

Given a Priestley space $(X, \leq, \mathcal{T})$, by $\mathcal{T}^+$ we denote the subtopology of $\mathcal{T}$ generated by all $\mathcal{T}$-clopen (closed and open) increasing subsets of X.

For every natural number n, we denote by $\mathbf{n}$ the set $\{0, 1, \ldots, n-1\}$.

First consider the s.a.t $\mathbf{S}^\Omega = (S^\Omega, J^\Omega, \eta, (\ )^\omega)$ associated to the category $Palg\Omega$ of all partial Ω-algebras.

Every $S^\Omega \mathbf{n}$ is the poset of all initial segments over $\mathbf{n}$ (a subset $\mathbf{n}_i$ of the set of Ω-terms over $\mathbf{n}$ is an *initial segment* if $\mathbf{n} \subset \mathbf{n}_i$ and together with any term t it contains every subterm of t).

Every $\mathbf{n}_i$ is a carrier of a partial Ω-algebra $\underline{\mathbf{n}}_i$ being a relative subalgebra ([2]) of the total Ω-algebra ΩX of Ω-terms over the alphabet X.

For any valuation $h : \mathbf{n} \longrightarrow U^\Omega A$ with a perfect extension $\tilde{h} : \underline{\mathbf{n}}_i \longrightarrow A$, $\mathbf{n}_i$ consists of all Ω-terms with variables in $\mathbf{n}$ defined in A at the valuation h.

$S^\Omega \mathbf{n}$ is an algebraic lattice. Hence $S^\Omega \mathbf{n}$ endowed with the *Lawson topology* (with an open subbase consistnig of sets $\uparrow \mathbf{n}_k = \{\mathbf{n}_j : \mathbf{n}_k \subset \mathbf{n}_j\}$ and $S^\Omega \mathbf{n} \setminus \uparrow \mathbf{n}_k$ for $\mathbf{n}_k$ compact in $S^\Omega \mathbf{n}$ ([1])) is a Priestley space.

Now assume that $\mathbf{V} \subset Palg\Omega$ is an elementary weak variety. Let $\mathbf{S}^V = (S^V, J^V, \hat{\eta}, (\ )^v)$ be the associated s.a.t..

Since $\mathbf{V} \subset Palg\Omega$, there is a morphism of s.a.ts $(w, \beta) : \mathbf{S}^\Omega \longrightarrow \mathbf{S}^V$.

We proved in [5] that every $w_n : S^V \mathbf{n} \longrightarrow S^\Omega \mathbf{n}$ reflects the order and every $\beta_i^n : J_n^\Omega(w_n(i)) = \mathbf{n}_{w_n(i)} \longrightarrow J_n^V(i)$ is surjective.

Thus every $S^V \mathbf{n}$ may be regarded as a subset of $S^\Omega \mathbf{n}$ with the induced order. More precisely, an initial segment $\mathbf{n}_i$ is in $S^V \mathbf{n}$ iff there exist a strong homomorphism from $\underline{\mathbf{n}}_i$ to A, where $A \in \mathbf{V}$. Then β_i^n is the first component of the (epi, monosource)- factorization of the cone
$(h : \underline{\mathbf{n}}_i \longrightarrow cod(h) : cod(h) \in \mathbf{V})$:

$$\underline{\mathbf{n}}_i \xrightarrow{\ \beta_i^n\ } \underline{\mathbf{n}}_i^V \xrightarrow{\ (m_h)\ } cod(h) \in \mathbf{V}$$

in $Palg\Omega$. Moreover, $\beta_i^n \cdot \eta_i^n = \hat{\eta}_i^n$.

We claim that $S^V \mathbf{n}$ is a closed subset of the Lawson space $S^\Omega \mathbf{n}$, i.e. any ultrafilter $\mathcal{U}$ on $S^\Omega \mathbf{n}$ containing $S^V \mathbf{n}$ has the (unique) limit point in $S^V \mathbf{n}$.

Consider the family $\{\hat{\eta}_i^n : \mathbf{n} \longrightarrow J_n^V(i) = U^V \underline{\mathbf{n}}_i^V : \mathbf{n}_i \in S^V \mathbf{n}\}$.

Let $A_\mathcal{U}$ be the ultraproduct of $\{\underline{\mathbf{n}}_i^V : \mathbf{n}_i \in S^V \mathbf{n}\}$ with respect to $\mathcal{U}$ and let $h : \mathbf{n} \longrightarrow U^V A_\mathcal{U}$ be the function determined by the family $\{\hat{\eta}_i^n : \mathbf{n}_i \in S^V \mathbf{n}\}$.

Finally, let $\tilde{h} : \underline{\mathbf{n}}_i \longrightarrow A_\mathcal{U}$ be the perfect extension of h in $Palg\Omega$. Since $A_\mathcal{U} \in \mathbf{V}$, $\mathbf{n}_i \in S^V \mathbf{n}$. Since $\mathbf{n}_i$ consists of terms defined in $A_\mathcal{U}$ at the valuation h, directly from the definition of ultraproduct we conclude that for every finite (i.e. compact) $\mathbf{n}_k \in S^\Omega \mathbf{n}$,

$$\mathbf{n}_k \subset \mathbf{n}_i \text{ iff } \uparrow \mathbf{n}_k \in \mathcal{U}$$

But this means that $\mathbf{n}_i$ is the limit point of $\mathcal{U}$ in the Lawson space considered. Thus $S^V \mathbf{n}$ - as a closed subset of $S^\Omega \mathbf{n}$ - is a Priestley space.

Let $\mathcal{T}_n$ denote the induced Priestley topology on $S^V \mathbf{n}$. The family of sets:

$$F(k) = \{\uparrow \mathbf{n}_k \cap S^V \mathbf{n} : \mathbf{n}_k \text{ is finite}\}$$

is a base of $\mathcal{T}_n^+$ ([6]). Put

$$\mathbf{n}_{(k)} = inf F(k) \text{ in } S^{\Omega}\mathbf{n}$$

and consider the cone:

$$\{ \mathbf{n}_{(k)} \xrightarrow{J_n^{\Omega}((k) \leq j)} \mathbf{n}_j \xrightarrow{\beta_j^n} J_{\mathbf{n}}^V(j) \quad : \mathrm{j} \in S^V \mathbf{n} \}$$

Let $\bar{J}_n^V(F(k))$ be the central object of the (epi,monosource)- factorization of this cone. This gives a function defined on a subbase of the topology $\mathcal{T}_n^+$. Next we extend the function $\bar{J}_n^V$ to a presheaf

$$\bar{J}_n^V : (\mathcal{T}_n^+)^{op} \longrightarrow \mathbf{Set}$$

in the usual way.

The straightforward consequence of the fact that $\mathbf{V}$ is elementary is that for every $\mathbf{n}_i \in S^V \mathbf{n}$,

$$J_n^V(i) = colim(\bar{J}_n^V(G) : \mathbf{n}_i \in G \in \mathcal{T}_n^+)$$

These observations lead to the following definition.

Definition 3.1. A finitary s.a.t $\mathbf{S} = (S, J, \eta, (\ )^*)$ is said to be elementary iff every spectrum $S\mathbf{n}$ may be endowed with a Priestley topology $\mathcal{T}_n$ subject to the following conditions:

i.) there is a presheaf $\bar{J}^n : (\mathcal{T}_n^+)^{op} \longrightarrow \mathbf{Set}$ such that for every $i \in S\mathbf{n}$:

$$J_n(i) = colim(\bar{J}_n(F) : F \in \mathcal{T}_n^+)$$

(i.e. $J_n(i)$ is the stalk of $\bar{J}_n$ at i .)

ii.) there is a natural transformation $\bar{\eta}^n : \Delta\mathbf{n} \longrightarrow \bar{J}_n$ such that for every $F \in \mathcal{T}_n^+$ and $i \in F$

$$\eta_i^n = \phi_{Fi} \cdot \bar{\eta}_F^n ,$$

(here and in what follows $\phi_{Fi} : \bar{J}_n(F) \longrightarrow J_n(i)$ denotes a suitable colimit embedding.)

iii.) for every $F \in \mathcal{T}_n^+$ and $f : \mathbf{m} \longrightarrow \bar{J}_n(F)$:
- the assignement $i \in F \longrightarrow \hat{f}(i) \in S\mathbf{m}$, where $(\phi_{Fi} \cdot f)^* : J_m(\hat{f}(i)) \longrightarrow J^n(i)$,
is a continuous and order preserving map,
- there is a natural transformation

$$\bar{f} = (\bar{f}_G : \bar{J}_m(G) \longrightarrow \bar{J}_n(\hat{f}^{-1}(G)) : G \in \mathcal{T}_m^+)$$

such that for every $i \in \hat{f}^{-1}(G)$ the diagram

$$\begin{array}{ccc}
J_m(\hat{f}(i)) & \xrightarrow{\quad (\phi_{Fi}\cdot f)^* \quad} & J_n(i) \\
\uparrow{\scriptstyle \phi_{G\hat{f}(i)}} & & \uparrow{\scriptstyle \phi_{\hat{f}^{-1}(G)i}} \\
\bar{J}_m(G) & \xrightarrow{\quad \bar{f}_G \quad} & \bar{J}_n\hat{f}^{-1}(G) \\
\uparrow{\scriptstyle \hat{\eta}_G} & & \uparrow \\
\mathbf{m} & \xrightarrow{\quad f \quad} & \bar{J}_n(F)
\end{array}$$

is commutative.

Hence for every elementary weak variety $\mathbf{V}$ the s.a.t. $\mathbf{S}^V$ is elementary (some details of the proof were omitted here).

We prove the converse.

Theorem 3.1. *For every elementary s.a.t.* $\mathbf{S}$ *there exists an elementary weak variety* $\mathbf{V}$ *together with a concrete isomorphism* $H : \mathbf{S}\text{-}Alg \longrightarrow \mathbf{V}$ *such that* $H(\mathbf{S}\text{-}Alg_p)$ *is the subcategory of strong homomorphisms.*

Proof. We follow the notation established above. For every $n \in N$ put

$$\Omega_n = \bigcup\{\bar{J}_n(F) : F \in \mathcal{T}_n^+\}$$

For every $\mathbf{S}$-algebra A, let

$$HA \;=\; (A = U^S A, (q^A)) \;,$$

where for any $q \in \bar{J}_n(F) \subset \Omega_n$ and $h : \mathbf{n} \longrightarrow A$ with the perfect extension $\tilde{h} : J_n(i) \longrightarrow A$;

$$q^A(h) \text{ is defined iff } i \in F \text{ and then } \; q^A(h) \;=\; \tilde{h}\cdot\phi_{Fi}(q)$$

The proof that H is a full embedding and $H(\mathbf{S}\text{-Alg})$ is a weak variety of partial Ω-algebras needs only a routine calculation (compare the proof of Theorem 3.5 in [4]).

We show that $H(\mathbf{S}\text{-Alg})$ is closed under formation of ultraproducts.

Let $\{\hat{A}^k : k \in K\} \subset \mathbf{S}\text{-Alg}$ and let $A_{\mathcal{U}}$ be the ultraproduct of the family of sets $\{A^k = U^S\hat{A}^k : k \in K\}$ with respect to a given ultraproduct $\mathcal{U}$. For every $k \in K$ let $\pi^k : A_{\mathcal{U}} \longrightarrow A^k$ be the obvious "ultraproduct projection".

For any function $h : \mathbf{n} \longrightarrow A_{\mathcal{U}}$ let $h_k = \pi^k \cdot h$ for every k in K. For each k, denote by $\tilde{h}_k : J_n(i_k) \longrightarrow \hat{A}^k$ the perfect extension of $(h_k, \hat{A}^k)$.

Let $\bar{\mathcal{U}}$ be the ultrafilter on $S\mathbf{n}$ obtained as the image of $\mathcal{U}$ under the assumption $\psi : k \in K \longrightarrow i_k \in S\mathbf{n}$

Thus, in particular, if $F \in \bar{\mathcal{U}}$ then $F_o = \{k \in K : i_k \in F\} \in \mathcal{U}$ and there is a function $\hat{h}_F : \bar{J}_n(F) \longrightarrow \prod(A^k : k \in F_o)$ such that the diagram

$$\begin{array}{ccc} \bar{J}_n(F) & \xrightarrow{\ \hat{h}_F\ } & \prod(A^k : k \in F_o) \\ \Big\downarrow \phi_{Fi_k} & & \Big\downarrow p^k(projection) \\ J_n(i_k) & \xrightarrow[\ \tilde{h}_k\ ]{} & A^k \end{array}$$

is commutative.

Now let i_h be a (unique) limit point of $\bar{\mathcal{U}}$ in $(S\mathbf{n}, \mathcal{T}_n)$. This means that for every clopen increasing set F,

$$i_h \in F \text{ iff } F \in \bar{\mathcal{U}}$$

Thus, because **S** is elementary, $J_n(i_h) = colim(\bar{J}_n(F) : F \in \bar{\mathcal{U}})$.

Hence there is a function

$$\bar{h} : J_n(i_h) \longrightarrow A_{\mathcal{U}}$$

such that for every $F \in \bar{\mathcal{U}}$ the diagram

$$\begin{array}{ccc} J_n(i_h) & \xrightarrow{\ \bar{h}\ } & A_{\mathcal{U}} \\ \Big\uparrow \phi_{Fi_h} & & \Big\downarrow \text{“projection”} \\ \bar{J}_n(F) & \xrightarrow[\ \hat{h}_F\ ]{} & \prod(A^k : k \in F_o) \end{array}$$

is commutative.

Notice that $\bar{h} \cdot \eta^n_{i_h} = h$

Given $w : \mathbf{m} \longrightarrow \mathbf{n}$ one may repeat this construction for the function $g = h \cdot w$. It is easily seen that

$$\bar{h} \cdot (\eta^n_{i_h} \cdot w)^* = \bar{g}$$

Thus, applying this construction to all embeddings $(y : Y \longrightarrow A_{\mathcal{U}} : Y \in FinA_{\mathcal{U}})$ we obtain the sequence:

$$i = (i_Y \in SY : Y \in FinA_{\mathcal{U}}) \in lim(SY : Y \in FinA_{\mathcal{U}}) = SA_{\mathcal{U}}$$

and the “compatible“ family of functions

$$(\bar{y} : J_Y(i_y) \longrightarrow A_{\mathcal{U}} : Y \in FinA_{\mathcal{U}})$$

S is finitary, hence those two sequences define an **S**-algebra

$$\hat{A}_{\mathcal{U}} = (A_{\mathcal{U}}, i, a : J_{A_{\mathcal{U}}}(i) \longrightarrow A_{\mathcal{U}})$$

where $a \cdot (\eta_i^{A_{\mathcal{U}}} \cdot y)^* = \bar{y}$ for each $Y \in FinA_{\mathcal{U}}$.

Notice that $\bar{h} : J_n(i_h) \longrightarrow U^S \hat{A}_{\mathcal{U}}$ is then the perfect extension of $(h, \hat{A}_{\mathcal{U}})$.

Notice also the following: for every $q \in \bar{J}_n(F) \subset \Omega_n$

$$q^{A_{\mathcal{U}}}(h) \text{ is defined iff } i_h \in F \text{ iff } F \in \bar{U} \text{ i.e.,}$$

$$q^{A_{\mathcal{U}}}(h) \text{ is defined iff } \{k \in K : q^{A^k}(h_k) \text{ is defined}\} \in U$$

and then

$$q^{A_{\mathcal{U}}}(h) = \bar{h} \cdot \psi_{Fi_h}(q)$$

Thus Ω-operations in $H(\hat{A}_{\mathcal{U}}) = (A_{\mathcal{U}}, (q^{A_{\mathcal{U}}}))$ are defined just in the same way as in the ultraproduct of the family $\{HA^k : k \in K\}$ in $Palg\Omega$. Hence $H(\hat{A}_{\mathcal{U}})$ *is* the ultraproduct of this family with respect tu $\mathcal{U}$. The proof is complete. □

Hence we may call the **S**-algebra $\hat{A}_{\mathcal{U}}$ the *ultraproduct of the family of* **S**-*algebras* $\{A^k : k \in K\}$ *in* **S**-*Alg*.

Lemma 3.2. *Let* $\mathbf{S} = (S, J, \ldots)$, *and* $\mathbf{S}' = (S', J', \ldots)$ *be elementary s.a.ts with the associated presheaves* $\bar{J}$ *and* $\bar{J}'$, *resp.*

A semiperfect functor H : **S**-*Alg* $\longrightarrow$ **S**$'$-*Alg preserves ultraproducts iff the associated morphism of s.a.ts* $(v, \gamma) : \mathbf{S}' \longrightarrow \mathbf{S}$ *(Lemma* 1.1*) has the following properties: for every* $n \in N$*:*

i.) $v_{\mathbf{n}} : S'\mathbf{n} \longrightarrow S\mathbf{n}$ *is a continuous map,*

ii.) *there exists a natural transformation* $\Gamma^n : \bar{J}'_n \longrightarrow \bar{J}_n \cdot v_n^{-1}$ *such that for every clopen increasing set* $F \subset S'\mathbf{n}$ *and* $i \in v_n^{-1}(F)$ *the square*

$$\begin{array}{ccc}
J'_n(v_n(i)) & \xrightarrow{\quad \gamma_i^n \quad} & J_n(i) \\
\Big\uparrow {\scriptstyle \phi'_{Fv_n(i)}} & & \Big\uparrow {\scriptstyle \phi_{v_n^{-1}(F)i}} \\
\bar{J}'_n(F) & \xrightarrow[\quad \Gamma_F^n \quad]{} & \bar{J}_n \cdot v_n^{-1}(F)
\end{array}$$

is commutative.

Proof. Proof is straightforward. □

4. Proof of Main Theorem

We shall need the following lemma:

Lemma 4.1. *The forgetful functor from the category of Priestley spaces and continuous order preserving maps into* **Pos** *has a left adjoint.*

Proof. Let $P = (P, \leq)$ be a poset. We construct a universal arrow $\tau_P : (P, \leq) \longrightarrow (Priest(P), \leq, \mathcal{F})$.

Let $In(P)$ be the poset of all increasing subsets of P and let $\mathcal{F}(P)$ be the algebraic lattice of all filters of $In(P)$. Let $\hat{\tau}_P : P \longrightarrow \mathcal{F}(P)$ assign to every x in P the filter $\{A \in In(P) : x \in A\}$. Every order preserving function $f : P \longrightarrow L$ into an algebraic lattice has a unique extension $f^o : \mathcal{F}(P) \longrightarrow L$: for every $f \in \mathcal{F}$ we put: $f^o(f) = sup(inf A : A \in f)$.

f^o preserves arbitrary infima and directed suprema, hence it is a continuous order preserving map between Lawson spaces $(\mathcal{F}(P)$ and L. From this it easily follows that the closure of $\hat{\tau}_P(P)$ in $\mathcal{F}(P)$ is the desired free Priestley space over the poset P (because every Priestley space is representable as a closed subset of some Lawson space). □

Corollary 4.2. *Let $G \subset Priest(P)$ be a clopen ,increasing set. Let $f : \tau_P^{-1}(G) \longrightarrow Q$ be an order preserving function. Then there exists a continuous order preserving function $f^P : G \longrightarrow Priest(Q)$ extending f.*

Proof. Notice that $G \subset \{f \in Priest(P) : \tau_p^{-1}(G) \in f\}$ and that the greater set is isomorphic to $Priest(\tau_P^{-1}(G))$ □

Let $\mathbf{S} = (S, J, \eta, (\ )^*)$ be a finitary s.a.t in **Set** . By Lemma 1.1, in order to prove main theorem it is enough to construct an elementary s.a.t. $\mathbf{S}^e = (S^e, J^e, \eta^e, (\ )^e)$ together with the "couniversal" morphism of s.a.ts $(w, \beta)\ : \mathbf{S}^e \longrightarrow \mathbf{S}$.

For every n in N we put:

$$S^e\mathbf{n} = Priest(S\mathbf{n})$$

By $\mathcal{F}_n$ we shall denote the topology on $Priest(S\mathbf{n})$. For every $G \in \mathcal{F}_n^+$, let

$$\bar{J}_n^e(G) \ = \ lim(J_n(i) : \tau_{Sn}(i) \in G)$$

and then $(\bar{\eta}_n^e)_G : \mathbf{n} \longrightarrow \bar{J}_n^E(G)$ is a function determined by the family $(\eta_i^n : \tau_{Sn}(i) \in G)$.
By $(p_k^G : \bar{J}_n^e(G) \longrightarrow J_n(k))$ we shall denote suitable limit projections.

Next we define J^e from the presheaf $\bar{J}_n^e$ using stalk-construction.

Let $f : \mathbf{m} \longrightarrow \bar{J}_n^e(G)$ and $F \in \mathcal{F}_m^+$.

Consider the diagram:

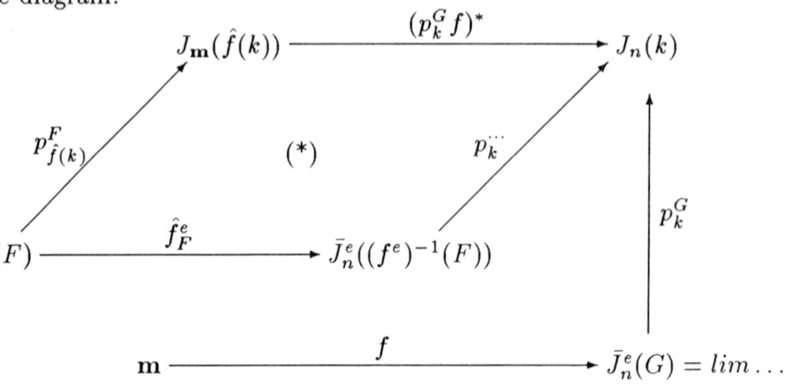

Since the function $\hat{f} : \tau_{Sn}^{-1}G \longrightarrow S\mathbf{m}$ (see Definition 3.1 iii.) is order preserving, hence by Corollary 4.2 there exists a continuous map $f^e : G \longrightarrow Priest(S\mathbf{m}) = S^e\mathbf{m}$. Then we define

$$\bar{f}^e_F : \bar{J}^e_m(F) \longrightarrow \bar{J}^e_n((f^e)^{-1}(F) = lim(J_n(j) : \tau_n(j) \in (f^e)^{-1}(F))$$

to be a unique function making the "square" $(*)$ commutative for every k such that $\tau_n(k) \in (f^e)^{-1}(f)$.

Now it is clear how to define the operator $(\ )^e$ (compare Definition 3.1 iv.).

We define the "couniversal" morphism $(w, \beta) : \mathbf{S}^e \longrightarrow \mathbf{S}$ as follows:

for every $n \in N$

$$w_n = \tau_{Sn} : S\mathbf{n} \longrightarrow S^e\mathbf{n} = Priest(S\mathbf{n})$$

while every β^n_i is a unique morphism making the diagram

$$\begin{array}{ccc}
\bar{J}^e_n(G) & \xrightarrow{\quad p^G_i \quad} & J_n(i) \\
\Big\downarrow {\scriptstyle \phi^e_{Gi}} & \nearrow {\scriptstyle \beta^n_i} & \\
J^e_n(\tau_{Sn}(i)) = \mathrm{colim}(\bar{J}^e_n(G) : \tau_{Sn}(i) \in G) & &
\end{array}$$

commutative for every $G \in \mathcal{F}_n$ containing i.

Let $(v, \gamma) : \mathbf{S}' \longrightarrow \mathbf{S}$ where $\mathbf{S}' = (S', J', \ldots)$ is elementary. Let $\mathcal{T}_n$ denote the associated Priestley topology on $S'\mathbf{n}$ while $\bar{J}'_n : (\mathcal{T}_n^+)^{op} \longrightarrow \mathbf{Set}$ the associated presheaf.

We define a morphism $(\bar{v}, \bar{\gamma}) : \mathbf{S}' \longrightarrow \mathbf{S}^e$ as folows: for every n, $\bar{v}_n : S^e\mathbf{n} = Priest(S\mathbf{n}) \longrightarrow S'\mathbf{n}$ is a unique continuous extension of v_n.

For any $A \in \mathcal{T}_n^+$ let $\Gamma^n_A : \bar{J}'_n(A) \longrightarrow \bar{J}^e_n(\bar{v}_n^{-1}(A)) = lim \ldots$ be the unique function making the diagram

$$\begin{array}{ccc}
J'_n(v_n(i)) & \xrightarrow{\quad \gamma^n_i \quad} & J_n(i) \\
{\scriptstyle \phi'_{Av_n(i)}} \Big\uparrow & & \Big\uparrow {\scriptstyle p_{v_n^{-1}(A)i}} \\
\bar{J}'_n(A) & \xrightarrow[\quad \Gamma^n_A \quad]{} & \bar{J}^e_n \cdot v_n^{-1}(A)
\end{array}$$

commutative for every i such that $\tau_S n(i) \in v_n^{-1}(A)$. Next for every $f \in S^e\mathbf{n}$ we define

$$\bar{\gamma}_{\mathrm{f}}^{n} : J'_n(\bar{v}_n(f) \longrightarrow J^e_n(f)$$

using stalk-construction in the obvious way.

Clearly, $(\bar{v}, \bar{\gamma}) \cdot (w, \beta) = (v, \gamma)$

The fact that the functor associated to the morphism $(\bar{v}, \bar{\gamma})$ preserves ultraproducts follows directly from Lemma 4.1.

The proof is complete.

References

1. G. Gierz,K. H.Hoffmann,K. Keimel,J. D.Lawson,M. Mislove, D. S.Scott, *A Compendium of Continuous Lattices*, Springer–Verlag, 1980.
2. G. Gratzer, *Universal Algebra 2nd ed.* Springer–Verlag, 1979.
3. S. S.Hong, *Categories in which every mono-source is initial*, Kyungpook Math.J.**15**, 1975, 133–139.
4. G. Jarzembski, *Finitary spectral algebraic theories*, J.Pure and Appl.Algebra **52**, 1988, 31–50.
5. G. Jarzembski, *Weak varieties of partial algebras*, Alg.Univ.**25**, 1988, 247–262.
6. G. Jarzembski,*Sheaves of finitely definable operations in weak varieties of partial algebras*, to apper in Alg. Univ.
7. A. H.Priestley,*Ordered sets and duality for distributive lattices*, Annals of Discrete Math.**23**, 39–60, 1984.

N. Copernicus University, Toruń, Poland
E-mail address: gj001@pltumk11.bitnet

Canadian Mathematical Society
Conference Proceeding
Volume **13**, 1992

REMARKS ON REPRESENTATIONS OF UNIVERSAL ALGEBRAS BY SHEAVES OF QUOTIENT ALGEBRAS

MICHAEL JOHNSON AND SHU-HAO SUN

ABSTRACT. In this paper, we give a sheaf representation for any universal algebra whose congruence lattice contains a particular type of subframe. This representation yields the following known representation theorems: Wolf-Maddana Swamy's representations for those algebras whose congruence lattices are distributive; Koh-Simmons' representation theorem for strongly harmonic rings which includes Mulvey's representations of Gelfand rings; and Georgescu and Voiculescu's representations of normal unital quantales which include Keimel's representations of F-rings and Cornish's representations of normal lattices.

1. INTRODUCTION

There are many sheaf representation theorems for a wide variety of algebras. One might hope to establish a unified proof of the sheaf representation theorems by searching for a sheaf representation of universal algebras. Non-trivial sheaf representation theorems are known for the class of algebras whose congruence lattices are distributive (see [16],[10]) and the class whose congruence lattices are normal commutative unital quantales (see [4]). Unfortunately such results do not include the sheaf representation theorems for rings with identity.

We present a sheaf representation for a class of universal algebras which specializes to give some of the known representation theorems including some representations of non-commutative rings. This solves the problem posed by Wolf [16], but it does not yield Grothendieck representations, nor some of the representations of Lambek [8]. This paper will not explore the applications of the individual representations, nor will we yet consider general categorical treatments.

For basic terminology the reader is referred to [5] and [1].

2. PRELIMINARIES

Let L be a complete lattice. Then $\mathcal{L} \subset L$ is called a subframe of L if $\mathcal{L}$ is closed under infinite joins and finite meets and $\mathcal{L}$ is itself a locale. Equivalently,

---

1991 *Mathematics Subject Classification.* Primary 1980 Mathematics Subject Classification (1985 Revision) 16A24 06B10.

The authors gratefully acknowledge the support of the Australian Research Council (ARC) and the advice of an anonymous referee.

This paper is in final form and no version of it will be submitted for publication elsewhere.

$\mathcal{L}$ is a subframe of L if it is a frame and the inclusion preserves finite meets and arbitrary joins (including the top and bottom elements).

For such an $\mathcal{L}$ let $Pt(\mathcal{L})$ denote the set of all prime elements of $\mathcal{L}$ endowed with the usual hull-kernel topology in which the open sets have the form $d(\theta) = \{\pi \in Pt(\mathcal{L}) \mid \theta \not\leq \pi\}$, for $\theta \in \mathcal{L}$.

Unless the context indicates otherwise, suppose henceforward that A is a universal algebra, that CA is its congruence lattice, and that $\mathcal{L}$ is a compact subframe of CA.

Let ω be the smallest (diagonal) congruence on A and write for each $\theta \in \mathcal{L}$

$$\theta^* = \bigvee\{\phi \in \mathcal{L} \mid \phi \cap \theta = \omega\}.$$

Then $\theta^* \in \mathcal{L}$ and $\theta^* \cap \theta = \omega$. For each $\pi \in Pt(\mathcal{L})$, let

$$r(\pi) = \bigcup\{\theta^* \mid \theta \not\leq \pi\}.$$

Notice that $\pi \supset r(\pi)$ and $\theta^* \subset \bigcap\{r(\pi) \mid \pi \in d(\theta)\}$.

Lemma 1. *For each $\theta \in \mathcal{L}$, $\theta = \bigcap\{\pi \mid \pi \in Pt(\mathcal{L})$ and $\theta \subset \pi\}$.*

Proof. See Cornish [3] Proposition 2.1. □

Lemma 2. *For each $\theta \in \mathcal{L}$, $\theta^* = \bigwedge_{\mathcal{L}}\{\pi \mid \pi \in d(\theta)\}$.*

Proof. Notice that

$$\begin{aligned} \theta \cap (\bigwedge_{\mathcal{L}}\{\pi \mid \pi \in d(\theta)\}) &= \theta \cap (\bigwedge_{\mathcal{L}}\{\pi \in Pt(\mathcal{L}) \mid \theta \not\leq \pi\}) \\ &= \bigwedge_{\mathcal{L}}\{\pi \in Pt(\mathcal{L}) \mid \theta \leq \pi\} \cap \bigwedge_{\mathcal{L}}\{\pi \in Pt(\mathcal{L}) \mid \theta \not\leq \pi\} \\ &= \omega \end{aligned}$$

(using Lemma 1). Thus we have $\theta^* \geq \bigwedge_{\mathcal{L}}\{\pi \mid \pi \in d(\theta)\}$. The reverse inequality is straightforward. □

Recall that an element x of a lattice L is said to be *compact* if for each family of elements of L whose join is above x there is a finite subfamily whose join is above x, and that a complete lattice L is called *algebraic* if each element x of L can be expressed as the join of a family of compact elements. For example any ideal lattice (two-sided or one-sided) of a ring is an algebraic complete lattice.

Lemma 3. *Let L be an algebraic complete lattice and $\mathcal{L}$ a subframe of L. Then $\mathcal{L}$ is a spatial locale.*

Proof. Let $a, b \in \mathcal{L}$ with $a \not\leq b$. Then there is a compact $c \leq a$ with $c \not\leq b$ since L is algebraic. Consider the set $Q(c) = \{x \in \mathcal{L} \mid c \not\leq x, b \leq x\}$. Then $Q(c)$ is non-empty since $b \in Q(c)$ and $Q(c)$ is closed under joins of chains in $Q(c)$ because c is compact. Hence by Zorn's Lemma there is a maximal element $m \in Q(c)$ with $b \leq m$. Of course, $c \not\leq m$.

It remains to show that m is a prime element in $\mathcal{L}$. Let $y_1, y_2 \in \mathcal{L}$ with $y_i \not\leq m$; then $c \leq y_i \vee m$ by the maximality of m. Furthermore, $c \leq (y_1 \vee m) \wedge (y_2 \vee m) = y_1 \wedge y_2 \vee m$, and hence $y_1 \wedge y_2 \not\leq m$. □

Remark 4. Let x and y be elements of a lattice L. We say that x is *way below* y if for any family of L whose join is above y there is a finite subfamily whose join is above x. A lattice L is called *continuous* if each element of L can be expressed as the join of those elements which are way below it. It is clear that every algebraic lattice is continuous.

The hypotheses of Lemma 3 are actually stronger than is required. Using our argument and a well-known characterization of continuous lattices, Lemma 3 holds for L a *continuous* complete lattice. We thank D.S. Zhao (Cambridge) for bringing this to our attention.

Corollary 5. *Every continuous locale is spatial.*

We will use the following Chinese Remainder Theorem [16].

Theorem 6. (CRT): *Let θ_i, $i = 1, 2, \ldots, n$ be mutually permutable congruences on A which generate a distributive sublattice $\mathcal{L}$ of CA. If a_i, $i = 1, 2, \ldots, n$ are elements of A such that $(a_i, a_k) \in \theta_i \vee \theta_k$ for all $i, k = 1, 2, \ldots, n$, then there is an element $a \in A$ such that $(a, a_i) \in \theta_i$ for all $i = 1, 2, \ldots, n$.*

Lemma 7. *Let $\theta_i \in \mathcal{L}$, $i = 1, 2, \ldots, n$ be mutually permutable with join $A \times A$; and $a_i \in A, i = 1, 2, \ldots, n$ such that $(a_i, a_j) \in (\theta_i \cap \theta_j)^*$ for all $i, j = 1, 2, \ldots, n$. Then there exists $a \in A$ such that $(a, a_i) \in \theta_i^*$ for all $i \leq n$.*

Proof. Since θ^* is the largest element in $\mathcal{L}$ with the property that $\theta \cap \theta^* = \omega$, we have $(\theta \cap \phi)^* \cap \theta \leq \phi^*$ for all $\theta, \phi \in \mathcal{L}$. Now let $i, j = 1, 2, \ldots, n$ and $\theta = \bigcap_{k=1}^{n} \{(\theta_i \cap \theta_k)^* \vee (\theta_k \cap \theta_j)^*\}$ and notice that $(a_i, a_j) \in \theta$. Then

$$\theta \cap \theta_k \subset ((\theta_i \cap \theta_k)^* \cap \theta_k) \vee ((\theta_k \cap \theta_j)^* \cap \theta_k) \subset \theta_i^* \vee \theta_j^*$$

for $1 \leq k \leq n$, which implies that $\theta = \theta \cap (\bigvee_{k=1}^{n} \theta_k) = \bigvee_{k=1}^{n} (\theta \cap \theta_k) \subset \theta_i^* \vee \theta_j^*$.

On the other hand, $\theta_i^* \vee \theta_j^* \subset (\theta_i \cap \theta_k)^* \vee (\theta_k \cap \theta_j)^*$ for $1 \leq k \leq n$ and hence $\theta_i^* \vee \theta_j^* \subset \theta$; so that $\theta = \theta_i^* \vee \theta_j^*$. Therefore, we have $(a_i, a_j) \in \theta_i^* \vee \theta_j^*$ for $1 \leq i, j \leq n$. Note that each $\theta_i^* \in \mathcal{L}$ and $\mathcal{L}$ is distributive. The lemma now follows from CRT (Theorem 6). □

3. A SHEAF REPRESENTATION

We will construct the sheaf $\mathcal{F}$ in the usual way: for each $\pi \in Pt(\mathcal{L})$, we denote by A_π the quotient algebra $A/r(\pi)$.

Let E be the disjoint union of the algebras A_π, $\pi \in Pt(\mathcal{L})$. Each $a \in A$ determines a Gelfand function $\widehat{a}\colon Pt(\mathcal{L}) \to E$ by defining $\widehat{a}(\pi) = [a]r(\pi)$ for all $\pi \in Pt(\mathcal{L})$.

We endow E with the finest topology for which all the maps $\widehat{a}$ are continuous.

Proposition 8. *The sets of the form $\widehat{a}(d(\theta))$ for $a \in A$ and $\theta \in \mathcal{L}$ form a basis of the topology on E and $\eta : E \to Pt(\mathcal{L})$ is a sheaf $\mathcal{F}$ of universal algebras. For each $a \in A$ the function $\widehat{a}$ is a global section and $a \mapsto \widehat{a}$ is an algebra homomorphism.*

Proof. First we show that the set

$$V = \{\pi \in Pt(\mathcal{L}) \mid (a,b) \in r(\pi)\}$$

is open for all $a, b \in A$: If $\pi \in V$, then $(a,b) \in \theta^*$ for some $\theta \in \mathcal{L}$ with $\theta \not\leq \pi$. Now $\pi \subset d(\theta) \subset V$ is easily seen. Next we see that for each $a, b \in A$,

$$\widehat{b}^{-1}(\widehat{a}(d(\theta))) = \{\pi \in Pt(\mathcal{L}) \mid (a,b) \in r(\pi)\} \cap d(\theta)$$

since $\pi \in \widehat{b}^{-1}(\widehat{a}(d(\theta)))$ if and only if $\widehat{b}(\pi) = \widehat{a}(\pi')$ for some $\pi' \in d(\theta)$ but this occurs if and only if $(a,b) \in \pi$ and $\pi = \pi' \in d(\theta)$. The rest of the proof is straightforward. □

We will require $\mathcal{L}$ to satisfy the following property (*):

For each finite set $\{\theta_i\}_1^n$ of $\mathcal{L}$ with $\bigvee \theta_i = A \times A$, there exists a finite set $\{\phi_i\}_1^n$ of $\mathcal{L}$ such that $\bigvee \phi_i = A \times A$, $\phi_i \leq \theta_i$, $i \leq n$ and

$$(\phi_i \cap \phi_j)^* \supset \bigcap \{r(\pi) \in Pt(\mathcal{L}) \mid \pi \in d(\theta_i) \cap d(\theta_j)\}$$

for all $i, j \leq n$.

Theorem 9. *Let A be a universal algebra with congruence lattice CA and let $\mathcal{L}$ be a compact subframe of CA which satisfies property (*) and is permutable. Then $a \mapsto \widehat{a}$ is an isomorphism from A onto the algebra of all sections of the sheaf $\mathcal{F}$ over $Pt(\mathcal{L})$ constructed above.*

Proof. The injectivity follows from Lemma 1. Now we show the surjectivity: Let σ be a section of $\mathcal{F}$. For each $\pi \in Pt(\mathcal{L})$, there is an $a_\pi \in A$ with $\sigma(\pi) = \widehat{a}_\pi(\pi)$. Then $\sigma^{-1}(\widehat{a}_\pi(Pt(\mathcal{L}))$ is a neighbourhood of π which contains some basic neighbourhood $d(\theta_\pi)$ of π, where $\theta_\pi \in \mathcal{L}$, so that $\sigma|d(\theta_\pi) = \widehat{a}_\pi|d(\theta_\pi)$ since they are sections. Thus $\{d(\theta_\pi) \mid \pi \in Pt(\mathcal{L})\}$ is an open cover of $Pt(\mathcal{L})$ and hence, since $Pt(\mathcal{L})$ is compact, there are $\theta_1, \theta_2, \ldots, \theta_n \in \mathcal{L}$ whose join is $A \times A$, and elements $a_1, a_2, \ldots, a_n \in A$ such that

$$\sigma|d(\theta_i) = \widehat{a}_i|d(\theta_i)$$

for $i = 1, 2, \ldots, n$. Furthermore, there exist $\{\phi_i\}_1^n$ satisfying (*).

Now for $i, k = 1, 2, \ldots, n$ we claim that $(a_i, a_k) \in (\phi_i \cap \phi_k)^*$: indeed, for all $\pi \in d(\theta_i \cap \theta_k) = d(\theta_i) \cap d(\theta_k)$ we have $\widehat{a}_i(\pi) = \sigma(\pi) = \widehat{a}_k(\pi)$, whence $(a_i, a_k) \in r(\pi)$ since $a_k \in [a_i]r(\pi)$; and hence $(a_i, a_k) \in \bigcap\{r(\pi) \mid \pi \in d(\theta_i \cap \theta_k)\} \subset (\phi_i \cap \phi_k)^*$ by (*). Now by Lemma 7, there is $a \in A$ such that $(a, a_i) \in {\phi_i}^*$, and hence in $\bigcap\{r(\pi) \mid \pi \in d(\phi_i)\}$ for $i = 1, 2, \ldots, n$ (using the definition of $r(\pi)$). Furthermore, $\widehat{a}|d(\phi_i) = \widehat{a}_i|d(\phi_i) = \sigma|d(\phi_i)$. As $\bigcup_{i=1}^n d(\phi_i) = d(A \times A) = Pt(\mathcal{L})$, we conclude that $\widehat{a} = \sigma$. □

Theorem 9 can be generalized as follows:

Theorem 10. *Let A be a universal algebra with congruence lattice CA and let $\mathcal{L}$ be a sublattice of CA satisfying* (0) *$\mathcal{L}$ and CA have the same top and bottom elements;* (1) *$\mathcal{L}$ itself is a compact spatial locale;* (2) *$\omega = \bigcap\{\pi \mid \pi \in Pt(\mathcal{L})\}$;* (3) *property (*) holds;* (4) *$\mathcal{L}$ is permutable. Then $a \mapsto \widehat{a}$ is an isomorphism from A onto the algebra of all sections of the sheaf $\mathcal{F}$ over $Pt(\mathcal{L})$.*

4. Quantales

To apply our sheaf representation to ring theory, we need the following notions (see [12] and [11]). Recall that a quantale L is a complete lattice with an associative binary operator $\cdot$ satisfying the infinite distributive law:

$$a \cdot (\bigvee S) = \bigvee \{a \cdot s \mid s \in S\};\ (\bigvee S) \cdot a = \bigvee \{s \cdot a \mid s \in S\}$$

for any $a \in L$ and $S \subset L$. We will often write as for $a \cdot s$.

A *unital quantale* is one in which the top element is also an identity with respect to $\cdot$ (a similar setting was called an *integral cl-groupoid* in [1]).

Let L be a unital quantale. An element $p \leq 1$ is called m*-prime* if for any $a, b \in L$ then $a \cdot b \leq p$ implies either $a \leq p$ or $b \leq p$, while $p \leq 1$ is called *maximal* if $p \leq a$ implies $a = p$ or $a = 1$.

Then for any element x of a unital quantale, x maximal implies x m-prime which implies that x is prime (in the lattice-theoretic sense).

Following Simmons([14]), a unital quantale which is an upper continuous lattice will be called a *carrier*. Examples of carriers include complete Heyting algebras, the ideal lattices of bounded distributive lattices and the 2-sided ideal lattices of rings, as well as the congruence lattice of any universal algebra with distributive congruences.

Note that subframes $\mathcal{L}$ of a carrier L correspond to order-preserving and top-preserving maps $g : L \to L$ by defining for any $a \in L$

$$g(a) = \bigvee \{x \in \mathcal{L} \mid x \leq a\}.$$

The following are basic examples of quantales with chosen subframes.

Example 11. Let $Id\,R$ be the set of 2-sided ideals of R. Following Borceux *et al* (see [2]), a 2-sided ideal I is called pure if $I \vee a^* = R$ for each $a \in I$, where a^* denotes the right annihilator of a. The set of all pure ideals of R is a subframe of $Id\,R$.

Example 12. Let $Id\,R$ be the set of 2-sided ideals of R. Following Simmons ([13]), a 2-sided ideal I is called *uniform virginal* if $I \vee (aR)^* = R$ where $(aR)^*$ is the right annihilator of the right ideal aR. Then the set of all uniform virginal ideals of R is a subframe of $Id\,R$.

Example 13. More generally, let L be any 2-sided quantale which is an upper continuous lattice. An element $a \in L$ is called right regular if $a = \bigvee \{x \in L \mid x^* \vee a = 1\}$, where $x^* = \bigvee \{y \in L \mid xy = 0\}$. Define $w : L \to L$ by sending each $a \in L$ to the join of all the right regular elements less than a. Then $w(L)$ is a subframe of L.

It is shown in [4] that, if L is a commutative continuous unital normal quantale, then w restricts to an isomorphism between $Max(L)$ and $Pt(w(L))$. Their proof does not appear to use commutativity but does make use of continuity. The scope of this result has been extended by Simmons [14] for 2-sided normal carriers with a compact top. It can also be extended to include 1-sided cases, see [15].

Example 14. Let R be a ring and $E(R)$ the set of all idempotent central elements of R; then $E(R)$ is boolean with respect to $a \wedge b = ab$, $a \vee b = a + b - ab$. Define $g : Id\,R \to Id\,R$ by $g(A) = (A \cap E(R))R$. Then $g(Id\,R)$ is a subframe of $Id\,R$.

Now we record some examples of subframes which satisfy property (*).

Example 15. Let A be a universal algebra with distributive congruence lattice CA then $\mathcal{L} = CA$ satisfies (*).

Example 16. Let A be an algebraic normal unital quantale with a compact top τ. For example, A could be the 2-sided ideal lattice of a strongly harmonic ring. Then $\mathcal{L} = w(A)$ as defined in Example 13 satisfies (*).

Proof. We will prove that for any finite open cover $\{d(a_i)\}_1^n$ of $Pt(\mathcal{L})$ there exists a finite open cover $\{d(b_i)\}_1^n$ such that each $d(b_i) \subset d(a_i)$ and $(b_i \wedge b_j)^* \supset \bigcap\{\pi \in Pt(\mathcal{L}) \mid \pi \in d(a_i \wedge a_j)\}$. To do this, define $\rho : A \to A$ by assigning to each $a \in A$ the join of those elements x below a such that $x^* \vee a = \tau$. Write $\tilde{\pi}$ for the unique maximal element containing π. It is known that $\rho(\tilde{\pi}) = \pi$, that $\rho = w$ and that $\rho(A)$ is a compact regular subframe of A (which implies $Pt(\mathcal{L})$ is a compact Haudorff space). So for any open cover of $Pt(\mathcal{L})$ there is a finite refinement $\{d(a_i)\}_1^n$. By regularity, there is a finite open cover $\{d(b_i)\}_1^n$ such that for each $i \leq n$, the closure $\overline{d(b_i)} \subset d(a_i)$, and hence

$$\overline{d(b_i \wedge b_j)} = \overline{d(b_i) \cap d(b_j)} \subset \overline{d(b_i)} \cap \overline{d(b_j)} \subset d(a_i) \cap d(a_j) = d(a_i \wedge a_j).$$

Now:

(a) $\pi \in d(a)$ if and only if $\tilde{\pi} \in \tilde{d}(a)$, where the $\tilde{d}(x)$, $x \in \mathcal{L}$, are the open sets for the usual topology on the space of maximal elements of A.

Proof of (a): If $a \not\leq \pi$, then $a \not\leq \tilde{\pi}$. Otherwise $a \leq \tilde{\pi}$ implies $a = \rho(a) \leq \rho(\tilde{\pi}) = \pi$ which is a contradiction. The converse is trivial.

(b) $\tilde{\pi} \in \overline{\tilde{d}(a)}$ implies $\rho(\tilde{\pi}) \in \overline{d(a)}$.

Proof of (b): Consider any neighbourhood $d(b)$ of $\pi = \rho(\tilde{\pi})$, i.e., $b \in \mathcal{L}$ with $b \not\leq \pi$. By (a), $\tilde{d}(b)$ is a neighbourhood of $\tilde{\pi}$. So there exists $m \in \tilde{d}(b) \cap \tilde{d}(a) = \tilde{d}(b \wedge a)$ which implies $\rho(m) \in d(b \wedge a)$ using (a) again. That is to say $d(a) \cap d(b) \neq \emptyset$. Thus $\pi \in \overline{d(a)}$.

Now write $b_{i,j}$ for $b_i \wedge b_j$ and $a_{i,j}$ for $a_i \wedge a_j$. We will first show that

$$\bigcap\{\pi \mid \pi \in \overline{d(b_{i,j})}\} \subset \rho(\bigcap\{\tilde{\pi} \mid \tilde{\pi} \in \overline{\tilde{d}(b_{i,j})}\}).$$

It suffices to show that for each compact element $c \leq \bigcap\{\pi \mid \pi \in \overline{d(b_{i,j})}\}$,

$$c^* \vee \bigcap\{\tilde{\pi} \mid \tilde{\pi} \in \overline{\tilde{d}(b_{i,j})}\} = \tau.$$

If not, there is a maximal element m_0 containing $c^* \vee \bigcap\{\tilde{\pi} \mid \tilde{\pi} \in \overline{\tilde{d}(b_{i,j})}\}$ since τ is compact. In particular, $m_0 \supset \bigcap\{\tilde{\pi} \mid \tilde{\pi} \in \overline{\tilde{d}(b_{i,j})}\}$ which implies that $m_0 \in \{\tilde{\pi} \mid \tilde{\pi} \in \overline{\tilde{d}(b_{i,j})}\}$ since $\overline{\tilde{d}(b_{i,j})}$ is closed. Hence $\rho(m_0) \in \overline{d(b_{i,j})}$ by (b) above. On the other hand, $c \leq \bigcap\{\pi \mid \pi \in \overline{d(b_{i,j})}\}$ and so $c \leq \rho(m_0)$ which implies $c^* \vee m_0 = \tau$ (since c is compact) and $m_0 = \tau$ which is a contradiction.

Now the proof is completed as follows:

$$\bigcap\{\pi \mid \pi \in \overline{d(b_{i,j})}\} \subset \rho(\bigcap\{\tilde{\pi} \mid \tilde{\pi} \in \overline{\tilde{d}(b_{i,j})}\})$$

$$= w(\bigcap\{\tilde{\pi} \mid \tilde{\pi} \in \overline{\tilde{d}(b_{i,j})}\}) \subset w(\bigcap\{w(\tilde{\pi}) \mid \tilde{\pi} \in \tilde{d}(b_{i,j})\})$$

$$= w(\bigcap\{\pi \mid \pi \in d(b_{i,j})\}) = (b_{i,j})^* \text{ by Lemma 2}.$$

Finally, we have

$$\bigcap\{\pi \mid \pi \in d(a_{i,j})\} \subset \bigcap\{\pi \mid \pi \in \overline{d(b_{i,j})}\} \subset (b_{i,j})^* . \quad \square$$

Remark 17. In this example, we can weaken the assumption that the lattice is algebraic to require just that the lattice is continuous. The proof of the more general case is similar to that given above. Using results from [15] we can also weaken the assumption that the quantale is unital to require only that it be 1-sided.

5. Applications

We now use Theorem 9 to derive several known sheaf representation theorems. We begin by applying Lemma 2 to obtain the following Corollary.

Corollary 18. *Let A be a universal algebra whose congruence lattice is distributive. Put $CA = \mathcal{L}$. Then we have Swamy-Wolf's sheaf representation theorem* [9],[16].

Corollary 19. . *Let A be an universal algebra of some type with the following properties:*

(i) *the congruence lattice CA is a normal unital quantale with a compact top element.*

(ii) *the virginal elements of CA permute.*

Then A is isomorphic to the algebra of global sections of a sheaf on the compact Hausdorff space $Max(CA)$.

The proof of Corollary 19 follows from Theorem 9 and Example 16. Once again the result may be extended to 1-sided quantales.

This Corollary unifies the following representation theorems: Georgescu and Voiculescu's representation of normal unital quantales [4], Keimel's sheaf representations for F-rings [6], Koh-Simmons's representations for strongly harmonic rings [7], including Mulvey's representations for Gelfand rings [10], and Cornish's sheaf representations for normal lattices [3].

Finally, if we restrict our attention to representations of rings (rather than general universal algebras) we can eliminate property (*).

Let R be a ring (with an identity). Write $Id\,R$ for the set of all 2-sided ideals of R. A subset $\mathcal{L} \subset Id\,R$ is said to be a *subframe* of $Id\,R$ if for any $I, J \in \mathcal{L}$, $IJ = I \cap J$ and $\mathcal{L}$ is closed under arbitrary sums and finite intersections. A subframe $\mathcal{L}$ of $Id\,R$ is called *small* if for each maximal ideal M of R, $\mathcal{L}(M) = \sum\{I \in \mathcal{L} \mid I \subset M\} \subset \sum\{LAnn <a> \mid a \notin M\}$, where $<a>$ is the principal ideal generated by a and $LAnn <a>$ denotes the left annihilator ideal of $<a>$. Then we have the following:

Theorem 20. *Let R be any ring with identity and $\mathcal{L}$ be a small subframe of $Id\,R$. Then R is isomorphic to the ring of global sections of a sheaf $\mathcal{F}$ whose base space is $Pt(\mathcal{L})$.*

Proof. The sheaf is built by taking as each stalk a quotient R/π, where $\pi \in Pt(\mathcal{L})$. The smallness condition ensures that the canonical projection is a local homeomorphism. The injectivity follows from Lemma 1. Now we show the surjectivity: Let σ be a section of $\mathcal{F}$. For each $\pi \in Pt(\mathcal{L})$, there is an $a_\pi \in A$ with $\sigma(\pi) = \widehat{a}_\pi(\pi)$. Then $\sigma^{-1}(\widehat{a}_\pi(Pt(\mathcal{L}))$ is a neighbourhood of π which contains some basic neighbourhood $d(I_\pi)$ of π, where $I_\pi \in \mathcal{L}$, so that $\sigma|d(I_\pi) = \widehat{a}_\pi|d(I_\pi)$ since they are sections. Thus $\{d(I_\pi) \mid \pi \in Pt(\mathcal{L})\}$ is an open cover of $Pt(\mathcal{L})$ and hence there are $I_1, I_2, \ldots, I_n \in \mathcal{L}$ whose join is R, and elements $a_1, a_2, \ldots, a_n \in R$ such that

$$\sigma|d(I_i) = \widehat{a}_i|d(I_i)$$

for $i = 1, 2, \ldots, n$. Furthermore, there exist $e_i \in I_i$ satisfying $e_1 + e_2 + \cdots + e_n = 1$.

Let $a = \sum_1^n a_i e_i$. We claim $\widehat{a} = \sigma$. In fact, for each $i \leq n$ and each $\pi \in Pt(\mathcal{L})$, if $\pi \notin d(I_i)$, i.e., $I_i \subset \pi$, then

$$[(\sigma - \widehat{a}_i)\widehat{e}_i](\pi) \subset \widehat{e}_i(\pi) \in I_i/\pi \subset \pi/\pi = 0;$$

if $\pi \in d(I_i)$, then $\sigma(\pi) = \widehat{a}_i(\pi)$ which implies

$$[(\sigma - \widehat{a}_i)\widehat{e}_i](\pi) = 0.$$

Thus we have shown that

$$\sum_i^n [(\sigma - \widehat{a}_i)\widehat{e}_i](\pi) = 0,$$

that is, $\widehat{a} = \sigma$. □

Generalizations and applications of Theorem 20 will be taken up elsewhere.

References

1. G. Birkhoff, *Lattice Theory*, Coloq. Publ., vol XXV, Providence, 1967.
2. F. Borceux, H. Simmons and G. Van Den Bossche, *A sheaf representation for modules with applications to Gelfand rings*, Proc. London. Math. Soc. (3) **48** (1984), 230–246.
3. W.H. Cornish, *The Chinese remainder theorem and sheaf representations*, Fund. Math. **96** (1977), 177–187.
4. G. Georgescu and I. Voiculescu, *Some abstract maximal ideal-like spaces*, Algebra Universalis **26** (1989), 90–102.
5. P.T. Johnstone, *Stone spaces*, Cambridge University Press, Cambridge, 1982.
6. K. Keimel, *The representation of lattice ordered groups and rings by sections in sheaves*, Lecture Notes in Mathematics **248** (1971).
7. K. Koh, *On a representation of strongly harmonic rings by sheaves*, Pacific J. Math. **41** (1972), 459–468.
8. J. Lambek, *Remarks on representations of modules by sheaves of factor modules*, Can. Math. Bull. **14** (1971), 359–368.
9. U. Maddana Swamy, *Representation of universal algebras by sheaves*, Proc. Amer. Math. Soc. **45** (1974), 55–58.
10. C.J. Mulvey, *A generalization of Gelfand duality*, J. Algebra **56** (1977), 499–505.
11. ——— *&*, Supplemento ai Rendiconti del Circolo Matematico di Palermo **II, 12** (1986), 99–104.
12. S. Niefield and K. Rosenthal, *Constructing locales from quantales*, Math. Proc. Camb. Phil. Soc. **104** (1988), 215–234.

13. H. Simmons, *Sheaf representations of stringly harmonic rings*, Proc. R. Soc. Edinburgh (Section A) **99** (1985), 249–268.
14. ——— *Compact Representations — The Lattice Theory of Compact Ringed Spaces*, J. Algebra **126** (1989), 493–531.
15. S.H. Sun, *Remarks on normality and regularity of quantales*, in preparation.
16. A. Wolf, *Sheaf representation of arithmetical algebras*, Memoirs of the Amer. Math. Soc. **148** (1974), 87–93.

School of Mathematics and Computing, Macquarie University, Australia
E-mail address: mike@macadam.mpce.mq.edu.au

School of Mathematics and Statistics, Sydney University, Australia
E-mail address: sun_s@maths.su.oz.au

Canadian Mathematical Society
Conference Proceeding
Volume 13, 1992

Categorical Aspects of Conformal Field Theory

V. LYUBASHENKO

ABSTRACT. We define a modular functor [19] as a 2-functor. The projective flat connection of [19,21] permits us to define a tensor category. This category can be non-semisimple, which is a new feature. A homotopical version of a modular functor is constructed via obtained tensor category.

AKNOWLEDGEMENTS. This work is inspired by papers of V.Drinfeld [2], D.Kazhdan and G.Lusztig [7] and G.Segal [19]. I would like to thank T.Fox, A.Joyal and R.Street for stimulating discussions. I am grateful to Centre Interuniversitaire en Etudes Categoriques for their kind hospitality and the opportunity to work on the subject[1].[2]

1 The 2-Category of Riemann Surfaces

We follow Segal [19] with minor modifications. Modular functor is defined as a 2-functor from the bicategory of Riemann surfaces to the bicategory of vector spaces.

Let $\mathcal{C}_0 = \{A, B, C, \ldots\}$ be an index set. Define a weak 2-category $\mathcal{S}$ of Riemann surfaces. By definition,

$$\mathcal{S}_0 = \{(A_1, \ldots, A_k) | A_i \in \mathcal{C}_0\}$$

is the set of objects (0-morphisms) of $\mathcal{S}$. The set of 1-morphisms $\mathcal{S}_1((A_1, \ldots, A_k), (B_1, \ldots, B_l))$ consists of Riemannian surfaces Σ with real-analytic parametrization of the boundary circles $\mu_m : S^1 \to \partial\Sigma$, $m \in [-k, l] - \{0\}$, $m \in \mathbf{Z}$, where μ_m preserves the orientation for $m > 0$ and inverts the orientation for $m < 0$. We add also such limit cases of Riemann surfaces as infinitely thin

[1]1991 Mathematics Subject Classification. Primary 18D05, 81T40.

[2]The detailed version of this paper will be submitted for publication elsewhere.

annulus: $\mathcal{O}_{\mu_{-1},\mu_1}(A,B)$ = the set $\mathcal{O} = S^1$ with real-analytic parametrization $\mu_{-1}, \mu_1 : S^1 \to \mathcal{O}$, preserving the orientation. We consider labels $A, B, \ldots \in \mathcal{C}_0$ as attached to boundary circles.

The set of 2-morphisms is $\mathcal{S}_2(\Sigma, \Sigma')$ = complex isomorphisms $f : \Sigma \to \Sigma'$, coherent with the labelling and parametrization of the boundary:

$$\mu'_m = f \circ \mu_m : S^1 \to \Sigma'.$$

The role of identity 1-morphisms is played by infinitely thin annulus $\mathcal{O} = \mathcal{O}_{id,id}$ with $\mu_{-1} = \mu_1 = id_{S^1}$. For $(A_1, \ldots, A_k)$ the identity morphism $\mathrm{Id}_{(A_1,\ldots,A_k)}$ equals $\coprod_1^k \mathcal{O}_{id,id}(A_i, A_i) = k\mathcal{O}$.

Composition $f *_1 g$ of two 2-morphisms is a usual composition of conformal isomorphisms. The composition $\Sigma *_0 \Sigma' = \Sigma \bigcup \Sigma'$ of $\Sigma \in \mathcal{S}_1(A,B)$, $\Sigma' \in \mathcal{S}_1(B,C)$ is defined by gluing Σ and Σ' at boundary points $\mu_m(z)$ and $\mu'_{-m}(z)$, $z \in S^1$, $m > 0$. Here A is an abbreviation for $(A_1, \ldots, A_k)$. Similarly, $f *_0 g$ is a glueing of two conformal isomorphisms f, g.

These operations make $\mathcal{S}$ into weak 2-category or bicategory [1]. Moreover, it is a symmetric monoidal weak 2-category (see [6]) with tensor product on objects defined via

$$(A_1, \ldots, A_k) \otimes (B_1, \ldots, B_l) = (A_1, \ldots, A_k, B_1, \ldots, B_l).$$

The symmetry $A \otimes B \to B \otimes A$ is a 1-morphism

$$s : (A_1, \ldots, A_k, B_1, \ldots, B_l) \to (B_1, \ldots, B_l, A_1, \ldots, A_k)$$

constructed from "identical" $\mathcal{O}$'s.

2 The 2-Category of Vector Spaces

We define a weak 2-category $\mathcal{V}$. It differs from the 2-category 2-Vect of Kapranov and Voevodsky [6]. Its objects are $\mathcal{V}_0$ = $\mathbf{C}$-vector spaces $A, B, \ldots$. Its 1-morphisms are $\mathcal{V}_1(A,B) = \{\xi : X \to L(A,B)\}$, where X is a $\mathbf{C}$-vector space mapped via ξ to the space of linear mappings $L(A,B) = \mathrm{Hom}_{\mathbf{C}}(A,B)$. Note that if we restrict to $X = \mathbf{C}$, $\mathcal{V}_1(A,B)$ would coincide with $L(A,B)$.

We define for $\xi : X \to L(A,B)$, $\eta : Y \to L(A,B)$ the set of 2-morphisms

$$\mathcal{V}_2(\xi, \eta) = \{f : X \to Y \mid \xi = \eta \circ f : X \to L(A,B)\}.$$

For an object $A \in \mathcal{V}_0$ there is an identity 1-morphism $\mathbf{C} \to L(A,A)$, $1 \mapsto 1_A$. A composition $*_1$ of 2-morphisms is the usual composition of linear maps. A composition $\xi *_0 \eta$ of 1-morphisms $\xi : X \to L(A,B)$ and $\eta : Y \to L(B,C)$ is defined as

$$X \otimes Y \xrightarrow{\xi \otimes \eta} L(A,B) \otimes L(B,C) \xrightarrow{\circ} L(A,C).$$

Finally, composition $f *_0 g$ of 2-morphisms $f : X \to Y$, $g : U \to W$ equals $f \otimes g : X \otimes U \to Y \otimes W$.

The category $\mathcal{V}$ thus defined is a symmetric monoidal weak 2-category (see [6]), when equipped with usual tensor product of vector spaces. Tensor product of 1-morphisms $\xi : X \to L(A,B)$ and $\eta : Y \to L(C,D)$ is $\xi\otimes\eta : X\otimes Y \to L(A\otimes C, B\otimes D)$. The 1-morphisms of associativity and symmetry are given by

$$\mathbf{C} \to L(A\otimes(B\otimes C),(A\otimes B)\otimes C), \quad 1 \mapsto a,$$

$$\mathbf{C} \to L(A\otimes B, B\otimes A), \quad 1 \mapsto P,$$

where a and P are usual associativity and permutation.

3 Modular 2-Functor

A modular functor is defined as a 2-functor $Z : \mathcal{S} \to \mathcal{V}$ "depending continuously on conformal structure". More precisely, to each label $A \in \mathcal{C}_0$ is assigned a vector space $Z(A)$, and

$$Z(A_1,\ldots,A_k) = Z(A_1)\otimes\ldots\otimes Z(A_k)$$

(we shall abbreviate this to $Z(A)$). To a 1-morphism Σ from $A = (A_1,\ldots,A_k)$ to $B = (B_1,\ldots,B_l)$ a modular functor assigns a map

$$\xi_\Sigma : Z(\Sigma;A;B) \to L(Z(A),Z(B)),$$

satisfying $\xi_{\Sigma\coprod\Sigma'} = \xi_\Sigma\otimes\xi_{\Sigma'}$. To a 2-morphism $f : \Sigma \to \Sigma'$ corresponds an isomorphism

$$Z(f) : Z(\Sigma) \to Z(\Sigma'),$$

such that $\xi_\Sigma = \xi_{\Sigma'}\circ Z(f)$, and $f \mapsto Z(f)$ is a representation of the mapping class groupoid. Again we have $Z(f\coprod g) = Z(f)\otimes Z(g)$.

An essential part of the functor is a multiplication (2-morphism)

$$m_{\Sigma,\Sigma'} : Z(\Sigma)\otimes Z(\Sigma') \to Z(\Sigma\cup\Sigma'),$$

given for any composable surfaces Σ and Σ'. The diagram

$$\begin{array}{ccc} Z(\Sigma;A;B)\otimes Z(\Sigma';B;C) & \stackrel{m}{\to} & Z(\Sigma\cup\Sigma';A;C) \\ \downarrow & & \downarrow \\ L(Z(A),Z(B))\otimes L(Z(B),Z(C)) & \to & L(Z(A),Z(C)) \end{array}$$

must commute and the multiplication m must be associative and functorial in 2-morphisms.

We have also a 2-morphism $\eta : \mathrm{Id}_{Z(A)} \Longrightarrow Z(\mathrm{Id}_A)$ for any $A \in \mathcal{S}_0$ with $\mathrm{Id}_A = k\mathcal{O}$. That is, the composition

$$\mathbf{C} \stackrel{\eta}{\to} Z(\mathrm{Id}_A;A;A) \to L(Z(A),Z(A))$$

coincides with mapping $1 \mapsto 1_{Z(A)}$. Finally, η must be a unity of multiplication m.

Moreover, we assume that $Z(A)$ are separable Hilbert spaces, the tensor product for them is the Hilbert tensor product, and for any Σ, which does not contain components like $\mathcal{O}$, the corresponding 1-morphism is

$$Z(\Sigma; A; C) \to N(Z(A), Z(C))$$

where N means nuclear operators[3], and for $\mathcal{O}$

$$Z(\mathcal{O}; A; C) \to B(Z(A), Z(C))$$

where B means bounded operators. For any surface $\tilde{\Sigma} : A \to B$, obtained by a sewing $g : \Sigma \to \tilde{\Sigma}$ from a surface $\Sigma : A \coprod C \to B \coprod C$ (without $\mathcal{O}$-components) we assume the existence of a sewing map $Z(g) : Z(\Sigma) \to Z(\tilde{\Sigma})$ making the diagram

$$\begin{array}{ccc} Z(\Sigma; A, C; B, C) & \xrightarrow{Z(g)} & Z(\tilde{\Sigma}; A; B) \\ \downarrow & & \downarrow \\ N(Z(A)\otimes Z(C), Z(B)\otimes Z(C)) & \xrightarrow{\mathrm{Tr}_{Z(C)}} & N(Z(A), Z(B)) \end{array}$$

commute[4]. Again sewing must be associative.

4 The Sewing Axiom

From a modular functor we construct a $\mathbf{C}$-linear category $\mathcal{C}$ with the set of objects $\mathcal{C}_0$ and the set of morphisms $\mathcal{C}(A, B) = Z(\mathcal{O}; A; B)$, where $\mathcal{O} = \mathcal{O}_{id,id}$. The composition of morphisms is given by m, for $\mathcal{O} \bigcup \mathcal{O} = \mathcal{O}$. The unity is $\eta : \mathbf{C} \to Z(\mathcal{O}; A; A)$.

The associativity of m implies that $Z(\Sigma; A_1, \dots, A_k; B_1, \dots, B_l)$ is a functor

$$Z(\Sigma) : \mathcal{C}^{op} \times \ldots \times \mathcal{C}^{op} \times \mathcal{C} \times \ldots \times \mathcal{C} \to \mathbf{C}\text{-vect}.$$

In examples coming from physics [16,22] we have that $\mathcal{C}$ is a semisimple category with finite number of simple objects. In this paper we *assume* that $\mathcal{C}$ is abelian and that the functor $Z(\Sigma)$ is left exact for any Σ.

Proposition 4.1 *Let $g : \Sigma \to \tilde{\Sigma}$ sew n pairs of boundary circles. Then for any $f \in Z(n\mathcal{O}; P; Q)$ the diagram*

$$Z(\Sigma; A, Q; f, B) \quad \begin{array}{ccc} Z(\Sigma; A, Q; P, B) & \xrightarrow{Z(\Sigma; A, f; P, B)} & Z(\Sigma; A, P; P, B) \\ \downarrow & & \downarrow \\ Z(\Sigma; A, Q; Q, B) & \xrightarrow{Z(g)} & Z(\tilde{\Sigma}; A; B) \end{array} \quad Z(g)$$

commutes.

---

[3] An operator $X : H \to H'$ is nuclear, if $\|X\|_1 = \mathrm{tr}_H \sqrt{X^* X}$ is finite. See [3] and references therein.

[4] For any $X \in N(Z(A)\otimes Z(C), Z(B)\otimes Z(C))$ the inequality $\|\mathrm{Tr}_{Z(C)} X\|_1 \le \|X\|_1$ follows from the inequality $|\mathrm{tr}_{Z(A)\otimes Z(C)} Y| \le \|Y\|_1$ for an operator $Y = U \otimes 1 \circ X : Z(A)\otimes Z(C) \to Z(A)\otimes Z(C)$, where unitary operator $U : Z(B) \to Z(A)$ is such that $\mathrm{Tr}_{Z(C)} Y = U \mathrm{Tr}_{Z(C)} X \ge 0$.

Proof follows from the associativity of $Z(g)$.

Now we can formulate the "sewing axiom".

Axiom a)*Assume that the boundary of any connected component of $\tilde{\Sigma}$ is not empty. Then the functor*

$$Z(\tilde{\Sigma}; A; B) : (\mathcal{C}^{op})^k \times \mathcal{C}^l \longrightarrow \mathbf{C}\text{-vect}$$

together with $Z(g)$ is a coend of the functor

$$Z(\Sigma; A, P; P, B) : (\mathcal{C}^{op})^{k+n} \times \mathcal{C}^{n+l} \longrightarrow \mathbf{C}\text{-vect},$$

viewed as a bifunctor

$$(\mathcal{C}^n)^{op} \times (\mathcal{C}^n) \longrightarrow \text{Left\_exact\_functors}((\mathcal{C}^{op})^k \times \mathcal{C}^l \to \mathbf{C}\text{-vect}),$$

$$Z(\tilde{\Sigma}; A; B) = \int^{P \in \mathcal{C}^n} Z(\Sigma; A, P; P, B).$$

b)*A functor $\mathcal{C}^{op} \to \mathbf{C}$-vect, $A \mapsto Z(D; A; \ )$, with parametrization of the boundary of the disk D given by $z \mapsto z^{-1}$, is assumed to be representable, i.e. there is an object $I \in \mathcal{C}$ and isomorphism $Z(\mathcal{O}; A; I) \simeq Z(D; A; \ )$. A connected surface Σ without boundary can be represented as a glueing of some surface Σ_1 with a disk D. Then*

$$\begin{aligned} Z(\Sigma_1; \ ; I) \ \simeq \ & \int^A Z(\Sigma_1; \ ; A) \otimes Z(\mathcal{O}; A; I) \simeq \\ & \int^A Z(\Sigma_1; \ ; A) \otimes Z(D; A; \ ) \to Z(\Sigma; \ ; \ ) \end{aligned}$$

is an isomorphism.

For definition of a coend see [14]. For semisimple $\mathcal{C}$ a coend is a direct sum over all simple objects of $\mathcal{C}$, thus, this axiom takes the standard form of [19,21].

5 Flat Connection

Continuous dependence on conformal structure means that $Z(\Sigma; A; B)$ is the fiber, over a point $\Sigma \in \mathcal{T}$, of a vector bundle. Here the Teichmüller space $\mathcal{T}$ consists of all complex structures and parameterizations of the boundary modulo homeomorphisms of the surface, isotopic to identity. More precisely, $\mathcal{T} = \{(\Sigma, [\psi])\}/\sim$, where $\psi : \Sigma \to F$ is a homeomorphism of a Riemann surface to a fixed oriented surface F, $[\psi]$ is its isotopy class. We will often drop ψ in the following.

Segal [19] (see also [21]) showed the existence of a projectively flat connection on the bundle Z for semisimple $\mathcal{C}$. The essential part of this construction is the action of the semigroup of annuli on positive energy modules over an affine Lie algebra. This can be realized also for some non-semisimple categories of modules.

Hence, we will *assume* for some $\mathcal{C}$ the existence of a projectively flat connection, which for any path Σ_t, $t \in [0,1]$ yields a linear map

$$\Gamma(\Sigma_t) : Z(\Sigma_0; A; B) \to Z(\Sigma_1; A; B).$$

In fact, we need a connection which can be pull down to the space of moduli $\mathcal{M}$. So, we assume that for any two paths Σ_t and Σ'_t, which are isomorphic via a continuous family of complex isomorphisms $g_t : \Sigma_t \to \Sigma'_t$, the diagram

$$\begin{array}{cccc} & Z(\Sigma_0; A; B) & \xrightarrow{\Gamma(\Sigma_t)} & Z(\Sigma_1; A; B) & \\ Z(g_0) & \downarrow & & \downarrow & Z(g_1) \\ & Z(\Sigma'_0; A; B) & \xrightarrow{\Gamma(\Sigma'_t)} & Z(\Sigma'_1; A; B) & \end{array}$$

commutes. We need also a locality condition on the connection by which we mean the commutativity of the diagram

$$\begin{array}{cccc} & Z(\Sigma'; A; B) \otimes Z(\Sigma_0; B; C) & \xrightarrow{1 \otimes \Gamma(\Sigma_t)} & Z(\Sigma'; A; B) \otimes Z(\Sigma_1; B; C) & \\ m & \downarrow & & \downarrow & m \\ & Z(\Sigma' \bigcup \Sigma_0; A; C) & \xrightarrow{\Gamma(\Sigma' \bigcup \Sigma_t)} & Z(\Sigma' \bigcup \Sigma_1; A; C) & \end{array}$$

for any composable Σ' and Σ_t. Similarly for composable Σ_t and Σ''. In particular, using $\Sigma' = \mathcal{O}$ or $\Sigma'' = \mathcal{O}$, we see that $\Gamma(\Sigma_t; A; B)$ is functorial in $A_i, B_j \in \mathcal{C}$.

These properties can be expressed better by the following *sewing axiom for connections*:

i) $\Gamma(\Sigma_t \coprod \Sigma'_t) = \Gamma(\Sigma_t) \otimes \Gamma(\Sigma'_t)$,

ii) For any continuous family of sewings $g_t : \Sigma_t \to \tilde{\Sigma}_t$ the diagram

$$\begin{array}{cccc} & Z(\Sigma_0; A, C; C, B) & \xrightarrow{\Gamma(\Sigma_t)} & Z(\Sigma_1; A, C; C, B) & \\ Z(g_0) & \downarrow & & \downarrow & Z(g_1) \\ & Z(\tilde{\Sigma}_0; A; B) & \xrightarrow{\Gamma(\tilde{\Sigma}_t)} & Z(\tilde{\Sigma}_1; A; B) & \end{array}$$

commutes.

Moreover, we assume that on $\mathcal{T}_0$, the Teichmüller space of surfaces of genus 0, the connection is flat. Example is given by Knizhnik-Zamolodchikov equations [8]. Flat connections naturally arise in the context of framed Riemann surfaces [24] or extended surfaces [9,23].

6 The Tensor Structure of the Category $\mathcal{C}$

Now we construct a tensor product in the category $\mathcal{C}$ using a method similar to [2,7].

Let P_l denote pants $S^2 \setminus 3D^2$ equipped with a hyperbolic metrics of the curvature -1 such that boundary circles are geodesics of the length l (see e.g. [4]). Boundary circles (one incoming and two outgoing) are parametrized proportionally to length, and marked points correspond to $0 \in \mathbf{R}/\mathbf{Z} \simeq S^1$.

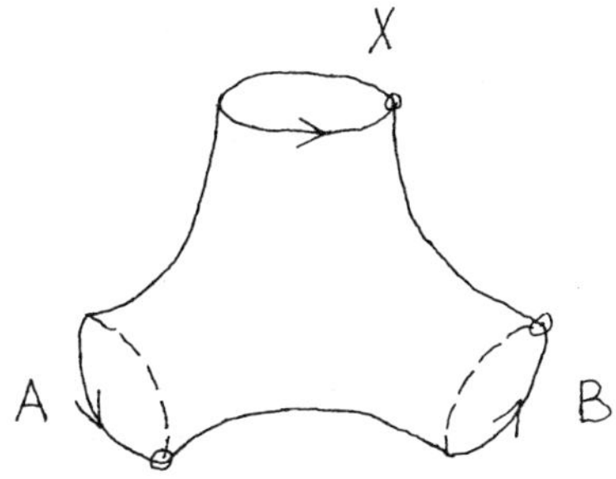

These pants will be refered to as $(P_l; X; A, B)$. Assume the following

Axiom *For any $A, B \in \mathcal{C}$ the functor $X \mapsto Z(P_l; X; A, B)$ is representable, i.e. there exists a representing object $A \otimes_l B \in \mathcal{C}$ and an isomorphism*

$$Z(P_l; X; A, B) \simeq \mathcal{C}(X, A \otimes_l B).$$

Thus a tensor product $\otimes_l : \mathcal{C} \times \mathcal{C} \to \mathcal{C}$ is defined. The 1-pameter family P_l, $l \in [l_1, l_2] \subset]0, \infty[$ defines via Γ the isomorphism

$$\mathcal{C}(X, A \otimes_{l_1} B) \simeq Z(P_{l_1}; X; A, B) \xrightarrow{\Gamma(P_l)} Z(P_{l_2}; X; A, B) \simeq \mathcal{C}(X, A \otimes_{l_2} B)$$

Thus, there exists an isomorphism of functors

$$\phi_{l_1, l_2} : A \otimes_{l_1} B \to A \otimes_{l_2} B.$$

In the following we fix l and denote $\otimes = \otimes_l$ for objects of $\mathcal{C}$. Standard choice [7,21] corresponds to $l \searrow 0$.

Consider a 1-parameter family of surfaces $PP_{\lambda, \lambda'} \simeq S^2 \setminus 4D^2$ with metrics of curvature -1 where lengths of geodesic boundary circles equal l and the lengths of two geodesics shown at figure are λ, λ'.

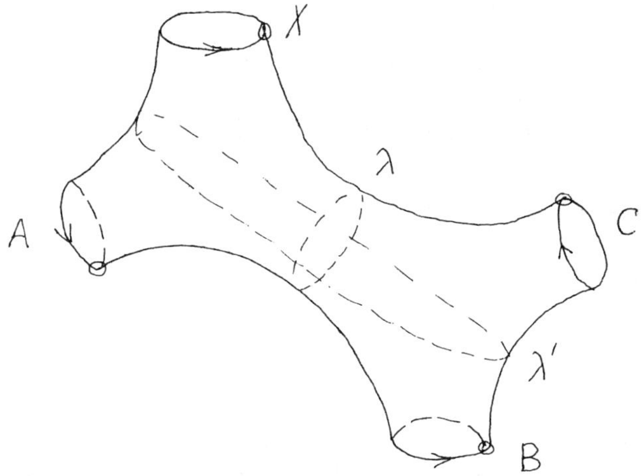

Varying λ we connect $PP(\lambda = l)$ to $PP(\lambda' = l)$ by a path. We have

$$Z(PP(\lambda = l); X; A, B, C) = \int^{D\in\mathcal{C}} Z(P; X; A, D)\otimes Z(P; D; B, C) =$$

$$\int^{D\in\mathcal{C}} \mathcal{C}(X, A\otimes D)\otimes\mathcal{C}(D, B\otimes C) \quad = \mathcal{C}(X, A\otimes(B\otimes C))$$

and

$$Z(PP(\lambda' = l); X; A, B, C) = \mathcal{C}(X, (A\otimes B)\otimes C).$$

The mapping $\Gamma(PP_{\lambda,\lambda'})$ (fusion matrix) yields a functorial isomorphism

$$\mathcal{C}(X, A\otimes(B\otimes C)) \simeq \mathcal{C}(X, (A\otimes B)\otimes C)$$

which defines an isomorphism of associativity

$$a : A\otimes(B\otimes C) \to (A\otimes B)\otimes C.$$

Considering a 2-parameter family $PPP \simeq S^2 \setminus 5D^2$ we can make a pentagon and conclude that a satisfy the Mac Lane–Stasheff pentagon commutativity condition. Thus $(\mathcal{C}, \otimes, a)$ is a monoidal category. Different l's give equivalent categories.

Recall that the functor $\mathcal{C}^{op} \to \mathbf{C}\text{-vect}$, $A \mapsto Z(D; A;\ )$, where parametrization of the boundary of the disk D is given by $z \mapsto z^{-1}$, is assumed representable by an object $I \in \mathcal{C}$ with isomorphism $Z(\mathcal{O}; A; I) \simeq Z(D; A;\ )$. Such I is a unity object of $\mathcal{C}$. Indeed,

$$\mathcal{C}(X, A\otimes I) \;=\; Z(P; X; A, I) \;=\; \int^{B} Z(P; X; A, B)\otimes Z(\mathcal{O}; B; I) =$$

$$\int^{B} Z(P; X; A, B)\otimes Z(D; B;\ ) \;=\; Z(P\bigcup D; X; A) =$$

$$Z(\mathcal{A}; X; A) \xrightarrow{\Gamma(\mathcal{A}_t)} Z(\mathcal{O}; X; A) \;=\; \mathcal{C}(X, A)$$

Here $\mathcal{A}$ is an annulus obtained by glueing a disk to pants, $\mathcal{A}_t$ connects $\mathcal{A}$ to $\mathcal{O}$. Hence, there exists a functorial isomorphism $A\otimes I \to A$. Thus, I is an identity object.

Finally, we assume that the monoidal category $\mathcal{C}$ is rigid, i.e. for any $A \in \mathcal{C}$ there exist $A\check{} \in \mathcal{C}$ together with evaluation $ev : A\otimes A\check{} \to I$ and coevaluation $coev : I \to A\check{}\otimes A$ satisfying usual axioms (see e.g. [5]).

7 Modular Group Action

Recall that $\mathcal{T}$ consists of classes $(\Sigma, [\psi])$, $\psi : \Sigma \to F$. The group of homeomorphisms of F modulo isotopy, the mapping class group, acts on $\mathcal{T}$ on the right. Here we are interested in another group, isomorphic to mapping class group of F.

Let Σ be a Riemann surface from $\mathcal{S}_1$. A *modular group* $\mathrm{Mod}\Sigma$ is defined as the group of isotopy classes of homeomorphisms $g : \Sigma \to \Sigma$ which preserve orientation, parameterization and labelling of the boundary (and preserve additional structures if defined). We think of g as acting on the right on Σ.

Let $g \in \mathrm{Mod}\Sigma$ be represented by homeomorphism g. Choose an arbitrary homeomorphism $f : \Sigma \to F$. Then $X_0 = (\Sigma \xrightarrow{f} F)$, $X_1 = (\Sigma \xrightarrow{g} \Sigma \xrightarrow{f} F)$ are points of $\mathcal{T}$. Let $X_t \in \mathcal{T}$, $t \in [0,1]$ be a path joining them.

Theorem 7.1 *In the case of a flat connection the operator*

$$\Gamma(X_t) : Z(\Sigma; A; B) \to Z(\Sigma; A; B)$$

depends only on g and is denoted by $\pi(g)$. Furthermore, $\pi(hg) = \pi(h) \circ \pi(g)$, so π is a representation of $\mathrm{Mod}\Sigma$. *For a projectively flat connection $\pi(g)$ is defined up to a constant and π is a projective representation.*

Proof follows from the fact that $\mathcal{T}$ is simple connected.

For example, a Dehn twist in an annulus $\mathcal{A}$ gives a family $\mathcal{A}_t$ of surfaces with fixed complex structure for which parametrization $\phi(s)$ of the incoming circle is rotated by $t \in [0,1]$ (changed by $\phi(s+t)$). The corresponding map

$$\Gamma(\mathcal{A}_t) : Z(\mathcal{A}; A; B) \to Z(\mathcal{A}; A; B)$$

in particular case $\mathcal{A} = \mathcal{O}$ yields $\mathcal{C}(A,B) \to \mathcal{C}(A,B)$, defining via Nakayama's lemma a functorial automorphism $\nu_A : A \to A$ called the *ribbon twist* [18].

A modification of this construction yields braiding in the monoidal category $\mathcal{C}$. Indeed, consider a path P_l^t, $t \in [0, \frac{1}{2}]$, consisting of pants with parametrization of the boundary $\phi_i(x)$ changed by $\phi_i(x+t)$, $x \in \mathbf{R}/\mathbf{Z}$, $i = 1,2,3$. This path connects $(P_l; X; A, B)$ with $(P_l^{\frac{1}{2}}; X; A, B) \simeq (P_l; X; B, A)$ yielding a map

$$\mathcal{C}(X, A \otimes B) \simeq \mathcal{C}(X, B \otimes A),$$

which produce an isomorphism

$$c : A \otimes B \to B \otimes A.$$

Morphisms c and a satisfy two hexagon commutativity conditions, hence, c is a brading. Moreover, c and ν are related by

$$c_{A,B}^2 = \nu_{A \otimes B} \circ \nu_A^{-1} \otimes \nu_B^{-1},$$

thus, $(\mathcal{C}, \otimes, a, c, \nu)$ is a ribbon (or tortile [20]) category.

For a torus with one hole $(T \setminus D; X; \ )$ we have a realization of the space $Z(T \setminus D)$ via coend:

$$Z(T \setminus D; X; \ ) = \int^M Z(P; X, M; M) \quad = \int^M \mathrm{Hom}_{\mathcal{C}}(X \otimes M, M) \simeq$$

$$\int^M \mathrm{Hom}_{\mathcal{C}}(X, M \otimes M\check{\ }) \simeq \quad \mathrm{Hom}_{\mathcal{C}}(X, \int M \otimes M\check{\ }) = \mathrm{Hom}_{\mathcal{C}}(X, F),$$

where $\int^M$ means a coend with values in the category of left exact functors $\mathcal{C}^{op} \to$ $\mathbf{C}$-vect, and $F = \int^M M \otimes M\check{\ }$ (when it exists in $\mathcal{C}$) is a Hopf algebra in the tensor category $\mathcal{C}$. The algebra F was introduced in [15]. The modular group $\mathrm{Mod}(T \setminus D; X;\ )$ has generators S, T (some diffeomorphisms of a torus), subject to a relation

$$(ST)^3 = S^2.$$

There is also a relation

$$S^{-4} = \text{Dehn twist } \nu \text{ near the boundary of } T \setminus D.$$

Then by functoriality the corresponding action $\pi(S)$, $\pi(T)$ on the vector space $Z(T \setminus D; X;\ )$ is induced by morphisms

$$S : F \to F, \qquad T : F \to F, \qquad T = \int \nu \otimes 1$$

$$S : F = \int^M M \otimes M\check{} \simeq \int^M M \otimes I \otimes M\check{} \xrightarrow{\int 1 \otimes \mu \otimes 1} \int^{M,N} M \otimes N \otimes N\check{} \otimes M\check{} \longrightarrow$$

$$\xrightarrow{\int c^{-1} \otimes c^{-1}} \int^{M,N} N \otimes M \otimes M\check{} \otimes N\check{} \xrightarrow{\int \mathrm{ev}} \int^N N \otimes I \otimes N\check{} \simeq \int^N N \otimes N\check{} \simeq F$$

for some $\mu : I \to F$, which is an integral for the dual Hopf algebra $F\check{}$ (see [10,11,13]). A morphism S of that form is the general solution to Moore and Seiberg's equation $Sa = bS$ [16] (see [12]). The morphisms S, T also satisfy the relations

$$(ST)^3 = \lambda S^2, \qquad S^{-2} = \text{antipode of the algebra } F, \qquad S^4 = \nu^{-1}$$

(cf. [17]) giving the projective representation of the modular group for non-extended surfaces.

8 Summary

We see that modular functor Z under mild assumptions generates a ribbon tensor category $\mathcal{C}$ and a homotopical modular functor

$$\bar{Z} : \text{Surfaces modulo homotopy} \to \mathbf{C}\text{-Vector spaces}.$$

In [12] I will prove that in fact any ribbon category $\mathcal{C}$ (with some non-degeneracy conditions) generates such a homotopical modular functor $Z_{\mathcal{C}}$. It is unique up to a sign. Our functor $\bar{Z}$ is isomorphic to a functor $Z_{\mathcal{C}}$ provided that non-degeneracy conditions are satisfied.

References

[1] J. Benabou, Introduction to bicategories, *Lect. Notes in Math.* **47** (1967), 1–71.

[2] V.G. Drinfeld, Quasi-Hopf algebras. Algebra and Analysis **1** (1989), n^0 6.

[3] *Encyclopaedia of Mathematics* (english translation of Matematicheskaia entsiklopediia), M. Hazewinkel, ed., Volume 6, Kluwer Academic Publishers, 1990.

[4] J.L. Harer, The cohomology of the Moduli Space of Curves, *Lect. Notes in Math.* **1337** (1988), 138–221.

[5] A. Joyal and R. Street, Braided monoidal categories, Mathematics Reports 86008, Macquarie University, 1986.

[6] M.M. Kapranov and V.A. Voevodsky, Braided monoidal 2-categories, 2-vector spaces and Zamolodchikov tetrahedra equations, Preprint, September 1991.

[7] D. Kazhdan and G. Lusztig, Affine Lie Algebras and Quantum Groups, Preprint, MIT, 1991.

[8] V.G. Knizhnik and A.B. Zamolodchikov, Current Algebra and Wess-Zumino models in two dimensions, Nuclear Physics **B247** (1984), 83–103.

[9] M. Kontsevich, Rational Conformal Field Theory and Invariants of 3-dimensional Manifolds, Preprint, CPT-88/P.2189, December 1988.

[10] V.V. Lyubashenko, Tensor categories and RCFT I,II, Preprint, Kiev ITP-90-30E, ITP-90-59E, 1990.

[11] V.V. Lyubashenko, Modular Transformations for Tensor Categories, Preprint, Pisa, SNS, **124**, February 1992.

[12] V.V. Lyubashenko, Ribbon Categories as Modular Categories, Work in progress.

[13] V.V. Lyubashenko and S. Majid, Braided Groups and Quantum Fourier Transform, Preprint, DAMTP/91-26, July 1991.

[14] S. Mac Lane, *Categories for the working mathematician*, Springer-Verlag, 1971.

[15] S. Majid, Reconstruction theorems and Rational Conformal Field Theories, Int. J. Mod. Phys. A **6** (1991), 4359–4374.

[16] G. Moore and N. Seiberg, Classical and Quantum Conformal Field Theories, Comm. Math. Phys. **123** (1989), 177–254.

[17] K.-H. Rehren, Markov Traces as Characters for Local Algebras, Nuclear Physics B (Proc. Suppl.) **18B** (1990), 259–268.

[18] N. Reshetikhin, Quasitriangular Hopf algebras, solutions of the Yang-Baxter equations and link invariants, Algebra and Analysis **1** (1989), n^0 2, 169–194.

[19] G. Segal, The definition of Conformal Field Theory, Preprint.

[20] M.C. Shum, Tortile tensor categories, Macquarie preprint, 1989.

[21] A. Tsuchiya, K. Ueno and Ya. Yamada, Conformal Field Theory on Universal Family of Stable Curves with Gauge Symmetries, Advanced Studies in Pure Math. **19** (1989), 459–566.

[22] E. Verlinde, Fusion Rules and Modular Transformations in 2D Conformal Field Theory, Nuclear Physics **B300** [**FS22**] (1988), 360–376.

[23] K. Walker, On Witten's 3-manifold Invariants, Preprint, February 1991.

[24] E. Witten, Quantum Field Theory and the Jones Polynomial, Commun. Math. Phys. **121** (1989), 351–399.

Department of Mathematical Methods of Systems Analysis, Kiev Polytechnic Institute, 37 Peremogy prosp., Kiev, Ukraine 252056

Current address: Classe di Scienze, Scuola Normale Superiore, Pisa, Italy I-56126

E-mail address: lyubashenko@vaxsns.infn.it

Canadian Mathematical Society
Conference Proceeding
Volume **13**, 1992

COHERENCE THEOREMS AND CONFORMAL FIELD THEORY

SAUNDERS MAC LANE

1. COHERENCE THEOREMS

The study of certain quantum field theories has involved known coherence theorems asserting that "all diagrams" commute in certain monoidal categories. A monoidal category M is a category with a distinguished object I and a bifunctor $\otimes : M \times M \to M$ which is associative up to a natural isomorphism a, for objects A, B, and C of M.

$$a : A \otimes (B \otimes C) \to (A \otimes B) \otimes C,$$

and such that the following two composites are equal:

$$A \otimes (B \otimes (C \otimes D)) \xrightarrow{a} (A \otimes B) \otimes (C \otimes D) \xrightarrow{a} ((A \otimes B) \otimes C) \otimes D, \tag{1}$$

$$A \otimes (B \otimes (C \otimes D)) \xrightarrow{1 \otimes a} A \otimes ((B \otimes C) \otimes D) \xrightarrow{a} (A \otimes (B \otimes C)) \otimes D \xrightarrow{a \otimes 1} ((A \otimes B) \otimes C) \otimes D. \tag{2}$$

Together they form a (commuting) pentagon. Naturality of a and the commutativity of this pentagon imply that every (larger) diagram involving only instances of a will commute. This "coherence" theorem was proved in 1963 by J. D. Stasheff and by S. Mac Lane; references to these papers (and other early references) are contained in Mac Lane (1971). For the distinguished object I (the unit element) there are natural isomorphisms.

$$\lambda : I \otimes A \to A, \quad \rho : A \otimes I \to A$$

such that

$$I \otimes \lambda : A \otimes (I \otimes C) \to A \otimes C \tag{3}$$

is equal to

$$A \otimes (I \otimes C) \xrightarrow{a} (A \otimes I) \otimes C \xrightarrow{\rho \otimes I} A \otimes C. \tag{4}$$

1991 *Mathematics Subject Classification.* 1802, 18A17, 81T05, 81T30.

This paper is in final form; no version of it will be submitted for publication elsewhere.

Again, these conditions imply that every diagram involving only instances of a, λ, and ρ is commutative.

A monoidal category is often called a tensored category, especially in current papers in theoretical physics.

A symmetric monoidal category is a monoidal category with an additional natural isomorphism.

$$c : A \otimes B \to B \otimes A$$

for which $c^2 = 1$ and the following two composites are equal

$$A \otimes (B \otimes C) \xrightarrow{1 \otimes c} A \otimes (C \otimes B) \xrightarrow{a} (A \otimes C) \otimes B \xrightarrow{c \otimes 1} (C \otimes A) \otimes B, \quad (5)$$

$$A \otimes (B \otimes C) \xrightarrow{a} (A \otimes B) \otimes C) \xrightarrow{a} C \otimes (A \otimes B) \xrightarrow{a} (C \otimes A) \otimes B. \quad (6)$$

They constitute a hexagon. Again, the pentagon above and this hexagon condition, with $c^2 = 1$, imply that every diagram involving a and c must commute.

The proof, by Mac Lane in 1963, of this coherence theorem made essential use of the relations on the symmetric groups on n letters, when generated by transpositions of successive letters. For the unit 1, one must assume also the equality of

$$A \otimes I \xrightarrow{\rho} A, \quad A \otimes I \xrightarrow{c} I \otimes A \xrightarrow{\lambda} A$$

A typical example of a symmetric monoidal category is the category of all vector spaces over a (fixed) field, with the usual tensor product of such spaces and the usual natural associativity isomorphism; the field itself serves as the unit element. A more decisive example (for the use in theoretical physics) is the category of all representations on such vector spaces of a fixed (perhaps topological) group G.

A braided monoidal category is defined as above (for a symmetric monoidal category) but without the assumption that $c^2 = 1$ and with the added requirement that the following two composites

$$(C \otimes A) \otimes B \xrightarrow{c \otimes 1} (A \otimes C) \otimes B \xrightarrow{a^{-1}} A \otimes (C \otimes B) \xrightarrow{1 \otimes c} A \otimes (B \otimes C), \quad (7)$$

$$(C \otimes A) \otimes B \xrightarrow{a^{-1}} C \otimes (A \otimes B) \xrightarrow{c} (A \otimes B) \otimes C \xrightarrow{a^{-1}} A \otimes (B \otimes C) \quad (8)$$

are equal. This is just the equality of (5) and (6) but with c there replaced by c^{-1}. These braided categories replace the symmetric group used in (5) and (6) by Artin's braid group. The corresponding coherence theorem holds in this case.

2. Conformal Field Theories

In theoretical physics, the collision and eventual separation of two particles is often represented by a Feynman diagram such as

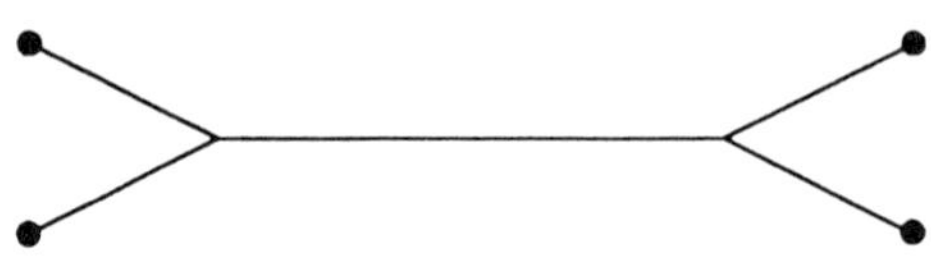

FIGURE 1

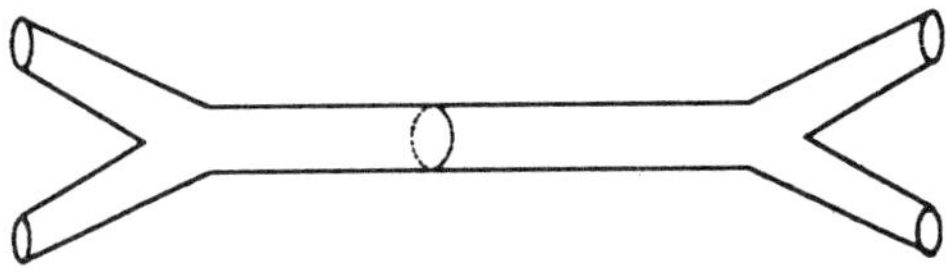

FIGURE 2

In string theory, each particle is replaced by a closed string (i.e., a circle) and the diagram becomes a branched tube, as in

The surface of this tube can be given a complex structure, so that the tube becomes a Riemann surface with the indicated circles as boundaries; the tube can also be considered as a cobordism from the left-hand bounding circles to those on the right. There are more elaborate such diagrams, involving successive collisions.

The study of such diagrams and their representations in string theory led G. Segal to make a corresponding formal definition of a conformal field theory (in lectures at the Institute for advanced Study, but fully published only after considerable delay).

In his lecture, a conformal field theory is a category M in which the objects are finite families of circles, while the morphisms are Riemann surfaces joining the domain family to the codomain family; they resemble cobordisms. Composition is given by glueing. This defines a monoidal category, in which the tensor product of two tubes means to place the tubes side by side. The commutativity isomorphism c twists the tube from circles A to circles A' over the tube from B to B'; schematically as in Figure 3.

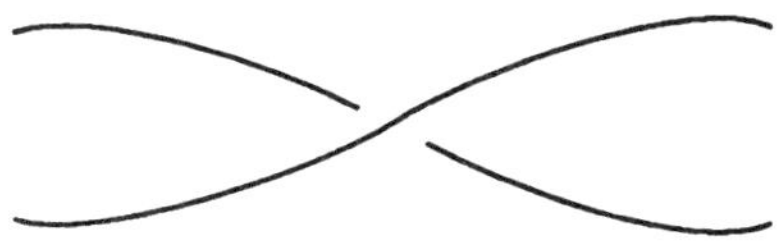

FIGURE 3

It is thus clear that c^2 is not the identity, and the connection to the braid group is evident.

Segal then considers representations of such categories M in the category with objects, Hilbert spaces, and morphisms, trace class mappings between Hilbert spaces. More recently in preprints, Louis Crane, Igor Frenkel, David Yetter, and others have replaced the (braided) categories used here by suitably braided 2-categories, for which the 2-cells are conformal maps between Riemann surfaces and "phase shifts" between trace class mappings of Hilbert space.

There are other closely related types of field theory, such as topological quantum theories, as defined for example by Witten [1988], where the objects of the category in question are closed oriented smooth manifolds of a fixed dimension d, and the morphisms are again suitable bordisms.

3. FUSION AND BRAIDING

In 1988, G. Moore and H. Seiberg [1989], in studying such conformal fields, proved theorems about their representations which involved exactly a pentagon and hexagons for what was then termed fusion (essentially, associativity) and braiding (commutativity). They gave proofs in this language for the coherence theorems; Witten observed that pentagons and hexagons had appeared in Mac Lane's coherence theorem. Shortly after this, S. Brustein, Y. Ne'eman, and S. Sternberg in a preprint later published in 1990 observed more exactly the connections between these coherence theorems.

Both these papers provide solutions for a form of the Yang-Baxter equation. This equation applies to a vector space V and a linear map

$$R : V \otimes V \to V \otimes V$$

which is required to satisfy

$$R_{12}R_{13}R_{23} = R_{23}R_{13}R_{12} \tag{9}$$

as an endomap of the triple tensor product $V \otimes V \otimes V$. Here R_{13}, for example, denotes the given map R acting on the first and third factors of $V \otimes V \otimes V$. This equation holds when $R = c$ is the commutativity isomorphism for V, an object in a symmetric or a braided monoidal category M. Indeed, the equation (9) is essentially the identity

$$(12)(23)(12) = (23)(12)(23)$$

valid for transpositions of adjacent letters in the symmetric group on n letters or in the braid group on n strings.

For some purposes, a Yang-Baxter equation (9) has added exponential numerical factors. It is often written as a condition on the matrix representing R relative to some basis.

The Zamolodchikov tetrahedral (or simplex) equation is analogous. It applies to an endomorphism S of a triple tensor product $V \otimes V \otimes V$ and requires the equality

$$S_{123}S_{124}S_{134}S_{234} = S_{234}S_{134}S_{124}S_{123} \tag{10}$$

on the four-fold tensor product. This equation is related to suitably defined braided monoidal 2-categories, as in current work by Kapranov and Voevodsky (1991); this involves more complex coherence conditions, and suggests that there are higher braiding conditions on 3-categories, etc. Such studies may require the use of pasting schemes, as in Power [1991] and Johnson [1989].

A Riemann surface with three bounding circles is sometimes called a "trunion" or a "pair of pants"; it represents the collision of just two particles to form a third (single) particle. Riemann surfaces with more boundary circles can be cut up into such trunions: all basic properties are assumed invariant under a change in the position of the cuts. The associativity isomorphism may be regarded as such a change—where a slice (34), cutting off circles 3 and 4, is tipped to a slice (23), as in the left hand passage of Figure 4 as in Walker [1990]. This whole figure represents the pentagon condition on such an operation a. Similarly, Figure 5 represents one of the two hexagonal conditions.

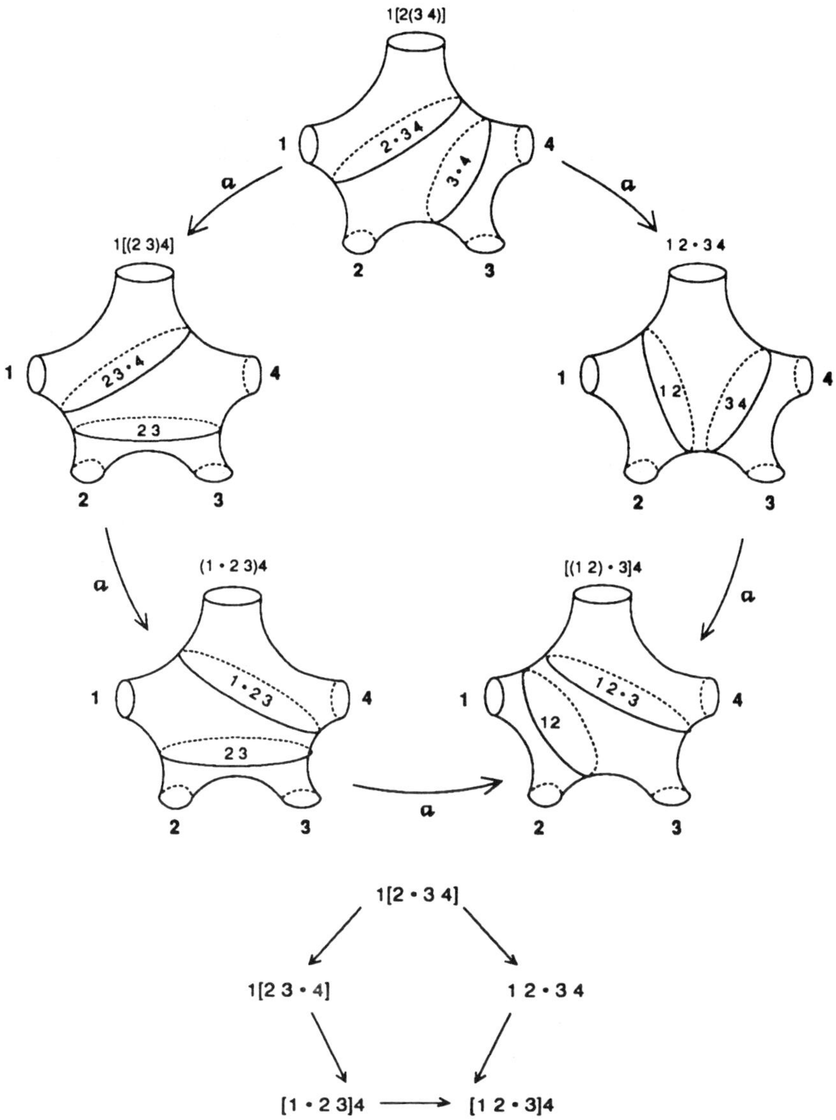

FIGURE 4. The Pentagon in Pants

4. C^* CATEGORIES

The Pontrjagin duality for compact abelian groups G takes on a more elaborate form when the group G is not abelian. The well-known Tannaka-Krein duality characterizes the dual of a compact group within the category of finite dimensional vector spaces. Recently, Doplicher and Roberts [1989] have developed a characterization of such a dual within abstract categories; this is again motivated by questions in quantum theory.

Consider the category $\operatorname{Rep} G$ of all continuous and unitary representations of the compact group G in a Hilbert space. This category is a C^*-category in the sense that each hom-set $\hom(A, B)$ has the structure of a Banach space, with

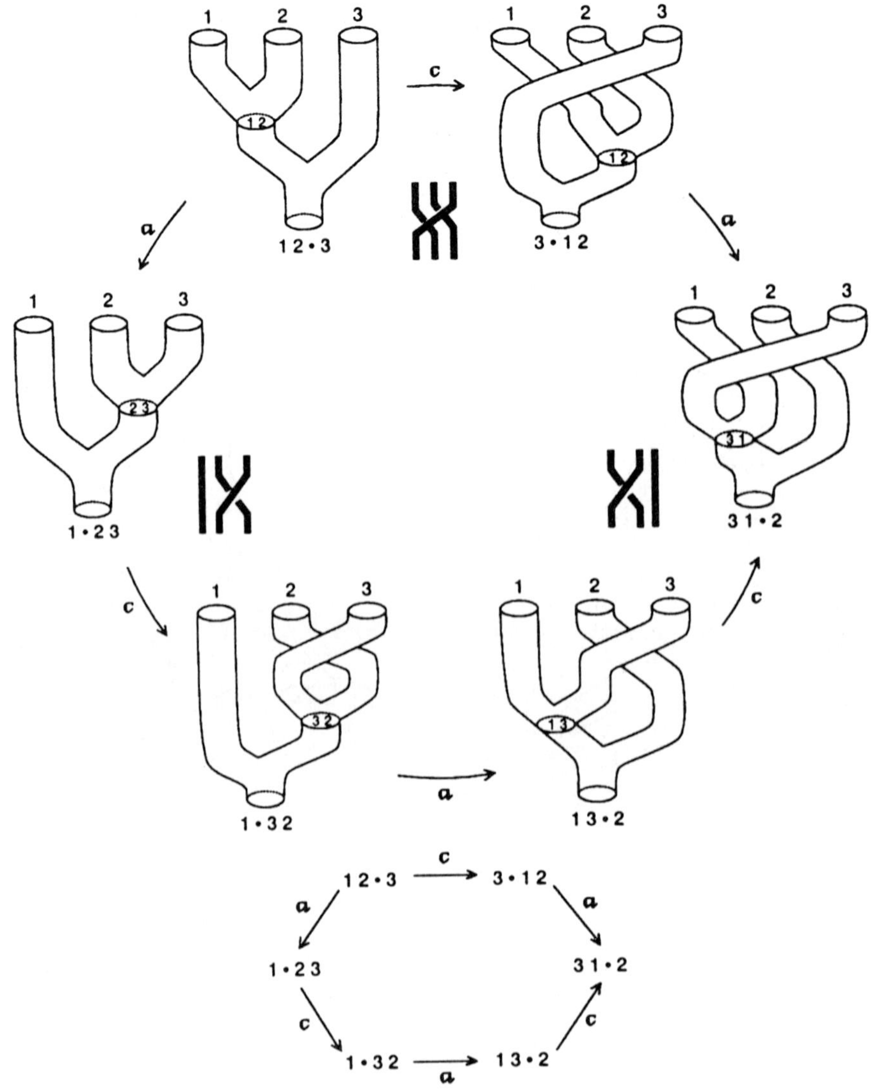

FIGURE 5. The Hexagon in Pants

norm $||f||$ for each $f : A \to B$ such that, for composites,

$$||g \circ f|| \le ||g|| \cdot ||f||.$$

Moreover, there is an operation of conjugation

$$* : \mathrm{hom}(A, B) \to \mathrm{Hom}(B, A)$$

with the property

$$||f^* \circ f|| = ||f||.$$

With a strictly associative tensor product, as for representations, such a category is called a strict symmetric monoidal C^*-category. It has three further special properties. First, $\hom(I, I)$ is the field of complex numbers. Second, any two objects A and B have a direct sum $A \oplus B$

$$A \xleftarrow{f} A \oplus B \xrightarrow{g} B$$

with $ff^* + gg^* = 1$. Third, subobjects exist; for any projection $f : A \to A$ there is a $g : A \to A$ with $f = g^* \circ g$. The essential result now is that any such category is a category $\operatorname{Rep} G$ for a unique G.

5. Tannakian Categories

A different extension of the Tannaka-Krein duality was initiated by Saavedra-Rivano [1972]. Certain imperfections in his work were subsequently corrected by P. Deligne and J. S. Milne [1982] They consider the category $\operatorname{Rep}(G)$ of all representation R of an affine group scheme G in a (complex) vector space W. This is an abelian compact closed category with the evident internal hom functor $[R, S]$, and with

$$[R, S] \otimes [R', S'] \cong [R \otimes R', S \otimes S'].$$

The forgetful functor ω from Rep G to the category of vector spaces is faithful, exact, closed, and respects isomorphisms and direct sums; this serves to characterize Rep G among such categories.

This result has been reformulated by Deligne, [1990], for the case of algebraic groups. There are also remarkable connections of category theory with Hopf algebras and the closely associated quantum groups. There are generalizations of braided categories to the so-called tortile tensor categories; the essential idea is to replace the strings of a braid by (possibly twisted) ribbons, as in Shum [1990].

6. Curves On A Three Fold

Initiatives from theoretical physics have also arisen in algebraic geometry. The classical result about the 27 lines on a cubic surface has extended to the study of lines and curves on the standard quartic three-fold in four-dimensional complex projective space, given by the equation, for a constant ψ,

$$x_1^2 + x_2^2 + x_3^2 + x_4^2 + x_5^2 - \psi x_1 x_2 x_3 x_4 x_5 = 0$$

in homogeneous coordinates x_1, x_2, x_3, x_4, x_5. Then J. Harris in 1975, using Galois theory, showed that there are 2875 lines on this three-fold. Subsequent results giving the number of rational quadratic curves on this three-fold and the more general studies of S. Mori are summarized in a paper of J. Kollar [1989]. Now a preprint by Candelas, de la Ossa, Green and Parker uses Calabi-Yau manifolds to get a conjecture as to the number of rational curves of any degree k on this three-fold. For $k = 5$ they obtain the number

$$n_5 = 22930\,58888\,87625.$$

These numbers are given up to $n = 10$, in which case there are 30 digits. This latter number is given exactly, although the argument depends on an unproven assumption about a certain prefactor.

The attached bibliography is a first indication of some of the many additional sources for these developments.

References

Atiyah, M., *Topological quantum field theories*, Inst. Hautes Etudes Sci. Publ. Math. **68** (1988), 175–186..

Brustein, R., Ne'eman, Y., and Sternberg, S., *Duality, crossing and Mac Lane's coherence*, Israel Journal of Mathematics **72** (1990), 19–37.

Candelas, P., de la Ossa, X. C., Green, P. S., and Parkes, L., *A pair of Calabi-Yau manifolds as an exactly soluble super conformal theory*, Univ. of Texas, (preprint, UTTG-25-19W)

Deligne, P., *Categories Tannakiens*, Grothendieck Festschrift, pp 114–196 (P. Cartier, et. al., eds.), vol. II, Birkhauser Boston, 1990..

Deligne, P. and Milne, J. S., *Tannakian Categories*, Hodge Cycles, Motives, and Shimura Varieties, Lecture Notes in Math., vol. 900, Springer Verlag, Heidelberg, 1982, pp. 101–128.

Doplicher, S., *Abstract compact group duals, operator algebras and quantum field theory*, Proc. Int. Cong. Math., Kyoto (1990) (to appear).

Doplicher, S. and Roberts, J. E., *A new duality for compact groups*, Invent. Math. **89** (1989), 157–218.

Harris, J., *Galois groups of enumerative problems*, Duke Math. J. **46** (1979), 685–724.

Johnson, M., *The combinatorics of n-categorical pasting*, J. Pure Appl. Algebra **62** (1989), 211–225.

Joyal, A. and Street, R., *Tortile Yang-Baxter operators in tensor categories*, J. Pure Appl. Algebra **71** (1991), 43–51.

Joyal, A. and Street, R., *Braided tensor categories*, Advances in Mathematics (to appear).

Joyal, A. and Street, R., *An Introduction to Tannaka duality and quantum groups*, Proceedings Como 1990, (A. Carboni et. al., eds.), vol. 1488, Springer Lecture Notes in Mathematics, 1991., pp. 413–492.

Kapranov, M. M. and Voevodsky, V. A., *Braided monoidal 2-categories, 2-vector spaces, and Zamolodchikov's tetrahedral equations*, (preprint) (1991).

Kollar, J., *The structure of algebraic three-folds: An Introduction to Mori's Program*, Bull. AMS **17** (1989), 211–273.

Mac Lane, S., *Categories for the Working Mathematician*, Springer Verlag, New York, 1971.

Majid, S., *Some physical applications of category theory*, Dept. Appl. Math. & Theoretical Physics, Cambridge, England (1991), (Preprint).

Majid, S., *Quasi-Triangular Hopf algebras and Yang-Baxter equations*, Int. J. Mod. Phys. **A5** (1990), 1–91.

Majid, S., *Tannaka-Krein theorem for quasi-Hopf algebras and other results*, Proc. Conference on Deformation Theory of Algebras and Quantization, 1990, (to appear).

Moore, G., and Seiberg, N., *Classical and quantum field theory*, Comm. Math. Phys. **123** (1989), 177–254.

Power, J., *Pasting diagrams in 2-categories*, Journal of Algebra **129** (1990), 439–445..

Rieffel, Mark A., *Category theory and quantum field theory*, (1990.), (preprint).

Saavedra Rivano, N., *Categories Tannakiennes*, SLN #285 418pp, Springer Verlag, 1972.

Segal, G., *The Definition of conformal field theory, Differential Geometry methods in theoretical physics*, Nat'l. Adv. Sci. Inst. Ser C Math. Phys. Sci., 1987, pp. 165–171; 250 Kluwer Acad. Publ. Dordrecht, 1988.

Shum, M. C., *Tortile tensor categories*, Macquarie Mathematical Reports, no. no 90 0047, (Apr. 1990).

Walker, Kevin, *On Witten's 3-manifold invariants*, (Preliminary version) Preprint 1990.

Witten, E., *Introduction to Cohomological Field Theories*, IASSNS-HEP, 90/57, (preprint).

Witten, E., *Topological quantum field theory*, Comm. Math. Phys. **117** (1988), 353–386.

Yetter, D. N., *Framed Tangles and a theorem of Deligne on braided deformation of Tannakian categories* (1990), (preprint).

Zamolodchikov, A. B., *Tetrahedral equations and integrable systems in three-dimensional space*, Soviet Physics, JETP **52** (1980), 325–336.

Department of Mathematics
The University of Chicago

Canadian Mathematical Society
Conference Proceeding
Volume **13**, 1992

BRAIDED GROUPS AND DUALS OF MONOIDAL CATEGORIES[1]

SHAHN MAJID

ABSTRACT Let $F : \mathcal{C} \rightarrow \mathcal{V}$ be a functor between monoidal categories. Previously we have shown that there is a dual monoidal category $\mathcal{C}^\circ$ of representations of $\mathcal{C}$ in $\mathcal{V}$ and a functor $\mathcal{C} \to \mathcal{C}^{\circ\circ}$ generalizing the Pontryagin duality of Abelian groups. If $\mathcal{V}$ is braided then there are also such notions as a coadjoint action of $\mathcal{C}$ on $\mathcal{C}^\circ$ and cross products by it. In another context we have shown in this situation (in the rigid case) that there is a Hopf algebra $\mathrm{Aut}\,(\mathcal{C}, F, \mathcal{V})$ in a cocompletion of $\mathcal{V}$ together with a functor $\mathcal{C} \to \mathrm{Aut}\,(\mathcal{C}, F, \mathcal{V})$-comodules (the category of comodules in $\mathcal{V}$). We report on these results and explore their relationship. We show that $\mathcal{C}^\circ \cong \mathrm{Aut}\,(\mathcal{C}, F, \mathcal{V})$-modules (modules in $\mathcal{V}$) and explain how these categorical duality constructions can be applied in an algebraic context to obtain new results about ordinary quasitriangular Hopf algebras (quantum groups), as well as their significance for physics in the context of quantum-gravity.

1 Introduction

This paper describes a construction of the author in [17] of the dual monoidal category of a monoidal category, and develops it further. We begin in Section 2 by recalling this construction and giving a number of details (including diagrammatic proofs) that were omitted in [17] for reasons of space. We then proceed in later sections to connect this construction with the author's recent work on braided groups and quantum braided groups[19], in the case when the monoidal category is braided. The braided group $\mathrm{Aut}(\mathcal{C})$ of a braided monoidal category $\mathcal{C}$ is a group-like object and expresses in its multiplication the monoidal product of $\mathcal{C}$. Section 3 begins by summarizing this construction (with full details to be found in the preprint [19]), and proceeds to show that in this case $\mathcal{C}^\circ$ can be identified with the category of $\mathrm{Aut}(\mathcal{C})$-modules. This thereby develops our

[1]1991 Math Subject Classification 18D10, 18D35, 16W30, 57M25, 17B37, 81R50, 83C45
SERC Research Fellow and Drapers Fellow of Pembroke College, Cambridge
This paper is in final form and no version of it will be submitted for publication elsewhere

0731-1036/92 \$1.00 + \$.25 per page

duality work of [17] in the braided group context, and is the modest new result of the paper. The results hold generally over a base category $\mathcal{V}$.

These categorical constructions are all fairly straightforward once the right definitions are found. Yet they have important and non-trivial applications. We focus on the applications to Drinfeld's theory of quasitriangular Hopf algebras (quantum groups). Section 4 begins by recalling these and then proceeds to describe some non-trivial corollaries of our general constructions from Sections 2 and 3. The main result is a new category-theoretic proof of the result of [22] that Drinfeld's quantum double $D(H)$ of a quantum group H is a semidirect product. This was proven by algebraic means, using an involved theorem of Radford [26]; we recover it now as a simple corollary of the main result of Section 3, applied in the case where $\mathcal{C}$ is the category of H-modules. The semidirect product explicitly takes the form $\mathrm{Aut}(\mathcal{C})\rtimes H$.

We note that a number of other applications of braided groups in this same context of quantum groups have recently been given in [20][21](with further applications in physics[15][16]). These applications use the general construction $\mathrm{Aut}(\mathcal{C}, F, \mathcal{V})$ (i.e., over base $\mathcal{V}$) as a way to shift the category in which an algebraic structure is defined. We begin with an algebraic structure of one sort, and use it to generate a category $\mathcal{C}$. By targeting $\mathcal{C}$ in another category $\mathcal{V}$ we can then transfer our original structure to one of the form $\mathrm{Aut}(\mathcal{C}, F, \mathcal{V})$ in $\mathcal{V}$. This process of 'transmutation' can be very useful in physics. For example, we may be presented with a group or quantum group in an ordinary sense but which lives more naturally (and looks simpler) when transmuted to the category of super-spaces, or vice-versa[15]. For a braided example, we have shown in [16] how the Weyl algebra of quantum mechanics of a one-dimensional particle lives more naturally in a braided category, where it corresponds simply to the braided line. These applications are all connected with braided monoidal categories and Yang-Baxter equations etc, which are quite a popular topic in physics and certain areas of mathematics. However, from our point of view they are a special case of more general duality considerations of quite wide validity.

It is not within our scope to discuss these detailed applications here. Instead, the paper concludes in Section 5 with a brief outline for category theorists of a vision of physics based on these duality considerations and category theory. This section is strictly optional but we hope it will be interesting for a reader who wants to know why the author, with a background in physics, should be led into category theory. Roughly speaking, one can argue[24] that complete theories of physics have two aspects, a combinatorial (quantum) and a geometrical, and these are in duality. During this conference I benefited from discussions with F.W. Lawvere and A. Kock who convinced me that there were some connections between these duality ideas and Lawvere's notions of 'extrinsic' and 'intrinsic' in topos theory. To explore such a connection would certainly be an interesting line for further work.

2 Duals of Monoidal Categories

In this section we recall the notion of duals of braided monoidal categories, with details of some of the proofs omitted in our report [17], and some further results. We use the standard definitions of a monoidal category and (weak) monoidal functor as in the text of Mac Lane[12]. Such a functor $F:\mathcal{C}\to\mathcal{V}$ means $F(X)\otimes F(Y)\cong F(X\otimes Y)$ by functorial isomorphisms which we denote $c_{X,Y}$, compatible with the associativity isomorphisms Φ in the two categories. These functorial associativity isomorphisms $\Phi_{X,Y,Z}: X\otimes(Y\otimes Z)\to(X\otimes Y)\otimes Z$ and $\Phi_{U,V,W}: U\otimes(V\otimes W)\to(U\otimes V)\otimes W$ will generally be suppressed for clarity, as also the isomorphisms $X\otimes\underline{1}\cong X$, $\underline{1}\otimes X\cong X$ for the unit object $\underline{1}$ (and similarly for the unit in $\mathcal{V}$). Throughout the paper, X, Y, Z etc denote general objects of $\mathcal{C}$ and U, V, W etc general objects of $\mathcal{V}$.

Definition 2.1 *[17] Let $F:\mathcal{C}\to\mathcal{V}$ be a monoidal functor between monoidal categories. We define a right $(\mathcal{C},F,\mathcal{V})$-module in $\mathcal{V}$ to be a pair (V,λ_V) where V is an object of $\mathcal{V}$ and $\lambda_V\in \mathrm{Nat}(V\otimes F, F\otimes V)$ is an invertible natural transformation $\lambda_{V,X}: V\otimes F(X)\to F(X)\otimes V$ obeying*

$$\lambda_{V,\underline{1}}=\mathrm{id},\qquad (\mathrm{id}\otimes\lambda_{V,Y})(\lambda_{V,X}\otimes\mathrm{id})=(c_{X,Y}^{-1}\otimes\mathrm{id})\lambda_{V,X\otimes Y}(\mathrm{id}\otimes c_{X,Y}).$$

This clearly 'represents' the monoidal product in a natural way, and generalizes the definition of representations of groups and Hopf algebras as we will see in the next section. The main results are

Theorem 2.2 *[17] The collection of right $(\mathcal{C},F,\mathcal{V})$-modules in Definition 2.1 form a monoidal category, denoted $\mathcal{C}^\circ$. The morphisms $(V,\lambda_V)\to(W,\lambda_W)$ are morphisms $\phi: V\to W$ in $\mathcal{V}$ such that the modules are intertwined in the form*

$$(\mathrm{id}\otimes\phi)\lambda_{V,X}=\lambda_{W,X}(\phi\otimes\mathrm{id}),\qquad \forall X \text{ in } \mathcal{C}.$$

The monoidal structure has identity $(\underline{1},\lambda_{\underline{1}})$, $\lambda_{\underline{1},X}=\mathrm{id}$ and product

$$(V,\lambda_V)\otimes(W,\lambda_W)=(V\otimes W,\lambda_{V\otimes W}),\ \ \lambda_{V\otimes W,X}=(\lambda_{V,X}\otimes\mathrm{id})(\mathrm{id}\otimes\lambda_{W,X}).$$

Proposition 2.3 *[17] In the setting of the preceding theorem, there is a monoidal functor $\psi:\mathcal{C}\to\mathcal{C}^{\circ\circ}$ defined tautologically by*

$$X\mapsto(F(X),\lambda_{\psi(X)}),\qquad \lambda_{\psi(X),(V,\lambda_V)}=\lambda_{V,X}^{-1}.$$

There is also a notion of left $(\mathcal{C},F,\mathcal{V})$-module, the collection of which is denoted ${}^\circ\mathcal{C}$ and a functor $\mathcal{C}\to{}^\circ(\mathcal{C}^\circ)$ given by $\lambda_{V,X}$ without inversion in Proposition 2.3. In these results it is clear that there are also monoidal functors $F^\circ:\mathcal{C}^\circ\to\mathcal{V}$ etc (given by the forgetful functor) so that $(\mathcal{C}^\circ,F^\circ,\mathcal{V})$ and $({}^\circ\mathcal{C},{}^\circ F,\mathcal{V})$ are dual systems of the same type as $(\mathcal{C},F,\mathcal{V})$. We can think of $\mathcal{V}$ as a fixed category over which $\mathcal{C},\mathcal{C}^\circ,{}^\circ\mathcal{C},\mathcal{C}^{\circ\circ}$ etc are functored. We consider a morphism between such triples as a monoidal functor between the first members that is compatible with the functors to $\mathcal{V}$. For example, the functor $\psi:\mathcal{C}\to\mathcal{C}^{\circ\circ}$ is such a morphism. Such triples and morphisms form a category MON/$\mathcal{V}$ which provides the setting in which we work. In such terms, Theorem 2.2 provides a contravariant self-adjoint endofunctor of MON/$\mathcal{V}$ and ψ in Proposition 2.3 then arises from the adjunction.

The proof of these results is in [17]. In fact, once the right statements are known, it is not hard to verify them by elementary and standard diagram filling. The results show clearly the similarity between the construction of $\mathcal{C}^\circ$ and the classical duality of Abelian groups, with their canonical inclusion $G \subseteq \hat{\hat{G}}$. However, by generalizing this Pontryagin duality to monoidal categories we extend it to non-Abelian groups also (as well as beyond these to include Hopf algebra duality). Many constructions for groups and Hopf algebras can be extended in the same way to the setting of monoidal categories over a base. An example of such a result, when the base is braided, is

Theorem 2.4 *[17] Given $(\mathcal{C}, F, \mathcal{V})$ where $\mathcal{V}$ is braided monoidal in the sense of [8] with braiding Ψ, there is a functor $\alpha : \mathcal{C} \times \mathcal{C}^\circ \to \mathcal{C}^\circ$ (compatible with F, F°) and isomorphisms of the form $\alpha(X, \alpha(Y, \ )) \cong \alpha(X \otimes Y, \ )$ for all X, Y in $\mathcal{C}$, i.e. the two functors*

$$\alpha(\mathrm{id} \times \alpha), \alpha(\otimes \times \mathrm{id}) : \mathcal{C} \times \mathcal{C} \times \mathcal{C}^\circ \to \mathcal{C}^\circ$$

are naturally equivalent. We call α the left coadjoint action *of $\mathcal{C}$ on $\mathcal{C}^\circ$. It is given tautologically by $\alpha(X, (V, \lambda_V)) = (F(X) \otimes V, \lambda_{\alpha(X,(V,\lambda_V))})$ with*

$$\lambda_{\alpha(X,(V,\lambda_V)),Y} = (\mathrm{id} \otimes \lambda_{V,X})(\lambda_{V,Y} \otimes \mathrm{id})(\mathrm{id} \otimes \Psi_{F(X),F(Y)})(\lambda^{-1}_{V,X} \otimes \mathrm{id}).$$

Proof This (and similar proofs) were omitted in [17] for lack of space. We give it now in detail as a demonstration of the new techniques needed when working with monoidal categories in this way. Recall first that a braided monoidal category $\mathcal{V}$ is a monoidal one equipped with a braiding or 'quasisymmetry' $\Psi_{V,W} : V \otimes W \to W \otimes V$ obeying two hexagon conditions and compatible with $\underline{1}$. We do not assume $\Psi_{W,V} \Psi_{V,W} = \mathrm{id}$. Omitting the associativity, the hexagons take the form $\Psi_{V,W \otimes U} = \Psi_{V,U} \Psi_{V,W}$ and $\Psi_{V \otimes W,U} = \Psi_{V,U} \Psi_{W,U}$. This is well known by now, as is the resulting coherence theorem for braided categories[8]. A useful technique for our proofs is a diagrammatic one. In this, all morphisms are written pointing downwards, Ψ and Ψ^{-1} are written as braid crossings and other morphisms are written in boxes or as nodes. Functoriality of Ψ, Ψ^{-1} says that such boxes or nodes can be translated through braid crossings. Recently, such short-hand notation has been given an ambient topological interpretation in [9]. Thus the braiding and Definition 2.1 take the form,

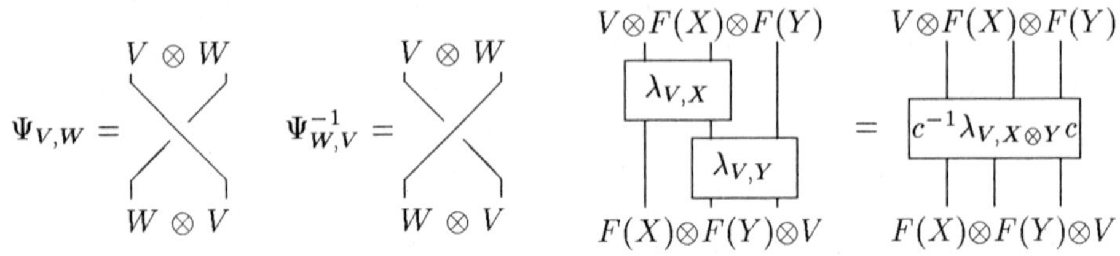

The proof that $\alpha(X, (V, \lambda_V))$ obeys Definition 2.1 is then easily seen from the diagram in Figure 1. The upper three boxes on the left are the definition of $\lambda_{\alpha(X,(V,\lambda_V)),Y}$ and the lower three of $\lambda_{\alpha(X,(V,\lambda_V)),Z}$. After cancelling inverses and using the braid relation for Ψ we obtain $c^{-1} \lambda_{\alpha(X,(V,\lambda_V)),Y \otimes Z} c$ as required. The action of α on morphisms is the obvious one given by $F \otimes \mathrm{id}$, with its result viewed as a morphism in $\mathcal{C}^\circ$. The proof that this functor α is a left action in the form stated is more complicated and shown in Figure 2. On the left is the

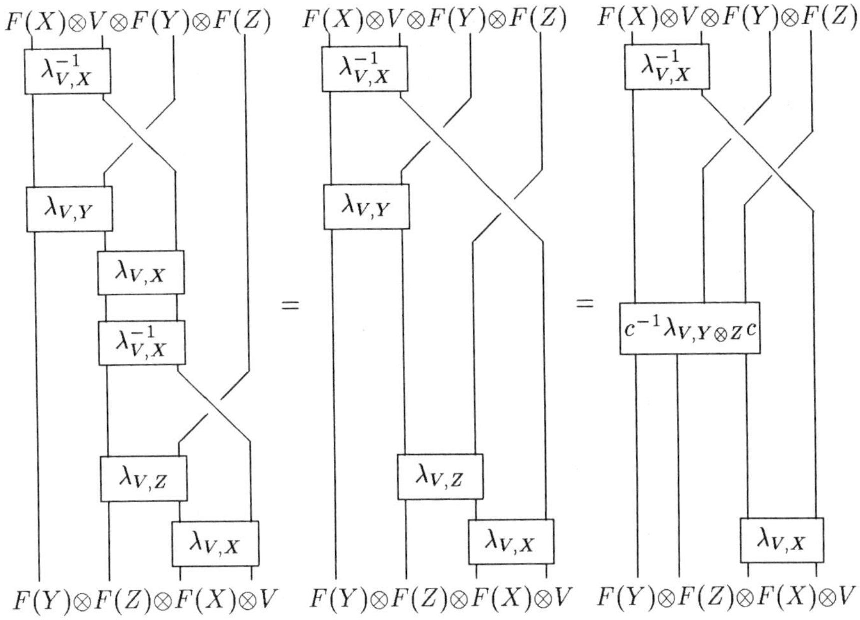

Figure 1: Proof that coadjoint action $\mathcal{C}\times\mathcal{C}^{\circ}\to\mathcal{C}^{\circ}$ is defined

definition of

$$\lambda_{\alpha(X,\alpha(Y,(V,\lambda_V))),Z}=\lambda_{\alpha(Y,(V,\lambda_V)),X}\lambda_{\alpha(Y,(V,\lambda_V)),Z}\Psi_{F(X),F(Z)}\lambda^{-1}_{\alpha(Y,(V,\lambda_V)),X}$$

which, after cancellation, the braid relations and Definition 2.1 for $\lambda_{V,X\otimes Y}$ computes to $c^{-1}\lambda_{\alpha(X\otimes Y,(V,\lambda_V)),Z}c$. Thus the natural equivalence required in the theorem is induced by c. Functoriality follows at once from that of c. This completes the proof.

The ordinary coadjoint action plays a fundamental role in many applications of groups in physics (in the realm of quantization), so it is interesting that it generalizes to our setting. An application in category theory was given in [17] where we used it to construct a double cross product $\mathcal{C}\bowtie\mathcal{C}^{\circ}$ in the case when $\mathcal{C}\cong\mathcal{C}^{\circ\circ}$. Here there is an action $\beta:\mathcal{C}\times\mathcal{C}^{\circ}\to\mathcal{C}$ analogous to the above and $\mathcal{C}_{\beta}\bowtie_{\alpha}\mathcal{C}^{\circ}$ is a double semidirect product by these two mutual coadjoint actions simultaneously (along the lines of [23, Sec. 3.2] for groups and Hopf algebras). When it exists, it is a monoidal category that non-trivially factorizes into the monoidal categories $\mathcal{C}$ and $\mathcal{C}^{\circ}$, and can be called the *outer quantum double* $D(\mathcal{C})$ of $\mathcal{C}$.

Returning now to the general situation, let us describe some results that do not have familiar group analogues, i.e. some genuinely new phenomena. One such canonical fact is that the dual in the special case $Z(\mathcal{C})=(\mathcal{C},\mathrm{id},\mathcal{C})^{\circ}$ is tautologically braided (with $\Psi_{(V,\lambda_V),(W,\lambda_W)}=\lambda_{V,W}$). This has been observed by Drinfeld[4] in another context, as well as independently in [10]. This $Z(\mathcal{C})$ can be called the *inner quantum double* and provides a way to generate braided monoidal categories from monoidal ones. We note also that our general $\mathcal{C}^{\circ}$ construction has subsequently been generalized still further in [11] in the setting

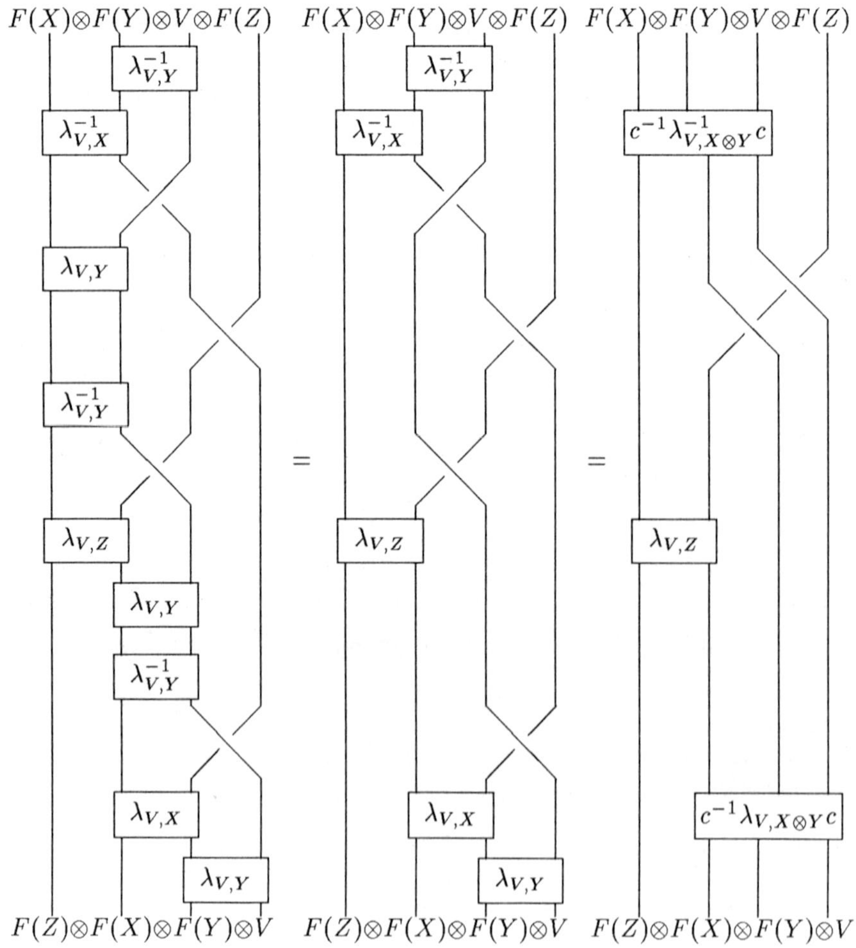

Figure 2: Proof that coadjoint action $\mathcal{C} \times \mathcal{C}^{\circ} \to \mathcal{C}^{\circ}$ is an action

of 2-categories. For a different type of result we observe also

Proposition 2.5 *Given $(\mathcal{C}, F, \mathcal{V})$ with $\mathcal{C}$ braided there is a monoidal functor $R : \mathcal{C} \to \mathcal{C}^{\circ}$ defined tautologically by*

$$R(X) = (F(X), \lambda_{R(X)}), \quad \lambda_{R(X),Y} = c^{-1}_{X,Y} F(\Psi_{X,Y}) c_{X,Y}, \qquad \forall X, Y \text{ in } \mathcal{C}.$$

Proof The output of R lies in $\mathcal{C}^{\circ}$ due to F applied to one of the hexagon identities for the braiding Ψ of $\mathcal{C}$, and is a monoidal functor due to F applied to the other. The other facts are equally easy.

3 Braided Groups

We have said that the duality constructions above have been modeled on Hopf algebras, with $\mathcal{C}^{\circ}$ considered heuristically as the modules of some Hopf algebra whose multiplication corresponds to the monoidal product of $\mathcal{C}$. In nice cases this can be made precise, i.e. the underlying Hopf algebra for the data $F : \mathcal{C} \to \mathcal{V}$ actually exists and $\mathcal{C}^\circ$ can be identified with its modules. This is the main result of the present section. The necessary Hopf algebra (in $\mathcal{V}$) that we need here has already been introduced in [19], cf[18] and we begin by briefly recalling it.

Recall that an algebra (or monoid) in a monoidal category is an object A and a morphism $A \otimes A \to A$ which is associative, and has a suitable unit morphism $\underline{1} \to A$. A coalgebra (comonoid) is the same thing with the arrows reversed. It has a comultiplication $\Delta : A \to A \otimes A$ and counit $\epsilon : A \to \underline{1}$. A bialgebra (bimonoid) is usually defined in a symmetric monoidal category and has both of these structures with the additional condition that Δ,ϵ are algebra homomorphisms. Here $A \otimes A$ has the tensor product algebra structure $(A \otimes A) \otimes (A \otimes A) \to A \otimes A$ defined by applying first the symmetry $\Psi_{A,A}$ in the middle two factors and then multiplications for A. This definition can be adopted also in the case when Ψ is a braiding not a symmetry, i.e. $A \otimes A$ is still associative. Bialgebras in braided monoidal categories were introduced in this way, and studied, in [18]. A Hopf algebra (Hopf monoid) is a bialgebra with an additional antipode morphism $S : A \to A$ playing the role of inversion. Finally, a (right) module of A is an object V and a morphism $V \otimes A \to V$ obeying the obvious axioms. A (right) comodule is an object V and a morphism $V \to V \otimes A$ obeying analogous axioms with the arrows reversed. It is clear that when the category is rigid, i.e. there are suitable dual objects, then A^* is also a Hopf algebra and comodules of A^* correspond to modules of A of appropriate (left or right) form.

Theorem 3.1 *[19], cf[18] Let $F : \mathcal{C} \to \mathcal{V}$ be as in Section 2, with $\mathcal{V}$ braided, rigid and cocomplete. Then there is a bialgebra* $\mathrm{Aut}(\mathcal{C}, F, \mathcal{V})$ *in $\mathcal{V}$, characterized as universal with the property that F factors through its category of right comodules in $\mathcal{V}$ via the forgetful functor. If $\mathcal{C}$ is also rigid then* $\mathrm{Aut}(\mathcal{C}, F, \mathcal{V})$ *is a Hopf algebra in $\mathcal{V}$.*

This $\mathrm{Aut}\,(\mathcal{C}, F, \mathcal{V})$ can be realized explicitly (cf [28][2]) as a representing object for a certain functor to Sets or, in the language of [12], the coend

$$\mathrm{Aut}\,(\mathcal{C}, F, \mathcal{V}) = \int^X F^*(X) \otimes F(X).$$

Here F^* is a contravariant functor provided by the rigidity in $\mathcal{V}$ and cocompleteness in needed for the existence of the coend to be assured. The $\mathrm{Aut}\,(\mathcal{C}, F, \mathcal{V})$ construction provides a functor from $\mathrm{MON}/\mathcal{V}$ to the category of bialgebras in $\mathcal{V}$ (giving Hopf algebras on rigid objects) and also behaves well under change of the base $\mathcal{V}$. In another variant we can fix $\mathcal{C}$ rigid instead of $\mathcal{V}$. Details are in [19].

The factorization of F in the theorem is via a monoidal functor $\mathcal{C} \to \mathcal{V}^{\mathrm{Aut}\,(\mathcal{C},F,\mathcal{V})}$ where $\mathcal{V}^{\mathrm{Aut}\,(\mathcal{C},F,\mathcal{V})}$ denotes the category of right $\mathrm{Aut}\,(\mathcal{C}, F, \mathcal{V})$ - comodules in $\mathcal{V}$. Thus $\mathcal{C}$ is mapped to the comodules of this underlying Hopf algebra. However, except in special cases (such as $\mathcal{V}$=Vec), this functor is not an equivalence. Thus we are not dealing here with a (braided version of) a full Tannaka-Krein reconstruction theorem as in [28][2][25]. Nevertheless, Aut $(\mathcal{C}, F, \mathcal{V})$ exists and contains a lot of information about the system $(\mathcal{C}, F, \mathcal{V})$. Indeed, we now show

Theorem 3.2 *In the situation of Theorem 3.1, there is an isomorphism of monoidal categories*

$$\mathcal{C}^\circ \cong \mathcal{V}_{\mathrm{Aut}(\mathcal{C},F,\mathcal{V})}$$

where $\mathcal{V}_{\mathrm{Aut}(\mathcal{C},F,\mathcal{V})}$ *denotes the category of right* $\mathrm{Aut}(\mathcal{C}, F, \mathcal{V})$*-modules in* $\mathcal{V}$.

Proof Firstly, rigidity of $\mathcal{V}$ means more precisely that for every object V there is a dual V^* and morphisms $\pi_V : \underline{1} \to V\otimes V^*$, $\mathrm{ev}_V : V^*\otimes V \to \underline{1}$ obeying some axioms. Omitting the associativity morphisms etc as usual, we require that $(\mathrm{id}\otimes \mathrm{ev}_V)(\pi_V\otimes \mathrm{id})$ and $(\mathrm{ev}_V\otimes \mathrm{id})(\mathrm{id}\otimes\pi_V)$ compose to the identity on V and V^* respectively. Using these maps it is not hard to see that there is a bijection

$$\hom(W\otimes U^*\otimes T, V) \cong \hom(W\otimes T, U\otimes V) \tag{1}$$

for any objects T, U, V, W in $\mathcal{V}$. The proof is similar to Lemma 2.1 of [19], for example if $f : W\otimes U^*\otimes T \to V$ is given, the corresponding morphism on the right is

$$W\otimes T \xrightarrow{\pi_U} W\otimes U\otimes U^*\otimes T \xrightarrow{\Psi_{W,U}} U\otimes W\otimes U^*\otimes T \xrightarrow{f} U\otimes V.$$

The automorphism bialgebra $A = \mathrm{Aut}(\mathcal{C}, F, \mathcal{V})$ was defined in [19], cf[18] as the representing object of a functor $V \mapsto \mathrm{Nat}(F, F\otimes V)$, i.e. there is a bijection $\hom(\mathrm{Aut}\,(\mathcal{C}, F, \mathcal{V}), V) \cong \mathrm{Nat}(F, F\otimes V)$. We then used rigidity of $\mathcal{V}$ in [19] and an observation like (1) (with W trivial) to give the coend realization mentioned above. We now use (1) in a similar way to conclude that

$$\hom(V\otimes \mathrm{Aut}\,(\mathcal{C}, F, \mathcal{V}), V) \cong \mathrm{Nat}(V\otimes F, F\otimes V)$$

for all objects V. This gives the correspondence that we need for the theorem. Explicitly, if $\alpha : V\otimes A \to V$ is a morphism the corresponding natural transformation is $\lambda_{V,X} = (\mathrm{id}\otimes\alpha)\Psi_{V,F(X)}(\mathrm{id}\otimes\beta_X)$ where $\beta_X : F(X) \to F(X)\otimes A$ is the natural transformation in $\mathrm{Nat}(F, F\otimes A)$ corresponding to the identity morphism $A \to A$ (it is the A-comodule associated to X in Theorem 3.1). Figure 3 then shows that α a right module corresponds to λ_V a right $(\mathcal{C}, F, \mathcal{V})$-module in the sense of Definition 2.1. We use the same notation as in Figures 1 and 2 above (with the morphisms $\alpha, \beta, \cdot$ now written as nodes rather than boxes). The left hand side is the definition of $c^{-1}\lambda_{V,X\otimes Y}c$ after using $c^{-1}\beta_{X\otimes Y}c = (\mathrm{id}\otimes\cdot)\Psi_{A,F(Y)}(\beta_X\otimes\beta_Y)$ (this is the definition of the multiplication in A from [19, Lemma 2.4]). The second equality is functoriality of Ψ and the right hand side is $\lambda_{V,Y}\lambda_{V,X}$. The two sides are equal as shown if (and only if) α is a right A-module. The proof can also be done by more conventional diagram-filling. This completes the proof of the theorem.

This representation of $\mathcal{C}^\circ$ in terms of $\mathrm{Aut}(\mathcal{C}, F, \mathcal{V})$ is of course fully functorial in $\mathcal{C}$ and has a number of corollaries for the work of Section 2, at least in those nice cases where $\mathrm{Aut}(\mathcal{C}, F, \mathcal{V})$ exists. For example,

Corollary 3.3 *Let* $F : \mathcal{C} \to \mathcal{V}$ *as in Section 2, with both* $\mathcal{C}, \mathcal{V}$ *rigid and* $\mathcal{V}$ *braided. Then* $\mathcal{C}^\circ$ *is rigid.*

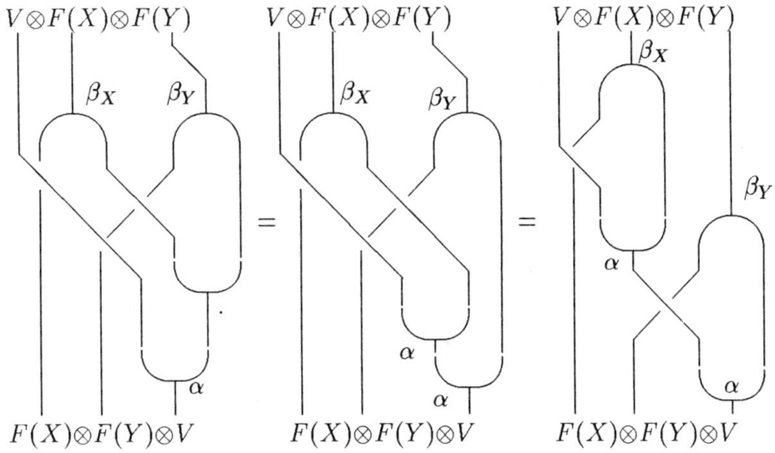

Figure 3: Diagram in proof of Theorem 3.2

Proof If Aut $(\mathcal{C}, F, \mathcal{V})$ actually lives in $\mathcal{V}$ (rather than a cocompletion), it is a Hopf algebra in $\mathcal{V}$ and hence its category of modules in $\mathcal{V}$ is also rigid. This is an elementary feature of the category of modules, or of comodules, of a Hopf algebra in a rigid braided monoidal category. The case of comodules (in the braided setting) was spelled out in [19, Lemma 1.1]. More generally we can argue by continuity. In fact the necessary maps π and ev are induced by those in $\mathcal{V}$ (viewed now as morphisms in $\mathcal{C}^{\circ}$) and hence serve quite generally.

In addition, if $\mathcal{C}$ is also braided then $\mathrm{Aut}(\mathcal{C}, F, \mathcal{V})$ is a (dual) quantum group, i.e. a dual quasitriangular bialgebra or Hopf algebra in $\mathcal{V}$, with various consequences for the structure of $\mathcal{C}^{\circ}$. We will discuss quantum groups in detail in the next section, in the simplest algebraic setting, and refer to [19] for the general case. Suffice it to say that 'dual quasitriangular' means that there is a morphism $\mathcal{R} : \mathrm{Aut}(\mathcal{C}, F, \mathcal{V}) \otimes \mathrm{Aut}(\mathcal{C}, F, \mathcal{V}) \to \underline{1}$ and $\mathrm{Aut}(\mathcal{C}, F, \mathcal{V})$ is 'commutative' in a certain sense (in the braided category) up to conjugation by this in the relevant convolution algebra. In addition, the morphism $\mathcal{R}$ in suitable conventions converts comodules to modules, along the lines of an extension of Proposition 2.5.

We mention that in general terms, the dual quasitriangular structure $\mathcal{R}$ of $\mathrm{Aut}(\mathcal{C}, F, \mathcal{V})$ measures the failure of F to respect the braidings: it is given by the ratio of the braidings in $\mathcal{C}$ and $\mathcal{V}$ and is trivial in the case Aut $(\mathcal{C})$ where $\mathcal{V}$ is a cocompletion of $\mathcal{C}$ and F is the identity. This Aut $(\mathcal{C})$ is therefore exactly 'commutative' in a certain braided sense and hence can be called a 'braided group' (in analogy with affine group schemes), while the general case can be called a 'quantum braided group'. See [19].

4 Applications to Quantum Groups

In this section we apply the category theoretic constructions above to obtain a result in the theory of quantum groups. Various applications of $\mathrm{Aut}(\mathcal{C}, F, \mathcal{V})$ and $\mathrm{Aut}(\mathcal{C})$ have already been made in this way in [20][21] as well as in [14][7].

Our new application is based on the connection with the duality of Section 2 as found in Theorem 3.2.

Briefly, an ordinary quantum group for us is a pair $(H, \mathcal{R})$ where H is a Hopf algebra over a field or commutative ring k, and $\mathcal{R} \in H\otimes H$ (the quasitriangular element) is invertible and obeys the axioms of Drinfeld[5]

$$\Delta^{\rm op} = \mathcal{R}(\Delta\ )\mathcal{R}^{-1}, \quad ({\rm id}\otimes\Delta)\mathcal{R} = \mathcal{R}_{13}\mathcal{R}_{12},\ (\Delta\otimes{\rm id})\mathcal{R} = \mathcal{R}_{13}\mathcal{R}_{23}.$$

Here $\Delta^{\rm op}$ denotes Δ followed by the usual permutation and $\mathcal{R}_{12} = \mathcal{R}\otimes 1, \mathcal{R}_{23} = 1\otimes\mathcal{R}$ (and similarly $\mathcal{R}_{13}$) as elements of $H\otimes H\otimes H$. If G is a group, its group algebra kG is trivially a quantum group with $\mathcal{R} = 1\otimes 1$ (it is cocommutative). Quantum groups are important in mathematics and physics because they provide, in their representation theory, a source of braided monoidal categories. The monoidal product (tensor product) of two H-modules is obtained by pulling back along the comultiplication Δ, in the usual way. The braiding Ψ is provided by the action of $\mathcal{R}$ followed by usual permutation of the underlying vector spaces.

If H is any Hopf algebra, say finite dimensional for simplicity, Drinfeld showed how to construct from it a quantum group $D(H)$. Most other quantum groups are quotients of such quantum doubles, so this is one of the key constructions of the theory. In the conventions of [22] it coincides as a linear space and coalgebra with $H^*\otimes H$, but has a modified algebra structure

$$(a\otimes h)(b\otimes g) = \sum b_{(2)}a\otimes h_{(2)}g < Sh_{(1)}, b_{(1)} >< h_{(3)}, b_{(3)} >,\ h, g \in H,\ a, b \in H^*$$

where $\Delta h = \sum h_{(1)}\otimes h_{(2)}$ etc [30] and $<\ ,\ >$ denotes evaluation. The quasitriangular structure is $\mathcal{R} = \sum_a (f^a\otimes 1)\otimes(1\otimes e_a)$ where $\{e_a\}$ is a basis of H and $\{f^a\}$ a dual one. The structure of $D(H)$ is a double cross product $H^{*\rm op} \bowtie H$[23, Sec. 4]. This is an example of the factorization of a Hopf algebra into two factors $H^{*\rm op}$ and H. It is clear that $D(H)$-modules are vector spaces on which both H and $H^{*\rm op}$ act, in a compatible way, (or on which H acts and also coacts, in a compatible way, i.e. crossed H-modules as in [31]). Moreover, if $\mathcal{C}$ is the category of H-modules then $Z(\mathcal{C}) = (\mathcal{C}, {\rm id}, \mathcal{C})^\circ$ in Section 2 is the category of $D(H)$-modules[4].

Knowing this, a corollary of Proposition 2.5 for example is that if H is itself a quantum group, then there is a Hopf algebra projection $p : D(H) \to H$ inducing the functor $\mathcal{C} \to \mathcal{C}^\circ$. It can be easily be computed in terms of $\mathcal{R}$, and was the starting point of [22]. Now, as a corollary of Theorem 3.2 we have,

Corollary 4.1 *Let H be a finite dimensional quasitriangular Hopf algebra (quantum group) and let $\mathcal{C}$ be the category of finite-dimensional H-modules. Then*

$$D(H) \cong {\rm Aut}(\mathcal{C}) \rtimes H$$

as a semidirect product algebra and semidirect product coalgebra by an action and coaction respectively of H.

Proof Identifying $\mathcal{C}^\circ$ as the category of $D(H)$-modules as explained, Theorem 3.2 now says that these correspond to Aut($\mathcal{C}$)-modules in $\mathcal{C}$. This means vector spaces V on which Aut($\mathcal{C}$) acts H-equivariantly. Here H acts on both V and Aut($\mathcal{C}$) as objects in the category. It is clear that such V are the same thing as Aut($\mathcal{C}$)$\rtimes H$-modules where the semidirect product is by the canonical action of H on Aut ($\mathcal{C}$). This establishes a correspondence between $D(H)$-modules and Aut($\mathcal{C}$)$\rtimes H$-modules, which one can easily see to be monoidal and compatible with the forgetful functors to Vec (i.e. it does not change the underlying vector space). As a result, by standard Tannaka-Krein ideas Aut($\mathcal{C}$)$\rtimes H$ is a Hopf algebra and isomorphic to $D(H)$. The coalgebra structure here comes out as a semidirect coproduct by the H-comodule structure induced by a functor from H-modules to H^*-modules (given by pull back along $\mathcal{R}$ as a map $H^* \to H$).

Let us note that the explicit form of Aut($\mathcal{C}$) in this example is known from [19]. Writing $A = H^*$ so that $\mathcal{C}$ is the category of A-comodules, we have from [19] that Aut ($\mathcal{C}$) $= \underline{A}$, realised on the same linear space and coalgebra as H^* but with a modified multiplication and a non-trivial H-module structure. Explicitly,

$$(\Delta\phi)(h\otimes g) = \phi(hg), \quad (g.\phi)(h) = \sum \phi((Sg_{(1)})hg_{(2)})$$

$$(\phi\psi)(h) = \sum (\mathcal{R}^{(2)}.\psi)(h_{(1)})\, \phi(h_{(2)}\mathcal{R}^{(1)})$$

for $h, g \in H$ and $\phi, \psi \in H^*$ and $\mathcal{R} = \sum \mathcal{R}^{(1)}\otimes\mathcal{R}^{(2)}$. This gives an algebra and coalgebra but not a Hopf algebra in the ordinary sense. This explains the structure of $D(H)$ found by algebraic means in [22]. There, using a theorem of Radford[26] it was shown that $D(H)\cong B\rtimes H$ where B was a mysterious algebra-coalgebra. We recognize it now as a Hopf algebra in a braided category. After this we can go further and express Radford's theorem itself in this language. It then reads: if $H_1 \to H$ is a Hopf algebra projection between ordinary Hopf algebras then there exists a bialgebra B in the category of $D(H)$-modules such that $H_1\cong B\rtimes H$ by the corresponding action and coaction of H.

Our Corollary 4.1 can also be viewed as an example of a general 'bosonization' theorem in [21]. Using ideas of transmutation mentioned in the introduction, this asserts that any Hopf algebra B in the braided monoidal category of H-modules of a quantum group H, can be converted by a tautological semidirect product to an ordinary Hopf algebra $B\rtimes H$. In these terms, Corollary 4.1 says that Drinfeld's celebrated quantum double $D(H)$ is the 'bosonization' of the braided group $\underline{A} = \mathrm{Aut}(\mathcal{C})$.

5 A Vision of Quantum-Gravity

This section aims to indicate how the above duality considerations fit into an approach to the unification of quantum theory and gravity in [23][24]. Such a unification is of course a fundamental problem in theoretical physics and I believe that category theory provides some fundamental insight into it.

The idea behind this application is very simple. Firstly, it should be stressed that theoretical physics is not so much concerned with 'what' is observed but with the question 'why is it observed', i.e. always seeks to explain structures in terms of still more fundamental structures or principles. This reductionist programme takes the naive view that there are indeed some fundamental laws of nature of which our experiments and observations are representations. Thus it is supposed that something is absolutely true, and that something else measures or observes it.

However, every mathematician knows that such evaluations should generally be thought of more symmetrically as a 'pairing' of one structure with another. An evaluation $f(x)$ can also be read $x(f)$ where f is an element of a dual structure. Since theoretical physics adopts the language of mathematics, such an 'observer-observed' reversed interpretation of the mathematical structure can always be forced, but will the dual interpretation also describe physics? The idea behind [23][24] was that the answer should be 'Yes', i.e. that a fundamental theory of physics is incomplete unless such a role-reversal is possible. We can go further and hope to fully determine the (supposed) structure of fundamental laws of nature among all mathematical structures by this self-duality condition. See [24] and the introduction of [23].

Certainly, such duality considerations in some form are evident in the context of quantum theory and gravity (and may also be relevant to recent developments of duality in string theory). The situation is summarized to the left in Figure 4. For example, Lie groups provide the simplest examples of Riemannian geometry, while the representations of related Lie groups provide the quantum numbers of elementary particles in quantum theory. Thus both quantum theory and non-Euclidean geometry are needed for a self-dual picture. Now, Hopf algebras (in the form of Hopf-von Neumann or Kac algebras) precisely serve to unify these mutually dual structures. Motivated by this we described in [24] systems of particles on homogeneous spaces where this was achieved. The algebra of observables was indeed a Hopf (von Neumann) algebra of self-dual type provided the homogeneous space and its dual one were suitably matched. The dual von Neumann algebra is built on the Banach space predual so that, roughly speaking, the role of observables and states in the quantum theory can be interchanged. For example, the quantum (von Neumann) algebra of observables of a $1+1$-dimensional differentiable dynamical system was shown to be a self-dual Hopf algebra if and only if the effective dynamics was that of a particle moving in the background of a 2-dimensional black hole-type metric.

The duality considerations of Section 2 and [17] are then motivated by the question of what is the right mathematical structure to unify more complex quantum and gravitational systems. We have established that (small) monoidal categories over a base make up a self-dual category, with the duality $\mathcal{C} \mapsto \mathcal{C}^{\circ}$ covering Hopf algebra duality (which in turn covers the duality operation on Abelian groups), and that one can begin to do physics in this setting (for example, in the form of the coadjoint action in Section 2). It should be mentioned here that the functors in Figure 4 are intended to be of the nature of embeddings and to respect the relevant duality functors. Moreover, the collection of monoidal categories over a base is large enough to receive various functors from both quantum systems (as in [3], as well as the modular functor in conformal field theory) and from Riemannian geometry. Precisely which functors we

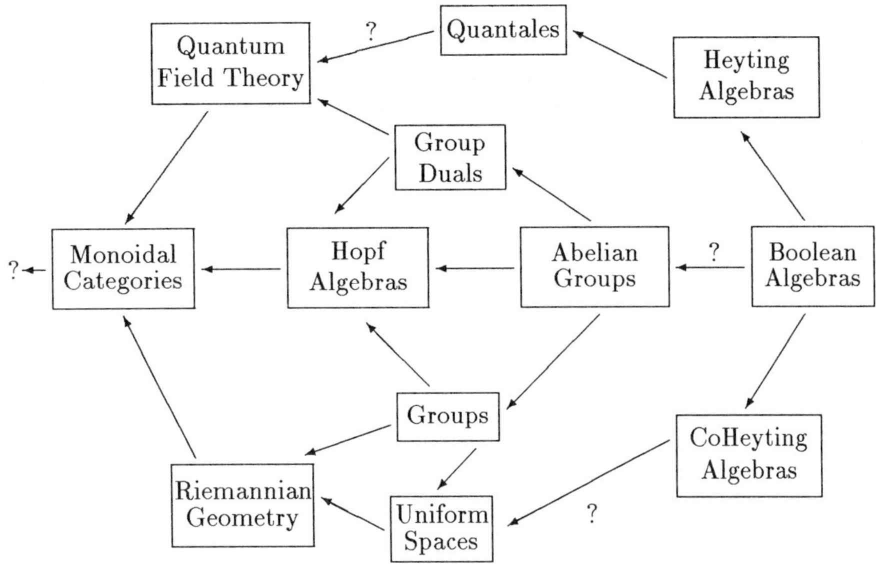

Figure 4: Representation-theoretic approach to quantum-gravity

should take here and the search for self-dual objects are topics for ongoing research. If they can be found then such self-dual objects would be interpretable as (possibly generalized) quantum theories and (possibly generalized) geometries, simultaneously. This is our approach to quantum-gravity.

Finally, we note that in addition to the 'end' of theoretical physics in the form of quantum-gravity, we can ask also about its 'birth'. We take the view that the simplest theories of physics are based on classical logic or, roughly speaking, Boolean algebras. Some physical (not mathematical) arguments in the last of [24] suggest that the relevant duality here is provided by complementation. The situation is summarized on the right in Figure 4. Boolean algebras are considered self-dual according to De Morgan's theorem. Going above the axis to Heyting algebras and beyond takes us into intuitionistic logic and presumably embeds (ultimately) into an axiomatic framework of quantum field theory. Dual to this, below the axis, F.W. Lawvere and his school have shown that coHeyting algebras can be viewed as the 'birth' of geometry[13]. Specifically, it has been observed that in a coHeyting algebra one can view $\partial a = a \wedge \tilde{\ } a$ as the 'boundary' of a, with the relevant derivation property[13]

$$\partial(a \wedge b) = (\partial a) \wedge b \vee a \wedge (\partial b).$$

Here $\tilde{\ } a$ is the smallest element such that $a \vee \tilde{\ } a = 1$. To develop this geometrical interpretation of cointuitionistic logic to embed (ultimately) in metric spaces and Riemannian (or Lorentzian) geometry is a task for further work.

Finally, we can ask if such problems as the unification of quantum theory and gravity have a useful parallel for the unification of (generalized) intuitionistic and (generalized) cointuitionistic logical systems. For example, this might be achieved using suitable forms of Hopf algebras or in the more general setting of monoidal categories. We note in this context that monoidal structures are now used routinely in various models of logic since the fundamental work of [6], see [29][1]. This unification is an interesting question suggested by the above. I would like to note the recent work [27] as a concrete step in this direction.

REFERENCES

[1] M. Barr. Accessible categories and models of linear logic *J. Pure Appl. Algebra*, 69:219-232, 1990.

[2] P. Deligne and J.S. Milne. Tannakian categories. *Springer Lect. Notes in Math.*, 900, 1982.

[3] S. Doplicher and J.E. Roberts. Why is there a field algebra with compact gauge group describing the superselection structure in particle physics? *Comm. Math. Phys.*, 131:51-107, 1990.

[4] V.G. Drinfeld. Private communication, February, 1990.

[5] V.G. Drinfeld. Quantum groups. In A. Gleason, editor, *Proceedings of the ICM*, pages 798–820, Rhode Island, 1987. AMS.

[6] J.-Y. Girard, Linear Logic, *J. Theor. Comp. Sci.*, 50:1-102, 1987; Towards a geometry of interaction, *Contemp. Math.*, 92:69-108, 1989.

[7] D.I. Gurevich and S. Majid. Braided groups of Hopf algebras obtained by twisting. *Preprint*, DAMTP/91-49, 1991.

[8] A. Joyal and R. Street. Braided monoidal categories. Mathematics Reports 86008, Macquarie University, 1986.

[9] A. Joyal and R. Street. The geometry of tensor calculus, I. *Adv. Math.*, 88:55–112, 1991.

[10] A. Joyal and R. Street. Tortile Yang-Baxter operators in tensor categories. *J. Pure Appl. Algebra*, 71:43-51, 1991.

[11] M.M. Kapranov and V.A. Voevodsky. Braided monoidal 2-categories and Zamalodchikov tetrahedra equations. Mathematics report, Harvard and Northwestern Universities, 1991.

[12] S. Mac Lane. *Categories for the Working Mathematician*. Springer, 1974. GTM vol. 5.

[13] F.W. Lawvere. Intrinsic boundaries in certain mathematical toposes exemplify 'logical' operators not passively preserved by substitution. Nov 1989.

[14] V.V. Lyubashenko and S. Majid. Braided groups and quantum Fourier transform. *Preprint*, DAMTP/91-26, 1991.

[15] A. Macfarlane and S. Majid, Quantum group structure in a fermionic extension of the quantum harmonic oscillator, *Phys. Lett. B*, 268:71–74, 1991; S. Majid and M.J. Rodriguez-Plaza, Universal R-matrix for non-standard quantum group and superization, *Preprint*, DAMTP/91-47, 1991.

[16] S. Majid. $\mathbb{C}$-statistical quantum groups and Weyl algebras. *Preprint*, DAMTP/91-46, 1991.

[17] S. Majid. Representations, duals and quantum doubles of monoidal categories (in Proc. Winter School Geom. Phys., Srni, January 1990). *Suppl. Rend. Circ. Mat. Palermo, Ser. II*, 26:197-206, 1991.

[18] S. Majid. Reconstruction theorems and rational conformal field theories, *Int. J. Mod. Phys. A*, 6(24):4359–4374, 1991; Some physical applications of category theory (in Proc. XIXth DGM, Rapallo , July 1990), *Springer Lect. Notes in Phys.* 375:131-142, 1991.

[19] S. Majid. Braided groups. *Preprint*, DAMTP/90-42, 1990.

[20] S. Majid. Transmutation theory and rank for quantum braided groups. *Preprint*, DAMTP/91-10, 1991.

[21] S. Majid. Cross products by braided groups and bosonization, 1991. To appear in *J. Algebra.*

[22] S. Majid. Doubles of quasitriangular Hopf algebras. *Comm. Algebra*, 19(11):3061–3073, 1991.

[23] S. Majid. Physics for algebraists: Non-commutative and non-cocommutative Hopf algebras by a bicrossproduct construction. *J. Algebra*, 130:17–64, 1990.

[24] S. Majid. Phd Thesis, Harvard University, 1988; Hopf-von Neumann algebra bicrossproducts, Kac algebra bicrossproducts, and the classical Yang-Baxter equations, *J. Funct. Analysis*, 95:291–319, 1991; Hopf algebras for physics at the Planck scale, *J. Classical and Quantum Gravity*, 5:1587–1606, 1988; Principle of representation-theoretic self-duality, *Phys. Essays*, 4(3):395–405, 1991.

[25] B. Pareigis. A non-commutative non-cocommutative Hopf algebra in nature. *J. Algebra*, 70:356, 1981.

[26] D. Radford. The structure of Hopf algebras with a projection. *J. Alg.*, 92:322–347, 1985.

[27] G.E. Reyes and H. Zolfaghari. Bi-Heyting algebras, topos and modalities. *Preprint*, 1991.

[28] N. Saavedra Rivano. Catégories Tannakiennes. *Springer Lect. Notes in Math*, 265, 1972.

[29] R.A.G. Seely. Linear logic, $*$-autonomous categories and cofree coalgebras. *Contemp. Math.*, 92:371-382.

[30] M.E. Sweedler. *Hopf Algebras.* Benjamin, 1969.

[31] D.N. Yetter. Quantum groups and representations of monoidal categories. *Math. Proc. Camb. Phil. Soc.*, 108:261–290, 1990.

Department of Applied Mathematics
& Theoretical Physics
University of Cambridge
Cambridge CB3 9EW, U.K.

Canadian Mathematical Society
Conference Proceeding
Volume **13**, 1992

A Quantisation of the Calculus of Relations

CHRISTOPHER J. MULVEY
AND
JOAN WICK PELLETIER

ABSTRACT. It is known that the lattice $\mathcal{Q}(X)$ of relations on a set X is a quantale with respect to composition of relations. This structure on $\mathcal{Q}(X)$ has been applied implicitly by Hoare and He in the case that X is a set of states of a machine to construct the weakest prespecification of one program with respect to another. It will be shown that this construction may be linearized to the case in which a Hilbert space H of states is introduced, corresponding to a quantization of the states of the machine. In this situation, the quantale $\mathcal{Q}(H)$ of linear relations on the Hilbert space H is considered. It may be shown that the quantale $\mathcal{Q}(H)$ is a particular case of a Gelfand quantale, first introduced in providing a concept of spectrum for non-commutative C^*-algebras. This motivates the identification of a category of Gelfand quantales, referred to as Hilbert quantales.

The calculus of relations provides a basic tool for describing and manipulating specifications of programs of a machine. A specification of a program may be represented by a relation R on the set X of states of the machine. The programs which satisfy this specification are those having the property that, if $y \in X$ is a terminal state corresponding to running the program with initial state $x \in X$, then $(x, y) \in R$. The fact that the set $\mathcal{Q}(X)$ of relations on a set X, partially ordered by inclusion of relations as subsets of the product $X \times X$, admits arbitrary joins, given by unions of subsets of the product, allows the construction of the weakest specification satisfied by any class of programs. This weakest specification may be obtained simply by taking the join of all specifications satisfied by each program in the class. In particular, it follows that, for any program Q and any specification R, there exists a weakest specification with the property that it is satisfied by precisely those programs P for which the program $P; Q$, namely P followed by Q, satisfies the specification R. The specification obtained in this way, denoted $Q \backslash R$, is called the weakest prespecification of the specification R given the program Q. This construction, together with others concerning the

1991 Mathematics Subject Classification. Primary 18F99.
The second author acknowledges support from the Natural Sciences and Engineering Research Council of Canada.
This paper is in final form and no version of it will be submitted for publication elsewhere.

calculus of relations in the context of specifications of programs, may be found in the fundamental paper of Hoare and He [3].

The present paper shows that these constructions can be carried through to the case in which the behaviour of the machine is quantised, by replacing the set X of states by a Hilbert space H of states, by applying the concept of a quantale, introduced in [4]. In making this generalisation, our motivation initially is no more than the realisation that in other non-deterministic situations the introduction of this quantisation, in which the lattice of subsets of the set X is replaced by that of closed subspaces of the Hilbert space H, is extremely productive. However, the structures that begin to appear once this generalisation is allowed to evolve indicate that there may already exist intrinsic reasons for extending the calculus of relations in this way. In particular, there emerges a natural relationship with the concept of Gelfand quantale [6] arising in the spectral theory of C^*-algebras [7] which itself seems to merit further investigation. This, together with the appearance of certain operations encountered already within the context of linear logic, then becomes the primary motivation for considering the quantisation of the calculus of relations which will be introduced.

1. Quantales of relations

DEFINITIONS. By a *quantale* Q will be meant a lattice having arbitrary joins $\bigvee$ together with an associative product & satisfying

$$a \,\&\, \bigvee b_i = \bigvee a\&b_i$$

and

$$\bigvee a_i \,\&\, b = \bigvee a_i\&b$$

for all $a, b, a_i, b_i \in Q$. The quantale Q will be said to be *unital* provided that there exists an element $e \in Q$ for which

$$e\&a = a = a\&e$$

for all $a \in Q$.

By an *involutive quantale* will be meant a quantale Q together with an involution, that is, with a unary operation $*$, satisfying

$$\begin{aligned} a^{**} &= a \\ (a\&b)^* &= b^*\&a^* \end{aligned}$$

and

$$(\bigvee a_i)^* = \bigvee a_i^*$$

for all $a, b, a_i \in Q$. In the event that Q is also unital, then necessarily

$$e^* = e.$$

By a *homomorphism of quantales* will be meant a $\bigvee$- and &-preserving mapping

$$\varphi : Q \to Q'.$$

The homomorphism φ will be said to be *unital* provided Q and Q' are unital and

$$\varphi(e) = e',$$

for e and e' the respective units of Q and Q'. If Q and Q' are involutive quantales, the homomorphism φ will be said to be a *homomorphism of involutive quantales* if φ preserves the involution.

EXAMPLES. 1. Any locale L provides a trivial example of a unital involutive quantale by taking the lattice L together with the product given by the meet $\wedge$, with unit given by the top element 1, and by defining $a^* = a$ for all $a \in L$.

2. Any C^*-algebra A with unit determines a unital quantale Max A by taking Max A to be the lattice of closed linear subspaces of A together with the product given by multiplication of subspaces followed by closure. The quantale Max A is also involutive with involution defined by

$$M^* = \{x^* \in A | x \in M\}$$

for any closed linear subspace M of A.

3. The set $\mathcal{Q}(X)$ of relations on a set X, partially ordered by inclusion of relations, is a unital quantale with multiplication given by

$$R\&S = \{(x,y) \in X \times X | \exists z \in X \ \ (x,z) \in R \text{ and } (z,y) \in S\}$$

for any $R, S \in \mathcal{Q}(X)$, and with the equality relation providing the unit. Moreover, $\mathcal{Q}(X)$ is involutive, with involution defined by

$$R^* = \{(y,x) \in X \times X | \ (x,y) \in R\}.$$

The quantale L obtained trivially from a locale, the quantale Max A which is the spectrum of a C^*-algebra A, and the quantale $\mathcal{Q}(X)$ of relations on a set X are each instances of the concept of a Gelfand quantale, in the sense now described:

DEFINITIONS. An element $a \in Q$ of a quantale Q is said to be *right-sided* if

$$a\&b \leq a$$

for all $b \in Q$. Similarly, $a \in Q$ is said to be *left-sided* if $b\&a \leq a$ for all $b \in Q$. The quantale Q is said to be *right-sided* (respectively, *left-sided*) if every $a \in Q$ is right-sided (respectively, left-sided). We observe that the subsets of right-sided and of left-sided elements of Q, $R(Q)$ and $L(Q)$ respectively, are quantales with respect to the operations of the quantale Q.

A quantale Q will be said to be a *Gelfand quantale* provided that Q is unital, involutive, and satisfies the condition

$$a\&a^*\&a = a$$

for all $a \in R(Q)$. We observe that this is equivalent to asking that

$$a \& a^* \& a = a$$

for all $a \in L(Q)$, since $a \in R(Q)$ if, and only if, $a^* \in L(Q)$.

For a C^*-algebra A the right-sided elements of the quantale Max A are the closed right ideals of A and the left-sided elements are the closed left ideals of A. The quantale Max A is, moreover, the motivating example of a Gelfand quantale, first introduced by Mulvey [5] to provide a concept of the spectrum for non-commutative C^*-algebras (see also [6,7]).

To see that Max A is a Gelfand quantale, let I be a closed right ideal and let $x \in I$. By the construction of approximate units in A, there exists a sequence (e_n) of self-adjoint elements of I for which

$$e_n x \longrightarrow x$$

in A. The closed linear subspace U generated by the (e_n) obtained for each $x \in I$ then satisfies

$$U^* = U \leq I \leq U \& I$$

by definition. Hence,

$$I \leq U \& I \leq U \& U \& I \leq I \& I^* \& I \leq I,$$

since I is a right ideal, so $I = I \& I^* \& I$.

In the case of $\mathcal{Q}(X)$, the quantale of relations on the set X, we observe that the right-sided elements are those relations R_A for $A \subseteq X$ given by

$$R_A = \{(x,y) \in X \times X | \, x \in X - A\}.$$

Clearly, relations of this form are right-sided. Moreover, if S is any right-sided relation, we must have

$$S \& 1 \leq S,$$

where 1 is the relation $\{(x,y) \in X \times X | \, x \in X, \; y \in X\}$. Hence, $S = R_A$, where

$$X - A = \{x \in X | \, (x,y) \in S \text{ for some } y \in X\}.$$

Similarly, the left-sided elements of $\mathcal{Q}(X)$ are those relations L_A of the type

$$L_A = \{(x,y) \in X \times X | \, y \in A\}.$$

It is the observation that $R_A^* = L_{X-A}$ and $L_A^* = R_{X-A}$ that permits us to establish the theorem:

THEOREM 1. *For any set X, the quantale $\mathcal{Q}(X)$ of relations on X is a Gelfand quantale.*

Given an involutive quantale Q, the quantale $R(Q)$ of right-sided elements of Q can be endowed with a negation operation $\perp$ by defining

$$a^{\perp} = \bigvee_{a^{*}\&b=0} b$$

for $a, b \in R(Q)$, that is, $a^{\perp}$ is the largest element of $R(Q)$ such that

$$a^{*}\&a^{\perp} = 0.$$

It is easy to see that

$$a \leq a^{\perp\perp}$$

and that

$$(\bigvee a_i)^{\perp} = \bigwedge a_i^{\perp}$$

for all $a, a_i \in R(Q)$. Similarly, $L(Q)$, the quantale of left-sided elements of Q, also possesses a negation operation $\perp$ defined by

$${}^{\perp}b = \bigvee_{a\&b^{*}=0} a$$

for $a, b \in L(Q)$; ${}^{\perp}b$ is the largest element of $L(Q)$ such that

$${}^{\perp}b\&b^{*} = 0.$$

For any right-sided element R_A of $\mathcal{Q}(X)$ one observes that

$$R_A^{*}\&R_{X-A} = \emptyset$$

and that R_{X-A} is the largest relation S having the property that

$$R_A^{*}\&S = \emptyset.$$

Hence,

$$R_A^{\perp} = R_{X-A}.$$

Similarly, one observes that

$${}^{\perp}L_A = L_{X-A}.$$

In general the negations on $R(Q)$ and $L(Q)$ do not provide a true complement in the lattice sense. However, there is a strong connection in the case of a Gelfand quantale between the notion of a negation and that of a complement as demonstrated by the Corollary to the following Proposition:

PROPOSITION. *For any Gelfand quantale Q, the Gelfand quantale $R(Q)$ of the right-sided elements has the following properties:*

(i) $a\&a = a$

(ii) $a \wedge b \leq a^*\&b$

for every $a, b \in R(Q)$.

PROOF. For any $a \in R(Q)$, we have

$$a\&a \leq a = a\&a^*\&a \leq a\&a,$$

which establishes (i). Moreover,

$$a = a\&a^*\&a \leq 1\&a^*\&a \leq a^*\&a,$$

since $a^*\&a \in L(Q)$. Thus, noting that the meet of right-sided elements is again right-sided, we have that

$$a \wedge b \leq (a \wedge b)^*\&(a \wedge b) \leq a^*\&b,$$

which establishes (ii).

COROLLARY. *If* Q *is a Gelfand quantale, then*

$$a \wedge a^{\perp} = 0$$

for all $a \in R(Q)$.

We note that the analogous results for $L(Q)$ are established similarly.

2. Quantales of Linear Relations

The aim is now to quantise the construction of the quantale $\mathcal{Q}(X)$ of relations on a set X to yield the quantale $\mathcal{Q}(H)$ of linear relations on a Hilbert space H. The quantale $\mathcal{Q}(H)$ will be shown to be once again a Gelfand quantale, albeit with, as will also be seen to be the case with the quantale $\mathcal{Q}(X)$, certain properties in addition which will be considered in detail in the next section. The observation that allows this quantisation to take place is the following concerning the quantale $\mathcal{Q}(X)$.

For any relation R on a set X, one may consider the mapping φ_R from the power set $\wp(X)$ of the set X to itself which assigns to each $A \in \wp(X)$ the subset

$$(A)\varphi_R = \{y \in X \mid \exists x \in A \ \ (x, y) \in R\}$$

of X. The mapping

$$\varphi_R : \wp(X) \rightarrow \wp(X)$$

is evidently a sup-preserving mapping from the sup-lattice $\wp(X)$ to itself. Moreover, it may be straightforwardly verified that any sup-preserving mapping

$$\varphi : \wp(X) \to \wp(X)$$

arises in this way from a unique relation on the set X. The multiplication of the quantale $\mathcal{Q}(X)$ is exactly that corresponding to composition of mappings, with mappings composed in the categorical sense to the right of their arguments, and the supremum is that corresponding to the pointwise ordering of mappings into the sup-lattice $\wp(X)$. The unit of the quantale $\mathcal{Q}(X)$, given by the equality relation on X, corresponds in this way to the identity mapping on $\wp(X)$.

The quantale $\mathcal{Q}(X)$ may therefore be identified with the quantale of sup-preserving mappings from the sup-lattice $\wp(X)$ to itself. The way forward to the quantisation of these ideas is now evident. The sup-lattice $\wp(X)$ of subsets of the set X may be replaced by the sup-lattice $\wp(H)$ of closed linear subspaces of the Hilbert space H to yield the following definition:

DEFINITION. By the *quantale* $\mathcal{Q}(H)$ *of linear relations* on a Hilbert space H will be meant the quantale of sup-preserving mappings from the sup-lattice $\wp(H)$ of closed linear subspaces of H to itself.

The sup-lattice $\wp(H)$ is canonically isomorphic to that of projections on the Hilbert space H. In particular, $\wp(H)$ is partially ordered by inclusion of closed linear subspaces, the supremum being given by taking the closure of the algebraic sum of the linear subspaces concerned. The quantale $\mathcal{Q}(H)$ has product given by composition of sup-preserving mappings and supremum defined pointwise with respect to the partial ordering on $\wp(H)$. Its unit is given by the identity operator.

Concerning this quantale, one may now prove the following:

THEOREM 2. *For any Hilbert space H, the quantale $\mathcal{Q}(H)$ of linear relations on H is a Gelfand quantale.*

Before doing so, and without any comment on the semantics that we have in mind, we make the immediate observation that the quantale structure of $\mathcal{Q}(H)$ leads to the existence of a quantised analogue of the construction of the weakest prespecification of Hoare and He ([3]). Namely, given any elements $\varphi, \psi \in \mathcal{Q}(H)$, there is an element $\varphi \backslash \psi \in \mathcal{Q}(H)$ with the property that it is the largest element ξ of $\mathcal{Q}(H)$ satisfying the condition that

$$\xi \& \varphi \leq \psi$$

in the quantale $\mathcal{Q}(H)$. Of course, the element is constructed exactly by taking the supremum of all elements of $\mathcal{Q}(H)$ satisfying this condition.

The sup-lattice $\wp(H)$ is rich in structure; in particular, it is *orthocomplemented*; that is, for every $M \in \wp(H)$ there exists a closed subspace $M^{\perp}$ of H such that

$$\begin{aligned} M^{\perp\perp} &= M \\ (\bigvee M_i)^{\perp} &= \bigwedge M_i^{\perp} \\ M \vee M^{\perp} &= H \\ M \wedge M^{\perp} &= 0. \end{aligned}$$

Using this structure, one can construct an operation * on $\mathcal{Q}(H)$ by defining

$$(M)\varphi^* = (\bigvee_{(N)\varphi \leq M^{\perp}} N)^{\perp}$$

for all sup-preserving mappings from $\wp(H)$ to itself and for all $M, N \in \wp(H)$. In other words, $(M)\varphi^*$ is the closed subspace of H of which the orthogonal complement $((M)\varphi^*)^{\perp}$ is the largest satisfying

$$((M)\varphi^*)^{\perp}\varphi \leq M^{\perp}.$$

If $N \leq ((M)\varphi^*)^{\perp}$, then $(N)\varphi \leq M^{\perp}$, so it is immediate that

$$N \leq ((M)\varphi^*)^{\perp} \Leftrightarrow (N)\varphi \leq M^{\perp}.$$

Thus, using the fact that $N \leq M \Leftrightarrow M^{\perp} \leq N^{\perp}$ we have that

$$\begin{aligned} N \leq ((M)\varphi^{**})^{\perp} &\Leftrightarrow (N)\varphi^* \leq M^{\perp} \\ &\Leftrightarrow M \leq ((N)\varphi^*)^{\perp} \\ &\Leftrightarrow (M)\varphi \leq N^{\perp} \\ &\Leftrightarrow N \leq ((M)\varphi)^{\perp}, \end{aligned}$$

which shows that $(M)\varphi^{**} = (M)\varphi$ for all $M \in \wp(H)$, and, hence, that $\varphi^{**} = \varphi$.

A similar use of the properties of $\perp$ and of * allows us to complete the verification of the fact that $*$ is an involution on $\mathcal{Q}(H)$, namely

$$\begin{aligned} (\varphi \& \psi)^* &= \psi^* \& \varphi^* \\ (\bigvee \varphi_i)^* &= \bigvee \varphi_i^*, \end{aligned}$$

for $\varphi, \varphi_i, \psi \in \mathcal{Q}(H)$.

In analysing the right-sided and left-sided elements of $\mathcal{Q}(H)$, we return to the analogy with the quantale $\mathcal{Q}(X)$ of relations on the set X, of which notion $\mathcal{Q}(H)$ is the quantisation. We have in that case described right-sided and left-sided relations as being of the form R_A and L_A, for $A \subseteq X$, where

$$R_A = \{(x, y) \in X \times X \mid x \in X - A\}$$

and

$$L_A = \{(x, y) \in X \times X \mid y \in A\}.$$

Under the identification of $\mathcal{Q}(X)$ with the quantale of sup-preserving mappings from $\wp(X)$ to itself, R_A and L_A clearly correspond, respectively, to the sup-preserving mappings λ_A and κ_A , where

$$(B)\lambda_A = \begin{cases} X & \text{unless} \\ \emptyset & B \subseteq A \end{cases}$$

and

$$(B)\kappa_A = \begin{cases} A & \text{unless} \\ \emptyset & B = \emptyset. \end{cases}$$

The obvious generalisation to $\mathcal{Q}(H)$ yields a description of the right- and left-sided elements. We define λ_M , for $M \in \wp(H)$, by

$$(N)\lambda_M = \begin{cases} H & \text{unless} \\ 0 & N \leq M, \end{cases}$$

for $N \in \wp(H)$. It is clear that λ_M is a sup-preserving mapping from $\wp(H)$ to itself and that

$$\lambda_M \& \varphi \leq \lambda_M$$

for all $\varphi \in \mathcal{Q}(H)$. Moreover, if φ is any right-sided element of $\mathcal{Q}(H)$, we must have

$$\varphi \& 1 \leq \varphi,$$

where

$$(N)1 = \begin{cases} H & \text{unless} \\ 0 & N = 0. \end{cases}$$

Then, clearly φ must be of the form λ_M , where

$$M = \bigvee \{N \in \wp(H) \mid (N)\varphi = 0\}.$$

Similarly, we define κ_M , for $M \in \wp(H)$, by

$$(N)\kappa_M = \begin{cases} M & \text{unless} \\ 0 & N = 0. \end{cases}$$

Sup-preserving mappings of the form κ_M are the left-sided elements of $\mathcal{Q}(H)$.

As in the case of $\mathcal{Q}(X)$, the verification that the involutive quantale $\mathcal{Q}(H)$ is a Gelfand quantale, that is, that

$$\lambda_M \& \lambda_M^* \& \lambda_M = \lambda_M$$

for all $M \in \wp(H)$, turns on the fact that $\lambda_M^* = \kappa_{M^\perp}$. For, by definition,

$$(N)\lambda_M^* = (\bigvee_{(L)\lambda_M \leq N^\perp} L)^\perp.$$

Since $(L)\lambda_M \leq N^\perp$ when $N = 0$ and L is any closed linear subspace of H or when $N \neq 0$ and $L \leq M$, we see that

$$(O)\lambda_M^* = H^\perp = 0$$

and for $N \neq 0$,

$$(N)\lambda_M^* = M^\perp,$$

which shows that

$$\lambda_M^* = \kappa_{M^\perp}.$$

This completes the proof of the theorem, since it is easy to see that

$$\lambda_M \& \kappa_{M^\perp} \& \lambda_M = \lambda_M.$$

We remark that from the above relation,

$$\kappa_M^* = \lambda_{M^\perp}^{**} = \lambda_{M^\perp},$$

as well.

Finally, we examine the negation operations on $R(\mathcal{Q}(H))$ and $L(\mathcal{Q}(H))$ to establish their connection with the orthocomplement on $\wp(H)$ and the involutive structure of $\mathcal{Q}(H)$. Recalling that for a right-sided element λ_A of $\mathcal{Q}(H)$, $\lambda_A^\perp$ is the largest element such that

$$\lambda_A^* \& \lambda_A^\perp = 0,$$

we see that $\lambda_{A^\perp} \leq \lambda_A^\perp$ since clearly

$$\lambda_A^* \& \lambda_{A^\perp} = \kappa_{A^\perp} \& \lambda_{A^\perp} = 0.$$

On the other hand, if $\varphi \in R(\mathcal{Q}(H))$ is such that

$$\kappa_{A^\perp} \& \varphi = 0,$$

it must be the case that $\varphi \leq \lambda_{A^\perp}$, so that

$$\lambda_A^\perp = \lambda_{A^\perp}.$$

A similar calculation shows that ${}^\perp\kappa_A = \kappa_{A^\perp}$ for any left-sided element κ_A of $\mathcal{Q}(H)$. Putting these results together with those above, we see that the involution and the negation are related in the following way:

$$\begin{aligned} \lambda_A^* &= {}^\perp\kappa_A \\ \kappa_A^* &= \lambda_A^\perp. \end{aligned}$$

3. Hilbert Quantales

It was seen in the last section that the construction of the quantale $\mathcal{Q}(X)$ of relations on a set X could be quantised to yield a Gelfand quantale $\mathcal{Q}(H)$ of linear relations on a Hilbert space H, thus permitting a quantised analogue of the weakest prespecification. The properties that were exploited in this quantisation were the identification of the quantale $\mathcal{Q}(X)$ with the quantale of sup-preserving mappings from $\wp(X)$ to itself and structures of the sup-lattices $\wp(X)$ and $\wp(H)$. In particular, it was the fact that $\wp(H)$, the sup-lattice of closed linear subspaces of H, is orthocomplemented that proved to be of great significance, replacing the complementation present in the sup-lattice $\wp(X)$ of subsets of the set X.

We now extract these salient facts and generalise the construction of the quantale of linear relations to a more abstract setting in which we begin by being provided with an orthocomplemented sup-lattice S. We recall that a sup-lattice S is said to be *orthocomplemented* if there exists a unary operation $\perp$ on S satisfying the conditions that:

$$\begin{aligned} s^{\perp\perp} &= s \\ (\bigvee s_i)^{\perp} &= \bigwedge s_i^{\perp} \\ s \vee s^{\perp} &= 1 \\ s \wedge s^{\perp} &= 0, \end{aligned}$$

for all $s, s_i \in S$.

DEFINITION. By the *quantale* $\mathcal{Q}(S)$ *of endomorphisms of the orthocomplemented sup-lattice* S will be meant the unital quantale of sup-preserving mappings from S to itself, with the supremum given by the pointwise ordering of mappings, with the multiplication corresponding to composition of mappings, and with the unit given by the identity mapping.

As in the case of $\mathcal{Q}(H)$, we can define an involution $*$ on $\mathcal{Q}(S)$ by defining

$$(s)\varphi^* = (\bigvee_{(t)\varphi \leq s^{\perp}} t)^{\perp},$$

for $\varphi \in \mathcal{Q}(S)$ and $s, t \in S$; that is, $(s)\varphi^*$ is the element of S of which the orthogonal complement is the largest element satisfying

$$((s)\varphi^*)^{\perp}\varphi \leq s^{\perp}.$$

Using the fundamental property

$$t \leq ((s)\varphi^*)^{\perp} \Leftrightarrow (t)\varphi \leq s^{\perp},$$

which is true for all $\varphi \in \mathcal{Q}(S)$ and $s, t \in S$, one can verify that $\mathcal{Q}(S)$ is an involutive quantale with $*$ as its involution, which is the first step towards establishing:

THEOREM 3. *For any orthocomplemented sup-lattice S, the quantale $\mathcal{Q}(S)$ of sup-preserving mappings from S to itself is a Gelfand quantale.*

The right- and left-sided elements of $\mathcal{Q}(S)$ can be described in a manner totally analogous to those of $\mathcal{Q}(H)$. We define for $s \in S$, the sup-preserving mappings λ_s and κ_s, given by

$$(t)\lambda_s = \begin{cases} 1 & \text{unless} \\ 0 & t \leq s \end{cases}$$

and

$$(t)\kappa_s = \begin{cases} s & \text{unless} \\ 0 & t = 0. \end{cases}$$

All right-sided elements of $\mathcal{Q}(S)$ are of the form λ_s for some $s \in S$, whilst mappings of the form κ_s characterize the left-sided elements.

The same properties relating λ, κ, the involution, and the negation operation available on $R(\mathcal{Q}(S))$ and $L(\mathcal{Q}(S))$, as defined in Section 1, pertain as well:

$$\begin{aligned} \lambda_s^* &= \kappa_{s^\perp} \\ \lambda_s^\perp &= \lambda_{s^\perp} \\ {}^\perp\kappa_s &= \kappa_{s^\perp}. \end{aligned}$$

These in turn imply:

$$\begin{aligned} \kappa_s^* &= \lambda_{s^\perp} \\ \kappa_s^* &= {}^\perp\kappa_s \\ \kappa_s^* &= \lambda_s^\perp. \end{aligned}$$

The final step in the proof of the theorem, namely that

$$\lambda_s \& \lambda_s^* \& \lambda_s = \lambda_s,$$

follows readily from these relations.

Since these quantales of the form $\mathcal{Q}(S)$ have arisen in considering generalisations of the quantale $\mathcal{Q}(H)$ of linear relations on a Hilbert space H, we make the following definition:

DEFINITION. A Gelfand quantale Q will be said to be a *Hilbert quantale* if it is isomorphic to a Gelfand quantale of the form $\mathcal{Q}(S)$ for some orthocomplemented sup-lattice S.

Our aim is now to characterise such quantales.

We recall from the Proposition and Corollary of Section 1 that whenever Q is a Gelfand quantale, then the quantale $R(Q)$ of right-sided elements of Q is an idempotent right-sided quantale for which the negation operation $\perp$, given by

$$a^{\perp} = \bigvee_{a^* \& b = 0} b,$$

satisfies

$$\begin{aligned} a &\leq a^{\perp\perp}, \\ (\bigvee a_i)^{\perp} &= \bigwedge a_i^{\perp}, \end{aligned}$$

and

$$a \wedge a^{\perp} = 0,$$

for each $a, a_i \in R(\mathcal{Q})$, but is not necessarily an orthocomplement. On the other hand, if Q is of the form $\mathcal{Q}(S)$ for some orthocomplemented sup-lattice S, then the negation $\perp$ on $R(\mathcal{Q}(S))$ is necessarily an orthocomplement in view of:

PROPOSITION. *For any orthocomplemented sup-lattice S, the mapping*

$$\chi : S \to R(\mathcal{Q}(S))$$

defined by

$$(s)\chi = \lambda_{s^{\perp}}$$

is an isomorphism of orthocomplemented sup-lattices.

PROOF. The map χ is clearly onto. It is one-one since if $s \leq t$ and $s \neq t$, then

$$(s^{\perp})\lambda_{s^{\perp}} = 0, \;\; (s^{\perp})\lambda_{t^{\perp}} = 1$$

and if $s \not\leq t$, then

$$(t^{\perp})\lambda_{s^{\perp}} = 1, \;\; (t^{\perp})\lambda_{t^{\perp}} = 0.$$

The mapping χ preserves order since

$$s \leq t \Rightarrow t^{\perp} \leq s^{\perp} \Rightarrow \lambda_{s^{\perp}} \leq \lambda_{t^{\perp}} \Rightarrow (s)\chi \leq (t)\chi.$$

Finally, we note that

$$(s^{\perp})\chi = \lambda_{s^{\perp\perp}} = (\lambda_{s^{\perp}})^{\perp} = (s)\chi^{\perp}.$$

COROLLARY. *For any orthocomplemented sup-lattice S, there is an isomorphism*

$$\sigma : \mathcal{Q}(S) \to \mathcal{Q}(R(\mathcal{Q}(S)))$$

of Gelfand quantales given by

$$(\varphi)\sigma = \chi^{-1}\varphi\chi$$

for each $\varphi \in \mathcal{Q}(S)$.

In the general case of an arbitrary Gelfand quantale, however, there already exist a mapping

$$\mu : Q \to \mathcal{Q}(R(Q))$$

given by $(a)\mu = \pi_a$, where

$$(b)\pi_a = a^* \& b,$$

which clearly preserves sups, & , and the unit. If we further assume that $R(Q)$ is orthocomplemented, then for μ to be a homomorphism of Gelfand quantales it is necessary and sufficient to verify that

$$(a^*)\mu = ((a)\mu)^*$$

for all $a \in Q$. If $b \in Q$, we have

$$(b)(a^*)\mu = (b)\pi_{a^*} = a^{**} \& b = a \& b$$

and

$$\begin{aligned} (b)((a)\mu)^* &= (b)\pi_a^* = (\bigvee_{(c)\pi_a \leq b^\perp} c)^\perp = (\bigvee_{a^* \& c \leq b^\perp} c)^\perp \\ &= (\bigvee_{b^* \& a^* \& c = 0} c)^\perp = (\bigvee_{(a \& b)^* \& c = 0} c)^\perp = (a \& b)^{\perp\perp}. \end{aligned}$$

What is proved is:

PROPOSITION. *For any Gelfand quantale* Q*,*

$$\mu : Q \to \mathcal{Q}(R(Q))$$

is a homomorphism of Gelfand quantales, if and only if, $R(Q)$ *is orthocomplemented.*

Thus, when Q is a Hilbert quantale, that is, a quantale isomorphic to one of the form $\mathcal{Q}(S)$ for some orthocomplemented sup-lattice S, the situation at which we have arrived is that there exist two mappings:

$$\sigma, \mu : Q \to \mathcal{Q}(R(Q)),$$

one of which has been shown to be an isomorphism. We claim that σ and μ are the same in this case.

To wit, identifying Q with the quantale $\mathcal{Q}(S)$ to which it is isomorphic, let $\varphi \in \mathcal{Q}(S)(= Q)$ and consider an arbitrary element λ_s of $R(Q)$. We have

$$\begin{aligned} (\lambda_s)(\varphi)\sigma &= (\lambda_s)\chi^{-1}\varphi\chi = (s^\perp)\varphi\chi \\ &= \lambda_{((s^\perp)\varphi)^\perp} \\ (\lambda_s)(\varphi)\mu &= (\lambda_s)\pi_\varphi = \varphi^* \& \lambda_s . \end{aligned}$$

For $t \in S$,

$$(t)\lambda_{((s^\perp)\varphi)^\perp} = \begin{cases} 1 & \text{unless} \\ 0 & t \leq ((s^\perp)\varphi)^\perp \end{cases}$$

and

$$(t)\varphi^* \& \lambda_s = \begin{cases} 1 & \text{unless} \\ 0 & (t)\varphi^* \leq s. \end{cases}$$

But $(t)\varphi^* \leq s \Leftrightarrow t \leq ((s^\perp)\varphi)^\perp$ by the fundamental property of the involution on $\mathcal{Q}(S)$.

Recalling that the canonical morphism $\mu : Q \to \mathcal{Q}(R(Q))$ is a homomorphism of Gelfand quantales if, and only if, $R(Q)$ is orthocomplemented, we have achieved the following characterisation of Hilbert quantales:

THEOREM 4. *A Gelfand quantale Q is a Hilbert quantale if, and only if, the canonical mapping*

$$\mu : Q \to \mathcal{Q}(R(Q))$$

is an isomorphism of Gelfand quantales.

REFERENCES

1. Giles, R., and Kummer, H., A non-commutative generalization of topology, Indiana Univ. Math. J. **21** (1971), 91-102.

2. Girard, J.-Y., Linear logic, Theoretical Computer Science **50** (1987), 1-102.

3. Hoare, C.A.R., and He, Jifeng, The weakest prespecification, Information Processing Letters **24** (1987), 127-132.

4. Mulvey, C.J., &, Rendiconti Circ. Mat. Palermo, 12, 99-104 (1986).

5. Mulvey, C.J., Quantales, Invited Lecture at the Curaçao Summer Conference on Locales and Topological Groups, Curaçao, 1989.

6. Mulvey, C.J., Gelfand quantales, *to appear.*

7. Mulvey, C.J., The spectrum of a C^*-algebra, *to appear.*

8. Rosenthal, K.I., *Quantales and their applications*, Pitman Research Notes in Mathematics Series, 234, Longman Scientific & Technical, Essex, U.K., 1990.

9. Rosicky, J., Multiplicative lattices and C^*-algebras, Cahiers de Top. et Géom. Diff. Cat. XXX-2 (1989), 95-110.

Christopher J. Mulvey
Mathematics Division, University of Sussex, Falmer, Brighton, BN1 9QH, United Kingdom
E-mail address: mmfc6 at vax2.sussex.ac.uk
and
Joan Wick Pelletier
Department of Mathematics and Statistics, York University, North York, Ontario, Canada, M3J 1P3
E-mail address: jwpell at vm2.yorku.ca

Canadian Mathematical Society
Conference Proceeding
Volume **13**, 1992

A Logical View on the Adjoint Functor Theorem

DUŠKO PAVLOVIĆ

ABSTRACT. We describe a logically motivated construction of a weak colimit from a solution set of a given diagram. (This construction is effective and it is used in a sequel to this paper for actual calculations in locally small fibred categories.)

1. Idea. Consider the following "ratios":

$$\frac{\bigvee D = \bigwedge\{x \geq D\}}{\varinjlim D \subseteq \varprojlim\{x \in \Sigma \mid D \to x\}} = \frac{\bigvee D = \neg\bigwedge\neg D}{\varinjlim D \subseteq \wp(\varprojlim \wp D)} \qquad (1$$

In the numerator on the left-hand side is the formula which expresses the least upper bound of a subset D of a complete (semi)lattice as the greatest lower bound of the set of all the upper bounds of D. The denominator represents the categorical generalisation of this formula: in principle, the General Adjoint Functor Theorem (Freyd 1964, exercise 3.J.; also in: MacLane 1971, V.6-8; Barr-Wells 1985, 1.9.). It tells that the colimit of a diagram D in a complete category $\mathcal{B}$ can be calculated as the limit of the diagram obtained by projecting on $\mathcal{B}$ the cocones from D to the elements of a *solution set* $\Sigma \subseteq |\mathcal{B}|$. On the right-hand side of (1), we have two very special cases of these constructions. The numerator shows a formula which holds in Boolean algebras – perhaps better known in its logical form:

$$\exists x.\, \delta(x) = \neg \forall x.\, \neg\delta(x) \qquad (2$$

The denominator presents the form in which a colimit is obtained in a topos $\mathcal{S}$, when Paré's theorem (1974) is applied. The setup of this formula is resumed on the following picture.

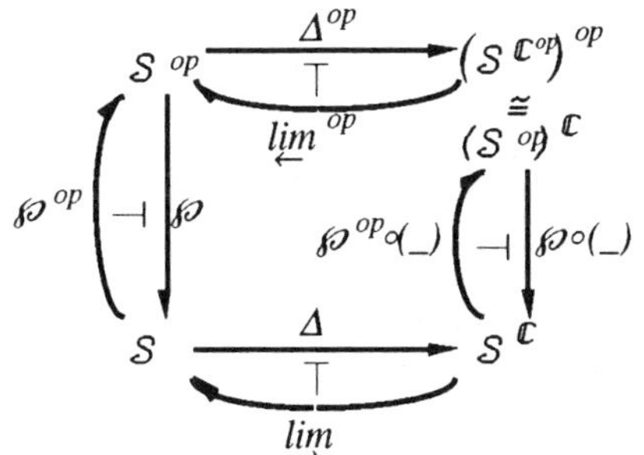

(3

1991 Mathematics Subject Classification. Primary 18A15, 18A40.

This paper is in final form and no version of it will be submitted for publication elsewhere.

While the General Adjoint Functor Theorem is basically an existential statement – which in principle may not allow an actual calculation – the special case captured in Paré's theorem is completely effective, probably the best way to calculate, say, the coequalizers in the category of sets.

Now note that (2) is a special case of the well-known formula of second-order logic:

$$\exists x.\delta(x) = \forall\xi(\forall x.\delta(x)\to\xi)\to\xi \qquad (4$$

where ξ varies over all propositions. This suggests a third "ratio", equal to those in (1). It relates the formulas that can be formulated in complete Heyting algebras and cartesian closed categories:

$$\ldots = \frac{\bigvee D = \bigwedge_{\xi}\bigwedge_{d}(d\to\xi)\to\xi}{\varinjlim D \subseteq \prod_{X\in\Sigma}[\varprojlim [D,X],\, X]} \qquad (1'$$

In the numerator, ξ runs over the whole algebra, and d over the subset D. As regards its generality and effectiveness, this "ratio" should be placed in between the previous two.

The main point of this note is the observation that the construction depicted in the denominator of (1') can be performed in arbitrary complete category $\mathcal{B}$ with small hom-sets. The setup for this echoes diagram (3). First of all, consider the functors:

$$\llcorner\_,\_\lrcorner := Hom\colon\ \mathcal{B}^{op}\times\mathcal{B}\to\underline{Set}\colon\ \langle A,B\rangle\mapsto\mathcal{B}(A,B) \qquad (5$$

$$\ulcorner\_,\_\urcorner\ :\quad \underline{Set}^{op}\times\mathcal{B}\to\mathcal{B}\colon\ \langle J,B\rangle\mapsto\prod_J B \qquad (6$$

Given a solution set Σ of a small diagram $D\colon\mathbb{C}\to\mathcal{B}$, define the functors

$$P\ :\ (\underline{Set}/\Sigma)^{op}\to\mathcal{B}\ :\ f\mapsto\prod_{X\in\Sigma}\ulcorner f^{-1}(X),\, X\urcorner \qquad (7$$

$$P_!\ :\ \mathcal{B}\to(\underline{Set}/\Sigma)^{op}\ :\ B\mapsto\Big(\coprod_{X\in\Sigma}\llcorner B,X\lrcorner\to\Sigma\Big) \qquad (8$$

(where $\coprod$ represents the disjoint union). They are adjoint – and here is the setup:

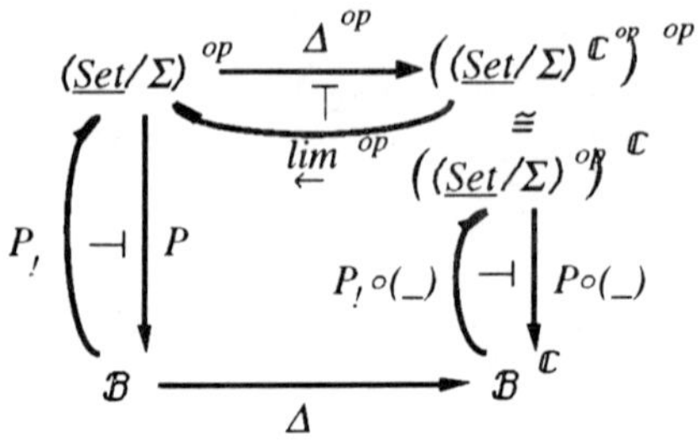

(9

The claim is now that sending $D \in |\mathcal{B}^{\mathbb{C}}|$ around this square gives its weak colimit.

2. Theorem. If Σ is the solution set of a small diagram $D: \mathbb{C} \longrightarrow \mathcal{B}$ in a complete category $\mathcal{B}$ with small hom-sets, then

$$\prod_{X \in \Sigma} \lceil \varprojlim \lfloor D, X \rfloor, X \rceil \tag{10}$$

is a weak colimit of D.

Proof. Note, first of all, that the elements of the set $L := \varprojlim \lfloor D, X \rfloor$ correspond to the cocones $D \longrightarrow X$. Thus, for every object $X \in |\mathcal{B}|$, there is in $\mathcal{B}$ a cocone

$$\delta_X := \langle \delta \rangle_{\delta \in L}: D \longrightarrow \prod_{\delta \in L} X = \lceil \varprojlim \lfloor D, X \rfloor, X \rceil \tag{11}$$

induced by the family of cocones $\{\delta: D \longrightarrow X\}_{\delta \in L}$. Clearly, for every set of objects $M \subseteq |\mathcal{B}|$, there is a cocone

$$\eta := \langle \delta_X \rangle_{X \in M} : \ D \longrightarrow \prod_{X \in M} \lceil \varprojlim \lfloor D, X \rfloor, X \rceil \tag{12}$$

We claim that this cocone is weakly initial when $M := \Sigma$ is a solution set. Suppose that a cocone $d: D \longrightarrow A$ is given. The assumption that Σ is a solution set for D now means that there is an object $S \in \Sigma$ and a cocone $s: D \longrightarrow S$ such that for some $h: S \longrightarrow A$ holds $d = h \circ s$. On the other hand, there is an obvious arrow

$$p := \pi_s \circ \pi_S: \ \prod_{X \in \Sigma} \lceil \varprojlim \lfloor D, X \rfloor, X \rceil \rightarrow \lceil \varprojlim \lfloor D, S \rfloor, S \rceil = \prod_{\delta \in L} S \rightarrow S$$

such that $s = p \circ \eta$. We have

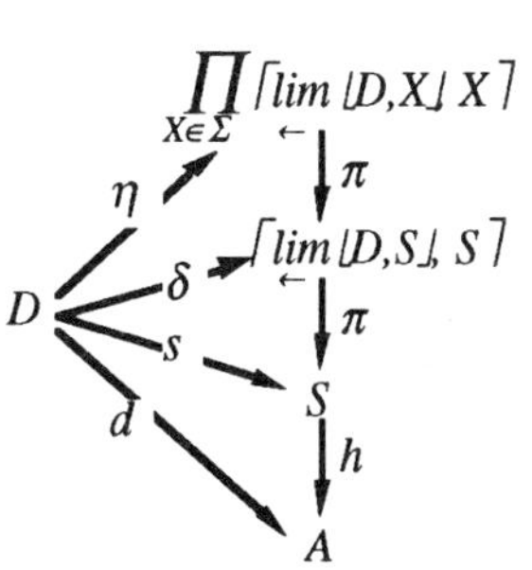

(13

We proved that η is weakly initial. •

3. Remark. The actual colimit Q of D can now be obtained from this weakly initial cocone $\eta: D \longrightarrow W$ by equalizing all the endomorphisms of W that stabilize η. In fact, it is enough to make the equalizer

$$Q \longrightarrow W \overset{\langle d_W \rangle_{v \in V}}{\underset{\langle v \rangle_{v \in V}}{\rightrightarrows}} \lceil V, W \rceil \tag{14}$$

where

$$V := \{v \in \lfloor W, W \rfloor \mid v \circ \eta = \eta\} \tag{15}$$

(This is a standard method, contained in almost every textbook on categories.)

4. Comments. By Freyd's definition, a solution set Σ of D is a set containing a coinitial set of upper bounds of D; it may also contain some superfluous objects. To produce a weak colimit, we must extract the actual upper bounds from it and make their product. In ordinary category theory – over $\underline{Set}$ – this is automatic, due to the comprehension principle. But when we work with categories over a base which is not topos, a construct like $\{x \in \Sigma \mid D \longrightarrow x\}$ may not be at our disposal. The usual proof does not go through. Of course, one could defer the problem by considering the "strict" solution sets which contain just the upper bounds – and still prove an adjoint functor theorem. But such a theorem wouldn't help us out when it comes, for instance, to calculating the colimits in a small complete category (which, in general setting, need not be a preorder!): comprehension is now necessary to make the solution sets.

The apparent value of theorem 2 is that it effectively calculates a weak colimit, seeking no help from the *deus ex machina* of comprehension. By the "logical" trick, it fabricates from a solution set Σ of D another solution set

$$\left\{ \lceil \underset{\leftarrow}{lim} \lfloor D, X \rfloor, X \rceil \mid X \in \Sigma \right\},$$

which consists of upper bounds only. For instance, when $\mathcal{B}$ is a complete preorder,

$$\underset{\leftarrow}{lim} \lfloor D, X \rfloor = \begin{cases} 1 & \text{if } X \geq D \\ \emptyset & \text{otherwise} \end{cases}$$

$$\lceil \underset{\leftarrow}{lim} \lfloor D, X \rfloor, X \rceil = \begin{cases} X & \text{if } X \geq D \\ \top & \text{otherwise} \end{cases}$$

To calculate the colimits in a small complete category, one can now simply take the whole category to be the solution set of every object. – This and some related themes will be studied in detail in a forthcoming paper (Pavlović 1992).

Theorem 2 considered only the colimits, i.e. the adjoints of the diagonal functors. The actual General Adjoint Functor Theorem – which provides a left adjoint for a general left exact functor – can be stated and proved along the same lines. We just formulate the part which extends theorem 2. The scheme is

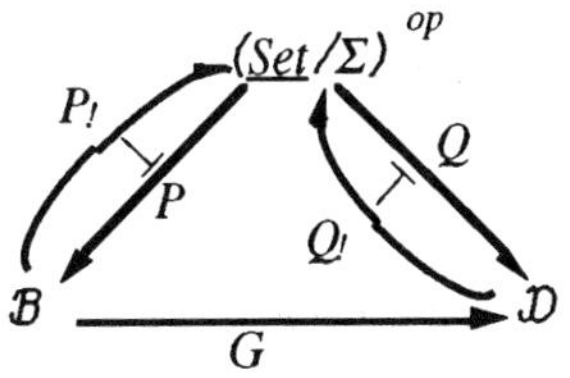

(16

with P and $P_!$ as in (7-8), $Q := GP$, and

$$Q_! \; : \; \mathcal{D} \longrightarrow (\underline{Set}/\Sigma)^{op} \; : D \mapsto \Big(\coprod_{X \in \Sigma} \lfloor D, GX \rfloor \longrightarrow \Sigma \Big)$$

There is now $\eta : D \longrightarrow GPQ_!D$, weakly initial among all arrows $D \longrightarrow GB$, for $B \in |\mathcal{B}|$.

5. Theorem. Let $\mathcal{B}$ be a complete category and $G: \mathcal{B} \longrightarrow \mathcal{D}$ a functor which preserves limits. If Σ is a solution set of $D \in |\mathcal{D}|$, then

$$\prod_{X \in \Sigma} \lceil \lfloor D, GX \rfloor, \, X \rceil \tag{17}$$

is weakly universal for G at D.

6. Remarks. A strongly universal object $G_!D := Q$ is obtained as in remark 3. If every D has a solution set, we get the object part of a functor $G_!$, left adjoint to G. The arrow part is determined by universality.

Finally, let us point out that a formula very similar to (17) appeared in Freyd's proof of his *Special* Adjoint Functor Theorem (op.cit., exercise 3.M.). Indeed, when Σ is not a solution set for D, but a cogenerating set of $\mathcal{B}$, scheme (16) takes up a quite different role. This time, the object $PQ_!D$ is not weakly universal, but its subobjects are shown to constitute a solution set for D – a *set*, provided that $\mathcal{B}$ is well powered. But this categorical construction seems to go beyond our logical intuitions.

References

Barr M., Wells C.
(1985) *Toposes, Triples and Theories*, Grundlehren der mathematischen Wissenschaften 278, Springer (Berlin)

Freyd P. J.
(1964) *Abelian Categories. An Introduction to the Theory of Functors*, Harper & Row (New York)

MacLane S.
(1971) *Categories for the Working Mathematician*, Springer (Berlin)

Pavlović D.
(1992) *On completeness and cocompleteness in and around small categories*, manuscript

Paré R.
(1974) Colimits in topoï, *Bull. of the A.M.S.* 80(3), 556-561

Author's address: DEPARTMENT OF MATHEMATICS AND STATISTICS
MCGILL UNIVERSITY
BURNSIDE HALL, MONTREAL, QUEBEC, CANADA H3A 2K6
E-mail: pavlovic@triples.Math.McGill.CA

Canadian Mathematical Society
Conference Proceeding
Volume **13**, 1992

Axiomatizing Algebraically Behaved Categories of Hausdorff Spaces

Günther Richter

Abstract. There is a whole realm of categories of Hausdorff spaces, the underlying set functors of which are of *descent type*, i.e. the corresponding Eilenberg–Moore comparison functors are fully faithful. The case of pointwise dense units of the adjunctions in question allows a purely categorical description of all such categories. Especially, this yields an axiomatic characterization of the category of *Stone spaces*, i.e. zero–dimensional compact T_0–spaces, and of the category of all compact Hausdorff spaces, which permits a new common approach to Stone duality as well as Gelfand duality.

0. Introduction

One of the fundamental differences between algebra and topology is that continuous bijections need not be isomorphisms. Therefore, and for related reasons, an arbitrary full subcategory **X** of the category **Top** of topological spaces and all continuous maps between them fails badly to be (dually) equivalent to a category of algebras and their homomorphisms, in general. Nevertheless there are a lot of remarkable exceptions, some of them given by topological

Mathematics Subject Classification (1991). Primary 18B30, 18C20, 54B30, 54D10, 54D30; Secondary 06E15, 46L05.
This paper is in final form and no version of it will be submitted for publication elsewhere.

properties like discreteness, indiscreteness or compactness, respectively. Moreover, the product closure of every strongly rigid, small, full subcategory **R** of the category $\mathbf{Top_2}$ of Hausdorff spaces together with the empty space is of *effective descent type* or *tripleable* [**3,6,8,10**] with respect to its underlying set functor, i.e. the corresponding Eilenberg–Moore comparison functor is an equivalence.

If the latter is supposed to be only full and faithful, i.e. in case of *descent type*, there exists a whole realm of examples. In fact, every right adjoint restriction of the underlying set functor to a full subcategory $\mathbf{X} \subseteq \mathbf{Top}$ with left adjoint F and counit ε yields various full subcategories **D** of descent type between the image $F(\mathbf{Set})$ of F and **X** by looking at certain $D \in \mathbf{X}$, for which εD is final [**11, 20**]. Note, that every εFT, T a set, is a retraction, hence final.

Section 1 contains a purely categorical description of a great number of categories **D** arising in this way, namely the case of pointwise dense units and $\mathbf{D} \subseteq \mathbf{Top_2}$. Pointwise density of the unit means, that closed sets are (homomorphic) images, which gives a minimal compatability between the topologies and the algebraic structures in question [**17**]. The second section specializes to $\mathbf{D} \subseteq \mathbf{Comp_2}$, the category of compact Hausdorff spaces. Axiomatizing the category $\mathbf{D} = \mathbf{Stone}$ of zero–dimensional compact T_0–spaces in section 3 permits a new appraoch to Stone duality. The same holds for $\mathbf{D} = \mathbf{Comp_2}$ and Gelfand duality in the last section.

1. Characterization

There are many more or less well–known descriptions of functors U, which are of descent type **[1,3,11,20]**. In general, there are no obvious relations to topoplogy. But if U is ranging in **Set** and if the corresponding monad preserves finite coproducts, then there is a very natural factorization of U via the underlying set functor of **Top** given by the images of underlying maps of morphisms with fixed range. In fact, for any object C in the domain of U,

$$\{\text{im } Uf \mid \text{range } f = C\}$$

is a topology of closed sets on the underlying set UC, the *image topology*. This is the main tool for the characterization below. Recall, that a full subcategory of **Top** is *nontrivial* if it contains a space with at least two points [7].

1.1. Theorem. *A category* $\mathbf{C}$ *is equivalent to a full, nontrivial subcategory* $\mathbf{D} \subseteq \mathbf{Top_2}$ *such that its underlying set functor is of descent type with pointwise dense unit iff the following axioms hold for a certain* $\mathbf{C}$*–object* $\mathbf{1}$*:*

(o) $\mathbf{C}$ *has all copowers* $I \cdot \mathbf{1} := \coprod_{i \in I} \mathbf{1}$, $I \in \mathbf{Set}$ [12].

(i) The canonical $\mathbf{C}$*–morphisms* $\varepsilon C : \mathbf{C}(\mathbf{1}, C) \cdot \mathbf{1} \longrightarrow C$ *(given by* $\varepsilon C u_f = f$, $u_f : \mathbf{1} \longrightarrow \mathbf{C}(\mathbf{1}, C) \cdot \mathbf{1}$ *coproduct injections) are coequalizers.*

(ii) (a) There are exactly two $\mathbf{C}$*–morphisms from* $\mathbf{1}$ *to* $\mathbf{2} := 2 \cdot \mathbf{1}$.

(b) $I \cdot \mathbf{1} \xleftarrow{\text{codiagonal}} 2 \cdot (I \cdot \mathbf{1}) = I \cdot \mathbf{2} \xrightarrow{\text{codiagonal}} \mathbf{2}$ *is a mono–source* [1] *for all sets* I.

(c) For any **C***–monomorphism* $m : \mathbf{2} \hookrightarrow C$ *there exists a* **C***–morphism* $d : \mathbf{C}(\mathbf{1}, C) \cdot \mathbf{1} \longrightarrow \mathbf{2}$ *with* $dn = id\ \mathbf{2}$ *for all liftings* $n : \mathbf{2} \longrightarrow \mathbf{C}(\mathbf{1}, C) \cdot \mathbf{1}$ *of* m *along* εC*, i.e.* $n\varepsilon C = m$.

Proof. Assume (o) – (ii) for $\mathbf{1} \in \mathbf{C}$. Axiom (o) means, that $U := \mathbf{C}(\mathbf{1}, -) : \mathbf{C} \longrightarrow \mathbf{Set}$ has a left adjoint F given by $FI = I \cdot \mathbf{1}$ with counit $\varepsilon = (\varepsilon C)$. The unit η of this adjunction is given by the canonical injections $\eta I(i) : \mathbf{1} \longrightarrow I \cdot \mathbf{1}$, $i \in I$. Condition (i) ensures, that U is of descent type **[3,11,20]**, and (ii)(a) forces η to be pointwise injective. Consequently, $\mathbf{C}(\mathbf{1}, \mathbf{2})$ contains precisely the two different coproduct injections $u, v : \mathbf{1} \longrightarrow \mathbf{2}$ which are coequalized by the codiagonal $\nabla : \mathbf{2} \longrightarrow \mathbf{1}, \nabla u = \mathrm{id}\ \mathbf{1} = \nabla v$. Especially, ∇ fails to be a monomorphism. But U reflects monomorphisms and limits, because it is of descent type. Consequently, $U\nabla$ is not injective, but surjective as retraction. This implies

$$|U\mathbf{1}| = |C(\mathbf{1}, \mathbf{1})| = 1 ,$$

hence U preserves the coproduct $u, v : \mathbf{1} \longrightarrow \mathbf{2}$, and $\mathbf{1}$ is terminal in $\mathbf{C}$.

Next consider any cosquare $k, l : C \longrightarrow C \amalg C$ in $\mathbf{C}$. The following diagram commutes

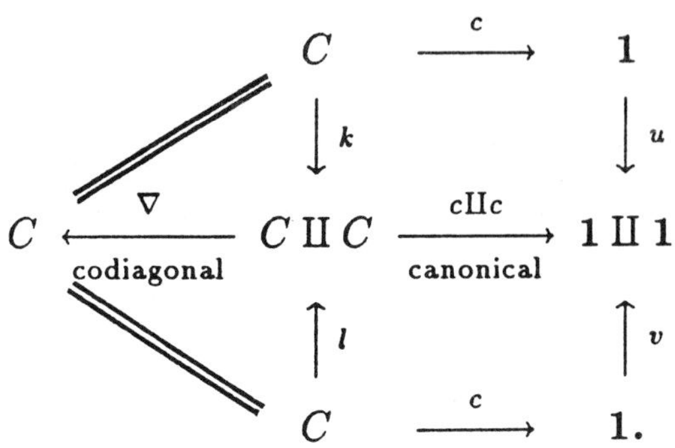

Both rectangles are pullbacks, if $(\nabla, c\amalg c)$ is a monosource. To show this, consider $\mathbf{C}$–morphisms $f : B \longrightarrow \mathbf{1}$, $g : B \longrightarrow C \amalg C$ with $uf = (c \amalg c)g$ and choose $h := \nabla g : B \longrightarrow C$. Now $ch = f$, because $\mathbf{1}$ is terminal, and

$$\left.\begin{array}{r} \nabla kh = h = \nabla g \\ (c \amalg c)kh = uch = uf = (c \amalg c)g \end{array}\right\} \Rightarrow kh = g \ .$$

Obviously, h is unique.

As a right adjoint, U preserves pullbacks. This implies

$$U(C \amalg C) = U(k)(UC) \dot\cup U(l)(UC) \ ,$$

hence U preserves the cosquare $C \amalg C$, especially for $C = FI = I \cdot \mathbf{1}$ using (ii)(b). The latter means that $UF : \mathbf{Set} \longrightarrow \mathbf{Set}$ preserves cosquares, thus arbitrary finite coproducts including the empty one by [**19, 4.1**].

A look at the corresponding category of Eilenberg–Moore algebras $\mathbf{Set}^{\mathbf{T}}$, $T := UF$, shows for every $\mathbf{C}$–object C that

$$\mathcal{C} = \{\text{im } Uf \mid \text{range } f = C\}$$

is closed under the formation of arbitrary intersections, because homomorphic images (of free algebras) coincide with subalgebras. Morevoer, the union $A \cup B$ of $A, B \in \mathcal{C}$ appears as image of $UFA \dot\cup UFB = UF(A \dot\cup B) = U(FA \amalg FB)$ and $\emptyset$ as image of $UF\emptyset = \emptyset$. Thus $\mathcal{C}$ is a topology of closed sets on UC, (the underlying mappings of) $\mathbf{C}$–morphisms are closed with respect to these topologies, and even continuous, because inverse images in $\mathbf{Set}^{\mathbf{T}}$ are images of certain free objects. This yields a factorization of U via the underlying set functor of $\mathbf{Top}$ as follows

$$\begin{array}{ccc} \mathbf{C} & \overset{H}{\dashrightarrow} & \mathbf{Top} \\ & {\scriptstyle U}\searrow \quad \swarrow {\scriptstyle \text{underlying set functor}} . & \\ & \mathbf{Set} & \end{array}$$

By construction, η is pointwise dense with respect to the image topologies [**18, 2.9**]. It remains to show, that H is full and $H(\mathbf{C}) \subseteq \mathbf{Top_2}$.

For the latter consider $x, y \in UC$ with $x \neq y$ for an arbitrary $\mathbf{C}$–object C. Using $\mathbf{1} = F1$ and $\mathbf{2} = 2 \cdot \mathbf{1}$, this yields a $\mathbf{C}$–monomorphism $m : \mathbf{2} \hookrightarrow C$ with underlying image $\{x, y\}$. By (ii)(c), there exists a $\mathbf{C}$–morphism $d : FUC \longrightarrow \mathbf{2}$ such that for arbitrary $s \in (U\varepsilon C)^{-1}(x)$, $t \in (U\varepsilon C)^{-1}(y)$ and the corresponding lifting $n : \mathbf{2} \longrightarrow FUC$ of m along εC with underlying image $\{s, t\}$ the equality $dn = \mathrm{id}\, \mathbf{2}$ holds. This means $(U\varepsilon C)^{-1}(x) \subseteq S$, $(U\varepsilon C)^{-1}(y) \subseteq T$, where $S \dot{\cup} T = UFUC$ denotes the decomposition given by d. S, T are closed as well as their images $A := (U\varepsilon C)(S)$, $B := (U\varepsilon C)(T)$. Moreover, $A \cup B = UC$ and $y \notin A$, because $y = (U\varepsilon C)(s)$ for some $s \in S$ would imply $S \cap (U\varepsilon C)^{-1}(y) \neq \emptyset$. Analogously one gets $x \notin B$, hence $UC \setminus A$ and $UC \setminus B$ separate x and y.

Now one gets $HF \overset{\eta}{\dashv} (H(\mathbf{C}) \xrightarrow{\text{underlying set}} \mathbf{Set})$ with counit $H\varepsilon$, where $H(\mathbf{C}) \subseteq \mathbf{Top_2}$ may be considered as full subcategory. This means that for any map $f : X \longrightarrow UC$ (= underlying set of HC) there is exactly one continuous extension $\bar{f} : HFX \longrightarrow HC$ of f along ηX [**18, 2.10**]. Especially, every continuous map g :

$HFX \longrightarrow HC$ appears as $g = Hh$ for some **C**–morphism $h : FX \longrightarrow C$.

To prove fullness of H, consider arbitrary objects B, C in **C** and any continuous map $c : HB \longrightarrow HC$. Then $g := cH\varepsilon B : HFUB \longrightarrow HC$ is continuous, hence $g = Hh$ for some **C**–morphism $h : FUB \longrightarrow C$ and $Uh = cU\varepsilon B$. Descent type of U forces εB to be final **[11, 20]**. Consequently, there exists a **C**–morphism $b : B \longrightarrow C$ with $Ub = c$ or $Hb = c$, respectively.

For the converse, consider $\mathbf{C} = \mathbf{D} \subseteq \mathbf{Top_2}$ as above. Then (o), (i), and (ii)(a) are obvious choosing $I \cdot \mathbf{1} = FI$, where F denotes the left adjoint of the underlying set functor of **D**, which is naturally isomorphic to $\mathbf{D}(\mathbf{1}, -)$. Note, that $F1$ is a singleton and $F2 \cong 2 = \{0, 1\}$, the two point discrete space. For (ii)(b) observe that for any space X

$$X \xleftarrow{\text{codiagonal}} X\dot{\cup}X \xrightarrow{\text{canonical}} 2$$

is a product in **Top**, where the restrictions of the canonical morphism to the summands are constant and distinct if X is nonempty. This yields a continuous map $m : F(I\dot{\cup}I) \longrightarrow FI\dot{\cup}FI$ for any set I such that

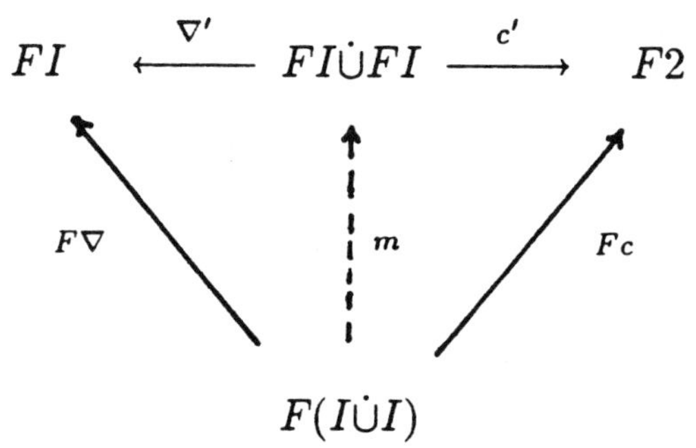

commutes, where ∇, ∇' denote the codiagonals and c, c' are canonical, respectively. Note, that $F\nabla$ and Fc are just the codiagonals mentioned in (ii)(b). Obviously, $(F\nabla, Fc)$ is a mono–source iff m is a monomorphism. To prove the latter, consider the coproduct injections $i, j : I \longrightarrow I\dot{\cup}I$, $v, w : FI \longrightarrow FI\dot{\cup}FI$, and the continuous map $s : FI\dot{\cup}FI \longrightarrow F(I\dot{\cup}I)$ with $sv = Fi$, $sw = Fj$. Let η denote the unit and ε the counit of the adjunction in question, respectively. For simplicity assume $\eta 2 = \mathrm{id}\ 2$. Now diagram chasing in

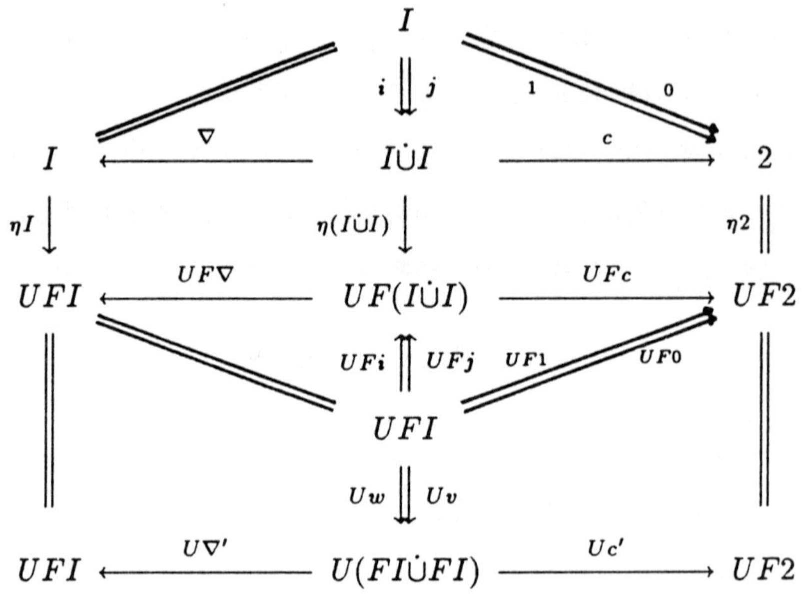

yields $ms = \mathrm{id}(F(I)\dot{\cup}F(I))$, because (∇', c') is a mono–source, (v, w) an epi–sink,

$$\nabla' msv = F\nabla Fi = \mathrm{id}\ FI = \nabla' v$$

$$c'msv = FcFi \ = F0 \quad = c'v \ ,$$

and the analogous equalities for w instead of v hold. Moreover,

$$U(sm)\eta(I\dot{\cup}I)i = U(sm)UFi\eta I = U(smsv)\eta I$$
$$= U(sv)\eta I = UFi\eta I = \eta(I\dot{\cup}I)i$$

together with the corresponding equalities for j instead of i shows $sm = \mathrm{id}\ F(I\dot{\cup}I)$.

For the last remaining step recall, that every FI is extremally disconnected **[17, 2.9]**. For any **C**–monomorphism $m : \{0,1\} \hookrightarrow C$ there are disjoint open neighbourhoods V_0 of $m(0)$ and V_1 of $m(1)$ in C, and $\varepsilon C^{-1}(V_i)$, $i = 0,1$, are disjoint open sets in FUC. Now $\varepsilon C^{-1}(V_1)^-$ is closed–open and defines a continuous map

$$d : FUC \longrightarrow 2$$

with $d(\varepsilon C^{-1}(V_i)) = \{i\}$, $i = 0,1$. Any lifting $n : 2 \longrightarrow FUC$ of m along εC fulfills

$$n(i) \in \varepsilon C^{-1}(m(i)) \subseteq \varepsilon C^{-1}(V_i) \ ,$$

hence $dn(i) = i$, $i = 0,1$.

□

1.2. Remarks.

(1) Using results from **[17]**, there is a shorter proof for 1.1.(ii)(b), because the hull $\hat{\mathbf{C}} \subseteq \mathbf{Top}$ of $F : \mathbf{Set} \longrightarrow \mathbf{C} \subseteq \mathbf{Top}$ contains the **Top**–coproducts $C\dot{\cup}C \cong C \times 2$ for any **C**–object C.

(2) If $\mathbf{B} \subseteq \mathbf{C}$ is *coseparating*, i.e. $(C \xrightarrow{f} B)_{f\in\mathbf{C}(C,B),B\in\mathbf{B}}$ is a mono–source for every **C**–object C, then the case $C \in \mathbf{B}$ suffices in 1.1.(ii)(b). Nothing has to be shown if **2** coseparates

C, because every **C**–monomorphism $m : \mathbf{2} \hookrightarrow \mathbf{2}$ is an isomorphism as well as $\varepsilon\mathbf{2}$.

(3) It will be shown elsewhere that image topologies are avaiblable for any right adjoint $U : \mathbf{C} \longrightarrow \mathbf{Set}$, the corresponding monad of which preserves finite coproducts.

(4) The image topologies on the spaces in $\mathbf{D} \subseteq \mathbf{Top_2}$ as in Theorem 1.1 yield a concrete equivalence $\mathbf{D} \simeq \mathbf{D}' \subseteq \mathbf{Top_2}$, full, such that continuous maps in $\mathbf{D}'$ are closed [**18**].

2. Compactness

The axioms (o) – (ii) mentioned in the previous section may be modified or replaced by others in order to get special kinds of subcategories $\mathbf{D} \subseteq \mathbf{Top_2}$. This section deals with the case $\mathbf{D} \subseteq \mathbf{Comp_2}$. Using **C**–morphisms $d : C \longrightarrow \mathbf{2}$, which correspond to closed–open sets, there is no problem to state compactness for the topologies generated by them in a categorical way. Unfortunately, these topologies are coarser than the image topologies, in general. Fortunately, they turn out to coincide for $C = I \cdot \mathbf{1}$. Note, that the categorical description of $\mathbf{Comp_2}$ and its regular epireflective subcategories in [**19**] uses the topologies generated by closed–open sets.

The "singleton" **1** forces the image topologies to be T_1. Using this and continuity as well as closedness of the canonical **C**–morphisms εC on gets compactness and Hausdorffness for all **C**–objects from the special objects $I \cdot \mathbf{1}$. The full subcategory generated by them turns out to be coseparated by **2**. As usual, $\beta : \mathbf{Set} \longrightarrow \mathbf{Comp_2}$ denotes the Stone–Čech compactification of discrete spaces (sets).

2.1. Definition. Let $\mathbf{C}$ be a category, $\mathbf{1}$ a $\mathbf{C}$–object, $\mathbf{2} := 2 \cdot \mathbf{1}$ in $\mathbf{C}$, and $o \in \mathbf{C}(\mathbf{1}, \mathbf{2})$. A subset $\mathcal{U} \subseteq \mathbf{C}(C, \mathbf{2})$ is called a *cover* (with respect to $o, \mathbf{1}$) if for all $\mathbf{C}$–morphisms $x : \mathbf{1} \longrightarrow C$ there exists $d \in \mathcal{U}$ with $dx = o$.

2.2. Theorem. *A category* $\mathbf{C}$ *is equivalent to a full subcategory* $\mathbf{D} \subseteq \mathbf{Comp_2}$ *with* $\beta(\mathbf{Set}) \subseteq \mathbf{D}$ *iff it fulfills the following axioms with respect to a certain* $\mathbf{C}$*–object* $\mathbf{1}$*:*

(o) $\mathbf{C}$ *has all copowers* $I \cdot \mathbf{1}$, $I \in \mathbf{Set}$.

(i) The canonical $\mathbf{C}$*–morphisms* $\varepsilon C : \mathbf{C}(\mathbf{1}, C) \cdot \mathbf{1} \longrightarrow C$ *are coequalizers.*

(ii) *(a) There are exactly two* $\mathbf{C}$*–morphisms from* $\mathbf{1}$ *to* $\mathbf{2} := 2 \cdot \mathbf{1}$.

(b) $I \cdot \mathbf{1} \xleftarrow{\text{codiagonal}} 2 \cdot (I \cdot \mathbf{1}) = I \cdot \mathbf{2} \xrightarrow{\text{codiagonal}} \mathbf{2}$ *is a monosource for all sets* I.

(c)' $\mathbf{2}$ *coseparates the full subcategory* $\mathbf{C}' \subseteq \mathbf{C}$ *given by the copowers of* $\mathbf{1}$.

(iii) For some fixed $o \in \mathbf{C}(\mathbf{1}, \mathbf{2})$ *and any set* I*, every cover* $\mathcal{U} \subseteq \mathbf{C}(I \cdot \mathbf{1}, \mathbf{2})$ *contains a finite (sub-)cover.*

Proof. Obviously, the axioms (o) – (iii) hold for $\mathbf{C} = \mathbf{D} \subseteq \mathbf{Comp_2}$, full, with $\beta(\mathbf{Set}) \subseteq \mathbf{D}$, because $I \cdot 1 = \beta I$ in this case. For the converse, Theorem 1.1 and 1.2.(2) apply to $\mathbf{C}'$, hence $\mathbf{C}' \simeq \mathbf{D}' \subseteq \mathbf{Top_2}$, full, nontrivial, with underlying set functor of descent type, and pointwise dense unit η. For simplicity assume, that $2 = \{0, 1\}$ is contained in $\mathbf{D}'$. Now show, that the image of every continuous map $f : I \cdot \mathbf{1} \longrightarrow 2^J$, I, J nonempty sets, $I \cdot \mathbf{1}$ in $\mathbf{D}'$, is closed with respect to the product topology on 2^J. This implies closedness of the canonical injection $I \cdot \mathbf{1} \hookrightarrow 2^{\mathbf{D}'(I\cdot\mathbf{1},2)}$ with respect to the image topology on $I \cdot \mathbf{1}$, which proves compactness for $I \cdot \mathbf{1}$.

For that consider $y \in 2^J \backslash f(I \cdot \mathbf{1})$, $d : 2 \longrightarrow 2$ with $d(0) = 1$, $d(1) = 0$, the product projections $p_j : 2^J \longrightarrow 2$, $j \in J$, and $d_j : I \cdot \mathbf{1} \longrightarrow 2$, $j \in J$, given by

$$d_j := \begin{cases} dp_j f & \text{for } p_j(y) = 0 \\ p_j f & \text{for } p_j(y) = 1 \ . \end{cases}$$

There is no $x \in I \cdot \mathbf{1}$ with $d_j(x) = 1$ for all $j \in J$, otherwise $d_j(x) = p_j f(x) = 1$ iff $p_j(y) = 1$, hence $f(x) = y$, a contradiction. Consequently, for every $x \in I \cdot \mathbf{1}$ there exists $j \in J$ with $d_j(x) = 0$, and, by axiom (iii), there is a finite subset $F \subseteq J$ with the corresponding property. Consider the closed subset

$$A := \bigcup_{j \in F} p_j^{-1}(2 \setminus \{p_j(y)\}) \subseteq 2^J \ .$$

Then $y \notin A$, because $y \notin p_j^{-1}(2 \backslash \{p_j(y)\})$ for all $j \in J$, but $f(I \cdot \mathbf{1}) \subseteq A$. Otherwise, $f(x) \notin A$ would imply $f(x) \notin p_j^{-1}(2 \backslash \{p_j(y)\})$, hence $p_j f(x) = p_j(y)$, thus $d_j(x) = 1$ for all $j \in F$.

Using density of ηI and compactness of $I \cdot \mathbf{1}$, the image of the canonical injection

$$I \cdot \mathbf{1} \hookrightarrow 2^{\mathbf{D}'(I\cdot\mathbf{1},2)} \cong 2^{2^I}$$

turns out to be the closure of the image of the canonical map $I \hookrightarrow 2^{2^I}$, which is just on of the usual constructions of βI. Hence $\beta I = I \cdot \mathbf{1}$ up to an isomorphism.

Assumptions (0), (i), and (ii)(a), (b) are sufficient for the construction of the image topology on $\mathbf{C}(\mathbf{1}, C)$ for every $\mathbf{C}$-object C. As mentioned above, these topologies inherit compactness from the ones on $\mathbf{C}(\mathbf{1}, I \cdot \mathbf{1})$ via the counit ε as well as Hausdorffness. The latter forces the factorization of $\mathbf{C}(\mathbf{1}, -)$ via $\mathbf{Comp_2}$ to be full just as in the proof of Theorem 1.1.

□

2.3. Remark. Up to equivalence, the full, reflective subcategories of $\mathbf{Comp_2}$ containing a two element discrete space are characterized by (o) – (iii) as above and the existence of coequalizers. The latter forces the Eilenberg–Moore comparison functor to be right adjoint.

3. Stone spaces and duality

Most approaches to Stone duality use the well–known and easy to prove fact, that the maximal ideals in any Boolean algebra B are just the kernels of all Boolean homomorphisms $h : B \longrightarrow \{0, 1\}$. This has remarkable consequences, for instance, that all Boolean monomorphisms are equalizers. This is immediate as soon as Stone duality is established. Alternatively, there is a direct proof **[14]**, which refers to surjectivity of Boolean epimorphisms **[13]**. But characterizing monomorphisms in the category **Bool** of Boolean algebras first seems to be more convenient.

3.1. Proposition. *Boolean monomorphisms are equalizers.*

Proof. Clearly, monomorphisms in **Bool** are injective. Therefore, consider some Boolean subalgebra $A \subseteq B$, $x \in B \setminus A$, and prove existence of Boolean homomorphisms $h_x, k_x : B \longrightarrow \{0, 1\}$ which concide on A and differ on x. Then the embedding $A \hookrightarrow B$ is just the equalizer of $h, k : B \longrightarrow \{0,1\}^{B \setminus A}$ with components h_x or k_x, respectively.

First of all consider the special case $A = \{0,1\}$, take maximal ideals $M, N \subseteq B$ with $x \in M$, $x' \in N$, and the corresponding homomorphisms $B \longrightarrow \{0,1\}$.

In the second case assume $x\!\uparrow \cap A = \{1\}$, where $x\!\uparrow := \{b \in B \,|\, x \leq b\}$ denotes the principal filter generated by x, and take M as above. Then $A \cap M$ is a maximal ideal in A (given by the restriction of the corresponding homomorphism $B \longrightarrow \{0,1\}$ to A). Now $A \cap M = A \cap I$, $I \subseteq B$ the ideal generated by $A \cap M$, and the embedding $A \hookrightarrow B$ yields an injection as follows

$$\{0,1\} \simeq A/A \cap M \hookrightarrow B/I \ .$$

This is just the previous case, because $x \in M$ implies $x' \notin I \subseteq M$ and $x \notin I$ by $x\!\uparrow \cap A \cap M = \emptyset$.

For arbitrary A consider the ideal

$$J := \{d \in B \mid d \leq y < x' \quad \text{for some} \quad y \in A\} \subseteq B$$

and the images $\hat{A}, \hat{x}$ of A, x in $\hat{B} := B/J$. Recall

$$b \equiv c \ (mod \ J) \Leftrightarrow b \vee d = c \vee d \quad \text{for some} \quad d \in J$$

[4], hence $a \not\equiv x' \, (mod \ J)$ for all $a \in A$. In fact, for any $d \in J$ there is $y \in A$ with $d \leq y < x'$, thus

$$a \vee d = x' \vee d = x' \Rightarrow x' = x' \vee y = a \vee y \in A \ ,$$

which contradicts $x \notin A$, consequently, $\hat{x} \notin \hat{A}$. Moreover, for any $a \in A$ and some $d \in J$

$$\begin{aligned} & a' \wedge x' \equiv a' \ (mod\ J) \Rightarrow (a' \wedge x') \vee d = a' \vee d \\ \Rightarrow \quad & (a' \vee d) \wedge x' = a' \vee d \Rightarrow a' \leq a' \vee d \leq x' \Rightarrow a' \in J \\ \Rightarrow \quad & a' \equiv 0 \ (mod\ J)\ , \end{aligned}$$

hence $\hat{x}\!\uparrow \cap \hat{A} = \{\hat{1}\}$ as in the previous case.

□

3.2. Corollary. *Boolean epimorphisms are surjective.*

Proposition 3.1 establishes axiom (i) for $\mathbf{C} = \mathbf{Bool}^{op}$, the dual of **Bool**. The distinguished object **1** in **C** is just $\{0,1\}$, and its square $\Diamond = \{0,1\}^2$ plays the role of **2**.

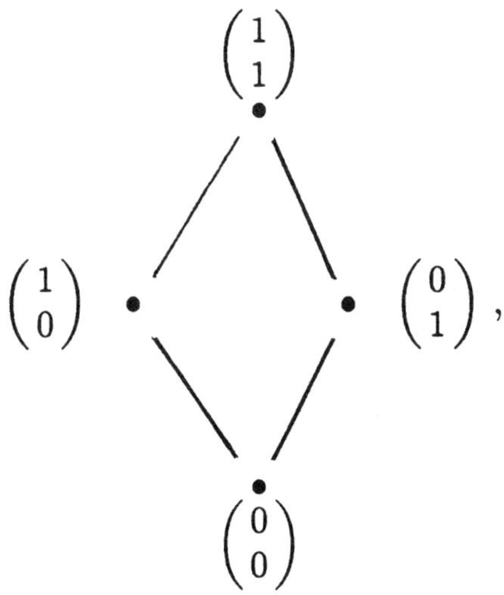

$\Diamond = \{0,1\}^2$.

3.3. Lemma. $B \xrightarrow[\text{diagonal}]{\Delta} B \times B \xleftarrow[\text{canonical}]{c} \Diamond$ *is an epi–sink in* **Bool** *for all Boolean algebras* B.

Proof. For every $\binom{a}{b} \in B \times B$ the following holds

$$\begin{aligned}\binom{a}{b} &= [\binom{a}{a} \wedge \binom{1}{0}] \vee [\binom{b}{b} \wedge \binom{0}{1}] \\ &= [\Delta(a) \wedge c\binom{1}{0}] \vee [\Delta(b) \wedge c\binom{0}{1}] \,.\end{aligned}$$

□

3.4. Lemma. *Let B be a Boolean algebra and $S \subseteq B$ a subset which fails to be contained in any maximal ideal $M \subseteq B$. Then S contains a finite subset F which fails to be contained in any M as well.*

Proof. The ideal generated by S is

$$\{b \in B \mid b \leq \vee F \text{ for some finite } F \subseteq S\} \,.$$

By assumption on S, it equals B, hence $\vee F = 1$ for some finite $F \subseteq S$.

□

3.5. Theorem. $\mathbf{Bool}^{op} \simeq$ Stone.

Proof. $\mathbf{C} := \mathbf{Bool}^{op}$ fulfills the axioms mentioned in Theorem 2.2 with respect to $\mathbf{1} = \{0, 1\}$, because

(o) **Bool** is complete.

(i) the kernel of $B \longrightarrow \{0,1\}^{\mathbf{Bool}(B,\{0,1\})}, b \mapsto (h(b))_{h \in \mathbf{Bool}(B,\{0,1\})}$ equals 0, thus it is a Boolean monomorphism, hence an equalizer by Proposition 3.1.

(ii) $\mathbf{2} = \Diamond$ is just the free Boolean algebra on one generator, which separates **Bool**. This proves (a) and (c)'. For (b) see Lemma 3.3.

(iii) $\mathbf{Bool}(\Diamond, B) \cong B$ by $\mathbf{Bool}(\Diamond, B) \ni d \mapsto d\binom{0}{1}$, hence a cover $\mathcal{U}$ corresponds to a subset $S \subseteq B$, while **C**–morphisms x with domain **1** correspond to maximal ideals M. Choose $o \in \mathbf{Bool}(\Diamond, \{0,1\})$ with $o\binom{1}{0} = 0$. This implies $o\binom{0}{1} = 1$. Now $xd = o$ in **Bool** means

$$d\binom{0}{1} \notin M := \ker x \ .$$

Consequently, $S := \{d\binom{0}{1} \mid d \in \mathcal{U}\}$ fails to be contained in any maximal ideal $M \subseteq B$ and Lemma 3.4 applies.

This proves $\mathbf{Bool}^{op} \simeq \mathbf{D} \subseteq \mathbf{Comp_2}$, full, with $\beta(\mathbf{Set}) \subseteq \mathbf{D}$. But, as mentioned above for the dual, **D** is coseparated by a two element discrete space, thus $\mathbf{D} \subseteq \mathbf{Stone}$. Finally, completeness yields $\mathbf{D} \simeq \mathbf{Stone}$, because **Stone** is the limit closure of (the two element discrete space in) **D**, and the underlying set functor of **D** preserves limits. This forces the canonical map from a limit in **D** to the corresponding one in **Stone** to be bijective, thus an isomorphism.

□

3.6. Remark. Up to equivalence, **Stone** is characterized by the axioms (o), (i), (ii)(a), (ii)(b), (iii) as in Theorem 2.2 together with

(ii) (c)" **2** coseparates **C** and

(iv) **C** is complete.

The latter may be weakened using special limits, because $A = A^- \subseteq \{0,1\}^I$ with respect to the product topology iff A is the equalizer of some pair $\{0,1\}^I \rightrightarrows \{0,1\}^J$ of continuous maps. Alternatively, axiom (iv) may be replaced by the existence of coequalizers as mentioned at the end of section 2. The duals of the axioms in question characterize **Bool** by Theorem 3.5. There is an alternative description in **[16]**, which may be used for an axiomatic approach to Stone duality as well. Details will appear elsewhere.

4. Gelfand duality for compact spaces

In this section, Theorem 2.2 is used to establish the part of Gelfand duality that can be described without the Gelfand Representation Theorem. Let $\mathbf{C}^*$ denote the full and isomorphism–closed subcategory of the category $\mathbf{R}$–$\mathbf{Alg}^\bullet$ of all unitary, commutative $\mathbf{R}$–algebras given by the function spaces

$$C(X) := \{f \mid f : X \longrightarrow \mathbf{R} \text{ continuous }\},\ X \in \mathbf{Comp_2}\ ,$$

and $\mathbf{C} := \mathbf{C}^{*op}$. As function space of a singleton, $\mathbf{R}$ is a $\mathbf{C}$–object which plays the role of $\mathbf{1}$. The maximal ideals in $C(X)$ are just the kernels of homomorphisms $C(X) \longrightarrow \mathbf{R}$, which coincide with the projections $p_x(f) := f(x)$.

$$M \subsetneq C(X) \text{ maximal ideal } \Leftrightarrow$$
$$M = \{f \in C(X) \mid f(x) = 0\} \text{ for some (unique) } x \in X\ .$$

This is originally due to M.H. Stone. For an easy and short proof see for instance [**21, 1.22**]. Especially, there are exactly two homomorphisms $\mathbf{R} \times \mathbf{R} \longrightarrow \mathbf{R}$.

4.1. Lemma. *$C(\beta I)$ is the I–the copower of $\mathbf{R}$ in $\mathbf{C}$ for any set I.*

Proof. $C(\beta I) \cong C^*(I) := \{b \mid b : I \longrightarrow \mathbf{R} \text{ bounded }\} \subseteq \mathbf{R}^I$. Consider $\mathbf{C}^*$-morphisms $h_i : C(X) \longrightarrow \mathbf{R}$, $i \in I$, $h := (h_i)_{i \in I} : C(X) \longrightarrow \mathbf{R}^I$, and $f \in C(X)$. Every h_i coincides with some projection p_{x_i}, thus $h(f) = (h_i(f))_{i\in I} = (f(x_i))_{i \in I}$ is bounded, and h factorizes via $C^*(I)$.

□

4.2. Lemma. *The canonical* $\mathbf{C}$*-morphisms* εC *are coequalizers.*

Proof. Consider again the dual situation, namely the canonical embedding

$$e : C(X) \hookrightarrow C(\beta(\text{ hom}(C(X), \mathbf{R})))$$

with $e(f)(h) = h(f) = f(x)$ for $h = p_x \in \text{hom}(C(X), \mathbf{R})$. Let $s : \beta(\text{ hom}(C(X), \mathbf{R})) \longrightarrow X$ denote the continuous extension of the surjection $h = p_x \mapsto x$, which is a coequalizer in $\mathbf{Comp_2}$. Applying $C := \mathbf{Top}(-, \mathbf{R})$ to s yields an equalizer $C(s)$ in $\mathbf{C}^*$. But $C(s) = e$, because for $h = p_x$ and any f

$$C(s)(f)(h) = (fs)(h) = f(s(p_x)) = f(x) = h(f) = e(f)(h).$$

□

4.3. Lemma. $A \xrightarrow[\text{diagonal}]{\Delta} A \times A \xleftarrow[\text{canonical}]{c} \mathbf{R} \times \mathbf{R}$ *is an epi-sink in* $\mathbf{R}$-$\mathbf{Alg}^\bullet$ *for all unitary, commutative* $\mathbf{R}$*-Algebras* A.

Proof. For every $\binom{a}{b} \in A \times A$ the following holds

$$\begin{pmatrix} a \\ b \end{pmatrix} = \begin{pmatrix} a \\ a \end{pmatrix} \cdot \begin{pmatrix} 1 \\ 0 \end{pmatrix} + \begin{pmatrix} b \\ b \end{pmatrix} \cdot \begin{pmatrix} 0 \\ 1 \end{pmatrix} = \Delta(a) \cdot c \begin{pmatrix} 1 \\ 0 \end{pmatrix} + \Delta(b) \cdot c \begin{pmatrix} 0 \\ 1 \end{pmatrix} .$$

□

Recall that $\mathbf{R}$–algebra homomorphisms $h : \mathbf{R} \times \mathbf{R} \longrightarrow A$ are just the idempotents in A, because $h\binom{1}{0} \in A$ is idempotent and vice versa, every idempotent $e \in A \setminus \{0\}$ yields a linear map $h : \mathbf{R} \times \mathbf{R} \longrightarrow A$ choosing

$$h\begin{pmatrix}1\\0\end{pmatrix} := e \;, h\begin{pmatrix}0\\1\end{pmatrix} := 1 - e \;,$$

which turns out to be a homomorphism of $\mathbf{R}$–algebras.

4.4. Lemma. $\mathbf{R} \times \mathbf{R}$ *separates the full subcategory of* $\mathbf{C}^*$ *given by the objects* $C(\beta I)$, I *a set.*

Proof. It suffices to consider distinct homomorphisms $f, g : C(\beta I) \longrightarrow \mathbf{R}$, which are projections, $f = p_x$, $g = p_y$, $x, y \in \beta I$, $x \neq y$. Distinct points in βI can be separated by a closed–open subset $S \subseteq \beta I$, $x \in S$, $y \notin S$. Now x, y are separated by $e : \beta I \longrightarrow \mathbf{R}$ with $e(S) = \{1\}$ and $e(\beta I \backslash S) = \{0\}$, which is continuous and idempotent in $C(\beta I)$. It corresponds to a homomorphism $h : \mathbf{R} \times \mathbf{R} \longrightarrow C(\beta I)$ as mentioned above and

$$fh\begin{pmatrix}1\\0\end{pmatrix} = f(e) = p_x(e) = e(x) = 1 \neq e(y) = gh\begin{pmatrix}1\\0\end{pmatrix} \;.$$

$\square$

4.5. Lemma. *Let* $E \subseteq C(X)$ *be a subset which fails to be contained in any maximal ideal* $M \subseteq C(X)$. *Then* E *contains a finite subset* F *which is not contained in any* M *as well.*

Proof. By assumption on E, the ideal generated by E contains the unit 1. Consequently, there is a finite subset $F \subseteq E$ such that

$$1 = \sum_{e \in F} e \cdot f_e$$

for some $f_e \in C(X)$, $e \in F$.

$\square$

Lemma 4.5 implies axiom (iii) for **C** just as in the proof of Theorem 3.5:

Choose $o \in \mathbf{C}^*(\mathbf{R} \times \mathbf{R}, \mathbf{R})$ with $o\binom{1}{0} = 0$. This implies $o\binom{0}{1} = 1$. Now $xd = o$ in $\mathbf{C}^*$ for $x : C(X) \longrightarrow \mathbf{R}$ means

$$d\begin{pmatrix} 0 \\ 1 \end{pmatrix} \notin M := \ker\ x\ .$$

Consequently, $E := \{d\binom{0}{1} \mid d \in \mathcal{U}\}$ fails to be contained in any maximal Ideal $M \subsetneq C(X)$.

4.6. Theorem. $\mathbf{C}^* \simeq \mathbf{Comp_2}^{op}$.

Proof. By Theorem 2.2, $\mathbf{C}^{*op} = \mathbf{C} \simeq \mathbf{D} \subsetneq \mathbf{Comp_2}$, where $\hom(C(X), \mathbf{R})$ together with the image topology is a $\mathbf{D}$–object for any compact Hausdorff space X. Now the bijections

$$j : X \longrightarrow \hom(C(X), \mathbf{R})$$

$$x \longmapsto p_x$$

turn out to be closed, hence isomorphisms, thus $\mathbf{D} \simeq \mathbf{Comp_2}$. In fact, for $A = A^- \subseteq X$, the inclusion $i : A \hookrightarrow X$, the projections $q_a : C(A) \longrightarrow \mathbf{R}$, $a \in A$, and any $f \in C(X)$ the following holds

$$\begin{aligned} j(a)(f) &= p_a(f) = f(a) = (f \circ i)(a) \\ &= q_a(f \circ i) = (q_a \circ C(i))(f) \\ &= \hom(C(i), \mathbf{R})(q_a)(f)\ . \end{aligned}$$

This proves $j(A) = \text{im hom}(C(i), \mathbf{R})$ with $C(i) : C(X) \longrightarrow C(A)$ in $\mathbf{C}^*$ given by $C(i)(f) = f \circ i$.

□

4.7. Remark.

As expected, the equivalence stated above is given by $\text{hom}(-, \mathbf{R})$ or $C : \mathbf{Comp_2}^{op} \to \mathbf{C}^*$ in the reverse direction, respectively. Clearly, this axiomatic approach to the duality between $\mathbf{Comp_2}$ and $\mathbf{C}^*$ is not the shortest one. But it separates purely categorical aspects from others and clarifies similarities of Stone and Gelfand dualities in this way. Moreover, Theorem 2.2 yields a categorical characterization of $\mathbf{Comp_2}$ (up to equivalence) by the axioms (o) – (iii), the existence of coequalizers, and effectiveness of equivalence pairs [5]. The dual properties characterize $\mathbf{C}^*$ and without the latter the coreflective subcategories of $\mathbf{C}^*$ containing $\mathbf{R} \times \mathbf{R}$.

References

[1] J. Adámek, H. Herrlich, G.E. Strecker, *Abstract and concrete categories*, Wiley, New York (1990).

[2] B. Banaschewski, G. Bruns, *The fundamental duality of partially ordered sets*, Order **5** (1988), 61–74.

[3] M. Barr, C. Wells, *Toposes, triples and theories*, Springer–Verlag, New York, Berlin, Heidelberg, Tokyo (1985).

[4] G. Birkhoff, *Lattice theory*, 3rd ed., AMS Colloquium Publications **25** (1967).

[5] J. Duskin, *Variations on Beck's tripleability theorem*, Proc. Midwest Category Seminar, III, Springer Lecture Notes in Math. **106** (1969), 74–129.

[6] H. Herrlich, *On the concept of reflections in general topology*, In: J. Flachsmeyer, H. Poppe, F. Terpe (eds.), Contributions to extension theory of topological structures, VEB Deutscher Verlag der Wissenschaften, Berlin (1969), 105–114.

[7] H. Herrlich, G.E. Strecker, *Algebra ∩ Topology = Compactness*, Gen. Top. Appl. **1** (1971), 283-287.

[8] H. Herrlich, G.E. Strecker, *Category theory*, 2nd ed., Heldermann Verlag, Berlin (1982).

[9] P.T. Johnstone, *Stone spaces*, Cambridge University Press (1982).

[10] V. Kaman, M. Rajagopalan, *Constructions and applications of rigid spaces, I*, Advances in Math. **29** (1978), 89–130.

[11] G.M. Kelly, A.J. Power, *Adjunctions whose counits are coequalizers, and presentations of finitary enriched monads*, The University of Sydney, Pure Math. Research Reports **22** (1990).

[12] F.W. Lawvere, *Functorial semantics of algebraic theories*, Proc. N.A.S. **50** (1963), 869–873.

[13] F.E.J. Linton, *Injective Boolean σ-Algebras*, Arch. Math. **17** (1966), 383-387.

[14] F.E.J. Linton, *Applied functorial semantics, I*, Annali di Matematica pura ed applicata **86**, no.4 (1970), 1–14.

[15] J.W. Negrepontis, *Duality in analysis from the point of view of triples*, Journal of Algebra **19** (1971), 228–253.

[16] A. Obtulowicz, K. Sokolnicki, *On the algebraic theory of Boolean algebras*, Bull. Acad. Pol. Sci., Sér. sci. math., astr. et phys. **26**, no.6 (1978), 483–487.

[17] G. Richter, *A characterization of the Stone-Čech Compactification*. In: J. Adámek, S. Mac Lane (eds.), Categorical Topology, Prague, Aug 22–27, 1988, World Scientific Publishing Company, Singapore (1989), 462–476.

[18] G. Richter, *Characterizations of algebraic and varietal categories of topological spaces*, Topology Appl. **42** (1991), 109–125.

[19] G. Richter, *Axiomatizing the category of compact Hausdorff spaces*, In: H. Herrlich, H.-E. Porst (eds.): Category Theory at Work, Heldermann Verlag, Berlin (1991), 199–216.

[20] W. Tholen, *Relative Bildzerlegungen und algebraische Kategorien*, Inaugural-Dissertation, Universität Münster (1974).

[21] R.C. Walker, *The Stone-Čech Compactification*, Springer-Verlag, Berlin, Heidelberg, New York (1974).

Günther Richter, Universität Bielefeld, Fakultät für Mathematik
Universitätsstraße 25, D-4800 Bielefeld 1

Canadian Mathematical Society
Conference Proceeding
Volume **13**, 1992

Relational Databases and Indexed Categories

ROBERT ROSEBRUGH AND R. J. WOOD

ABSTRACT. A description of relational databases in categorical terminology given here has as intended application the study of database dynamics, in particular we view (i) updates as database objects in a suitable category indexed by a topos; (ii) L-fuzzy databases as database objects in sheaves. Indexed categories are constructed to model the databases on a fixed family of domains and also all databases for a varying family of domains. Further, we show that the process of constructing the relational completion of a relational database is a monad in a 2-category of functors.

Introduction

We use the term *relation* for a subobject of a finite product of objects in a category. Following the relational database literature, we use the term *domain* for an object of the ambient category (and warn readers that these are *not* the ordered objects which go by the name "domain" elsewhere in theoretical Computer Science.) A *relational database*, as defined by E. F. Codd [3], is first of all a family of relations (or tables) on a family of domains. A heavily used example of domain is the set of character strings over an alphabet. Thus domains should be logically permitted to be infinite, though in practice they are always finite sets, e.g. character strings up to a fixed maximum length. The theory of databases as families of relations views domains simply as discrete objects. We adopt that point of view for this paper though the domains of practice usually have at least an order structure.

A very brief example will serve to illustrate the concepts mentioned so far. We introduce three domains, `name`, `address`, `phone`, which can each be viewed as sets of character strings satisfying appropriate constraints. An example of a database on this family of domains is the family of two relations, `address-book`, `phone-book`, where `address-book` is a subobject of `name` × `address`, and `phone-book` is a subobject of `name` × `phone`. Clearly, the storage and manipulation of databases is an important part of computing practise.

Research partially supported by grants from NSERC Canada. Diagrams typeset using Catmac.

This paper is in final form and no verison of it will be submitted for publication elsewhere.
1991 Mathematics Subject Classification. Primary 68P15; Secondary 18C20, 18D30.

The theory of relational databases is well-developed, and the relational model for databases is now the most widely implemented. Earlier database paradigms ("network" and "hierarchical") are still found in many older systems. They are not as amenable to theoretical treatment, do not provide a portable conceptual structure and are of decreasing interest. Moreover, there is active current research on enhancements and extensions of the relational model. Current editions of the texts by Date [4] or Ullman [18] contain pointers to this work.

The theory of families we use is the theory of *indexed categories* as studied by Paré and Schumacher [13]. Indexed categories are a widely used categorical tool, but have only begun to be explicitly used in theoretical computer science relatively recently [6,7,12,16]. The "relational algebra" of relational database theory involves operations which are set-theoretic and other operations which can be defined by a language involving only constants, variables of domain or relation type and equality. An objective of this article is to construct the relational completion of a database as the action of a monad, so that relationally complete databases are algebras for this monad. Section 1 gives some examples, and then the construction of a required family. In Section 2 we describe databases as families of relations in an $\mathbf{S}$-indexed category $\boldsymbol{A}$ and construct an indexed category of databases for a fixed family of domains. We then return to examples, including updates and fuzzy databases. Section 3 considers the effect of varying domains and attributes and finds an indexed category of all databases in an indexed category. In Section 4 we construct the relational completion monad. We find that the endo-functor part of the monad is an endo-functor on the fibration which arises from the indexed category of databases. Finally, we observe that relationally complete databases are 2-categories of relations.

1. The Setting

We will freely use the notion of indexed category, so we first describe the basic language of indexed categories. We begin with a base category, $\mathbf{S}$, which is required to have finite limits. Moreover, for our description of database objects, $\mathbf{S}$ must allow construction of free monoids. It suffices to assume that $\mathbf{S}$ is an elementary topos with natural numbers object N. Appropriate examples of $\mathbf{S}$ include the category of sets and functions, $\mathbf{set}$, any topos of diagrams (or presheaves), or any Grothendieck topos. A topos which will interest us below is $\mathbf{set}^{\mathbf{2}}$, the topos whose objects are functions in $\mathbf{set}$ and whose arrows are commutative squares.

An $\mathbf{S}$-*indexed category* $\boldsymbol{A}$ is given by a category $\mathbf{A}^I$ for each object I in $\mathbf{S}$ and a functor $\alpha^* : \mathbf{A}^I \longrightarrow \mathbf{A}^J$ for each arrow $\alpha : J \longrightarrow I$ in $\mathbf{S}$. These *substitution functors* α^* are subject to isomorphisms making them compatible with identities and composition in $\mathbf{S}$, and coherent with associativity. For example, if $\beta : K \longrightarrow J$ is also in $\mathbf{S}$, then there is a canonical isomorphism $(\alpha\beta)^* \cong \beta^*\alpha^*$. For a complete description see [13].

We will often want $\boldsymbol{A}$ to be just $\mathbf{S}$ with a canonical indexed structure. We denote it by $\boldsymbol{S}$ with $\mathbf{S}^I$ defined to be the slice category $\mathbf{S}/I$, and the required substitutions defined by pullbacks. We detail two examples of $\boldsymbol{S}$ now.

EXAMPLE 1. When we take $\mathbf{S}$ to be the category $\mathbf{set}$, we find that the $\mathbf{set}$-indexed category $\boldsymbol{set}$ has, for any set I, ordinary I-indexed families of sets as its I-indexed families. This follows since **set**/I has functions with codomain I as objects. Such a function, $x : X \longrightarrow I$ say, may be identified with a family of sets $< X_i >_{i \in I}$ defined by $X_i = x^{-1}(i)$, and conversely. In fact *any* category is **set**-indexed, again taking I-indexed families to be just ordinary families of objects.

EXAMPLE 2. When **S** is $\mathbf{set}^{\mathbf{2}}$ we get a more interesting indexing. The indexing objects are now functions in **set**, e.g. $I : I_0 \longrightarrow I_1$, and an I-indexed family X, being an arrow of $\mathbf{set}^{\mathbf{2}}$, is a pair of functions $x_0 : X_0 \longrightarrow I_0$, $x_1 : X_1 \longrightarrow I_1$ making a commutative square in **set**, $x_1 X = I x_0$. Substitutions are defined by pullback which are computed point-wise.

EXAMPLE 3. Another example of a $\mathbf{set}^{\mathbf{2}}$-indexed category arises when we allow the object X above to be replaced by a *partial* function, which we will denote $X : X_0 \longleftarrow X_d \xrightarrow{X} X_1$. Thus, when $I = 1$ we obtain the category whose objects are partial functions and whose morphisms, from X to Y say, are pairs of functions $f_0 : X_0 \longrightarrow Y_0$ and $f_1 : X_1 \longrightarrow Y_1$ such that the restriction of f_0 to X_d factors through Y_d by f_d and $Y f_d = f_1 X$. An I-indexed family is a pair $X_0 \xrightarrow{x_0} I_0$, $X_1 \xrightarrow{x_1} I_1$ so that $x_1 X = I x_d$ — with x_d the restriction of x_0 to X_d. A morphism in I-indexed families is a pair (f_0, f_1) of functions so that in

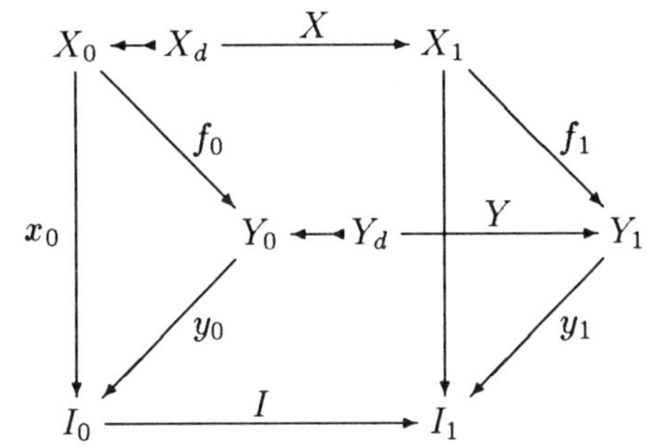

f_0 restricts to $f_d : X_d \longrightarrow Y_d$ and $f_1 X = Y f_d$. Substitution is still accomplished by pointwise pullback, including on the domain of full definition, X_d. We denote the resulting indexed category by $\mathbf{set}^{pf}$.

We want to define a relational database to be a J-indexed family of relations in $\boldsymbol{A}$ on some I-indexed family of domains, say A. A central feature of indexed category theory is that it identifies a J-indexed family of structures as a structure in the category of J-families, e.g. a J-indexed family of groups is a group in J-families. Similarly, a J-indexed family of relations is a *single* relation in the category $\mathbf{A}^J$ of J-indexed families. To define relational database in $\mathbf{A}$ we need to be able to say when a J-indexed family of relations is a family of subobjects of finite products of domains in A. In order to make this requirement precise, we will need some notation and some hypotheses on $\boldsymbol{A}$. The remainder of this section provides this background.

For an object I of **S** we denote the free monoid on I by $M(I)$. Henceforth *we assume that* $M(I)$ *exists in* **S**. It is well known that $M(I)$ exists in any topos with a natural numbers object [10,13]. We also need to assume, and do so for the remainder of this paper that $\boldsymbol{A}$ *has finite products.* This requires that each $\mathbf{A}^I$ has finite products preserved by the α^*. If A is an object of $\mathbf{A}^I$, we will need the $M(I)$-indexed family of "finite products of members of A", denoted $P_0(A)$. When **S** is **set** the family desired has as fibre over a word $w = i_1 i_2 \ldots i_k \in M(I)$, the (finite) product whose description is $A^w = A_{i_1} \times A_{i_2} \times \ldots \times A_{i_k}$. We conclude this section by finding sufficient conditions for the existence of $P_0(A)$.

Under suitable hypotheses, the required family of finite products can be constructed as a solution to a "recursion problem" [15] for the indexed functor "crossing with A". We recall that, for **S**-indexed categories **A** and **B**, an *indexed functor* $F : \mathbf{A} \longrightarrow \mathbf{B}$ is a family of functors $F^I : \mathbf{A}^I \longrightarrow \mathbf{B}^I$, one for each I in **S**. Further, for any arrow $\alpha : J \longrightarrow I$ in **S**, we must have the squares

$$\begin{array}{ccc} \mathbf{A}^I & \xrightarrow{F^I} & \mathbf{B}^I \\ {\scriptstyle \alpha^*}\downarrow & & \downarrow{\scriptstyle \alpha^*} \\ \mathbf{A}^J & \xrightarrow[F^J]{} & \mathbf{B}^J \end{array}$$

commuting up to coherent isomorphism. A *recursion problem* on **A** is a pair (Φ, C_0) with Φ an indexed endofunctor of **A** and C_0 in $\mathbf{A}^1$. The recursion problem (Φ, C_0) *has a solution* if there is an object C in $\mathbf{A}^N$ such that $0^*C = C_0$ and $s^*C = \Phi^N C$.

An hypothesis we shall need on **A** is that it has an indexed functor $E : \mathbf{A} \longrightarrow \mathbf{S}$ with small fibres. An indexed functor E has *small fibres,* when the objects of $\mathbf{A}^I$ whose image under E^I is a given object of $\mathbf{S}^I$ form a family indexed by an object of $\mathbf{S}/I$. An indexed functor with codomain $\mathbf{S}$ and small fibres was called an "e-functor" in [15]. The name refers to "elements" since the idea is that E^I gives a (very rough) idea of the cardinality of an object in $\mathbf{A}^I$. Examples include the identity functor on $\mathbf{S}$ and forgetful functors from (the **S**-indexed) category of groups (in **S**.) More interesting, any category of sheaves has an e-functor to **set** given by taking the union of all sections.

In the proposition which follows, we use a basic technique of indexed category theory "localizing" at $M(I)$. That is, we use the $\mathbf{S}/M(I)$-indexed category $\mathbf{A}^{M(I)}$, whose σ-indexed families, for $\sigma : J \longrightarrow M(I)$, are defined by $(\mathbf{A}^{M(I)})^\sigma = \mathbf{A}^J$. The hypotheses on the indexed category **A** remain true for $\mathbf{A}^{M(I)}$, and allow us to construct the required object there. We denote the inclusion of generators I in the monoid $M(I)$ by η. The internal sum of A in $\mathbf{A}^I$ along η is denoted $\Sigma_\eta A$. When **S** is **set**, $\Sigma_\eta A$ is the $M(I)$ family with fibre A_i over i and empty elsewhere.

PROPOSITION 1. *Let* **S** *be a topos with NNO and* **A** *an* **S***-indexed category with an e-functor* $E : \mathbf{A} \longrightarrow \mathbf{S}$. *Suppose that* **A** *has finite products and that the canonical arrows* $E^I(A \times B) \longrightarrow E^I(A) \times E^I(B)$ *are all monic. Suppose that* A *is in* $\mathbf{A}^I$ *and* $\Sigma_\eta A$ *exists. Then the recursion problem* $(\Phi, 1)$*, where* $\Phi : \mathbf{A}^{M(I)} \longrightarrow \mathbf{A}^{M(I)}$ *is defined by* $\Phi(C) = C \times \Sigma_\eta A$*, has a solution.*

PROOF. We first localize E to $E^{M(I)} : \mathbf{A}^{M(I)} \longrightarrow \mathbf{S}^{M(I)}$, and note that the canonical arrows are still monic. Now we can invoke Proposition 2.2 of [15] as soon as we show that Φ is mono-bounded, i. e. for all C in $\mathbf{A}^{M(I)}$ there is a B in $\mathbf{S}/M(I)$ such that

(i) $E^{M(I)}(C) \rightarrowtail B$;

(ii) for all C' in $(A^{M(I)})^\sigma$, if $E^\sigma(C') \rightarrowtail B$ then $E^\sigma(\Phi^\sigma C') \rightarrowtail \sigma^* B$.

We let $B = \Omega^{(E(C)+E(\Sigma_\eta A)+2)^N}$ where Ω is the subobject classifier in $\mathbf{S}/M(I)$. Then using the coproduct injection and singleton, $E(C) \rightarrowtail B$ is clear and

$$E(\Phi C') = E(C' \times \Sigma_\eta A) \rightarrowtail E(C') \times E(\Sigma_\eta A) \rightarrowtail B \times B \rightarrowtail B.$$

The last monomorphism, whose existence is guaranteed by Lemma 2.2.3, Chapter 5 of [13], localizes to $(\mathbf{S}/M(I))^\sigma$ giving (ii). ∎

2. A Categorical Model for Databases

For this section we assume that $\mathbf{S}$ is a topos with NNO and $\boldsymbol{A}$ is $\mathbf{S}$-indexed and has finite products and an e-functor to $\boldsymbol{S}$. The *scheme* of a relation in a database is the list of attributes (= names of domains) appearing in the relation, that is, if A in $\mathbf{A}^I$ is an I-indexed family viewed as a family of domains, the scheme of a relation is in $M(I)$. A database with J relations and (database) signature $\sigma : J \longrightarrow M(I)$ is as follows:

DEFINITION 1. *Let A in $\mathbf{A}^I$ be a family of domains. An* A-database *with* database signature σ, *where $\sigma : J \longrightarrow M(I)$, is a relation $R : R_0 \rightarrowtail \sigma^* P_0(A)$ in $\mathbf{A}^J$.*

REMARKS. The signature σ of a database should not be confused with the scheme of a relation in the database. When $\mathbf{A} = \boldsymbol{S} = \mathbf{set}$, the schemes of the relations in the database, R_j for $j \in J$, are elements of $M(I)$ given by the signature, $\sigma : J \longrightarrow M(I)$.

As J and σ vary we obtain databases with varying families of relations. From this variation we will find a category of databases below.

The category $\mathbf{S}/M(I)$ is the base for indexing below, but we wish to also note that the full subcategory of $\mathbf{set}/M(I)$ whose objects are those $J \longrightarrow M(I)$ with J finite is of important practical interest, for these objects are signatures of databases in $\mathbf{set}$ with finitely many relations. However, this subcategory fails to have a terminal object. (In fact the terminal object in $\mathbf{S}/M(I)$ is $1_{M(I)} : M(I) \longrightarrow M(I)$, whose domain is essentially never finite.) The subcategory is thus unsuitable as a base category for indexing as arguments by localization become impossible. ∎

For any object $\sigma : J \longrightarrow M(I)$ of $\mathbf{S}/M(I)$, we define $\mathcal{D}(A)^\sigma$ to be the partial order, hence category, of subobjects of $\sigma^* P_0(A)$. If $\alpha : \sigma' \longrightarrow \sigma$ is an arrow of $\mathbf{S}/M(I)$, i.e. $\sigma\alpha = \sigma' : J' \longrightarrow M(I)$, define $\alpha^* : \mathcal{D}(A)^\sigma \longrightarrow \mathcal{D}(A)^{\sigma'}$ using substitution in $\boldsymbol{A}$.

PROPOSITION 2. *Assume that all the functors α^* preserve monomorphisms. The $\mathcal{D}(A)^\sigma$ determine an $\mathbf{S}/M(I)$-indexed category denoted $\mathcal{D}(A)$.*

PROOF. We merely note that to $R : R_0 \rightarrowtail \sigma^* P_0(A)$, we have $\alpha^* R$ defined by

$$\alpha^* R_0 \rightarrowtail \alpha^* \sigma^* P_0(A) \cong \sigma'^* P_0(A).$$

Checking functorality of α^* and the equations is routine. ∎

REMARKS. The indexed category of databases which we denote $\mathcal{D}(A)$ above is locally a poset; in fact, it is the indexed category $\boldsymbol{sub}(P_0(A))$ described in [13], and when $\boldsymbol{A} = \boldsymbol{S}$ it is internally complete and cocomplete. Below we will consider subcategories arising by restricting the morphisms of $\mathcal{D}(A)$. ∎

Except for Example 3, where we will comment in more detail, we may take the identity functor as e-functor in the examples below. Thus all conditions of Proposition 1 are easily met. The substitution functors below are all defined by pulling back, so the conditions of Proposition 2 are met in all of the examples.

EXAMPLE 1 - CONTINUED. Once again, let $\mathbf{S} = \mathbf{set}$ and $\boldsymbol{A} = \boldsymbol{set}$. We note that for any family $< A_i >_{i \in I}$ of sets, $P_0(A)$ is the family of finite products of the A_i. When I, all of the A_i and J are finite, a database according to Definition 1 coincides exactly with the usual definition of a relational database with I domains $< A_i >_{i \in I}$ and J relations. That is, we have not lost anything by considering the setting proposed. We have gained a definition of morphism of databases — here just an inclusion, and further a definition of substitution along a function $\phi : K \longrightarrow J$, say. This substitution with ϕ monomorphic, for example, defines a database from a subset of the current relations. It might define an authorization class for example.

EXAMPLE 2 - CONTINUED. For an object A in $(\mathbf{set}^{\mathbf{2}})^I$, the object $P_0(A)$ is constructed "pointwise" i.e.

$$\begin{array}{ccc} P_0(A_0) & \xrightarrow{P_0(A)} & P_0(A_1) \\ {\scriptstyle p_0}\downarrow & & \downarrow{\scriptstyle p_1} \\ M(I_0) & \xrightarrow[M(I)]{} & M(I_1) \end{array}$$

where to $(a_{i_1}, \ldots, a_{i_k}) \in p_0^{-1}(i_1 \ldots i_k)$ we have

$$P_0(A)(a_{i_1}, \ldots, a_{i_k}) = (A(a_{i_1}), \ldots, A(a_{i_k})) \in p_1^{-1}(M(I)(i_1 \ldots i_k))$$

Thus a database object with $J : J_0 \longrightarrow J_1$ relations arises from a diagram such as the following:

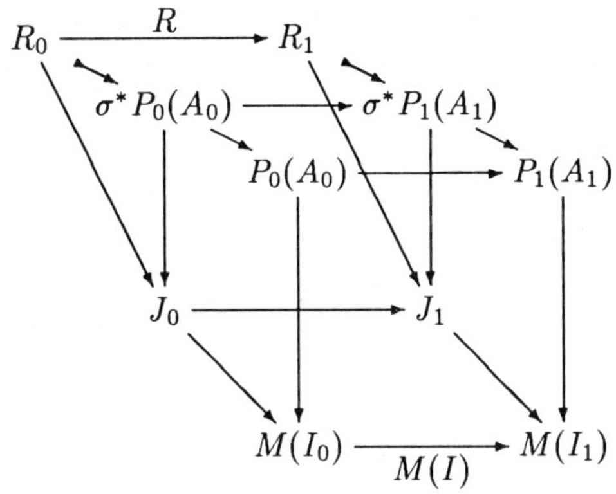

That is, it consists of an A_0 database, R_0, in ***set*** with J_0 relations, an A_1 database, R_1, in ***set*** with J_1 relations and a function $R : R_0 \longrightarrow R_1$ so that a tuple $\mathbf{a} = (a_{l_1}, \ldots, a_{l_k})$ in the j'th relation of R_0 has $R(\mathbf{a})$ in the $J(j)$'th relation of R_1. Thus R itself is a rather general morphism of databases in that the domains, relations and attributes are allowed to be fully variable. This notion of database morphism will be encountered again in the next section. Restricting A and I to be identity functions, and similarly J, provides a simpler notion of database morphism. We exploit such restrictions in the next example.

EXAMPLE 3 - CONTINUED. Again letting $\mathbf{S} = \mathbf{set^2}$, but now with $\boldsymbol{A} = \boldsymbol{set}^{pf}$, we obtain rather complicated objects similar to those in Example 2 as databases, but restricting attention to special cases will provide a description of the notion of "update". First, in order to guarantee that $P_0(A)$ exists we define an e-functor $E : \boldsymbol{A} \longrightarrow \boldsymbol{S}$ as in Proposition 1. If $a_0 : A_0 \longrightarrow I_0, a_1 : A_1 \longrightarrow I_1$ is an object of $\mathbf{A}^I$ with domain $A_0 \longleftarrow A_d \longrightarrow A_1$, we define $E^I(a_0, a_1)$ by

$$\begin{array}{ccc} A_0 & \xrightarrow{A^+} & A_1 + I_1 \\ {\scriptstyle a_0}\downarrow & & \downarrow{\scriptstyle a_1^+} \\ I_0 & \xrightarrow[I]{} & I_1 \end{array}$$

where A^+ is the canonical extension of $A : A_d \longrightarrow A_1$ given by viewing A as an I_1-indexed family of functions in **set** viz,

$$\langle \Sigma_{I(i_0)=i_1} (A_d)_{i_0} \longrightarrow (A_1)_{i_1} \rangle_{i_1 \in I_1}$$

and since $(A_d)_{i_0} \rightarrowtail (A_0)_{i_0}$ we may extend the I_1-family canonically to

$$\langle \Sigma_{I(i_0)=i_1} (A_0)_{i_0} \longrightarrow (A_1)_{i_1} + 1 \rangle_{i_1 \in I_1}.$$

Then a_1^+ is defined on $A_1 + I_1$ using a_1 and the identity on I_1. The E^I define an indexed functor $E : \boldsymbol{A} \longrightarrow \boldsymbol{S}$ with small fibres. We omit the technicalities of proving this. With this definition of E it is easy to see that

$E^I(A \times B) \rightarrowtail E^I(A) \times E^I(B)$. Moreover, $\Sigma_\eta A$ exists as internal sums are inherited from $\mathbf{set^2}$. So the requirements of Proposition 1 are met. The experienced reader will have noted that $P_0(A)$ could have been constructed directly in the case at hand. The point of the construction given is that it can be carried out for $\mathbf{S}^{pf}$ for any topos $\mathbf{S}$ with NNO. In that case, $a_1^+ : A_1 + I_1 \longrightarrow I_1$ above is replaced with the partial morphism classifier [8] for a_1 in $\mathbf{S}/I_1$.

An update of a database which does not change the current family of relations is one or several of three possible actions on elements (called "tuples") of the current relations:
(i) delete a tuple from a relation
(ii) add a tuple to a relation
(iii) change the values of some components of a tuple.
In fact, the third action can clearly be accomplished by a combination of the first two actions, namely delete the tuple in its current state and add it in the new state. Thus, an update can be accomplished by specifying a subset of current tuples which will be unchanged by the update (i.e. deleting the tuples to be changed or deleted) and specifying tuples to be added. In short, it is specified by a partial monomorphism whose domain is the current set of tuples, whose subset of total definition is the set of unchanged tuples, and whose function part is a monomorphism on that subset. Now let us describe the database objects in $\mathbf{set}^{pf}$ which express updates. For simplicity, we assume that the family of domains and the database signature are constant. That is, we assume that I and J in $\mathbf{set^2}$ are identity functions, and A in $(\mathbf{set}^{pf})^I$ has a fully defined identity function as its domain. In this special case, consider the diagram below for a database r with signature $\sigma : J \longrightarrow M(I)$. In the diagram following we denote the domain of $\sigma^* P_0(A)$ by P.

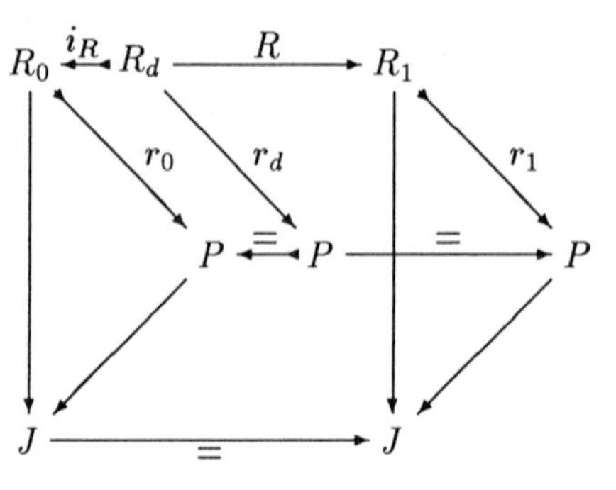

We see that since r_0 and i_R are monic, r_d is monic and then R is too as $r_1 R = r_d$. The conclusion is:

a database object in $\mathbf{set}^{pf}$ *is an update of a database object in* $\mathbf{set}$.

It follows from this conclusion that we can reason about updates by reasoning about mere database objects in $\mathbf{set}^{pf}$.

EXAMPLE 4. Let us denote by ω the category freely generated by the graph

$$\mathbf{0} \longrightarrow \mathbf{1} \longrightarrow \mathbf{2} \longrightarrow \ldots.$$

$\mathbf{set}^{\omega}$ is a topos whose objects are diagrams

$$X : X_0 \xrightarrow{x_0} X_1 \xrightarrow{x_1} X_2 \xrightarrow{x_2} \ldots$$

with X_i sets and x_i functions, and which can be thought of as X varying in discrete time steps encoded in the x_i. An analysis similar to that in Example 2 shows that a database object in $\mathbf{set}^{\omega}$ is a sequence $R_0, R_1, \ldots$ of databases in **set**, and the general morphisms of them discussed in Example 2. Restrictions to constant attribute families and relation signatures (the σ's of Definition 1) again provide a description of a database's variation through discrete time steps.

An important class of elementary topos is the categories of sheaves on a topological space, or more generally a locale. All base categories $\mathbf{S}$ above are of this form. Categories of fuzzy sets, after appropriate completion for fuzzy equality, are categories of sheaves on a locale, as Barr [1] has explained.

EXAMPLE 5. Let L be a locale. Following Barr, we will denote by L^+ the locale L with a new bottom element adjoined. Then (the completion for fuzzy equality of) L-fuzzy sets is the category of sheaves on L^+, $\mathbf{sh}(L^+)$. It is thus appropriate to define an *L-fuzzy database* to be a database object in $\mathbf{sh}(L^+)$. With this definition, we note that Examples 1 and 2 above are special cases; Example 4 can be modified to $\mathbf{set}^{\omega+1}$, where $\omega + 1$ is $\mathbf{0} \longrightarrow \mathbf{1} \longrightarrow \ldots \omega$.The topos $\mathbf{set}^{(\omega+1)^+}$ is of the form $\mathbf{sh}(L^+)$. It is worth emphasizing that, with the definition we give here, an L-fuzzy database has not only fuzzy relations, but also fuzzy domains and fuzzy index objects for its family of relations. Compare [14].

3. More indexed categories of databases

We have concentrated on a fixed (I-indexed) object A in the discussion above. It is clear that variation of A within $\mathbf{A}^I$, and further, variation of I, will define more indexed categories of databases. We have already seen such morphisms for the case $\mathbf{S} = \mathbf{set}$. They appeared as objects with variable attribute families and indexes for them, together with their associated databases in $\mathbf{set}^{\mathbf{2}}$. We indicate a few definitions:

Let $f : B \longrightarrow A$ be a morphism of $\mathbf{A}^I$. The arrow f determines an arrow of $\mathbf{A}^{M(I)}$ which we denote $P_0(f) : P_0(B) \longrightarrow P_0(A)$. For signatures $\sigma : J \longrightarrow M(I)$ and $\tau : K \longrightarrow M(I)$ we have a morphism between them in $\mathbf{S}/M(I)$ when there is $\nu : J \longrightarrow K$ satisfying $\tau\nu = \sigma$. Now we define a category $\mathcal{D}(\mathbf{A})^I$ as follows. Objects are triples (A, σ, R) with R an A-database with signature σ. An arrow from (A, σ, R) to (B, τ, R') is a pair $(f : A \longrightarrow B, \nu : \sigma \longrightarrow \tau)$ such that $\sigma^* P_0(f)$ restricted to R factors through $\nu^* R'$ as in the following (recall that we assume substitution preserves monics):

$$\begin{array}{ccc}
\sigma^* P_0(A) & \xrightarrow{\sigma^* P_0(f)} & \nu^* \tau^* P_0(B) = \sigma^* P_0(B) \\
\uparrow & & \uparrow \\
R & \xrightarrow[\phi]{} & \nu^* R'
\end{array}$$

Thus, when $\mathbf{S} = \mathbf{set}$ for example, we are requiring that for any $r \in R, \phi_j(r) = (\sigma^* P_0(f))_j(r)$ lies in $(\nu^* R')_j = R'_{\nu(j)}$. Since f determines ϕ, our description of arrows is appropriate.

Now suppose $\alpha : J \longrightarrow I$ is an arrow of $\mathbf{S}$. We obtain a functor $\alpha^* : \mathcal{D}(\mathbf{A})^I \longrightarrow \mathcal{D}(\mathbf{A})^J$ when we define

$$\alpha^*(A, \sigma, R) = (\alpha^*(A), \sigma', R')$$

where $\alpha^*(A)$ refers to the indexed structure on $\mathbf{A}$, $R' = (M(\alpha)')^* R$, and the following is a pullback:

$$\begin{array}{ccc}
J' & \xrightarrow{M(\alpha)'} & J \\
{\scriptstyle \sigma'} \downarrow & & \downarrow {\scriptstyle \sigma} \\
M(J) & \xrightarrow[M(\alpha)]{} & M(I)
\end{array}$$

PROPOSITION 3. *The categories $\mathcal{D}(\mathbf{A})^I$ and functors α^* determine an $\mathbf{S}$-indexed category, denoted $\mathcal{D}(\mathbf{A})$.*

PROOF. All that remains after the construction above is to verify that the $\mathcal{D}(\mathbf{A})^I$ are categories and that the α^* are functorial and compatible with identities and composition. This is a good exercise in the definitions of indexed category. ∎

4. Relational Completion is a Monad

The information obtained by querying a relational database is presented as relations. The relations which present the information derived are obtained by operating on the "base tables" or stored relations. The operations are collectively known as the "relational algebra" and depend on the schemes of stored relations as well as on the constants in the family of domains for the database. Simple examples of the operations are "list tuples in either of two relations" (union), "list the cartesian product of two relations", "list tuples whose value in one attribute equals their value in another" (selection.) The reader will have noted that combining the latter two sorts of operations allows formation of pullbacks. We will say that a database is *relationally complete* if it is closed under the operations of relational algebra. Of course, a relationally complete database can never be physically stored, having at least countably many relations. Thus, the ability to present any relation in the relational completion of a physically stored database is an important objective for a database system. Our objective in this section is to show that constructing the relational completion of database objects on a family of domains is a monad in a suitable 2-category, and that relationally complete databases are precisely algebras for this monad. Until further notice, *we assume that our databases are in* **set**, though much of what we say holds for databases in indexed categories satisfying the hypotheses of Proposition 1 by having $\boldsymbol{A} = \boldsymbol{S}$. Various families of binary and unary operations on the relations in a database have been called the "relational algebra". We describe a simple family of them namely two boolean operations, cartesian product, projection and selection. We remark that the other common operations, with the exception of difference, may be derived from those mentioned below [11]. We are unable to handle the binary difference of relations operation in our setup. Unlike the other operations of relational algebra we use here, it is not monotonic. This spoils functoriality of relational completion and that is too high a price to pay. We make some further remarks about this difficulty below.

The boolean operations are the simplest to describe. To do so, suppose that $R : R_0 \rightarrowtail \sigma^* P_0(A)$ is an A-database with signature $\sigma : J \longrightarrow M(I)$ and r, s are relations in R with the same scheme i.e. for some $j_1, j_2 : 1 \longrightarrow J$ we have $r = j_1^*(R_0)$, $s = j_2^*(R_0)$ and $\sigma j_1 = \sigma j_2$ so $r, s \rightarrowtail (\sigma j_1)^* P_0(A)$. Then $r \cup s, r \cap s \rightarrowtail (\sigma j_1)^* P_0(A)$ are relations with scheme σj_1. Next we consider cartesian product. If $r \rightarrowtail (\sigma j_1)^* P_0(A)$ and $s \rightarrowtail (\sigma j_2)^* P_0(A)$ then $r \times s \rightarrowtail (\sigma j_1 \cdot \sigma j_2)^* P_0(A)$ where $\sigma j_1 \cdot \sigma j_2$ denotes product in M(I) of σj_1, and σj_2.

Projection and selection require a little more care. Suppose that $j : 1 \longrightarrow J$ and that $r \rightarrowtail (\sigma j)^* P_0(A)$. Since σj is a word of length $n : 1 \longrightarrow N$, say, we may view the word as $\sigma j : [n] \longrightarrow I$. Now let $\varphi : [n'] \rightarrowtail [n]$ be one-one. The *projection of r along φ*, denoted $\pi_\varphi r$ is the (n'-ary) relation obtained by "projecting on the n' columns specified by φ." Finally, for j and r as above, and viewing σj as $\sigma j : [n] \longrightarrow I$, suppose that $\sigma j(i_1) = \sigma j(i_2) \in I$ and $c \in A_{\sigma j(i_1)}$. The selection operators $S_{i_1 = i_2}(r)$ and $S_{i_1 = c}(r)$ are defined by equalizers with codomain r as follows:

$$S_{i_1=i_2}(r) \rightarrowtail r \longrightarrow (\sigma j)^* P_0(A) \underset{\pi_{i_2}}{\overset{\pi_{i_1}}{\rightrightarrows}} A_{\sigma j(i_1)}$$

and

$$\begin{array}{ccccccc} S_{i_1=c}(r) & \rightarrowtail & r & \rightarrowtail & (\sigma j)^* P_0(A) & \xrightarrow{\pi_{i_1}} & A_{\sigma j(i_1)} \\ & & & \searrow & & \nearrow_{c} & \\ & & & & 1 & & \end{array}$$

Before proceeding, to establish notation, we give a brief description of 2-categories. The standard reference is [9]. A *2-category* $\mathbf{B}$ consists of a class of objects, denoted $\mathbf{B}_0$, and for each pair B, C of objects, a *hom category*, denoted $\mathbf{B}(B, C)$, whose objects are called 1-cells, and whose arrows are called 2-cells. We denote a 2-cell by $\sigma : f \Longrightarrow g$ in $\mathbf{B}(B, C)$. Composition of 2-cells in $\mathbf{B}(B, C)$

is called *vertical composition*. There is also a *horizontal composition* of 2-cells defined by a functor $\mathbf{B}(C,D) \times \mathbf{B}(B,C) \longrightarrow \mathbf{B}(B,D)$ for each triple B, C, D of objects of $\mathbf{B}$. An identity 1-cell in $\mathbf{B}(B,B)$ for each object B completes the necessary data. The data are required to satisfy associativity of horizontal composition, neutrality of identity arrows for horizontal composition and an interchange law linking horizontal and vertical composition. We generally denote vertical composition by juxtaposition or $\cdot$, and horizontal composition by $\circ$. The interchange law, for example, states that for

$$B \;\underset{\Downarrow\sigma'}{\overset{\Downarrow\sigma}{\longrightarrow}}\; C \;\underset{\Downarrow\tau'}{\overset{\Downarrow\tau}{\longrightarrow}}\; D$$

we have $(\tau' \cdot \tau) \circ (\sigma' \cdot \sigma) = (\tau' \circ \sigma') \cdot (\tau \circ \sigma)$. The motivating example of a 2-category is **cat**, the 2-category with categories as objects, functors as 1-cells and natural transformations as 2-cells. Another important example is **rel**, the 2-category of relations. The objects of **rel** are sets. If B and C are sets, then the category $\mathbf{rel}(B,C)$ is the partial order, qua category, of relations from B to C, so vertical composition is trivial and horizontal composition is the usual composition of relations. It is defined on 2-cells since if $R \subseteq R' : B \longrightarrow C$ and $S \subseteq S' : C \longrightarrow D$ are relations then $SR \subseteq S'R'$. The interchange law follows since all equations among 2-cells in **rel** are trivial: inclusion either holds or it doesn't.

We next describe the 2-category $\mathbf{cat^2}$ in which we will work below. The objects of $\mathbf{cat^2}$ are functors. If $A : \mathbf{A}_0 \to \mathbf{A}_1$ and $B : \mathbf{B}_0 \to \mathbf{B}_1$ are functors a 1-cell $F : \mathbf{A} \to \mathbf{B}$ is a triple $\mathbf{F} = (F_0, F_1, \Phi)$ where $F_0 : \mathbf{A}_0 \to \mathbf{B}_0$ and $F_1 : \mathbf{A}_1 \to \mathbf{B}_1$ are functors and $\Phi : BF_0 \cong F_1 A$ is a natural isomorphism. A 2-cell $\tau : \mathbf{F} \Longrightarrow \mathbf{G} : \mathbf{A} \to \mathbf{B}$, with $\mathbf{G} = (G_0, G_1, \Gamma)$, is a pair of transformations $\tau_0 : F_0 \Longrightarrow G_0$ and $\tau_1 : F_1 \Longrightarrow G_1$ such that $\Gamma \cdot (\tau_1 \circ A) = (B \circ \tau_0) \cdot \Phi$. Verification that the obvious compositions of 1- and 2- cells provide a 2-category is left to the reader.

For any $\mathbf{S}$-indexed category $\mathbf{A}$ there is a functor with codomain $\mathbf{S}$ called the Grothendieck construction for $\mathbf{A}$. We denote this functor $G_A : gr(\mathbf{A}) \to \mathbf{S}$ and briefly describe it. The objects of $gr(\mathbf{A})$ are pairs (I, A) with A an object of $\mathbf{A}^I$. The hom-sets are defined by the formula $gr(\mathbf{A})((I,A),(J,A')) = \left\{(\alpha,\varphi) \middle| I \xrightarrow{\alpha} J, A \xrightarrow{\varphi} \alpha^* A'\right\}$. The functor G_A is given on objects and arrows by projecting on I and α. In fact, G_A is a fibration [5].

We can now proceed to describe the relational completion monad. We recall that a monad in a 2-category is a 1-cell with a monoid structure given by "unit" and "multiplication" 2-cells [17]. Thus in $\mathbf{cat^2}$ a monad is an arrow of $\mathbf{cat^2}$ together with unit and multiplication 2-cells. We note that a monad on $A : \mathbf{A}_0 \longrightarrow \mathbf{A}_1$ in $\mathbf{cat^2}$ is specified by a monad M_0 in **cat** on $\mathbf{A}_0$, a monad M_1 in **cat** on $\mathbf{A}_1$ and a 2-cell $M : M_1 A \cong AM_0$. These must satisfy equations involving the structures of the given monads in **cat** and M.

The domain of our monad $\mathbf{C}$ is the Grothendieck construction for the $\mathbf{S}/M(I)$-indexed category $\mathcal{D}(A)$, which is denoted $G_{\mathcal{D}(A)} : gr\mathcal{D}(A) \longrightarrow \mathbf{S}/M(I)$.

Thus, the monad $\mathbf{C} = (C^0, C^1, \Gamma)$ requires functors $C^0 : gr\mathcal{D}(A) \longrightarrow gr\mathcal{D}(A)$, $C^1 : \mathbf{S}/M(I) \longrightarrow \mathbf{S}/M(I)$ and a transformation $\Gamma : C^1 G_{\mathcal{D}(A)} \Longrightarrow G_{\mathcal{D}(A)} C^0$.

The second functor C^1 is inductively defined and we describe it first. Its action on a database signature $\sigma : J \longrightarrow M(I)$ produces the database signature we denote $C^1(\sigma) : \sigma^1 \longrightarrow M(I)$. This is generated, beginning from the set of relation names J, by freely adjoining to σ^1 all expressions of relational algebra as described above for the family of domains A. These expressions have appropriate schemes given by $C^1(\sigma)$. The inductive definition of σ^1 (and simultaneously $C^1(\sigma)$) follows:

σ^1 is the smallest set satisfying clauses 1. – 5. below subject to identifications generated by equations 1. – 7.:

1. $\forall j \in J, j \in \sigma^1$ (with scheme $\sigma(j)$)
2. if s_1, $s_2 \in \sigma^1$ (with scheme $C^1(\sigma)(s_1) = C^1(\sigma)(s_2)$) then $(s_1 \cup s_2)$, $(s_1 \cap s_2) \in \sigma^1$ (with scheme $C^1(\sigma)(s_1)$)
3. if s_1, s_2 in σ^1 (with schemes $C^1(\sigma)(s_1)$ and $C^1(\sigma)(s_2)$ then $(s_1 \times s_2) \in \sigma^1$ (with scheme $C^1(\sigma)(s_1) \cdot C^1(\sigma)(s_2)$)
4. if $s \in \sigma^1$ (with scheme $C^1(\sigma)(s) : [n] \longrightarrow I$) and $\varphi : [n'] \longrightarrow [n]$, then $(\pi_\varphi s) \in \sigma^1$ (with scheme $C^1(\sigma)(s) \cdot \varphi$)
5. if $s \in \sigma^1$, $C^1(\sigma)(s)(i_1) = C^1(\sigma)(s)(i_2)$ and if $c \in A_{C^1(\sigma)(s)(i_1)}$, then $(S_{i_1=i_2}(s))$ and $(S_{i_1=c}(s))$ are in σ^1 (with schemes $C^1(\sigma)(s)$.)

The following equations (which have no effect on schemes) apply to the "words" generated above. Let $s, s_1, s_2, s_3 \in \sigma^1$ be such that the operations mentioned below are defined, then

1. $(s_1 \times s_2) \times s_3 = s_1 \times (s_2 \times s_3)$
2. $(s_1 \cup s_2) \cup s_3 = s_1 \cup (s_2 \cup s_3)$ $(s_1 \cup s_2) = (s_2 \cup s_1)$ $(s_1 \cup s_1) = s_1$
   $(s_1 \cup s_2) \cap s_3 = (s_1 \cap s_3) \cup (s_2 \cap s_3)$ $(s_1 \cup s_2) \cap s_2 = s_1$
   and the dual equations
3. if $\varphi : [n'] \rightarrowtail [n]$ and $\psi : [n''] \rightarrowtail [n']$ then $\pi_\psi \pi_\varphi s = \pi_{\varphi\psi} s$
4. $S_{i=j}(S_{i=j}s) = S_{i=j}s = S_{j=i}s$ $S_{i=i}s = s$ $S_{i=j}(S_{k=l}s) = S_{k=l}(S_{i=j}s)$
   $S_{i=c}(S_{i=j}s) = S_{i=j}(S_{i=c}s)$ $S_{i=c}(S_{i=c}s) = S_{i=c}s$
5. $s_1 \times (s_2 \cup s_3) = (s_1 \times s_2) \cup (s_1 \times s_3)$ $s_1 \times (s_2 \cap s_3) = (s_1 \times s_2) \cap (s_1 \times s_3)$
6. $\pi_\varphi(s_1 \cup s_2) = \pi_\varphi s_1 \cup \pi_\varphi s_2$ $\pi_\varphi(s_1 \cap s_2) = \pi_\varphi s_1 \cap \pi_\varphi s_2$
7. $S_{i=j}(s_1 \cup s_2) = S_{i=j}s_1 \cup S_{i=j}s_2$ $S_{i=j}(s_1 \cap s_2) = S_{i=j}s_1 \cap S_{i=j}s_2$

The definition of C^1 on an arrow $\tau : \sigma_1 \longrightarrow \sigma_2$ of $\mathbf{S}/M(I)$ is that induced by the action of τ from generators of σ_1^1 to those of σ_2^1. The transformations making C^1 a monad on $\mathbf{S}/M(I)$ are easily described. The unit is simply inclusion of the generators, while the multiplication simply rewrites a "word of words" in

$(C^1(\sigma))^1$ as that element of σ^1 obtained by treating operations on elements of σ^1 as operations defining an element of σ^1.

Next we decribe C^0. The image under C^0 of an object (σ, R) of $gr\mathcal{D}(A)$ is $(C^1(\sigma), R^0)$ where R^0 is the σ^1-indexed family of relations where, for $s \in \sigma^1$, R^0_s is the relation obtained by applying the relational operations in the construction of s to the corresponding relations of R. Extending C^0 to arrows of $gr\mathcal{D}(A)$ presents no difficulties. Once again, the monad unit is inclusion of generators and the multiplication arrow on the family of relations component merely applies the appropriate relational operations.

The non-monotonicity of binary difference of relations would prevent the definition of C^0 indicated above. One way to get around this problem is to use updates rather than simply inclusions as arrows in a modified $\mathcal{D}(A)$. The relationally complete databases are then those algebras for the resulting monad which have fully defined structural arrows. We have chosen to leave out difference in order that a "relationally complete database" is precisely an algebra for the monad $\mathbf{C}$ we have described.

Now the definition of Γ is straightforward. For any object (σ, R) of $gr\mathcal{D}(A)$, the natural isomorphism required has components the identity on $C^1(\sigma)$. The reader will have no trouble verifying that $\mathbf{C}$ is a monad on $G_{\mathcal{D}(A)}$ in $\mathbf{cat}^{\mathbf{2}}$.

The category $gr\mathcal{D}(A)^{\mathbf{C}^0}$ of Eilenberg-Moore algebras [17] for the monad $\mathbf{C}^0$ has objects pairs (σ, R) where σ is the database signature of a relationally complete database with relations R. The Eilenberg-Moore algebras $\mathbf{set}/M(I)^{\mathbf{C}^1}$ for $\mathbf{C}^1$ has objects database signatures with interpretations for the operations of relational algebra. There is a functor between these categories $G^{\mathbf{C}}_{\mathcal{D}(A)}$: $gr\mathcal{D}(A)^{C^0} \longrightarrow \mathbf{set}/M(I)^{C^1}$ which forgets the relations.

PROPOSITION 4. *The functor* $G^{\mathbf{C}}_{\mathcal{D}(A)}$ *is the Eilenberg-Moore object for the monad* $\mathbf{C}$ *on* $G_{\mathcal{D}(A)}$ *in* $\mathbf{cat}^{\mathbf{2}}$.

PROOF. This is an instance of a more general result which asserts that Eilenberg-Moore algebras for monads in $\mathbf{cat}^{\mathbf{2}}$ are computed, as here, by taking algebras for the domain and codomain monads and the induced functor between these categories of algebras. ∎

For the balance of this section we will assume that the indexing category $\mathbf{S}$ is an elementary topos with NNO and that $\mathbf{A} = \mathbf{S}$. Before describing the 2-category arising from a relationally complete database, we note that to any word $w : 1 \longrightarrow M(I)$ of $M(I)$ we can associate a "diagonal" relation $\delta_w : \Delta_w \rightarrowtail (ww)^* P_0(A) \cong w^* P_0(A) \times w^* P_0(A)$. We take Δ_w to be $w^* P_0(A)$, and define δ_w by projections to $w^* P_0(A)$ which are both the identity.

Now let $R : R_0 \rightarrowtail \sigma^* P_0(A)$ be a datbase with database signature $\sigma : J \longrightarrow M(I)$. Provided that R is closed under projection, selection and cartesian product, we can construct a 2-category which we will denote $\mathbf{R}$. We define the objects of $\mathbf{R}$ to be $M(I)$. Letting $v, w : 1 \longrightarrow M(I)$ be objects of $\mathbf{R}$, we define

1-cells by the formula

$$\mathbf{R}(v,w) = \begin{cases} \{(v,R_j,w) \in R \mid \sigma(j) = vw\} & \text{if } v \neq w \\ \{(v,R_j,w) \in R \mid \sigma(j) = vw\} \bigcup \{\Delta_w\} & \text{if } v = w \end{cases}$$

The two-cells are given by inclusion.

PROPOSITION 5. *Let ρ be a database closed under selection, projection and cartesian product. The structure $\mathbf{R}$ is a 2-category with composition of 1-cells given by relational composition.*

PROOF. We need only observe that if $r \in \mathbf{R}(v,w)$ and $s \in \mathbf{R}(u,v)$ then their relational composite (with a slight abuse of our notation for projection and selection) is

$$rs = \pi_{uw}(s \bowtie r) := \pi_{uw}(S_{r.v=s.v}(s \times r))$$

so that rs is in R. Now, with identities given by the Δ's, we have that R is a sub-2-category of $\mathbf{rel}(\mathbf{S})$ and the required equations all follow from the latter's structure. ∎

COROLLARY 1. *If ρ is a relationally complete database, then $\mathbf{R}$ is a 2-category of relations.* ∎

We end this section with two remarks:
(i) the results raise the question whether a relationally closed database is a "2-category of relations" in the sense of Carboni and Walters [2] (= the 2-category $\mathbf{rel}(\mathbf{E})$ for a regular category $\mathbf{E}$) – the answer appears to be no;
(ii) the setting we have constructed will be useful for the expression of functional dependencies which are important in database design – the statement that there is a functional dependency becomes the categorical statement that a 1-cell in $\mathbf{R}(v,w)$ always has a right adjoint.

Conclusion

We conclude this article with some directions for further work. While the theory of indexed categories gives a satisfactory account of the theory of relational databases, viewing databases as 2-categories of relations begs the investigation of databases on a different data structure which appears in a bicategories closely related to 2-categories of relations, namely bicategories of profunctors. A (binary) relation from set A_1 to set A_2 is determined by stating whether or not an element of A_1 is related to an element of A_2. A profunctor generalizes this in two ways. First, A_1 and A_2 are categories and there is a set (not just a boolean truth value) relating an object of A_1 to an object of A_2. Second, this doubly indexed family of sets is acted on by arrows of the categories A_1 and A_2. Replacing the discrete objects used as domains in the relational theory by categories, and hence considering the profunctor data structure, may provide a suitable theoretical foundation for object-oriented databases. This observation is currently under investigation.

References

1. M. Barr, Fuzzy set theory and topos theory. Bull. Can. Math. Soc. **29** (1986), 501–508.

2. Aurelio Carboni and R. F. C. Walters, Cartesian bicategories 1. Journal of Pure and Applied Algebra **49** (1987), 11–32.

3. E. F. Codd, A relational model for large shared data banks. Comm. ACM **13** (6) (1970), 377–387.

4. C. J. Date, *Introduction to Database Systems, Fifth edition.* Addison-Wesley, 1990.

5. J. W. Gray, Fibered and Cofibered Categories, *Proceedings of the Conference on Categorical Algebra, La Jolla*, pages 21–83, Springer, 1966.

6. B. Hilken and D. E. Rydeheard, Indexed categories for program development. Cahiers de topologie et géometrie différentielle catégoriques, XXXII (1991), 165–185.

7. M. Hyland and A. M. Pitts. The Theory of Constructions: Categorical Semantics and Topos-Theoretical Models, *Categories in Computer Science and Logic,* Contemporary Math No. 92, American Mathematical Society, 1989.

8. Peter T. Johnstone, *Topos Theory*, Academic Press, 1977.

9. G. M. Kelly and R. Street, Review of the elements of 2-categories, *Lecture Notes in Math. 420,* pages 75–103. Springer-Verlag, 1974.

10. B. LeSaffre, *Structures algebriques dans les topos elementaires.* PhD thesis, U. de Paris 7, 1974.

11. D. Maier, *The Theory of Relational Databases*, Computer Science Press, 1983.

12. Eugenio Moggi, A category-theoretic account of program modules, Mathematical Structures in Computer Science **1** (1991), 103–139.

13. R. Paré and D. Schumacher, Abstract Families and the Adjoint Functor Theorems, *Indexed Categories and their Applications,* pages 1–125. Volume 661 of *Lecture Notes in Math.*, Springer-Verlag, 1978.

14. K. Raju and A. Majumdar, Fuzzy functional dependencies and losslessjoin decomposition of fuzzy relational database systems. ACM Transactions on Database Systems, **13** (1988), 129–166.

15. Robert Rosebrugh, On defining objects by recursion in a topos, Journal of Pure and Applied Algebra, **20** (1981), 325–335.

16. R. A. G. Seely, Categorical semantics for higher order polymorphic lambda calculus, Journal of Symbolic Logic, **52** (1987), 4.

17. R. Street, The formal theory of monads, Journal of Pure and Applied Algebra, **2** (1972), 149-168.

18. J. D. Ullman, *Principles of Database and Knowledge Base Systems*, Volumes 1 and 2, Computer Science Press, 1988.

DEPARTMENT OF MATHEMATICS AND COMPUTER SCIENCE
MOUNT ALLISON UNIVERSITY
SACKVILLE, N.B. CANADA
RROSEBRUGH@MTA.CA

and

DEPARTMENT OF MATHEMATICS, STATISTICS AND COMPUTING SCIENCES
DALHOUSIE UNIVERSITY
HALIFAX, N.S. CANADA
RJWOOD@CS.DAL.CA

Canadian Mathematical Society
Conference Proceeding
Volume 13, 1992

LINEAR INDEXED QUANTALES AND QUANTALIC HYPERDOCTRINES

Kimmo I. Rosenthal

INTRODUCTION

Quantales are partially ordered algebraic structures, which are complete lattices equipped with a sup-preserving multiplication. Examples include frames, various lattices of multiplicative ideals from algebra, as well as relations on a set. Quantales are of interest in theoretical computer science in several different contexts. The partially ordered semantics of Girard's linear logic [G1] involves certain quantales, called Girard quantales, (see [Y], [R1], and [R2]). Quantales have also been systematically used in studying process semantics [AV] and the theory of categories enriched in quantales provides a way of considering automata ([BK], [KR], [R5]).

The category theoretical approach to logic has become increasingly important as a tool in analyzing logical concepts and in formulating appropriate semantics for different logical settings that may arise. Various partially ordered structures, looked at categorically, play a key semantic role. If one considers propositional logic and the entailment relation, then partially ordered models will provide the necessary semantics, e.g. frames for intuitionistic logic; Girard quantales for linear logic. However, when considering predicate logic, one needs to deal with types or sorts and a more complicated syntax. To deal with types categorically, there is the seminal work of F.W. Lawvere [L1], [L2]. For intuitionistic logic, the theory of indexed frames and the resulting hyperdoctrines are described and motivated in the highly readable lectures of A. Pitts [P], where the connection with first order predicate logic is explained. If one wishes to analyze linear logic in a similar fashion, one is forced to deal with the notion of indexed quantales.

1991 Mathematics Classification. Primary 18 A 15, B 35, D 99 ; 03 G 30.

Research supported by NSF-RUI Grant No.9002364.

This paper is in final form and will not be published elsewhere.

In this paper, we begin to develop the theory of indexed quantales and quantalic hyperdoctrines. Our goal is to provide the correct definitions and formulate the necessary adjustments in going from frames to quantales and to lay the foundation for further investigation along these lines. We shall look at several examples in order to illuminate these definitions. Also, we shall consider indexed Girard quantales and propose a definition of linear hyperdoctrine. In looking at some simple, but important examples, we shall see some of the linear structure, such as linear negation, is preserved by the transition maps. However, the transition maps do not, in general, preserve the multiplicative conjunction. This is reasonable to expect from the point of view of quantale theory, however it is problematic from the perspective of modelling predicate linear logic. This is an area in need of further investigation and will be discussed in the paper. We view this paper as a first step, which can lead to a more detailed analysis of these ideas.

The author would like to thank the referee for some insightful comments.

1. Quantales

We begin with a brief review of quantales. For a comprehensive introduction to quantales see Rosenthal [R1].

DEFINITION 1.1. A quantale is a complete lattice Q together with an associative binary operation $\circ$ satisfying the following for all $a \in Q$ and $\{b_\alpha\} \subseteq Q$.
$a \circ (sup_\alpha b_\alpha) = sup_\alpha (a \circ b_\alpha)$ and $(sup_\alpha b_\alpha) \circ a = sup_\alpha (b_\alpha \circ a)$

It follows that because $a \circ -$ and $- \circ a$ preserve sups, they have right adjoints, which are denoted $a \rightarrow_l -$ and $a \rightarrow_r -$ respectively (referred to as left and right residuation). Q is called *unital* iff $\exists\ 1 \in Q$ with $1 \circ a = a = a \circ 1$ for all a.

EXAMPLES: (1) Let M be a monoid. Consider $P(M)$, the power set of M. $P(M)$ is a quantale with operation $A \circ B = \{ab | a \in A \text{ and } b \in B\}$. Sups are unions and $A \rightarrow_r B = \{m \in M | am \in B \text{ for all } a \in A\}$ and $A \rightarrow_l B = \{m \in M | ma \in B \text{ for all } a \in A\}$. $P(M)$ is unital with unit $\{e\}$, where $e \in M$ is the identity. In fact, $P(M)$ is the free unital quantale on the monoid M.

(2) Let X be a set and let Rel(X) denote the set of relations on X. Rel(X) is a quantale under the operation of relational composition. If $R, S \in$ Rel(X), then $R \circ S = \{(x,z) \in X \times X \mid \exists y \text{ with } (x,y) \in S \text{ and } (y,z) \in R\}$. Again, unions play the role of sups and the residuations are provided by $R \rightarrow_r S = \{(x,y) \mid (y,z) \in R \Rightarrow (x,z) \in S \text{ for all } z \in X\}$ and similarly $R \rightarrow_l S = \{(x,y) | (z,x) \in R \Rightarrow (z,y) \in S \text{ for all } z \in X\}$.

If R is a relation on X, R^{o} denotes its dual relation. Rel(X) is also unital with the diagonal relation Δ as the unit.

Further examples include various ideal lattices from ring theory with the operation $\circ$ of ideal multiplication, as well as the lattice of closed right ideals of a C^* -algebra with $\circ$ being the closure of the ideal product. (See [R1]).

DEFINITION 1.2. If Q and S are quantales, a function f: $Q \to S$ is a *homomorphism* iff it preserves sups and $\circ$. If Q and S are unital and f preserves identity elements, then f is called a *unital homomorphism*.

DEFINITION 1.3. Let Q and S be quantales. A function $f : Q \to S$ is called a *closed map* of quantales iff $f(a) \circ f(b) \leq f(a \circ b)$ for all $a, b \in Q$.

Suppose Q and S are unital with identities 1_Q and 1_S.

Then f is a *unital closed map* iff it is closed and $1_Q \leq f(1_S)$.

PROPOSITION 1.1. Let $f : Q \to S$ be a homomorphism of quantales. Then, its right adjoint $f_* : S \to Q$ is a closed map of quantales. (If f is a unital homomorphism, then f_* is a unital closed map).

Proof: See [R1; Prop. 2.3.3].

Unital quantales are the partially ordered examples of monoidal biclosed categories and unital closed maps are just closed functors, in the sense of Lawvere [L3]. As shown there, closed functors do not preserve the monoidal structure of a closed category. This is borne out by many examples and thus, the notion of quantale homomorphism may be too restrictive.

In the examples we consider, the closed maps arise as right adjoints of quantale homomorphisms and hence automatically preserve all infs. In considering Girard quantales and linear indexed quantales in §3, the closed maps will also have right adjoints and thus preserve all sups as well. We shall denote the category of quantales and closed maps by *Quant*.

Other considerations in the theory of quantales, such as the notion of quantic nucleus ([NR], [R1]), will not be discussed here.

2. Indexed quantales

Indexing plays an important role in categorical logic. These ideas originate in the work of F.W.Lawvere [L1]. Propositional logics can be modelled as categories, in fact partially ordered ones. The notion of indexed category combines this idea with the well known fact that algebraic theories (with sorts and terms) can be modelled by categories $\mathbf{C}$ with products. Predicate logics are then modelled by categories $X(C)$ of propositional logics indexed by $C \in \mathbf{C}$. Predicates describing properties of terms can be combined using the structure of the $X(C)$ and quantification can also be described. This is best expressed through Lawvere's notion of hyperdoctrine [L2]. The case of intuitionistic logic and indexed frames is clearly described in the lecture notes of A. Pitts [P]. We shall develop the notions of indexing and hyperdoctrine for quantales and try to see how much of the theory goes through and what modifications need to be made. Then we shall consider the case of linear logic and its partially ordered models, namely Girard quantales, and try to develop the notion of a linear hyperdoctrine together with several key examples.

We begin with a discussion of indexed quantales and shall first develop their theory.

DEFINITION 2.1. Let $\mathbf{C}$ be a category. A $\mathbf{C}-$*indexed quantale* is a functor $X\colon \mathbf{C}^{op} \to Quant$.

Thus X consists of the following data :

(1) a quantale $X(C)$ for each object $C \in \mathbf{C}$.

(2)If $\alpha : C \to D$ is a morphism of $\mathbf{C}$, there is a closed map of quantales $X(\alpha) : X(D) \to X(C)$.

(3) $X(i_C) = i_X(C)$, where $i_{(\ )}$ denotes identity morphisms

(4) $X(\beta \circ \alpha) = X(\alpha) \circ X(\beta)$, whenever α and β are composable morphisms of $\mathbf{C}$.

As discussed at the end of §1, requiring that $X(\alpha)$ be a quantale homomorphism seems to be too restrictive and fails to hold in the naturally arising examples described below. (When discussing nuclei, a quantic nucleus is just a closed map with respect to the quantale operation $\circ$, rather than a homomorphism, as in the frame case. See [NR], [R1]).

DEFINITION 2.2. If $X\colon \mathbf{C}^{op} \to Quant$ and $Y\colon \mathbf{C}^{op} \to Quant$ are $\mathbf{C}$-indexed quantales, then a $\mathbf{C}$-*indexed morphism* is a natural transformation F from X to Y such that $F_I\colon X(I) \to Y(I)$ is a closed map of quantales for all $I \in \mathbf{C}$.

EXAMPLES: (1) Let $\mathbf{C}$ be the category of monoids Mon. For a monoid M, we can take $X(M) = P(M)$, the power monoid of M. If α is a monoid homomorphism, let $X(\alpha) = \alpha^{-1}$, the inverse image map. The related example of taking $\mathbf{C}$ to be the category Grp of groups, will also be of interest to us.

In the same vein, this example can be modified by taking $X(M) = \mathrm{Idl}(M)$, the quantale of ideals of M. Another modification is to take $\mathbf{C}$ to be the allegory of monoids, replacing monoid homorphisms with relational morphisms.

(2) Let $\mathbf{C}$ be the category Rng of rings with unit. If R is a ring, let $X(R) = \mathrm{Idl}(R)$, the quantale of two-sided ideals of R. (We could equally well consider $\mathrm{RIdl}(R)$ (the quantale of right ideals) or $\mathrm{LIdl}(R)$ (the quantale of left ideals)). As above, if α is a ring homomorphism, we can take $X(\alpha) = \alpha^{-1}$.

(3) Let Q be a quantale and let consider $Q-CAT$, the category of Q-enriched categories and Q-functors. If A is a Q-category, let $X(A) = Bim(A)$, which consists of the bimodules $\theta : A \Rightarrow A$. This is a quantale under bimodule composition. (For background on the theory of bimodules, see [L3] or [CW]).

If B is also a Q-category and $f : A \to B$ is a Q-functor, then f gives rise to an adjoint pair of bimodules $f_* : A \Rightarrow B$ defined by $f_*(b,a) = B(b, f(a))$ and $f^* : B \Rightarrow A$ defined by $f^*(a,b) = B(f(a), b)$.

Define $X(f) : Bim(B) \to Bim(A)$ by $X(f)(\tau) = f^* \circ \tau \circ f_*$. This is a closed map of quantales and gives us the required structure for an indexed quantale. (Note that if f and g are composable Q-functors, then we have $(g \circ f)^* = f^* \circ g^*$ and $(g \circ f)_* = g_* \circ f_*$.)
(4) Let $\mathbf{C}$ = *Sets*, the category of sets and functions. If A is a set, let $X(A)$ = Rel(A), the quantale of relations on A. If $f : A \to B$ is a function, define $X(f)$: Rel(B) $\to$ Rel(A) by $X(f)(R) = f^o \circ R \circ f$. Using ordered pairs, $X(f)(R) = \{(a, a') | (f(a), f(a')) \in R\}$.

More generally, $\mathbf{C}$ could be a Grothendieck topos, or we could consider distributive categories of relations [CW]. Even more generally, by considering the notion of *quantaloid*, (see [R3], [R4], [R5]), the last two examples are special cases of the same situation.

3. Girard quantales and linear indexed quantales

J.Y. Girard has developed a logical system called linear logic, which holds much promise for playing a significant role in theoretical computer science ([G1], [Sc]). The partially ordered semantics of linear logic is modelled by a certain class of quantales called Girard quantales. Actually, one does not need the completeness requirements of quantales to model linear logic (just as Heyting algebras are sufficient for modelling intuitionistic logic). Related structures other than quantales are also under investigation (see [de P], [L]). The connection between linear logic and quantale theory was first made precise by [Y]. Girard quantales were further investigated by [R2].

Since Girard quantales are the partially ordered models of linear logic, it is natural to try to investigate to what extent the notions of indexing and hyperdoctrines can be developed for quantales and to look for examples of indexed Girard quantales and investigate their properties relative to the notion of hyperdoctrine. Seely presents some discussion of what properties would be desirable in indexed Girard categories in [Se], however we shall see that in all the naturally arising partially ordered examples, our transition maps will not in general preserve the quantale operation $\circ$ strictly, although they will preserve linear negation.

We are now ready to present the definition of Girard quantale. Various results will be stated without proof. For details, the reader can consult [R1] and [R2].

DEFINITION 3.1. Let Q be a quantale. Let $d \in Q$.

d is called a *dualizing element* iff $(a \to_l d) \to_r d = a = (a \to_r d) \to_l d$ for all $a \in Q$.

d is called *cyclic* iff $a \to_r d = a \to_l d$ for all $a \in Q$.

(If d is cyclic, we shall drop the subscripts $\to_r$ and $\to_l$).

DEFINITION 3.2. A quantale G is called a *Girard quantale* iff it has a cyclic dualizing element d.

For example, in a power monoid $P(M)$, a subset $D \subseteq M$ is cyclic precisely when $ab \in D$ iff $ba \in D$ for all $a, b \in M$ holds. Letting $P(M)_D = \{A \in P(M) \mid (A \to D) \to D = A\}$, then $P(M)_D$ is a Girard quantale with cyclic dualizing element D. It is shown in [R2] that every Girard quantale is isomorphic to one of this form.

We shall use $a^\perp$ to denote $a \to_l d = a \to_r d = a \to d$. Note that $a^{\perp\perp} = a$ and $a \leq b^\perp$ iff $b \leq a^\perp$. $(\ )^\perp$ is the operation of "linear negation" in Girard's linear logic . (A Girard quantale is a frame iff it is a complete Boolean algebra, with dualizing element 0).

The following proposition lists some basic facts about Girard quantales (proofs can be found in [R1], [R2]).

PROPOSITION 3.1. Let G be a Girard quantale with cyclic dualizing element d. Let $a, b \in G$. Then,

(1) $a \to_l b = (a \circ b^\perp)^\perp$
(2) $a \to_r b = (b^\perp \circ a)^\perp$
(3) $a \circ b = (a \to_l b^\perp)^\perp$
(4) $b \circ a = (a \to_r b^\perp)^\perp$
(5) G is a unital quantale with unit $d^\perp$.

We now wish to present some natural non-commutative examples of Girard quantales, which arise from the classical operation of complementation by a dualization.

PROPOSITION 3.2. Let X be a set. Then $\mathrm{Rel}(X)$, the quantale of relations on X, is a Girard quantale.

Proof: Let $\Delta = \{(x, x) \mid x \in X\}$ be the diagonal relation. We shall use $(\ )^c$ to denote the complement of a relation (viewed as an element of $P(X \times X)$). Let $D = \Delta^c$. Then, if R is a relation on X, $R \to_r \Delta^c = \{(x, y) \mid (y, z) \in R \Rightarrow (x, z) \in \Delta^c\} = \{(x, y) \mid (y, z) \in R \Rightarrow z \neq x\} = (R^\circ)^c$. A similar calculation shows that $R \to_l \Delta^c = (R^\circ)^c$ as well, and thus Δ^c is cyclic. It follows that $(R \to_r \Delta^c) \to_l \Delta^c = (R^\circ)^c \to_l \Delta^c = R$, since $(\ )^\circ$ and $(\ )^c$ commute with each other and $(R^\circ)^\circ = R$ and $(R^c)^c = R$. Thus, Δ^c is a cyclic dualizing element, proving that $\mathrm{Rel}(X)$ is a Girard quantale.

PROPOSITION 3.3. Let G be a group. Then, $P(G)$ is a Girard quantale.

Proof: Let $D = \{e\}^c$ (e the identity of G). If $A \subseteq G$, then $A \to_l D = \{b \in G \mid bA \cap \{e\} = \emptyset\} = \{b \in G \mid b^{-1} \in A^c\}$. Denote this by $A^\perp$; easy calculations show that $A^\perp = A \to_r D$ and that $(A^\perp)^\perp = A$ and thus D is a cyclic dualizing element and $P(G)$ is Girard.

In both examples, we have complete Boolean algebras as underlying sets, an involution $(\ )^\circ$, and linear negation arises from the involution followed by taking the set-theoretic complement.

A third example arises from considering the notion of an order ideal as developed by Carboni and Street [CS]. This example is a variant of our first example in that it involves relations.

DEFINITION 3.3. Let X be a preordered set (i.e. a set equipped with a reflexive and transitive relation $\leq$). An order ideal of X is a relation R on X satisfying:

(1) $(x,y) \in R$ *and* $x' \leq x \Rightarrow (x',y) \in R$

(2) $(x,y) \in R$ *and* $y \leq y' \Rightarrow (x,y') \in R$

Let $I(X)$ denote the set of order ideals of X. It is not hard to show that $I(X)$ is a quantale under relational composition.

PROPOSITION 3.4. Let X be a preordered set. Then $I(X)$, the quantale of order ideals of X, is a Girard quantale.

Proof: From the above remarks, we know that $I(X)$ is a quantale. Let $D = \{(x,z) | z \leq x \textit{ fails}\}$. If $R \in I(X)$, then $R \rightarrow_r D = \{(x,y) | (y,z) \in R \textit{ implies } (x,z) \in D\}$. A straightforward calculation shows that $R \rightarrow_r D = \{(x,y) | (y,x) \in R^c\}$. Denote this by $R^\perp$. Then $R^\perp = R \rightarrow_l D$ yielding cyclicity and $(R^\perp)^\perp = R$ finishing the proof that $I(X)$ is Girard.

This last proposition can be proved as a special case of a much more general construction involving enriched categories and bimodules. (See [R4] for this, as well as an extension of the concept of Girard quantale to quantaloids).

The following observation about Girard quantales and adjoints will prove to be useful when we start considering hyperdoctrines.

PROPOSITION 3.5. Let G and H be Girard quantales and let $f : G \rightarrow H$ be an order-preserving function, which preserves linear negation, i.e. $f(a^\perp) = f(a)^\perp$. Then, f has a left adjoint iff it has a right adjoint.

Proof: Suppose f has a left adjoint $\lambda : H \rightarrow G$. If $\rho : H \rightarrow G$ is defined by $\rho(a) = (\lambda(a^\perp))^\perp$, then using properties of linear negation and adjointness, we have $b \leq \lambda(a^\perp)^\perp$ iff $\lambda(a^\perp) \leq b^\perp$ iff $a^\perp \leq f(b^\perp) = f(b)^\perp$ iff $f(b) \leq a$. Thus, ρ is the right adjoint of f.

On the other hand, given ρ right adjoint to f, define $\lambda(a) = \rho(a^\perp)^\perp$ and perform a similar calculation to show that λ is left adjoint to f.

To define the linear version of indexed quantale, we need to replace the notion of quantale by that of Girard quantale. The quantale operation will not be strictly preserved, however it would certainly be desirable to have our transition maps at least preserve linear negation.

DEFINITION 3.4. A $\mathbf{C}$-indexed quantale L: $\mathbf{C}^{op} \rightarrow \textit{Quant}$ is *linear* iff

(1) $L(C)$ is a Girard quantale for all $C \in \mathbf{C}$.

(2) If $f : C \rightarrow D$ is a morphism in $\mathbf{C}$, then $L(f) : L(D) \rightarrow L(C)$ is a closed, unital map of quantales.

(3) If $f : C \to D$ is a morphism in **C**, then $L(f) : L(D) \to L(C)$ preserves linear negation
(i.e. $L(f)(a^{\perp}) = (L(f)(a))^{\perp}$ for $a \in L(D)$).

Since the transition maps are unital and preserve linear negation, we have $L(f)(d_D) \leq d_C$, where d_D and d_C are the respective cyclic, dualizing elements of $L(D)$ and $L(C)$.

Since in our examples, $L(f)$ will be the right adjoint of a quantale homomorphism, by Prop. 3.5. it will preserve all the lattice structure (i.e. both infs and sups).

DEFINITION 3.5. If L: $\mathbf{C}^{op} \to Quant$ and K: $\mathbf{C}^{op} \to Quant$ are linear **C**-indexed quantales, then a morphism $F : L \to K$ of **C** -indexed quantales is *linear* iff $F_I : L(I) \to K(I)$ preserves linear negation for all $C \in \mathbf{C}$.

PROPOSITION 3.6. Rel: $Sets^{op} \to Quant$, the set indexed quantale of relations, is a linear indexed quantale.

Proof: From Prop. 3.2., Rel(X) is a Girard quantale for all sets X. If $f : X \to Y$ is a function and $R \in$ Rel(Y), then Rel(f)(R) $= f^{o} \circ R \circ f$. $\Delta_X \subseteq \{(x, x') \mid f(x) = f(x')\} = f^{o} \circ \Delta_Y \circ f$ implies that Rel(f) is unital. We have that Rel(f)($R^{\perp}$) $= \{(x, x') | (f(x), f(x')) \in (R^{o})^{c}\} = \{(x, x') \mid (f(x'), f(x)) \in R^{c}\}$. But, a moment's reflection shows that this equals (Rel(f)(R))$^{\perp}$.

Analogously, we obtain the following.

PROPOSITION 3.7. $P : Grp^{op} \to Quant$, the group indexed quantale of power sets, is a linear indexed quantale.

Proof: Given a homomorphism $f : G \to H$, then $f^{-1} : P(H) \to P(G)$ is closed and $\{e_G\} \subseteq f^{-1}(e_H)$ proves f^{-1} is unital. If $A \subseteq H$, then $f^{-1}(A^{\perp}) = \{x \in G \mid f(x^{-1}) \in A^{c}\} = \{x \in G \mid x^{-1} \in f^{-1}(A)^{c}\} = f^{-1}(A)^{\perp}$, proving that inverse image preserves linear negation.

The case of order ideals proceeds similarly. Let *Preord* denote the category of preordered sets and order preserving maps. For $f : X \to Y$ order preserving, define $f^{*} : I(Y) \to I(X)$ by $f^{*}(S) = \{(x_1, x_2) \mid (f(x_1), f(x_2)) \in S\}$. f^{*} is a closed unital map of quantales.

We can now define an indexed quantale **I**: $Preord^{op} \to Quant$ by sending $X \in Preord$ to the quantale $I(X)$ of order ideals of X and letting $I(f) = f^{*}$ for f order preserving. The following is straightforward.

PROPOSITION 3.8. $\mathbf{I} : Preord^{op} \to Quant$ is a linear indexed quantale.

The fact that the transition maps are merely closed maps of quantales, rather than homomorphisms, is problematic for the modelling of predicate linear logic. We shall discuss this briefly at the end of §4. We do wish to point out that in our three examples, if f is a surjective function, then $L(f)$

will in fact be a quantale homomorphism. Thus, as long as one is considering surjections, then all of the linear structure will be preserved.

4. Quantalic hyperdoctrines and a definition of linear hyperdoctrine

The notion of a linear hyperdoctrine of quantales extends that of a linear indexed quantale L: $\mathbf{C}^{op} \to Quant$ in that we require that certain completeness properties hold for L. This completeness is relative to the indexing by $\mathbf{C}$ and thus we shall begin by discussing completeness for indexed quantales. In logical terms, this will correspond to having existential and universal quantification for variables ranging over our sorts (objects of $\mathbf{C}$). This can be done by requiring certain adjoints to exist. This approach is due to Lawvere [L1], [L2]. We shall follow the presentation outlined by Pitts [P] and transcribe the results for indexed frames to the setting of indexed quantales.

Let $Q : \mathbf{C}^{op} \to Quant$ be a $\mathbf{C}$-indexed quantale, where the category has finite products. If $C \in \mathbf{C}$, we can form a new $\mathbf{C}$-indexed quantale Q^C, by defining $Q^C(I) = Q(C \times I)$ for $I \in \mathbf{C}$, and if $\alpha : I \to J$ is a morphism of $\mathbf{C}$, then $Q^C(\alpha) : Q(C \times J) \to Q(C \times I)$ is given by $Q^C(\alpha) = Q(1 \times \alpha)$.

We have a morphism $\Delta_C : Q \to Q^C$ of $\mathbf{C}$-indexed quantales given by $\Delta_{C,I} : Q(I) \to Q^C(I)$, where $\Delta_{C,I} = Q(\pi_2)$. The following proposition follows easily from the fact that Q is a linear $\mathbf{C}$-indexed quantale.

PROPOSITION 4.1. Let $Q : \mathbf{C}^{op} \to Quant$ be a linear $\mathbf{C}$-indexed quantale and let $C \in \mathbf{C}$. Then, Q^C is a linear $\mathbf{C}$-indexed quantale and $\Delta_C : Q \to Q^C$ is a linear morphism of linear $\mathbf{C}$-indexed quantales.

We now shall define the notion of a stable adjoint of a morphism of $\mathbf{C}$-indexed quantales. Due to the nice duality features of Girard quantales, as exemplified by Prop. 1.5., it will turn out that for linear indexed quantales, a morphism will have a stable left adjoint iff it has a stable right adjoint.

DEFINITION 4.1. Let Q and S be $\mathbf{C}$-indexed quantales and let $f : Q \to S$ be a morphism of $\mathbf{C}$-indexed quantales.
A morphism of $\mathbf{C}$-indexed quantales $g : S \to Q$ is the *stable left adjoint of f* iff

(1) $g_I : S(I) \to Q(I)$ is left adjoint to $f_I : Q(I) \to S(I)$ for all $I \in \mathbf{C}$.
(2) if $\alpha : I \to J$ is a morphism of $\mathbf{C}$, then $Q(\alpha) \circ g_I = g_J \circ S(\alpha)$.

A morphism of $\mathbf{C}$-indexed quantales $h : S \to Q$ is the *stable right adjoint of f* iff

(1) $h_I : S(I) \to Q(I)$ is right adjoint to $f_I : Q(I) \to S(I)$ for all $I \in \mathbf{C}$.
(2) if $\alpha : I \to J$ is a morphism of $\mathbf{C}$, then $Q(\alpha) \circ g_I = g_J \circ S(\alpha)$.

(2) is often referred to as the Beck (or Beck-Chevalley) condition. First, we note that Proposition 1.5 extends to the setting of linear indexed quantales.

PROPOSITION 4.2. Let $f : Q \to S$ be a linear morphism of linear **C**-indexed quantales. Then, f has a stable left adjoint iff it has a stable right adjoint.

Proof: Suppose $g : Q \to S$ is the stable left adjoint of f. Define $h : S \to Q$ by $h_I(s) = g_I(s^{\perp})^{\perp}$. By Prop. 1.5., h_I is right adjoint to f_I for all I. It remains to prove stability. Let $\alpha : I \to J$ be a morphism of **C**. By the definition of linear indexed quantale, both $Q(\alpha)$ and$S(\alpha)$ preserve linear negation. So, $Q(\alpha)(h_J(s)) = Q(\alpha)(g_J(s^{\perp})^{\perp}) = (Q(\alpha)(g_J(s^{\perp}))^{\perp} = (g_I(S(\alpha)(s^{\perp}))^{\perp} = (g_I(S(\alpha)(s)^{\perp}))^{\perp} = h_I(S(\alpha)(s))$, proving that h is in fact a stable right adjoint. The converse follows in an analogous fashion.

We shall now define indexed joins and meets.

DEFINITION 4.2. Let $Q : \mathbf{C}^{op} \to Quant$ be a **C**-indexed quantale, where **C** is a category with finite products. Let $C \in \mathbf{C}$.

Q has $C-$ *indexed meets* iff the morphism $\Delta_C : Q \to Q^C$ has a stable right adjoint $\forall_C : Q^C \to Q$.

Q has $C-$ *indexed joins* iff the morphism $\Delta_C : Q \to Q^C$ has a stable left adjoint $\exists_C : Q^C \to Q$.

From Prop. 4.1. it follows that if Q is a linear **C** -indexed quantale, then it has C-indexed meets iff it has C-indexed joins for all $C \in \mathbf{C}$.

PROPOSITION 4.3. Rel: $Sets^{op} \to Quant$ has C-indexed meets and joins for all sets C.

Proof: It suffices to prove that Δ_C: Rel $\to$ Rel^C has a stable left adjoint $\exists_C$: Rel$^C \to$ Rel. Since $\Delta_{C,I}$ = Rel(π_2), where $\pi_2 : C \times I \to C$ is the projection, it has a left adjoint $\exists_{C,I}$: Rel($C \times I$) $\to$ Rel(I) given by $\exists_{C,I}(S) = \{(x, x') \in I \times I \mid \exists c, c' \in C \; with \; (c, x) \sim_S (c', x')\}$. Denote Rel($\alpha$) by α^{ast} , if $\alpha : X \to Y$ is a function . If $S \in$ Rel($X \times Y$), we must show $\alpha^*(\exists_{C,Y}(S)) = \exists_{C,X}((1 \times \alpha)^*)$.One can easily calculate that $(x, x') \in \alpha^*(\exists_{C,Y}(S))$ iff $\exists c, c' \in C$ with $(c, \alpha(x)) \sim_S (c', \alpha(x'))$. Also, $(1 \times \alpha)^* = \{((c, x), (c', x')) \mid (c, \alpha(x)) \sim_S (c', \alpha(x')\}$ and thus $(x, x') \in \exists_{C,X}((1 \times \alpha)^*$ iff $\exists c, c' \in C$ with $(c, \alpha(x)) \sim_S (c', \alpha(x'))$ iff $(x, x') \in \alpha^*(\exists_{C,Y}(S))$. This proves the stability of the left adjoint as desired, and hence Rel has C-indexed meets for all sets C. By Prop. 4.2. it follows that Rel has C-indexed joins.

The following proposition can also be established with straightforward arguments analogous to the proof above.

PROPOSITION 4.4. (1) $P : Grp^{op} \to Quant$ has C-indexed meets and joins for all groups C.

(2) $\mathbf{I}$: $Preord^{op} \to Quant$ has X-indexed meets and joins for all preordered sets X .

We shall now define linear hyperdoctrines, slightly strengthening the definition of linear indexed quantale.

DEFINITION 4.3 A *first order linear hyperdoctrine* consists of a pair $(\mathbf{C}, G)$, where $\mathbf{C}$ is a category having finite products and $G : \mathbf{C}^{op} \to \mathit{Quant}$ is a linear indexed quantale such that

(1) If $\alpha : C \to D$ is a morphism of $\mathbf{C}$, $G(\alpha) : G(D) \to G(C)$ is the right adjoint of a quantale homomorphism $G(C) \to G(D)$

(2) G has C-indexed meets and joins for all $C \in \mathbf{C}$.

We immediately obtain the following from Prop. 3.1, 3.2., and 3.3 and Prop. 4.3. and 4.4.

THEOREM 4.1. (1) (*Sets*, Rel) is a first order linear hyperdoctrine.

(2) (*Grp*, *P*) is a first order linear hyperdoctrine.

(3) (*Preord*, **I**) is a first order linear hyperdoctrine.

The use of closed maps, as opposed to homomorphisms, while being the natural notion from a quantale theoretic perspective (as substantiated by all of the examples following Def. 2.1.), is of concern with regard to the proper translation of predicate linear logic into the language of hyperdoctrines. It is important that the so-called Frobenius Reciprocity condition hold. As pointed out by Pitts [P], Frobenius Reciprocity is equivalent to the pullback (transition) maps preserving implication (residuation), which by Prop. 3.1. is definable in terms of $\circ$ and negation. Thus, it would seem desirable to have $\circ$ preserved, yet in our examples this only happens for surjections. This is clearly a question that needs further analysis; perhaps there are natural modifications of the examples or of the transition maps.

REFERENCES

[AV] Abramsky, S. and Vickers, S., Quantales, observational logic, and process semantics, Imperial College Research Report No. DC 90/1, 1990.

[BK] Betti, R. and Kasangian, S., Tree automata and enriched category theory, Rend. Istit. Mat. Univ. Trieste 17, no. 1-2, 1985, 71-78.

[CS] Carboni, A. and Street, R., Order ideals in categories, Pac. J. of Math. Vol. 124, No. 2., 1986, 275-288.

[CW] Carboni, A. and Walters, R.F.C., Cartesian bicategories I, J. Pure Appl. Alg. 45, 1987, 127-141.

[de P] de Paiva, V., Lineales, (talk at Category Theory '91, Montreal, June, 1991)

[G1] Girard, J.Y., Linear logic, Theor. Comp. Sci. 50, 1987, 1-102.

[KR] Kasangian, S. and Rosebrugh, R., Glueing enriched modules and the composition of automata, Cah. de Top. et Geom. Diff. Cat. XXXI (4), 1990, 283-291.

[L] Lambek, J., From categorial grammar to bilinear logic, (preprint), 1991

[L1] Lawvere, F.W., Adjointness in foundations, Dialectica 23, 1969, 281-296.

[L2] Lawvere, F.W., Equality in hyperdoctrines and comprehension schema as adjoint functors, In A. Heller, ed. Categorical Algebra and its Applications Proc. N.Y. Symp., Amer. Math. Soc., 1970,1-14.

[L3] Lawvere, F.W., Metric spaces, generalized logic, and closed categories, Rend. Sem. Mat. e Fis. Milano, 1973, 135-166.

[P] Pitts, A., Notes on categorical logic, manuscript, Cambridge University Computer Laboratory, 1989.

[NR] Niefield, S.B. and Rosenthal, K.I., Constructing locales from quantales, Math. Proc. Camb. Phil. Soc. 104, 1988, 215-234.

[R1] Rosenthal, K.I., Quantales and their Applications, Pitman Research Notes in Math. No. 234, Longman, Scientific, and Technical, 1990.

[R2] Rosenthal, K.I., A note on Girard quantales, Cah. de Top. et Geom. Diff. Cat. XXXI (1), 1990, 3-12.

[R3] Rosenthal, K.I., Free quantaloids, J. Pure Appl. Alg. 72, 1991, 67-82.

[R4] Rosenthal, K.I., Girard quantaloids, to appear in Math. Structures in Computer Science

[R5] Rosenthal, K.I., Quantaloidal nuclei, the syntactic congruence and tree automata, J. Pure Appl. Alg. 77, 1992, 189-205.

[Sc] Scedrov, A. , A brief guide to linear logic, Bull. E.A.T.C.S. 41, 1990, 154-165.

[Se] Seely, R., Linear logic, *-autonomous categories, and cofree coalgebras, In J.W. Gray and A. Scedrov, eds. Categories in Comp. Sci. and Logic, Cont. Math. Vol. 92, Amer. Math. Soc., 1989, 371-382.

[Y] Yetter, D., Quantales and (non-commutative) linear logic, J. Symb. Logic 55 (1), 1990, 41-64.

Department of Mathematics, Union College, Schenectady, N.Y. 12308
E-mail address: RosenthK@gar.union.edu

Canadian Mathematical Society
Conference Proceeding
Volume 13, 1992

EFFECTIVE DESCENT MORPHISMS AND EFFECTIVE EQUIVALENCE RELATIONS

M. SOBRAL and W. THOLEN

ABSTRACT: A characterization of effective descent morphisms is given for categories with pullbacks and coequalizers, in terms of effectiveness of certain naturally arising equivalence relations and stability under pullback of the regular epimorphisms which are their coequalizers. Stability properties of the class of all effective descent morphisms are also derived, in particular, its stability under pullback.

1. Introduction

Let $\mathcal{C}$ be a category with pullbacks, and let $P : \mathcal{C}^2 \longrightarrow \mathcal{C}$ be the functor *"codomain"* of the category of morphisms of $\mathcal{C}$. The fibres of P are the slice categories $\mathcal{C}/_B$, for every $\mathcal{C}$-object B, and for every $\mathcal{C}$-morphism $p : E \longrightarrow B$ there is an associated adjunction $p! \dashv p^* : \mathcal{C}/_B \longrightarrow \mathcal{C}/_E$.

In setting up the notation, we recall that the change-of-base functor p^* is defined by $p^*(D,\delta) = \left(E \times_B D, \pi_1\right)$ and $p^*(f) = 1 \times_B f$

(1)

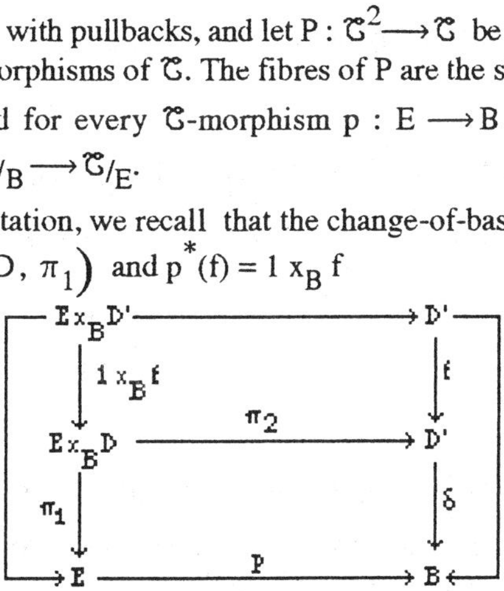

1991 Mathematics Subject Classification. 18A20, 18A30, 18C20.

Partial financial support by INIC/CMUC is greatfully acknowledged by the first author. The second author acknowledges partial financial support by NSERC and from a NATO Collaborative Grant.

This paper is in final form and no version of it will be submitted for publication elsewhere.

with π_1, π_2 the pullback projections. The functor p! is defined by composition: $p!\,(C,\gamma) = (C, p\gamma)$ and $p!(g) = g$. The unit η of the adjunction is the unique morphism for which the diagram

(2)

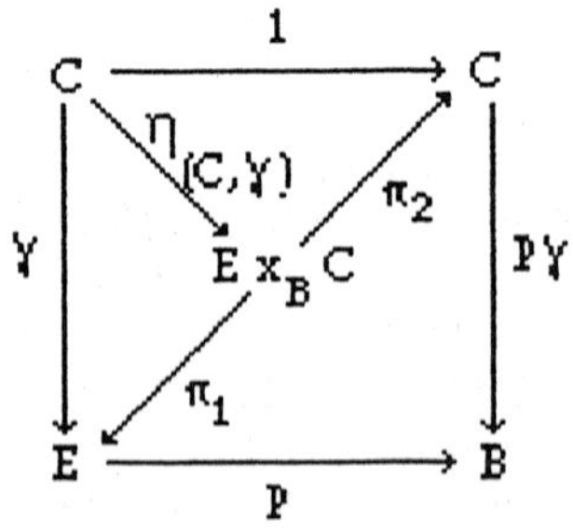

commutes; the counit ε is defined by $\varepsilon_{(D,\delta)} = \pi_2$.

The functor P is a bifibration which satisfies *Chevalley's* condition ([1], 3.2). *Bénabou* and *Roubaud* showed in [1] that descent data for functors fulfilling the above conditions can be interpreted as the structure maps of the algebras over a monad. In particular, for the functor P, descent data for a $\mathcal{C}$-object (C,γ) over E relative to a morphism $p : E \longrightarrow B$ is a morphism $\xi : E \times_B C \longrightarrow C$ over E such that $(C, \gamma\, ;\, \xi)$ is a $\mathbb{T}$-algebra for the monad $\mathbb{T}$ induced in $\mathcal{C}/_E$ by the adjunction $p! \dashv p^*$, i.e. ξ is a $\mathcal{C}$-morphism for which the diagrams

(3)

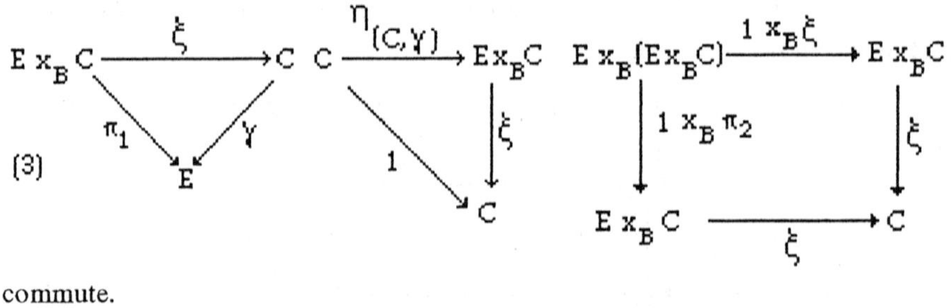

commute.

Let $\Phi : \mathcal{C}/_B \longrightarrow (\mathcal{C}/_E)^{\mathbb{T}}$ be the *Eilenberg-Moore* comparison functor. The descent problem for $\mathbb{T}$-algebras can be summarized as follows: given a $\mathbb{T}$-algebra $(C, \gamma\, ;\, \xi)$ when does there exist a $\mathcal{C}/_B$-object (D,δ) such that $(C, \gamma\, ;\, \xi)$ is isomorphic to $\Phi(D,\delta)$?

A $\mathcal{C}$-morphism p is called a *descent morphism* if Φ is full and faithful, i.e. if p^* is premonadic, and p is an *effective descent morphism* if Φ is an equivalence, i.e. if p^* is monadic ([6]).

Criteria for premonadicity and monadicity of p^* were investigated by *Janelidze* and *Tholen* [7], especially for $\mathcal{C}$ = Top, the category of topological spaces and continuous maps. In [12], *Reiterman* and *Tholen* gave a complete (but complicated) characterization of the effective descent morphisms in Top, in terms of convergence structures. They also proved that the class

$$\mathbb{M} := \{p : p \text{ is an effective descent morphism in } \mathfrak{C}\}$$

is properly contained in the class

$$\mathbb{Q} := \{p : p \text{ is a descent morphism in } \mathfrak{C}\}.$$

The latter class coincides with the class of universal regular epimorphisms in $\mathfrak{C}$, i.e. those regular epimorphisms p which are stable under pullback (so that $\pi_2 \colon E \times_B D \longrightarrow D$ is a regular epimorphism for every map $\delta \colon D \longrightarrow B$). $\mathbb{Q}$ therefore enjoys a number of pleasant stability properties, in particular: $\mathbb{Q}$ is stable under composition and under pullback; see [7], 1.2. One of the aims of this paper is to investigate stability properties of the class $\mathbb{M}$. We prove that $\mathbb{M}$ is always stable under pullback, and give a sufficient condition for $\mathbb{M}$ to be closed under composition; see Section 3 below.

We believe that stability of $\mathbb{M}$ under pullback is shown here for the first time. The proof relies on a characterization of effective descent morphisms in terms of effective equivalence relations.

Let us first recall the known result in the area of general solutions for the descent problem, as given by *Moerdijk* [10] and slightly improved in [12]: any class $\mathbb{O}$ of morphisms is contained in $\mathbb{M}$, provided it satisfies the following conditions:

I - $\mathbb{O}$ contains all isomorphisms and is closed under composition with isomorphisms.

II - $\mathbb{O} \subset \text{Reg Epi}(\mathfrak{C})$, the class of regular epimorphisms.

III $\mathbb{O}$ is stable under pullback.

IV - The coequalizer of a pair of $\mathbb{O}$-morphisms over some $\mathfrak{C}$-object B exists and is preserved by $p^* \colon \mathfrak{C}/_B \longrightarrow \mathfrak{C}/_E$ for each $\mathbb{O}$-morphism $p \colon E \longrightarrow B$.

Condition IV certainly holds when each parallel pair (f,g) of $\mathbb{O}$-morphisms has a universal coequalizer, i.e. if $X \underset{g}{\overset{f}{\rightrightarrows}} X' \overset{q}{\longrightarrow} Q$ is a coequalizer diagram then its pullback $X \times_Q Y \rightrightarrows X' \times_Q Y \longrightarrow Y$ along an arbitrary morphism $h \colon Y \longrightarrow Q$ is again a coequalizer diagram, which is equivalent to Axiom 1 in [10].

In [9] *Joyal* and *Tierney* proved that open surjections are effective descent morphisms in the category of locales. Using the above criterion, a much simpler proof of this fact was given in [10].

In this paper, by analysing the particular coequalizers whose preservation by the change-of-base functor p^* is essential in this context, we derive necessary and sufficient conditions for a morphism to be an effective descent morphism in terms of equivalence relations naturally arising from every $\mathbb{T}$-algebra (with $\mathbb{T}$ the monad as above). Our main result says that a regular epimorphism p is an effective descent morphism if and only if these equivalence relations are effective and their coequalizers are universal regular epimorphisms (cf. Thm. 2.5).

2. Main Theorem and Some Corollaries

Throughout this paper $\mathfrak{C}$ will denote a category with pullbacks and coequalizers. Then also each sliced category $\mathfrak{C}/_B$ has coequalizers. Hence the comparison functor

$\Phi : \mathfrak{C}/_B \longrightarrow \left(\mathfrak{C}/_E\right)^{\mathbb{T}}$ has a left adjoint L, defined on objects by $L(C, \gamma ; \xi) = (Q,\delta)$, with (Q,δ) the codomain of the coequalizer of the pair $\left(\xi, \pi_2\right)$ in $\mathfrak{C}/_B$.

The comparison functor is full and faithful if and only if the counit ε of the adjunction $p! \dashv p^*$ is pointwise a regular epimorphism (see e.g. [2] 3.2, Thm 9). Since $\varepsilon_{(D,\delta)} = \pi_2$, of (1), one has:

2.1. PROPOSITION ([7] 2.2) *A morphism of $\mathfrak{C}$ is a descent morphism if and only if it is a universal regular epimorphism.*

In particular, $\mathbb{D} \subseteq \text{Reg Epi}(\mathfrak{C})$.

In order to consider equivalence relations, it is convenient (but not essential) to assume that $\mathfrak{C}$ has binary products. Recall that a pair $\left(f,g : R \rightrightarrows X\right)$ of $\mathfrak{C}$-morphisms $\left(\text{or the induced } < f,g > : R \longrightarrow X \times X\right)$ is called an equivalence relation on X (see, e.g., [8] 0.43) if

(i) (f,g) is a mono-pair, i.e. $< f,g >$ is a monomorphism.

(ii) the diagonal $\Delta : X \longrightarrow X \times X$ factors through $< f,g >$.

(iii) there exists a morphism $t : R \longrightarrow R$ such that $f.t = g$ and $g.t = f$.

(iv) for the pullback diagram

$$\text{(4)} \qquad \begin{array}{ccc} A & \xrightarrow{\;g'\;} & R \\ {\scriptstyle f'}\downarrow & & \downarrow{\scriptstyle f} \\ E & \xrightarrow[\;g\;]{} & X \end{array}$$

$< ff',gg' > : A \longrightarrow X \times X$ factors through $< f,g >$.

Hence by (ii), (iii) and (iv) , $< f,g >$ is reflexive, symmetric and transitive, respectively.

The kernel pair of a morphism always yields an equivalence relation; an equivalence relation is said to be *effective* if it is the kernel pair of a morphism.

2.2. LEMMA. *For every $\mathbb{T}$-algebra $(C, \gamma ; \xi)$, $\left(\pi_2 , \xi\right)$ is an equivalence relation on C.*

Proof: We shall check in detail conditions (i) - (iv) above. (i) Let $q = \mathrm{coeq}(\pi_2, \xi)$ and (p_1, p_2) be its kernel pair. Consider the following diagram in $\mathcal{C}$

(5)

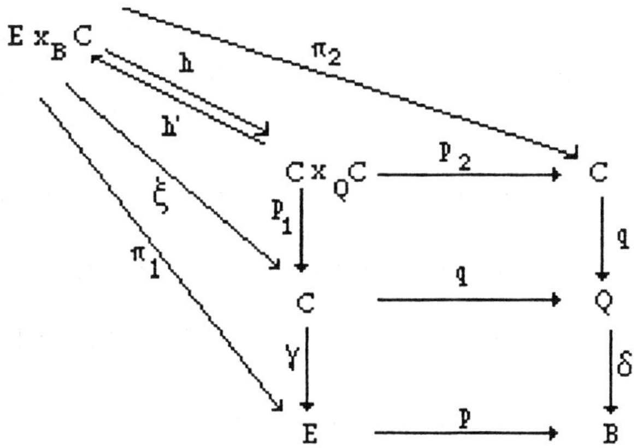

where h, h' are the uniquely determined morphisms with $p_1 . h = \xi$, $p_2 . h = \pi_2$ and $\pi_1 . h' = \gamma\, p_1$, $\pi_2 . h' = p_2$, respectively. Since h'h = 1, the morphism

(6)

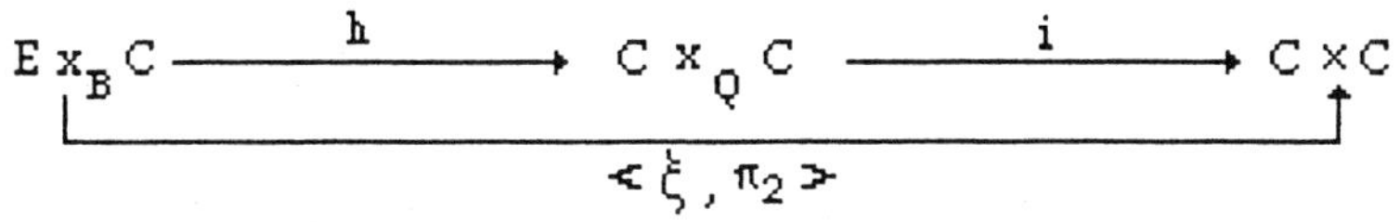

is a monomorphism.

(ii) $< \xi, \pi_2 > . \eta_{(C,\gamma)} = \Delta$.

(iii) Consider the following diagram in $\mathcal{C}$

(7)

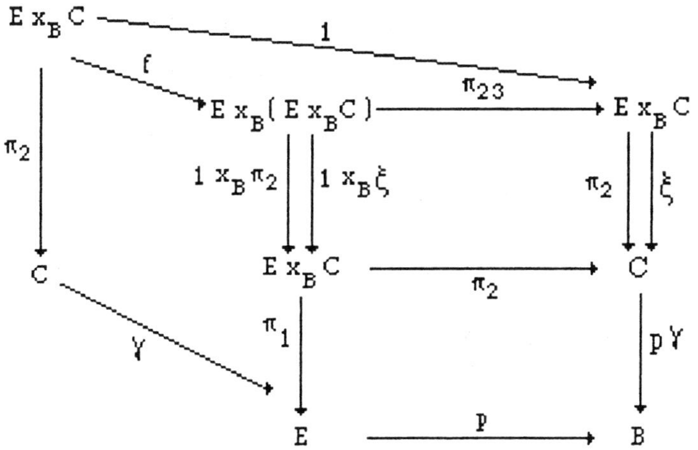

where f is the unique morphism such that $\pi_{23} . f = 1$ and $\pi_1 . 1 \times_B \pi_2 . f = \gamma\, \pi_2$. Note that $\left(\hat{\pi}_1 = \pi_1 . 1 \times_B \pi_2 = \pi_1 . 1 \times_B \xi\,,\, \pi_{23}\right)$ is the pullback of $\left(p,\, p\, \gamma\, \pi_2 = p\, \gamma\, \xi\right)$.

For $t = 1 \times_B \xi . f$, $\pi_2 . t = \xi$. It remains to be shown that $\xi . t = \pi_2$. By applying the projections π_1, π_2 one easily shows that $1 \times_B \pi_2 . f = \eta_{(C,\gamma)} . \pi_2$, hence

$$\xi . t = \xi . 1 \times_B \xi . f = \xi . 1 \times_B \pi_2 . f = \xi . \eta_{(C,\gamma)} . \pi_2 = \pi_2 .$$

(Note that, necessarily, t is an isomorphism since tt = 1)

(iv) The pullback of $\left(\xi, \pi_2\right)$ is $\left(\pi_{23}\,,\, 1 \times_B \xi\right)$, and one has

$$< \xi, \pi_2 > . 1 \times_B \pi_2 = < \xi . 1 \times_B \xi\,,\, \pi_2 . \pi_{23} > .$$

□

Now we shall look for conditions under which these equivalence relations are effective. We shall make use of the following fact:

2.3 LEMMA. *For every $\mathbb{T}$-algebra $(C, \gamma\,;\, \xi)$, $\left(\xi,\, \gamma\, \pi_2\right)$ is a monopair in $\mathcal{C}$*

Proof: Let $\gamma\, \pi_2 . a = \gamma\, \pi_2 . b$ and $\xi . a = \xi . b$. Then there exist uniquely determined morphisms h_a and h_b such that $\pi_{23} . h_a = a$, $\hat{\pi}_1 . h_a = \gamma\, \pi_2 . a$ and $\pi_{23} . h_b = b$, $\hat{\pi}_1 . h_b = \gamma\, \pi_2 . b$, with the notations of 2.2 (iii).

Using again the universal property of pullbacks, the equalities

$$1 \times_B \xi . h_a = 1 \times_B \xi . h_b$$
$$1 \times_B \pi_2 . h_a = \eta_{(C,\gamma)} . \pi_2 . a$$

and

$$1 \times_B \pi_2 . h_b = \eta_{(C,\gamma)} . \pi_2 . b$$

are easily verified. It follows that

$$\xi . 1 \times_B \xi . h_a = \xi . 1 \times_B \pi_2 . h_a = \xi . \eta_{(C,\gamma)} . \pi_2 . a = \pi_2 . a$$

and, similarly, $\xi . 1 \times_B \xi . h_b = \pi_2 . b$. Hence $\pi_2\, a = \pi_2\, b$. Since, in addition, $\pi_1\, a = \gamma\, \xi\, a = \gamma\, \xi\, b = \pi_1\, b$ we conclude that $a = b$.

□

Recall that the unit α of the adjunction $L \dashv \Phi$ is given by the unique $\mathbb{T}$-morphism $\alpha_{(C,\gamma;\xi)} : (C, \gamma ; \xi) \longrightarrow \Phi L\ (C, \gamma ; \xi)$ such that $\alpha_{(C,\gamma;\xi)} \cdot \xi = 1 \times_B q$. Equivalently, it is the unique morphism in $\mathcal{C}$ for which the upper-left triangles in the diagram

(8)

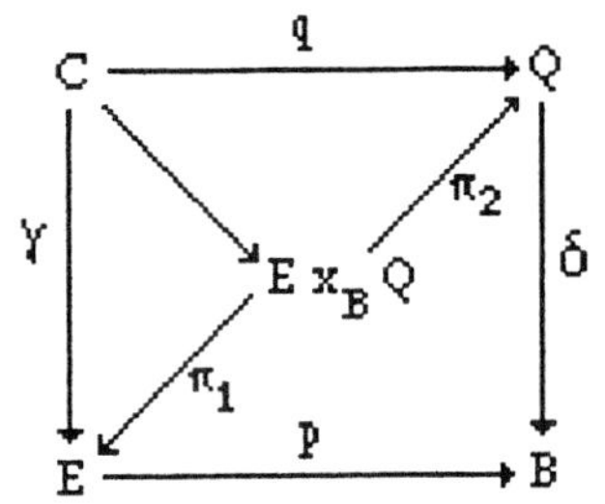

commute. Hence $\alpha_{(C,\gamma;\xi)}$ is a monomorphism if and only if (γ , q) is a monopair.

2.4 PROPOSITION. *For a $\mathbb{T}$-algebra $(C, \gamma ; \xi)$ the following are equivalent:*

(i) (π_2 , ξ) is a kernel pair in $\mathcal{C}$

(ii) $\alpha_{(C,\gamma;\xi)}$ is a monomorphism.

Proof: (i) $\Rightarrow$ (ii) Let (π_2, ξ) be a kernel pair; then it is the kernel pair of its coequalizer. If the $\mathcal{C}$-morphism underlying $\alpha_{(C,\gamma;\xi)}$ coequalizes $f, g : D \longrightarrow C$ then $q.f = q.g$. Hence there exists a unique morphism $d : D \longrightarrow E \times_B C$ such that $\pi_2 . d = f$ and $\xi . d = g$.

Since $\gamma\, \pi_2\, d = \gamma\, f = \gamma\, g = \gamma\, \pi_2\, \eta_{(C,\gamma)}\, g$ and $\xi . d = g = \xi . \eta_{(C,\gamma)} . g$, by 2.3, one concludes that $\eta_{(C,\gamma)} . g = d$. Thus $f = \pi_2 . d = \pi_2 . \eta_{(C,\gamma)} . g = g$.

(ii) $\Rightarrow$ (i) Let (γ , q) be a monopair. In 2.2 (i), since $\gamma . \xi . h' = \gamma . p_1$ and $q . \xi . h' = q . p_1$, we have that $\xi . h' = p_1$ and so $p_1 . h h' = \xi h' = p_1$. But the equality $p_2\, h h' = p_2$ always holds. Hence, one concludes $hh' = 1$. Since it is also a split monomorphism (2.2 (i)), h is an isomorphism.

□

We have shown that (π_2, ξ) is a kernel pair for each $\mathbb{T}$-algebra $(C, \gamma ; \xi)$ if and only if the left adjoint of the comparison functor is faithful.

Now we are ready to give the following characterization of effective descent morphisms.

2.5 THEOREM *A regular epimorphism in $\mathcal{C}$ is an effective descent morphism if and only if, for each $\mathbb{T}$-algebra $(C, \gamma ; \xi)$, the equivalence relation (π_2, ξ) is effective and its coequalizer belongs to $\mathbb{Q}$.*

Proof: The necessity of these conditions is now obvious.

To prove that they are also sufficient we first remark that p belongs to $\mathbb{Q}$ since p is the coequalizer of (ζ, π_2), with $(E, 1_E ; \zeta)$ the terminal algebra, i. e. $\zeta = \pi_1$ with (π_1, π_2) the kernel pair of p. Hence, by 2.1, p is a descent-morphism which means that the counit of the adjunction $L \dashv \Phi$ is an isomorphism.

We now show that also the unit α is an isomorphism. For a $\mathbb{T}$-algebra $(C, \gamma ; \xi)$, if (π_2, ξ) is the kernel pair of q in $\mathfrak{C}$ then it is also its kernel pair in $\mathfrak{C}/_B$. Since the comparison functor $\Phi : \mathfrak{C}/_B \longrightarrow (\mathfrak{C}/_E)^{\mathbb{T}}$ is a right adjoint, it preserves limits, hence $(\Phi(\pi_2), \Phi(\xi)) = (1 \times_B \pi_2, 1 \times_B \xi)$ is the kernel pair of $\Phi(q) = 1 \times_B q$. Furthermore, $1 \times_B q$ is a regular epimorphism in $\mathfrak{C}$, since it is the pullback of the $\mathbb{Q}$-morphism q along $E \times_B Q \longrightarrow Q$, a fact that also holds in $(\mathfrak{C}/_E)^{\mathbb{T}}$. Hence $1 \times_B q$ is the coequalizer of its kernel pair $(1 \times_B \pi_2, 1 \times_B \xi)$ in the category of $\mathbb{T}$-algebras. Since ξ is the coequalizer of $(1 \times_B \pi_2, 1 \times_B \xi)$, it follows that $\alpha_{(C,\gamma;\xi)}$ is an isomorphism.

Hence Φ is an equivalence of categories, i.e. p is an effective descent morphism.

$\square$

There is a number of general criteria for monadicity given in [5] none of which seems to yield conveniently the characterization of 2.5. The point of 2.5 is that it minimizes the sufficient conditions for effective descent, whereas the conditions given in [5] make use of more general types of equivalence relations. However, we are grateful to the referee for the observation that Lemma 3.4 of [5] suffices to derive the second of the following two corollaries, as was shown previously in [3], Corollary 2.4, in the more general context of indexed categories. Here we invoke Theorem 2.5 and Lemma 2.2 above to obtain immediatly:

2.6 COROLLARY. *Let $\mathfrak{C}$ be a finitely complete category with coequalizers. If equivalence relations in $\mathfrak{C}$ are effective then p is an effective descent morphism if and only if, for each $\mathbb{T}$-algebra $(C, \gamma; \xi)$, the coequalizer of (π_2, ξ) belongs to $\mathbb{Q}$.*

2.7 COROLLARY. *For a finitely complete category $\mathfrak{C}$ with coequalizers such that regular epimorphisms are universal and equivalence relations are effective, every regular epimorphism is an effective descent morphism.*

For example, in any topos $\mathcal{E}$, colimits are universal (hence, in particular, Reg Epi $(\mathcal{E}) = \mathbb{Q}$), and equivalence relations are effective ([8] 1.51 and 1.23). Hence all (regular) epimorphisms are effective descent morphisms. This also holds in any Barr-exact category.

Let us now apply 2.5 when $\mathfrak{C}$ = Top is the category of topological spaces. First we remark that, for every $\mathbb{T}$-algebra $(C, \gamma; \xi)$, (π_2, ξ) is an effective equivalence relation. Indeed, by the proof of (i) in 2.2 it is clear that $E \times_B C$ has the initial topology with respect to (ξ, π_2), and so (ξ, π_2) is the kernel pair of q.

2.8 COROLLARY. *A quotient map p in Top is an effective descent morphism if and only if, for each $\mathbb{T}$-algebra $(C, \gamma; \xi)$, the coequalizer of the pair (π_2, ξ) belongs to $\mathbb{Q}$.*

Of course, 2.8 still holds if Top is replaced by a concrete category over Set whose underlying functor preserves pullbacks and coequalizers.

The continuous maps $p : E \longrightarrow B$ which belong to $\mathbb{Q}$ were characterized by *Day* and *Kelly* in [4]. The class of open surjections is strictly contained in $\mathbb{Q}$ ([7] 1.2).

Using 2.8, the proof of the fact that open surjections are effective descent morphisms in Top is immediate: the coequalizer of the two open surjections (π_2, ξ) belongs to $\mathbb{Q}$ since it is open.

3. The Class $\mathbf{M}$ of Effective Descent Morphisms

For a category $\mathfrak{C}$, the class $\mathbf{M}$ is a class of regular epimorphisms containing all split-epimorphisms (= retractions), hence

$$\text{Split Epi } (\mathfrak{C}) \subset \mathbf{M} \subset \mathbb{Q} \;;$$

see [7] 2.4 and 2.2. For $\mathfrak{C}$ = Top, both inclusions are strict: since open surjections belong to $\mathbf{M}$ (see the remarks following 2.8), not every effective descent morphism is split, and an example of a non-effective descent morphism is exhibited in [12].

Both the classes Split Epi ($\mathfrak{C}$) and $\mathbb{Q}$ are closed under composition and stable under pullback. We first prove that also $\mathbf{M}$ satisfies the second property, in every category with pullbacks and coequalizers.

3.1 THEOREM. *The class $\mathbf{M}$ is stable under pullback.*

Proof: Let p' be the pullback of p along σ, let $(C, \gamma; \xi)$ be a $\mathbb{T}^{p'}$-algebra for the monad induced by the adjunction $(p')_! \dashv (p')^*$ and let q be the coequalizer of (π_2, ξ) in $\mathfrak{C}$, hence also in $\mathfrak{C}/_A$. In the diagram

(9)

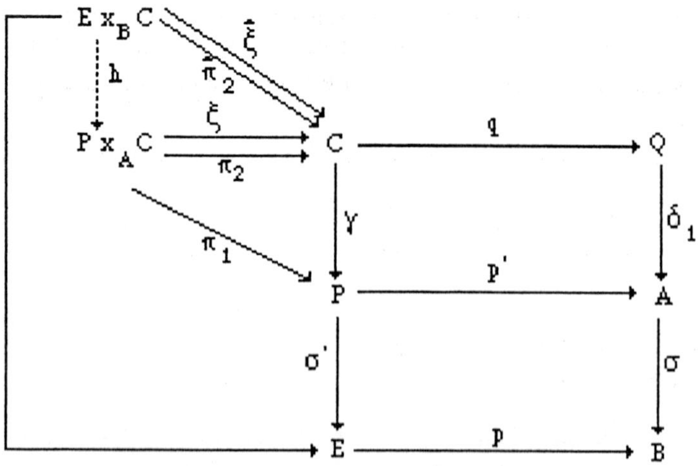

h is the unique isomorphism such that $\pi_2 \,.\, h = \hat{\pi}_2$ and $\sigma' \,.\, \pi_1 \,.\, h = \hat{\pi}_1$. Hence, it is clear that $\left(C,\ \hat{\gamma}\ ;\ \hat{\xi}\right)$ is a $\mathbb{T}^p$-algebra with $\hat{\gamma} = \sigma' \,.\gamma$ and $\hat{\xi} = \xi \,.\, h$.

Therefore, if p is an effective descent morphism, then $\left(\hat{\pi}_2\ ,\ \hat{\xi}\right)$ is a kernel pair since the same holds for $\left(\pi_2\ ,\ \xi\right)$, and $q = \text{coeq}\left(\pi_2\ ,\ \xi\right) = \text{coeq}\left(\hat{\pi}_2\ ,\ \hat{\xi}\right)$ belongs to $\mathbb{Q}$. By 2.5, we have therefore that $p' \in \mathbb{M}$.

□

The question of whether $\mathbb{M}$ is closed under composition is closely related to the question of whether $\mathbb{M}$ satisfies condition IV of the Introduction (so that $\mathbb{M}$ would be the largest class $\mathbb{O}$ satisfying conditions I-IV).

We first remind the reader of the following sufficient condition for compositivity of monadic functors. One says that $U : \mathcal{A} \longrightarrow \mathcal{X}$ satisfies the *homomorphism condition* if for every regular epimorphism $d : UB \longrightarrow D$ whose kernel pair is of the form (Uf, Ug), with $f,g : A \longrightarrow B$, there is a morphism $c : B \longrightarrow C$ and an isomorphism $i : UC \longrightarrow D$ with $i.\ Uc = d$. One then has:

3.2 PROPOSITION *([13]) If $\mathcal{X}$ has kernel pairs, then the composite of monadic functors $V : \mathcal{K} \longrightarrow \mathcal{A}$, $U : \mathcal{A} \longrightarrow \mathcal{X}$ satisfying the homomorphism condition is again a monadic functor satisfying the homomorphism condition.*

This proposition enables us to prove:

3.3 THEOREM. *For the statements (i), (ii), (iii) below, we have (i) ⇒ (ii) ⇒ (iii):*

(i) $\mathcal{M}$ satisfies condition IV of the Introduction.

(ii) The pullback of a regular epimorphism by an $\mathcal{M}$-morphism is again a regular epimorphism.

(iii) $\mathcal{M}$ is closed under composition.

Proof: (i) $\Rightarrow$ (ii) Consider a morphism $f : A \longrightarrow B$ and let c be the coequalizer of its kernel pair. If f is a regular epimorphism, then the morphism g with $g.c = f$ is an isomorphism. For $p : E \longrightarrow B$ in $\mathbf{M}$, by hypothesis p^* preserves the coequalizer diagram in $\mathcal{C}/_B$ since the projections of a kernel pair are split epimorphisms and so they belong to $\mathbf{M}$. Hence the pulback of f along p is a regular epimorphism.

(ii) $\Rightarrow$ (iii) According to 3.2, it suffices to show that p^* satisfies the homomorphism condition if $p \in \mathbf{M}$. Consider a parallel pair (a,b) of $\mathcal{C}/_B$-morphisms with codomain $f : A \longrightarrow B$, such that $\left(p^*(a), p^*(b)\right)$ is a kernel pair, for $p : E \longrightarrow B$ belonging to $\mathbf{M}$. Let c be the coequalizer of (a,b). By (ii), p^* (c) is a regular epimorphism, because p' belongs to $\mathbf{M}$.

(10) 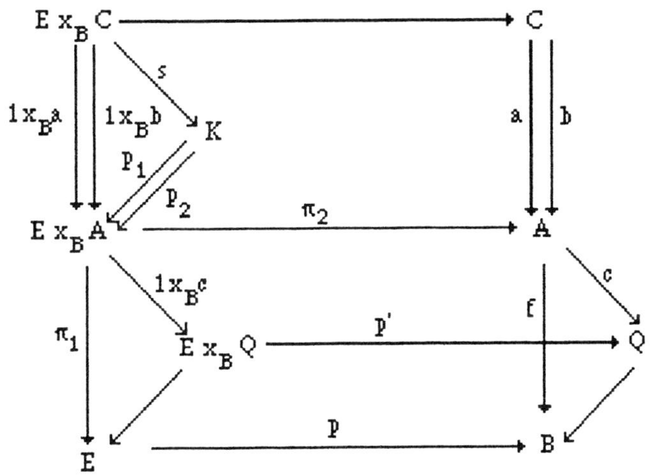

If $\left(p_1, p_2\right)$ is the kernel pair of $p^*(c) = 1 \times_B c$ then there exists a unique morphism s such that $p_1.s = p^*(a)$ and $p_2.s = p^*(b)$.

Let us prove now that $p^*(c)$ is the coequalizer of $\left(p^*(a), p^*(b)\right)$: if $g \cdot p^*(a) = g \cdot p^*(b)$ then $g \cdot p_1.s = g \cdot p_2.s$ implies that $g \cdot p_1 = g \cdot p_2$, because s is a (split) epimorphism, and so there exists a unique morphism h such that $h \cdot p^*(c) = g$.

If $d : E \times_B A \longrightarrow D$ is the coequalizer of $\left(p^*(a), p^*(b)\right)$ then, it is clear that, there exists a morphism c and an isomorphism i such that $i \cdot p^*(c) = d$.

□

3.4 REMARKS (1) For $\mathfrak{C}$ = Top, $\mathbf{M}$ is closed under composition. This is most easily concluded from the formal criterion for effective descent morphisms in Top given in 1.7 of [12]. Recently, J. Reiterman constructed an example showing that (ii) of 3.3 does not hold in Top (private communication). Consequently, (i) of 3.3 does not hold either in Top..

(2) The class $\mathfrak{Q}$ for $\mathfrak{C}$ = Top does not satisfy condition IV of the Introduction since, otherwise, one would have $\mathfrak{Q} = \mathbf{M}$ which was disproved in [12].

(3) Is there an example of a (decent) category $\mathfrak{C}$ for which $\mathbf{M}$ fails to be closed under composition?

Acknowledgement

The second author gratefully acknowledges valuable discussions with George Janelidze on the subject of the paper.

References

1. J. Bénabou and J. Roubaud, *Monades et descente*, Comptes Rendus Acad. Sc. Paris 270, A (1970) 96-98.
2. M. Barr and C. Wells, *Toposes, Triples and Theories*, Springer-Verlag (New York, Berlin, Heidelberg, Tokyo) 1985.
3. M. Bunge and R. Paré, *Stacks and equivalence of indexed categories*, Cahiers Top. Géom. Diff. Catégoriques 20 (1979) 373-399.
4. B.J. Day and G.M. Kelly, *On topological quotient maps preserved by pullbacks and products*. Proc. Cambridge Phil. Soc. 67 (1970) 553-558.
5. J. Duskin, *Variations on Beck's tripleability criterion*, in: Reports of the Midwest Cat. Sem. III, (Springer Lecture Notes in Math. 106, Berlin 1969) pp. 74-129.
6. A. Grothendieck, *Technique de descente et théorèmes d'existence en Géometrie Algébrique*, Séminaire Bourbaki (1959).
7. G. Janelidze and W. Tholen, *How algebraic is the change-of-base functor?* Proc. Conf. Category Theory, Como 1990 (Springer Lecture Notes in Math. 1488, Berlin 1991) pp. 174-186.
8. P. T. Johnstone, *Topos Theory*, Academic Press (London, New York, San Francisco) 1977.
9. A. Joyal and M. Tierney, *An extension of the Galois theory of Grothendieck*, Memoirs Amer. Math. Soc. 309 (1984).
10. I. Moerdijk, *Descent theory for toposes*, Bull. Soc. Math. Belgique 41 (1989) 373-391.
11. I. Moerdijk, *Descent for proper maps of topological spaces*, manuscript (1990).
12. J. Reiterman and W. Tholen, *Effective descent maps of topological spaces*, preprint (York University, 1991).
13. M. Pfender, *Kongruenzen, Konstruktion von Limiten und Cokernen und Algebraische Kategorien.*, Ph.D. Thesis (Technical University of Berlin, 1971).

Manuela SOBRAL
Departamento de Matemática, Universidade de Coimbra, 3000 Coimbra, PORTUGAL
E-mail address: sobral@ciuc2.uc.pt

Walter THOLEN
Department of Mathematics and Statistics, York University,Toronto,CANADA M3J 1P3
E-mail address: tholen@vm1.yorku.ca

Canadian Mathematical Society
Conference Proceeding
Volume **13**, 1992

Functorial Selection of Morphisms

VĚRA TRNKOVÁ

ABSTRACT. Simultaneous representations and slices of functors are investigated. Structural conditions for collections of slices are presented. Functors which admit a simultaneous representation into Unif $\to$ Top are characterized.

I. Introduction

I.1. The present paper investigates concepts associated with a commutative square of functors

$$\begin{array}{ccc} \mathcal{H}_1 & \xrightarrow{H} & \mathcal{H}_2 \\ {\scriptstyle G_1}\big\downarrow & & \big\downarrow{\scriptstyle G_2} \\ \mathcal{K}_1 & \xrightarrow{K} & \mathcal{K}_2 \end{array} \tag{1}$$

that have the following form: given functors $H : \mathcal{H}_1 \to \mathcal{H}_2$ and $K : \mathcal{K}_1 \to \mathcal{K}_2$, there exist functors $G_i : \mathcal{H}_i \to \mathcal{K}_i$ with properties P_i (for $i = 1, 2$) such that the square (1) commutes and has a property Q. Let us recall some of these concepts:

(1) *simultaneous representation* (see $[T_1]$), for which both G_1 and G_2 are full embeddings (=full and faithful functors);
(2) *strong embedding* (see [P]), where the functor G_1 is a full embedding and G_2 is faithful;
(3) *slice of the functor* K (see [ST]): both G_1 and G_2 are faithful and the square (1) a *subpullback*, that is, for every pair a, b of objects of $\mathcal{H}_1$, the corresponding square of hom-sets

$$\begin{array}{ccc} \mathcal{H}_1(a,b) & \xrightarrow{H} & \mathcal{H}_2(H(a),H(b)) \\ {\scriptstyle G_1}\big\downarrow & & \big\downarrow{\scriptstyle G_2} \\ \mathcal{K}_1(G_1(a),G_1(b)) & \xrightarrow{K} & \mathcal{K}_2(F(a),F(b)) \end{array}$$

(where $F = K \circ G_1 = G_2 \circ H$) is a pullback in *Set*.

1991 *Mathematics Subject Classification*. Primary 18A22, 18B15.
This paper is in final form and no version of it will be submitted for publication elsewhere.

I.2. Of these three concepts, the *strong embedding* is the oldest one; it was formulated at the time when categorists in Prague and elsewhere were intensively investigating full embeddings into various conventional concrete categories.

In a strong embedding, the faithful functor $G_2 : \mathrm{Set} \to \mathrm{Set}$ ensures that the full embedding $G_1 : \mathcal{H}_1 \to \mathcal{K}_1$ of a concrete category $(\mathcal{H}_1, H)$ into a concrete category $(\mathcal{K}_1, K)$ [where $H : \mathcal{H}_1 \to \mathrm{Set}$, $K : \mathcal{K}_1 \to \mathrm{Set}$ are their forgetful functors] maps $\mathcal{H}_1$-objects with the same underlying set X to $\mathcal{K}_1$-objects with the same underlying set $G_2X \supseteq X$, and that morphisms satisfy an analogous requirement. For conventional categories, the full embedding G_1 is then created by structures defined on these uniformly expanded sets. Strong embeddings induce finer distinctions between concrete categories than full embeddings do: for instance, while the category of graphs has a full embedding into the category Smg of semigroups [HL] and vice versa [PT], graphs cannot be strongly embedded into Smg or any other category of algebras. These and other older results about full and strong embeddings of concrete categories were summarized in the monograph [PT].

I.3. The notion of *simultaneous representation* originated from an inquiry about the effect of structure reduction on morphism selection. For instance, if M_0 denotes the endomorphism monoid of an algebra $A = (X, \alpha, \beta, \gamma)$ with three unary operations and M_1 is the endomorphism monoid of its reduct (X, α, β), are there any conditions (other than the obvious $M_0 \subseteq M_1$) that these monoids must satisfy?

Let $\mathrm{Alg}(1, 1, 1)$ (or $\mathrm{Alg}(1, 1)$) denote the category of universal algebras with three (or two) unary operations and let $K : \mathrm{Alg}(1, 1, 1) \to \mathrm{Alg}(1, 1)$ be the functor that forgets the last operation, that is, sends any $(X, \alpha, \beta, \gamma)$ to (X, α, β). A categorial generalization of the question about monoids is this: which functors $H : \mathcal{H}_1 \to \mathcal{H}_2$ have a simultaneous representation in K? Since every small category has a full embedding into both $\mathrm{Alg}(1, 1, 1)$ and $\mathrm{Alg}(1, 1)$, see [PT], the latter question, at least for **small** categories $\mathcal{H}_1$, $\mathcal{H}_2$, concerns their relationship, but **not** their separate representability in categories of algebras. For small categories $\mathcal{H}_1, \mathcal{H}_2$, the question is answered in [PST]. Given disjoint types Δ_0, Δ_1, and a functor $K : \mathrm{Alg}(\Delta_0 \cup \Delta_1) \to \mathrm{Alg}\,\Delta_0$ forgetting all Δ_1-operations, a functor $H : \mathcal{H}_1 \to \mathcal{H}_2$ between small categories has a simultaneous representation in K exactly when it is faithful and fulfills Isbell's Zig-Zag condition ([$I_{1,2}$], see also II.9 below) or, in the special case of a nullary type Δ_1, a stronger condition (p) (of II.6 below). Simultaneous representations in forgetful functors $K : V \to W$ for some subvarieties V of $\mathrm{Alg}(\Delta_0 \cup \Delta_1)$ and W of $\mathrm{Alg}\,\Delta_0$ and functors forgetting all Δ_1-operations are constructed in [LST].

I.4. Simultaneous representations in functors of general topology (such as the forgetful functor $U : \mathrm{Unif} \to \mathrm{Top}$, the Stone-Čech compactification $\beta : \mathrm{Tich} \to \mathrm{Comp}$, the compactly generated modification $\mathrm{CG} : \mathrm{Top} \to \mathrm{CGTop}$, the completely regular modification $\mathrm{CR} : \mathrm{Top} \to \mathrm{CRTop}$ and others) are investigated in [T_2], [T_3] and [TH]. In topology, the definition of simultaneous representation has to be modified a little: the functors G_1 and G_2 in the definition must be almost full embeddings, that is, faithful functors that are almost full (this means that for every $c, d \in \mathrm{obj}\,\mathcal{H}_i$, G_i maps the set $\mathcal{H}_i(c, d)$ onto the set of all *nonconstant* maps in $\mathcal{K}_i(G_i(c), G_i(d))$). This modification also applies to representations of individual categories in Top, its full subcategories Comp (= the compact Hausdorff spaces), Tich (= the Tichonov spaces), in Unif (= the category of uniform spaces and uniformly continuous maps), Metr (= the category of metric spaces and nonexpanding maps) and others, see [PT]. In contrast to algebraic results, which depend on the Zig-Zag condition, mere faithfulness of a functor $H : \mathcal{H}_1 \to \mathcal{H}_2$ between small categories suffices (and is necessary as well, of course) for its simultaneous representability in U, β, CG, CR, by [T_2], [T_3], [TH].

For simultaneous representations in functors between categories of algebras it is quite reasonable to consider only small categories $\mathcal{H}_1$, $\mathcal{H}_2$: every small category can be fully embedded into Alg(1, 1) and into Alg(2) (and into any Alg Δ with a larger type Δ), while the full embeddability of $\mathrm{Set}^{\mathrm{op}}$ (the category opposite to Set) into Alg(1, 1) is equivalent to a set-theoretical statement, see [PT]. This, however, is not the case in topology: every concretizable category (i.e., a category admitting a faithful functor into Set) can be almost fully embedded into Unif and into Top without any special set-theoretical assumption, by [K], see also [PT]. Hence there is no reason to restrict investigations of simultaneous representability of $H : \mathcal{H}_1 \to \mathcal{H}_2$ in U : Unif $\to$ Top to small $\mathcal{H}_1$, $\mathcal{H}_2$. Section III of this paper characterizes functors $H : \mathcal{H}_1 \to \mathcal{H}_2$ that can be simultaneously represented in U : Unif $\to$ Top, CG : Top $\to$ CGTop and CR : Top $\to$ CRTop.

I.5. *Slices of functors*, formally introduced in [ST], were used implicitly already in [PST] and [LST] to construct various simultaneous representations. To sketch an idea of these constructions, we begin with a commutative diagram

(2)
$$\begin{array}{ccccc} \mathcal{H}_1 & \xrightarrow{H} & \mathcal{H}_2 & \xrightarrow{G} & \mathcal{L} \\ {\scriptstyle H_1}\downarrow & & \downarrow{\scriptstyle H_2} & & \downarrow{\scriptstyle L} \\ \mathcal{T} & \xrightarrow{T} & \mathrm{Set} & \xrightarrow{F} & \mathrm{Set} \end{array}$$

whose left square is a subpullback of faithful functors and the right square is a strong embedding of the concrete category $(\mathcal{H}_2, H_2)$ into $(\mathcal{L}, L)$ (i.e. G is full). Having formed a pullback of L and FT, we obtain the square (3) below.

(3)
$$\begin{array}{ccccc} \mathcal{T}\dot{\times}\mathcal{L} & & \xrightarrow{P_2} & & \mathcal{L} \\ {\scriptstyle P_1}\downarrow & & & & \downarrow{\scriptstyle L} \\ \mathcal{T} & \xrightarrow{T} & \mathrm{Set} & \xrightarrow{F} & \mathrm{Set} \end{array}$$

Then a (unique) functor $\Phi : \mathcal{H}_1 \to \mathcal{T}\dot{\times}\mathcal{L}$ with $P_1\Phi = H_1$ and $P_2\Phi = GH$ is a full embedding (a verification of this is quite straightforward). Hence

$$\begin{array}{ccc} \mathcal{H}_1 & \xrightarrow{H} & \mathcal{H}_2 \\ {\scriptstyle \Phi}\downarrow & & \downarrow{\scriptstyle G} \\ \mathcal{T}\dot{\times}\mathcal{L} & \xrightarrow{P_2} & \mathcal{L} \end{array}$$

is a simultaneous representation of H in P_2. This, however, is not always precisely what is needed. For instance, if $(\mathcal{T}, T)$ is Alg Δ_0 and $(\mathcal{L}, L)$ is Alg Δ_1 (both with their usual forgetful functors), then the pullback of L and T gives a category Alg $\Delta_0 \star$ Alg Δ_1 which can be regarded as Alg$(\Delta_0 \cup \Delta_1)$ with the types Δ_0, Δ_1 disjoint. But, unfortunately, the pullback $\mathcal{T} \dot{\times} \mathcal{L}$ we have is that of L and FT. If $(\mathcal{T}, T)$ is Alg(1), this is easily remedied: we simply 'lift' the unary operation of Alg(1) by F to get a simultaneous representation of H in the forgetful functor Alg$(\Delta_0 \cup \{1\}) \to$ Alg Δ_0. In other concrete cases, a more sophisticated approach constructs (2) so that a strong embedding G of $(\mathcal{H}_2, H_2)$ into $(\mathcal{L}, L)$ is carried by a rather special functor F, and H_1 is carefully chosen so that it sends objects of $\mathcal{H}_1$ only to objects of $\mathcal{T}$ that admit 'lifting' by F. This procedure is used in [LST] to show, for instance, that any faithful functor $H : \mathcal{H}_1 \to \mathcal{H}_1$ between small categories that fulfills the Zig Zag condition has a simultaneous representation in Smg$\star$Alg(1, 1) $\to$ Alg(1, 1).

I.6. Another approach, described in [ST], calls for the left square in (2) to be a subpullback of faithful functors and disposes of the functor F in (2) altogether. Instead, it requires G to be a full embedding, L a faithful functor, and the existence of a monotransformation $\mu : H_2 \to LG$ [in the terminology of [PT], this means that the concrete category $(\mathcal{H}_2, H_2)$ has an *extension* into the concrete category $(\mathcal{L}, L)$; let us recall from [PT] that every small category $\mathcal{H}_2$ with *arbitrary* faithful $H_2 : \mathcal{H}_2 \to$ Set can be extended into a number of conventional concrete categories, such as Rings, Smg, Alg(1, 1), Graph and, replacing the fullness of G by the fullness up to constant morphisms, also into Top, Unif, Comp, Metr and others]. To compensate for this weaker requirement, we assume that the concrete category $(\mathcal{T}, T)$ admits free extensions of structures to any larger set [the categorial description : the functor $R : \mathcal{T} \to T \downarrow \mathrm{Id}_{\mathrm{Set}}$ sending any a to $(a, 1_{T(a)}, T(a))$ has a left adjoint L with $LR = \mathrm{Id}_{\mathcal{T}}$, $TL(a, m, X) = X$]. Under these conditions,

> if $H : \mathcal{H}_1 \to \mathcal{H}_2$ is a slice of $T : \mathcal{T} \to$ Set and $\mathcal{H}_2$ (with arbitrary faithful $H_2 : \mathcal{H}_2 \to$ Set) has an extension into $(\mathcal{L}, L)$ then H has a simultaneous representation in $\mathcal{T} \star \mathcal{L} \to \mathcal{L}$,

where $\mathcal{T} \star \mathcal{L}$ is the *pullback of T and L*; its objects are thus pairs of structures (i.e., a $\mathcal{T}$-structure and an $\mathcal{L}$-structure) on *the same set*. In [ST], such combinations of slices of forgetful functors associated with conventional concrete categories (see also Section II below) and earlier results (see [PT]) about extensions produced many new simultaneous representations (in functors such as Top$\star$Smg $\to$ Smg, Metr$\star$Rings $\to$ Rings, Top$\star$Top $\to$ Top, Graph$\star$Graph $\to$ Graph, ...).

This is one reason why functor slices may warrant further attention. Another reason are the slices themselves, for they give rise to a new classification of functors, particularly of faithful functors into Set. Section II explores this point of view.

II. Slices of functors

II.1. To illustrate this concept and explain its name, let us consider an example of categories Top and Pos (= partially ordered sets and monotone mappings).

As shown in [ST], the forgetful functor Top $\to$ Set is a slice of the forgetful functor Pos $\to$ Set, so that there are faithful functors Φ and F for which

$$\begin{array}{ccc} \text{Top} & \longrightarrow & \text{Set} \\ {\scriptstyle\Phi}\big\downarrow & & \big\downarrow{\scriptstyle F} \\ \text{Pos} & \longrightarrow & \text{Set} \end{array}$$

is a subpullback. We recall a well-known fact that any faithful functor F : Set $\to$ Set contains a copy of the identity functor I : Set $\to$ Set in the sense that there exists a monotransformation $I \to F$. Thus if $\Phi(X_i, t_i) = (F(X_i), \leq_i)$ for objects (X_i, t_i) of Top, $i = 1, 2$, then a mapping $f : (X_1, t_1) \to (X_2, t_2)$ is continuous exactly when $F(f) : (F(X_1), \leq_1) \to (F(X_2), \leq_2)$ is monotone and, since $f : X_1 \to X_2$ is the restriction of the mapping $F(f) : F(X_1) \to F(X_2)$ to $X_1 \subseteq F(X_1)$, it is natural to visualize the forgetful functor Top $\to$ Set as a 'sliced off' section of the forgetful functor Pos $\to$ Set.

II.2. We say that $H : \mathcal{H}_1 \to \mathcal{H}_2$ and $K : \mathcal{K}_1 \to \mathcal{K}_2$ are *slice equivalent* (or *s-equivalent*) if H is a slice of K and K is a slice of H.

For the sake of brevity, when discussing familiar concrete categories with standard forgetful functors into Set we shall say that these categories are s-equivalent whenever, in actual fact, the s-equivalence applies to their forgetful functors. Under this convention, many familiar concrete categories fall into several 'baskets' determined by the s-equivalence of their forgetful functors.

II.3. *The basket* $\mathbb{R}$ *contains following categories*:

$S(F)$ for any faithful functor F : Set $\to$ Set of either variance; we recall that objects of $S(F)$ are all pairs (X, R) of sets with $R \subseteq FX$, and that a mapping $f : X \to X'$ is a morphism of $S(F)$ from (X, R) to (X', R') exactly when $[F(f)](R) \subseteq R'$ for a covariant functor F while $[F(f)](R') \subseteq R$ in the contravariant case; the forgetful functor $S(F) \to$ Set sends (X, R) to the set X; in particular

Rel Δ^+ for an arbitrary *positive* type Δ^+, whose objects are relational systems of type Δ^+ and whose morphisms are all compatible maps; these are just categories $S(F)$ for which F is a coproduct of the appropriate covariant hom-functors;

Palg Δ^+ for an arbitrary *positive* type Δ^+; these are full subcategories of Rel Δ^+ consisting of partial algebras and their homomorphisms;

Pos — the category of all partially ordered sets and monotone mappings;

Top and all its full subcategories down to the category of all metrizable spaces;

Unif and all its full subcategories down to the category of all complete metrizable spaces;

Metr — the category of all metric spaces and maps which do not increase the distance and all its full subcategories down to the category of all complete metric spaces of diameter at most one.

Reasons for mutual s-equivalence of these categories are quite straightforward, see [ST].

II.4. *The basket* $\mathbb{A}$ *contains following categories*:

$A(F,G)$ for any nonconstant covariant functors F, G : Set $\to$ Set; the objects of $A(F,G)$ are all pairs (X,α) where X is a set and $\alpha : FX \to GX$ a mapping, and a mapping $f : X \to X'$ is a morphism of $A(F,G)$ from (X,α) to (X',α') exactly when $G(f)\alpha = \alpha' F(f)$; in particular

$\operatorname{Alg}\Delta^+$ — the category of all universal algebras of a positive type Δ^+ and all their homomorphisms; these are just categories $A(F,G)$ for which $G = I$ and F is a coproduct of the appropriate covariant hom-functors;

$\mathrm{Set}^{\mathbb{T}}$ — the category of all monadic algebras, for any *non-degenerate* monad $\mathbb{T} = (T,\eta,\mu)$ in Set (a monad $\mathbb{T}$ is non-degenerate iff its functor part T is neither the identity I nor a constant nor their coproduct).

The proof that all categories $\operatorname{Alg}\Delta^+$ with a positive Δ^+ are s-equivalent appears in [ST]; the fact that $A(F,G)$ are s-equivalent with any of them for any covariant nonconstant F, G is in [RST], and the result about $\mathrm{Set}^{\mathbb{T}}$ appears in [KST].

Problem: How should a 'non-degenerate comonad' be defined? Could it then be shown that, for any non-degenerate comonad $\mathbb{G} = (G,\varepsilon,\delta)$ in Set, the category of all $\mathbb{G}$-coalgebras belongs to the basket $\mathbb{A}$?

II.5. *The basket* $\mathbb{P}$ *contains following categories*:

$\mathrm{Set}^{\mathbb{T}}$ with $\mathbb{T} = (T,\eta,\mu)$, where T is the coproduct of the identity functor and a constant functor with a non-empty value;

$A(F,G)$ with covariant F, G : Set $\to$ Set, G nonconstant, F constant with a non-empty value; particularly

Set_* — the category of all pointed sets

Proofs are easy, see also [ST], [KST], [RST].

II.6. Let $\mathcal{R}$, $\mathcal{A}$, $\mathcal{P}$ denote the respective collections of slices of any, and hence each, forgetful functor of a category from a corresponding basket. It is easy to see that $\mathrm{Set}_* \to$ Set is a slice of Alg(1) $\to$ Set and that Alg(1) $\to$ Set is a slice of Rel(2) $\to$ Set, so that $\mathcal{P}$ is a subcollection of $\mathcal{A}$ and $\mathcal{A}$ is a subcollection of $\mathcal{R}$. These three slice collections differ from one another already on the level of one-object categories and, for a functor $H : \mathcal{H}_1 \to \mathcal{H}_2$ between small categories, [ST] shows that:

(R) H is in $\mathcal{R}$ iff it is faithful;

(A) H is in $\mathcal{A}$ iff H is faithful and obeys Isbell's Zig-Zag condition (cf. also II.9 below);

(P) H is in $\mathcal{P}$ iff it is faithful and satisfies the condition

(p) for $a, b, c \in \operatorname{obj}\mathcal{H}_1$, if $\mu \in \mathcal{H}_1(a,b)$, $\beta \in \mathcal{H}_2(H(b), H(c))$ and $\beta \circ H(\mu) = H(\sigma)$ for some $\sigma \in \mathcal{H}_1(a,c)$, then $\beta = H(\nu)$ for some $\nu \in \mathcal{H}_1(b,c)$.

Let $\mathcal{F}^{\mathrm{op}}$ denote the dual collection of all opposites $H^{\mathrm{op}} : \mathcal{H}_1^{\mathrm{op}} \to \mathcal{H}_2^{\mathrm{op}}$ to functors $H : \mathcal{H}_1 \to \mathcal{H}_2$ from a collection $\mathcal{F}$. While $\mathcal{R}$ and $\mathcal{A}$ are self-dual, that is $\mathcal{R} = \mathcal{R}^{\mathrm{op}}$ and $\mathcal{A} = \mathcal{A}^{\mathrm{op}}$, [ST], this is not the case for the collection $\mathcal{P}$, see (P) above. The collection $\mathcal{P}^{\mathrm{op}}$ thus produces a fourth basket, denoted $\mathbb{P}^{\mathrm{op}}$. This basket contains, for instance, all $A(F,G)$ with F, G covariant, F nonconstant and G constant with non-empty value, [RST] and, in particular, the subcategory

$\mathrm{Rel}_*(1)$ of $\mathrm{Rel}(1)$ whose morphisms, in addition to the single unary relation, preserve also its complement, see [ST].

II.7. Functors between small categories which are in $\mathcal{R}$, $\mathcal{A}$, $\mathcal{P}$, $\mathcal{P}^{\mathrm{op}}$ are described in II.6 above. Now we are going to characterize **all** functors in $\mathcal{R}$, $\mathcal{P}$, $\mathcal{P}^{\mathrm{op}}$.

Let $H : \mathcal{H}_1 \to \mathcal{H}_2$ be an arbitrary functor. For any $x \in \mathrm{obj}\,\mathcal{H}_2$, let $\mathrm{In}(x)$ denote the class of all pairs (a, m) where $a \in \mathrm{obj}\,\mathcal{H}_1$ and $m \in \mathcal{H}_2(H(a), x)$, and $\mathrm{Out}(x)$ the class of all pairs (n, b) with $b \in \mathrm{obj}\,\mathcal{H}_1$ and $n \in \mathcal{H}_2(x, H(b))$. Following [ASH], we consider the equivalence on $\mathrm{In}(x)$ defined by

$$\begin{aligned}(a, m) \sim (a', m') \equiv {} & \text{for every } (n, b) \in \mathrm{Out}(x), nm = H(\alpha) \text{ for} \\ & \text{some } \alpha \in \mathcal{H}_1(a, b) \text{ iff } nm' = H(\alpha') \text{ for some} \\ & \alpha' \in \mathcal{H}_1(a', b).\end{aligned}$$

The functor H is called *strongly small fibered* (abbr. SSF) if there is a set $S(x) \subseteq \mathrm{In}(x)$ such that every (a, m) in $\mathrm{In}(x)$ is equivalent to a member of $S(x)$, see [AHS]. In what follows, we shall assume that

α) any two distinct elements of $S(x)$ are non-equivalent, and
β) if $x = H(a)$, then $(a, 1_x)$ is in $S(x)$.

Lemma. *Let (1) be a subpullback. If G_1, K, G_2 are faithful, then H is faithful. If, moreover, K satisfies SSF or (p) or $(p)^{\mathrm{op}}$, then so does H.*

We omit the easy and straightforward proof of this lemma.

II.8.

Proposition. *A functor $H : \mathcal{H}_1 \to \mathcal{H}_2$ is in $\mathcal{R}$ (or $\mathcal{P}$ or $\mathcal{P}^{\mathrm{op}}$) if and only if $\mathcal{H}_2$ is concretizable (i.e., admits a faithful functor $\mathcal{H}_2 \to \mathrm{Set}$) and H is a strongly small fibered faithful functor (or also fulfills (p) or $(p)^{\mathrm{op}}$).*

Proof: 1) It is well-known and easy to verify that the forgetful functors $\mathrm{Rel}(1) \to \mathrm{Set}$, $\mathrm{Set}_* \to \mathrm{Set}$, $\mathrm{Rel}_*(1) \to \mathrm{Set}$ are SSF and faithful, and that $\mathrm{Set}_* \to \mathrm{Set}$ satisfies (p) and $\mathrm{Rel}_*(1) \to \mathrm{Set}$ satisfies $(p)^{\mathrm{op}}$. Hence, by the Lemma II.7, any H in $\mathcal{R}$ (or $\mathcal{P}$ or $\mathcal{P}^{\mathrm{op}}$) is SSF, faithful (and fulfills (p) or $(p)^{\mathrm{op}}$); it is clear that $\mathcal{H}_2$ must be concretizable.

2) Conversely, let $H : \mathcal{H}_1 \to \mathcal{H}_2$ be faithful and SSF. For every $x \in \mathrm{obj}\,\mathcal{H}_2$, let $S(x) \subseteq \mathrm{In}(x)$ be as in II.7. Define a functor $S : \mathcal{H}_2 \to \mathrm{Set}$ so that, for any morphism $f : x \to y$ and any $(a, m) \in S(x)$, $[S(f)]\,(a, m)$ is the unique element of $S(y)$ equivalent to (a, fm). Let $\mathcal{H}_2(S)$ denote the category of whose objects are all pairs (x, R) with $x \in \mathrm{obj}\,\mathcal{H}_2$, $R \subseteq S(x)$; a $\mathcal{H}_2$-morphism $f : x \to x'$ is a morphism $(x, R) \to (x', R')$ in $\mathcal{H}_2(S)$ iff $[S(f)]\,(R) \subseteq R'$. For any $c \in \mathrm{obj}\,\mathcal{H}_1$, set

$$R_c = \{(a, m) \in S(H(c)) | (a, m) \sim (a, H(\alpha)), a \in \mathrm{obj}\,\mathcal{H}_1, \alpha \in \mathcal{H}_1(a, c)\}$$

and define $\Psi : \mathcal{H}_1 \to \mathcal{H}_2(S)$ by

$$\Psi(c) = (H(c), R_c), \quad \Psi(\gamma) = H(\gamma).$$

Then Ψ is a full embedding, see [AHS]. Therefore, to prove that $H : \mathcal{H}_1 \to \mathcal{H}_2$ is in $\mathcal{R}$, it suffices to show that the forgetful functor $U : \mathcal{H}_2(S) \to \mathcal{H}_2$ (sending

(x, R) to x) belongs to this slice collection. Since $\mathcal{H}_2$ is concretizable there is a faithful functor $V : \mathcal{H}_2 \to \mathrm{Set}$; let F denote the coproduct of V and S and $\Phi : \mathcal{H}_2(S) \to \mathrm{Rel}(1)$ the functor sending any (x, R) to $(F(x), R)$ and any $f : (x, R) \to (x', R')$ to $F(f)$. It is easy to see that the square

$$\begin{array}{ccc} \mathcal{H}_2(S) & \xrightarrow{U} & \mathcal{H}_2 \\ {\scriptstyle \Phi}\downarrow & & \downarrow{\scriptstyle F} \\ \mathrm{Rel}(1) & \longrightarrow & \mathrm{Set} \end{array}$$

is a subpullback.

3) If H is not only faithful and SSF but also fulfills the condition $(p)^{\mathrm{op}}$, then the functor $\Psi : \mathcal{H}_1 \to \mathcal{H}_2(S)$ defined above preserves not only the unary relations R_c on $H(c)$ but also their complements, so it is in fact a full embedding of $\mathcal{H}_1$ into $(\mathcal{H}_2(S))_*$ (where $(\mathcal{H}_2(S))_*$ relates to $\mathcal{H}_2(S)$ as $\mathrm{Rel}_*(1)$ relates to $\mathrm{Rel}(1)$, see II.6). The remainder of the proof is analogous to that in 2), to produce a subpullback

$$\begin{array}{ccc} (\mathcal{H}_2(S))_* & \longrightarrow & \mathcal{H}_2 \\ {\scriptstyle \Phi}\downarrow & & \downarrow{\scriptstyle F} \\ \mathrm{Rel}_*(1) & \longrightarrow & \mathrm{Set} \end{array}$$

4) Finally, let $H : \mathcal{H}_1 \to \mathcal{H}_2$ be faithful, SSF and fulfill (p), and let $\mathcal{H}_2$ be concretizable. Then $\mathcal{H}_2^{\mathrm{op}}$ is also concretizable and $H^{\mathrm{op}} : \mathcal{H}_1^{\mathrm{op}} \to \mathcal{H}_2^{\mathrm{op}}$ is also faithful and SSF, see [AHS]. By 3), there are faithful $\Phi_1 : \mathcal{H}_1^{\mathrm{op}} \to \mathrm{Rel}_*(1)$ and $\Phi_2 : \mathcal{H}_2^{\mathrm{op}} \to \mathrm{Set}$ such that

$$\begin{array}{ccc} \mathcal{H}_1^{\mathrm{op}} & \xrightarrow{H^{\mathrm{op}}} & \mathcal{H}_2^{\mathrm{op}} \\ {\scriptstyle \Phi_1}\downarrow & & \downarrow{\scriptstyle \Phi_2} \\ \mathrm{Rel}_*(1) & \longrightarrow & \mathrm{Set} \end{array}$$

is a subpullback. There is a subpullback

$$\begin{array}{ccc} \mathrm{Rel}_*(1) & \longrightarrow & \mathrm{Set} \\ {\scriptstyle \Psi_1}\downarrow & & \downarrow{\scriptstyle \Psi_2} \\ \mathrm{Set}_* & \longrightarrow & \mathrm{Set} \end{array}$$

with faithful contravariant functors Ψ_1, Ψ_2, where Ψ_2 is the contravariant power-set functor and the unary relation R is interpreted as the distinguished element $R \in \Psi_1(X, R)$, see [ST].

II.9. Let us turn our attention to the collection $\mathcal{A}$. Since Isbell's Zig-Zag condition, $[I_1]$, $[I_2]$, is crucial here, we recall it first.

A functor $H : \mathcal{H}_1 \to \mathcal{H}_2$ fulfills the Zig-Zag condition (zz) if for every zig-zag line in $\mathcal{H}_1$

$$a \xrightarrow{\alpha} c_1 \xleftarrow{\gamma_1} d_1 \xrightarrow{\delta_1} c_2 \xleftarrow{\gamma_2} d_2 \xrightarrow{\delta_2} \dots \xleftarrow{\gamma_n} d_n \xrightarrow{\beta} b$$

whose H-image can be completed by $\mathcal{H}_2$-morphisms $m_i : H(a) \to H(d_i)$ and $p_i : H(c_i) \to H(b)$, $i = 1, \dots, n$ to a commutative diagram of $\mathcal{H}_2$-morphisms from $H(a)$ to $H(b)$ [this means $H(\gamma_1)m_1 = H(\alpha)$, $H(\gamma_{i+1})m_{i+1} = H(\delta_i)m_i$ and $p_i H(\gamma_i) = p_{i+1}H(\delta_{i+1})$ for $i = 1, \dots, n-1$ and $p_n H(\gamma_n) = H(\beta)$], the unique morphism $q : H(a) \to H(b)$ determined by this commutative diagram [i.e., $q = p_1 H(\alpha) = p_1 H(\gamma_1)m_1 = \cdots = H(\beta)m_n$] is an H-image of some $\mu \in \mathcal{H}_1(a, b)$.

Routine (but somewhat tedious) calculations result in these two claims:

a) if K in the subpullback (1) fulfills (zz), so does H;

b) $\mathrm{Alg}(1) \to \mathrm{Set}$ (hence also $A(F, G) \to \mathrm{Set}$, $\mathrm{Set}^{\mathbb{T}} \to \mathrm{Set}$) fulfills (zz).

Therefore,

if $H : \mathcal{H}_1 \to \mathcal{H}_2$ is in the collection $\mathcal{A}$, then necessarily
$\mathcal{H}_2$ is concretizable, H is faithful, SSF and fulfills (zz).

Are these conditions also sufficient? The general answer is unknown, but [R] answers this question affirmatively for $\mathcal{H}_2 = \mathrm{Set}$ (the proof in [R] is rather complicated and difficult!). Next, we show that the answer is affirmative also for any right adjoint H whatsoever – this very strong assumption eliminates the need for the condition SSF altogether.

II.10. The following lemma holds quite generally.

Lemma. *A right adjoint $H : \mathcal{H}_1 \to \mathcal{H}_2$ is of descent type if and only if it is faithful and fulfills (zz).*

Proof: Let $F : \mathcal{H}_2 \to \mathcal{H}_1$ be a left adjoint of H, $\mathbb{T} = (T, \eta, H\varepsilon F)$ the resulting monad, $U : \mathcal{H}_2^{\mathbb{T}} \to \mathcal{H}_2$ the forgetful functor of the category of $\mathbb{T}$-algebras, and $\Phi : \mathcal{H}_1 \to \mathcal{H}_2^{\mathbb{T}}$ its comparison functor. A routine verification shows that U always fulfills (zz). Thus, if H is of descent type (i.e. Φ is a full embedding, [BW]), then $U \circ \Phi = H$ is faithful and fulfills (zz).

Conversely, if H is faithful, then Φ is faithful. To show that H fulfills (zz) only when Φ is full, let $h : \Phi(a) \to \Phi(b)$ be a homomorphism of $\mathbb{T}$-algebras, that is, $h : H(a) \to H(b)$ is a $\mathcal{H}_2$-morphism satisfying $h \circ H(\varepsilon_a) = H(\varepsilon_b) \circ T(h)$ [where $\Phi(a) = (H(a), H(\varepsilon_a))$, $\Phi(b) = (H(b), H(\varepsilon_b))$]. We have to show that $h = \Phi(\alpha)$ (that is, $h = H(\alpha)$) for some $\alpha : a \to b$ in $\mathcal{H}_1$. Consider the zig-zag line in $\mathcal{H}_1$

$$a \xrightarrow{1a} a \xleftarrow{\varepsilon_a} FH(a) \xrightarrow{\varepsilon_b \circ F(h)} b\,.$$

Clearly, its H-image can be completed to a commutative diagram by $\eta_{H(a)} : H(a) \to HFH(a)$ and $h : H(a) \to H(b)$. By (zz), the unique morphism determined by the diagram, namely the morphism $h : H(a) \to H(b)$, is a H-image of some $\alpha : a \to b$.

II.11.

Proposition. *Let a right adjoint $H : \mathcal{H}_1 \to \mathcal{H}_2$ be faithful and fulfill (zz), and let $\mathcal{H}_2$ be concretizable. Then H is in $\mathcal{A}$.*

Proof: By the above lemma, it suffices to show that if $\mathbb{T} = (T, \eta, \mu)$ is a monad over a concretizable category $\mathcal{H}_2$, then the forgetful functor $H : \mathcal{H}_2^{\mathbb{T}} \to \mathcal{H}_2$ of the category of $\mathbb{T}$-algebras belongs to $\mathcal{A}$. Since $\mathcal{H}_2$ is concretizable, there exists faithful $V : \mathcal{H}_2 \to \mathrm{Set}$. Let F be the coproduct of V and VT. In a commutative square

$$\begin{array}{ccc} \mathcal{H}_2^{\mathbb{T}} & \xrightarrow{H} & \mathcal{H}_2 \\ {\scriptstyle \Psi}\downarrow & & \downarrow{\scriptstyle F} \\ \mathrm{Alg}(1) & \longrightarrow & \mathrm{Set} \end{array}$$

the functor Ψ sends a $\mathbb{T}$-algebra (a, α) to the unary algebra $(\bar{a}, \bar{\alpha})$, where $\bar{a}$ is a coproduct of $V(a)$ and $VT(a)$ and $\bar{\alpha}$ is the identity on $V(a)$ and it sends $VT(a)$ to $V(a)$ as $V(\alpha)$. We omit the routine verification that this commutative square is really a subpullback.

II.12. Baskets other than $\mathbb{R}$, $\mathbb{A}$, $\mathbb{P}$, $\mathbb{P}^{\mathrm{op}}$, particularly those between $\mathbb{A}$ and $\mathbb{R}$, are investigated in [RST].

III. Functor slices and simultaneous representations

III.1. Here we apply results of Section II to describe all functors that have simultaneous representations in the forgetful functor $U : \mathrm{Unif} \to \mathrm{Top}$, the compactly generated modification $\mathrm{CG} : \mathrm{Top} \to \mathrm{CGTop}$ and the completely regular modification $\mathrm{CR} : \mathrm{Top} \to \mathrm{CRTop}$.

Theorem. *A functor $H : \mathcal{H}_1 \to \mathcal{H}_2$ has a simultaneous representation in U : Unif $\to$ Top (or in CG : Top $\to$ CGTop or in CR : Top $\to$ CRTop) if and only if it belongs to the slice collection $\mathcal{R}$.*

Since the functors in $\mathcal{R}$ are fully characterized in II.8, this fully describes all functors that admit a simultaneous representation in U, CG, CR. Parts III.2–III.6 outline a proof of this result.

III.2. If $H : \mathcal{H}_1 \to \mathcal{H}_2$ has a simultaneous representation in U, then, by the definition, there are faithful and almost full (= full up to constant map, see I.4) functors G_1, G_2 such that the square

(4)
$$\begin{array}{ccc} \mathcal{H}_1 & \xrightarrow{H} & \mathcal{H}_2 \\ {\scriptstyle G_1}\downarrow & & \downarrow{\scriptstyle G_2} \\ \mathrm{Unif} & \xrightarrow{U} & \mathrm{Top} \end{array}$$

commutes. We have to show that H is in $\mathcal{R}$, that is, that $\mathcal{H}_2$ is concretizable and H is faithful and SSF. Only the last claim requires a little reasoning, the first two are evident. Since U preserves constant maps, the square (4) is a subpullback. Thus we need only show that U is SSF; we proceed as follows. If P is a uniform space and f is a map of P into a topological space X, let $\langle P, f\rangle$ denote the subset $f(P)$ of X endowed with the quotient-uniformity given by $P \xrightarrow{f} f(P)$ (the final structure of [AHS]). If $f : P \to X$ and $f' : P' \to X$ are maps such that $\langle P, f\rangle = \langle P', f'\rangle$, then (P, f) is equivalent to (P', f') in the sense of II.7. But for every topological space X there is only a set of all possible uniformities on subsets of X.

Similarly, if H admits a simultaneous representation in CG or in CR, then it belongs to $\mathcal{R}$. Proofs of these claims are quite analogous.

III.3. The proof of the converse requires more work.

Suppose that $H : \mathcal{H}_1 \to \mathcal{H}_2$ belongs to $\mathcal{R}$. We must construct a simultaneous representation of H in U (or CG or CR). Let P^- denote the contravariant power-set functor, i.e., let $P^-X = \{Z | Z \subseteq X\}$, $[P^-f](Z) = f^{-1}(Z)$; the category $S(P^-)$ and its forgetful functor $V : S(P^-) \to \mathrm{Set}$ are defined as in II.3. Since $\mathcal{R} = \mathcal{R}^{\mathrm{op}}$, there exist faithful functors $M_1 : \mathcal{H}_1^{\mathrm{op}} \to S(P^-)$ and $M_2 : \mathcal{H}_2^{\mathrm{op}} \to \mathrm{Set}$ such that the square

$$\begin{array}{ccc} \mathcal{H}_1^{\mathrm{op}} & \xrightarrow{H^{\mathrm{op}}} & \mathcal{H}_2^{\mathrm{op}} \\ {\scriptstyle M_1}\downarrow & & \downarrow{\scriptstyle M_2} \\ S(P^-) & \xrightarrow{V} & \mathrm{Set} \end{array}$$

is a subpullback. By III.3.3, III.4.3 and II.7.5 of [PT], every concrete category, and hence also the concrete category $(\mathcal{H}_2^{\mathrm{op}}, M_2)$, has an extension into $(S(P^-), V)$. Let

$$\begin{array}{ccc} S(P^-) \star S(P^-) & \xrightarrow{P_2} & S(P^-) \\ {\scriptstyle P_1}\downarrow & & \downarrow{\scriptstyle V} \\ S(P^-) & \xrightarrow{V} & \mathrm{Set} \end{array}$$

be a pullback, the objects of $S(P^-) \star S(P^-)$ thus can be regarded as triples $(X, \mathcal{Z}_1, \mathcal{Z}_2)$ with $\mathcal{Z}_1 \subseteq P^-X$, $\mathcal{Z}_2 \subseteq P^-X$ and

$$P_1(X, \mathcal{Z}_1, \mathcal{Z}_2) = (X, \mathcal{Z}_1), \quad P_2(X, \mathcal{Z}_1, \mathcal{Z}_2) = (X, \mathcal{Z}_2).$$

Since unary relations can be freely extended to any larger set (we simply take the same unary relation on the larger set), the procedure described in I.6 can be applied to the case $(\mathcal{T}, T) = (S(P^-), V) = (\mathcal{L}, L)$. We then get a simultaneous representation of H^{op} in P_2, that is, full embeddings $L_2 : \mathcal{H}_2^{\mathrm{op}} \to S(P^-)$ and $L_1 : \mathcal{H}_1^{\mathrm{op}} \to S(P^-) \star S(P^-)$ completing the commutative square

$$\begin{array}{ccc} \mathcal{H}_1^{\mathrm{op}} & \xrightarrow{H^{\mathrm{op}}} & \mathcal{H}_2^{\mathrm{op}} \\ {\scriptstyle L_1}\downarrow & & \downarrow{\scriptstyle L_2} \\ S(P^-) \star S(P^-) & \xrightarrow{P_2} & S(P^-) \end{array}$$

III.4. In III.5 below, we construct faithful almost full functors $K_1 : (S(P^-) \star S(P^-))^{\mathrm{op}} \to \mathrm{Unif}$ and $K_2 : (S(P^-))^{\mathrm{op}} \to \mathrm{Top}$ for which the square

$$\begin{array}{ccc} (S(P^-) \star S(P^-))^{\mathrm{op}} & \xrightarrow{P_2^{\mathrm{op}}} & S(P^-)^{\mathrm{op}} \\ {\scriptstyle K_1}\downarrow & & \downarrow{\scriptstyle K_2} \\ \mathrm{Unif} & \xrightarrow{U} & \mathrm{Top} \end{array}$$

commutes. The composite functors $G_1 = K_1L_1^{\text{op}}$ and $G_2 = K_2L_2^{\text{op}}$ are then faithful, almost full, and complete the commutative square (4); in other words, they constitute a simultaneous representation of H in U.

III.5. The construction of $K_1 : (S(P^-)\star S(P^-))^{\text{op}} \to \text{Unif}$ and $K_2 : S(P^-)^{\text{op}} \to \text{Top}$ is rather involved, but not quite as much as the simultaneous representation of a faithful functor between small categories in the functor $\beta : \text{Tich} \to \text{Comp}$ described in [T_3]. The present construction is, in fact, a simplification (with one trick added) of that from [T_3], so that we shall only discuss the necessary modifications of some parts of the text in [T_3]. To conform with the notation of [T_3], we denote $S(P^-)\star S(P^-)$ by $\mathcal{K}_1$, $S(P^-)$ by $\mathcal{K}_2$ and $P_2 : \mathcal{K}_1 \to \mathcal{K}_2$ by F (so that F sends any $(X, \mathcal{Z}_1, \mathcal{Z}_2)$ to $(X, \mathcal{Z}_2)$, as in [T_3]; in [T_3], $\mathcal{Z}_i \subseteq P^-X$ have to satisfy an additional condition, which is not needed here). We construct $K_1 : \mathcal{K}_1^{\text{op}} \to \text{Unif}$ and $K_2 : \mathcal{K}_2^{\text{op}} \to \text{Top}$ just as described in 3.1–3.5 of [T_3]: we construct the triangle spaces from collections of pairwise disjoint non-degenerate subcontinua of a Cook continuum just as in 3.3, use three triangle spaces and a subcontinuum D precisely as in 3.4, and then assemble the space $P(X; \mathcal{Z}_2)$ exactly as in 3.5. While the space $P(X, \mathcal{Z}_2)$ plays only an auxiliary role in [T_3], it is already our final space $K_2(X, \mathcal{Z}_2)$. Let us recall that it is obtained by the gluing together of the compact Hausdorff space D^X, the compact Hausdorff space $T^{(0)}$, many copies of the compact Hausdorff space $T^{(2)}$ (one copy for each $Z \in \mathcal{Z}_2$) and many copies of $T^{(1)} \setminus \{b^{(1)}\}$ (one copy for each $Z \subseteq X$). If a uniformity on each copy of $T = T^{(1)} \setminus \{b^{(1)}\}$ is given, we get (in the standard way) the uniformity on the whole $K_2(X, \mathcal{Z}_2)$. Now the trick (used already in [T_2]) is added: we choose two uniformities on T, say u_1 and u_2, both consistent with the topology, and such that the identity map $(T, u_2) \to (T, u_1)$ is uniformly continuous but its inverse $(T, u_1) \to (T, u_2)$ is **not**. Given an object $(X, \mathcal{Z}_1, \mathcal{Z}_2)$ of $\mathcal{K}_1$, we endow $K_2(X, \mathcal{Z}_2)$ by the uniformity equal to (T, u_1) on any copy of T corresponding to $Z \in \mathcal{Z}_1$ and equal to (T, u_2) on any other copy of T. The resulting uniform space is $K_1(X, \mathcal{Z}_1, \mathcal{Z}_2)$. Clearly, the forgetful functor Unif $\to$ Top sends $K_1(X, \mathcal{Z}_1, \mathcal{Z}_2)$ to $K_2(X; \mathcal{Z}_2)$. The proof that $K_1 : \mathcal{K}_1^{\text{op}} \to \text{Unif}$ and $K_2 : \mathcal{K}_2^{\text{op}} \to \text{Top}$ are almost full embeddings follows from the (evident formal modifications of) Lemmas 3.2, 3.3 and Propositions 3.4 and 3.5 of [T_3].

III.6. The construction of simultaneous representation of $\mathcal{K}_1^{\text{op}} \xrightarrow{F^{\text{op}}} \mathcal{K}_2^{\text{op}}$ in CG : Top $\to$ CGTop and in CR : Top $\to$ CRTop proceeds similarly to III.5; only the space $T = T^{(1)} \setminus \{b^{(1)}\}$ has to be replaced by the more complicated spaces from [TH]: here we have two distinct topologies t_1 and t_2 on T, t_1 being a CG- or CR- modification of t_2 and we take a copy of (T, t_1) for each $Z \in \mathcal{Z}_1$ and a copy of (T, t_2) for all other $Z \subseteq X$; the space $K_1(X, \mathcal{Z}_1, \mathcal{Z}_2)$ thus obtained is such that its compactly generated (or completely regular) modification proceeds only on the copies of T.

III.7. Let Top_* denote the category of all pointed topological spaces. The result below can be proved by a suitable modification of arguments from III.2–III.5.

Proposition. *A functor $H : \mathcal{H}_1 \to \mathcal{H}_2$ admits a simultaneous representation in the forgetful functor* $\text{Top}_* \to \text{Top}$ *if and only if it belongs to the slice collection*

$\mathcal{P}$.

Since the functors in $\mathcal{P}$ are fully characterized in II.8, this also gives a final result on simultaneous representability in $\mathrm{Top}_* \to \mathrm{Top}$.

Acknowledgment. I am indebted to J. Sichler for his comments on an earlier version of this paper.

References

[AHS] J.Adámek, H.Herrlich and G.Strecker, *The structure of initial completions*, Cahiers Topologie Géom. Différentielle **20** (1980), 333-352.

[BW] M.Barr and Ch.Wells, *Toposes, Triples and Theories*, Springer Verlag, New York Berlin Heidelberg Tokyo, 1985.

[C] H.Cook, *Continua which admit only the identity mapping onto non-degenerate subcontinua*, Fund. Math. **60** (1967), 241-249.

[FS] E.Fried and J.Sichler, *Homomorphisms of commutative rings with unit element*, Pacific J. Math. **45** (1973), 485–491.

[HL] Z.Hedrlín and J.Lambek, *How comprehensive is the category of semigroups?*, J. Algebra **11** (1969), 195-212.

[I_1] J.R.Isbell, *Epimorphisms and dominions*, in: Proc. Conf. Categorical Algebra, La Jolla, 1965, Springer Verlag, New York Heidelberg Berlin (1966), 232-246.

[I_2] J.R.Isbell, *Epimorphisms and dominions III*, Amer. J. Math. **90** (1968), 1025-1030.

[K] V.Koubek, *Each concrete category has a representation by T_2-paracompact topological spaces*, Comment. Math. Univ. Carolin. **15** (1974), 655-663.

[KST] V.Koubek, J.Sichler and V.Trnková, *Algebraic functor slices,*, in J. Pure Appl. Algebra (to appear).

[LST] W.A.Lampe, J.Sichler and V.Trnková, *Homomorphisms of unary algebras and of their expansion*, preprint 1991.

[PST] M.Petrich, J.Sichler and V.Trnková, *Simultaneous representations in categories of algebras*, Algebra Universalis **27** (1990), 426-453.

[P] A.Pultr, *Eine Bemerkung über volle Einbettungen von Kategorien von Algebren*, Math. Ann. **178** (1968), 78-82.

[PT] A.Pultr and V.Trnková, *Combinatorial, Algebraic and Topological Representations of Groups, Semigroups and Categories*, North Holland, 1980.

[R] J.Reiterman, *Concrete full embeddings into categories of algebras and coalgebras*, preprint 1991.

[RST] J.Reiterman, J.Sichler and V.Trnková, *On slice classification of functors*, in preparation.

[ST] J.Sichler and V.Trnková, *Functor slices and simultaneous representations*, Contributions to general algebra 7, Verlag Hölder-Pichler-Tempsky, Wien 1991 – Verlag B. G. Teubner, Stuttgart, 299-320.

[T_1] V.Trnková, *Simultaneous representations in categories*, in: Proc. Categ. Conf. Louvain-La-Neuve, Lect. N. in Math. 1348, Springer-Verlag, New York Heidelberg Berlin (1988), 217-239.

[T_2] V.Trnková, *Simultaneous representations by metric spaces*, Cahiers Topologie Géom. Différentielle **29** (1988), 217-239.

[T_3] V.Trnková, *Nonconstant continuous maps of spaces and of their β-compactifications*, Topology Appl. **33** (1989), 47-62.

[TH] V.Trnková and M.Hušek, *Nonconstant continuous maps of modifications of topologies*, Comment. Math. Univ. Carolin. **29** (1988), 747-765.

[W] O.Wyler, *Operational categories*, in: Proc. Conf. Categorical Algebra, La Jolla 1965, Springer Verlag, New York Heidelberg Berlin (1966), 295-316.

MÚ UK, Sokolovská 83, 186 00 Praha 8, Czechoslovakia
E-mail address: : UMMVT@csearn.bitnet